Soil Resilience and Sustainable Land Use

SOIL RESILIENCE AND SUSTAINABLE LAND USE

Proceedings of a Symposium held in Budapest, 28 September to 2 October 1992, including the Second Workshop on the Ecological Foundations of Sustainable Agriculture (WEFSA II)

Edited by

D.J. Greenland

Visiting Professor
Department of Soil Science
University of Reading and
Former Director
Scientific Services
CAB International

and

I. Szabolcs

Research Institute for Soil Science
and Agricultural Chemistry
Hungarian Academy of Sciences

CAB INTERNATIONAL

CAB INTERNATIONAL
Wallingford
Oxon OX10 8DE
UK

Tel: Wallingford (0491) 832111
Telex: 847964 (COMAGG G)
Telecom Gold/Dialcom: 84: CAU001
Fax: (0491) 833508

A catalogue entry for this book is available from the British Library

ISBN 0 85198 871 7

Typeset by Solidus (Bristol) Limited, UK
Printed and bound in Great Britain by Short Run Press Ltd

Contents

Preface

Amongst many topics discussed at the United Nations Conference on Environment and Development (UNCED) held in Rio de Janeiro in 1992 were 'Sustainable Agriculture and Rural Development' and 'Land Resource Planning and Management'. Both figured amongst the topics for 'priority action'. Discussion related to these topics centred on the problems of soil degradation associated with increasing intensity of land use, a necessary accompaniment to the growth of population, and the search for a higher standard of living.

Soils worldwide are being subjected to increasing degrees of stress – chemical, physical and biological. If they are to continue to support the growth of populations and to provide improving living standards for all, it is essential that they display resilience to these stresses.

Perhaps surprisingly, considering its importance, the topic of soil resilience has received rather little attention. Agriculture almost always applies a stress of some sort to the soil. This may be through nutrient removal, by reducing the return of organic matter to the soil and so reducing biological activity, or by the imposition of physical stress. Frequently all of these stresses arise, although their significance varies greatly according to local circumstances. The resilience of some soils allows them to recover naturally and rapidly. Others may require assistance in the form of fertilizers, soil amendments, or other action. How effective the interventions are in restoring soil productivity depends on the resilience of the particular soil. Fortunately the great majority of soils show significant resilience, but it is important to recognize the importance of this role: were it not for soil resilience the agricultural systems of the world would have collapsed long ago.

The present meeting was organized to draw attention to the importance of soil resilience and to review what is known of the subject. It was felt that recovery of soil health was often very dependent on soil organisms, and, following the findings of WEFSA I, the first Workshop on the Ecological Foundations of Sustainable Agriculture, which was concerned with the importance of biodiversity among invertebrates and microorganisms, special attention was given to that aspect.

The meeting was organized by the Hungarian Academy of Sciences and CAB International, in association with the International Society of Soil Science. Financial support came from UNEP, the European Environmental Research Organization (EERO), the OECD, the Ford Foundation, ICSU/CASAFA and the British Council. It was divided into: (i) a Symposium, at which the invited papers were presented and discussed, and at which a poster session was held when volunteered papers were presented; and (ii) a Workshop where the papers were reviewed and discussed, research priorities set, and recommendations to policy-makers and funding agencies developed.

This volume includes the invited papers, the Reports of the Working Groups, and the Recommendations. The papers presented in the poster session are appearing in a special edition of *Agrokemia es Talajtan* Volume 42, Numbers 1 and 2, 1993.

D.J. Greenland I. Szabolcs

May, 1993

Recommendations of the Second Workshop on the Ecological Foundations of Sustainable Agriculture (WEFSA II)

held in Budapest, 28 September to 2 October 1992

The participants of the Symposium on Soil Resilience and Sustainable Land Use appreciate the concerns and recommendations of the United Nations Conference on Environment and Development – AGENDA 21.

They also recognize that the global growth in produce from the land over the last century has increased despite losses from land degradation, and that this increase has exceeded population growth in most areas.

The current challenge is to maintain this growth in the well-endowed areas, and to use the natural resilience of soils to establish sustainable production in the unimproved and degraded systems in the face of continued population increase.

In this context the Symposium suggests that:

1. the global database of human-induced soil degradation developed by UNEP, ISSS and ISRIC, in collaboration with countries, should be complemented by a similar assessment of:

(a) areas with sustainable land management systems,

(b) areas where degraded lands have been rehabilitated,

(c) the resilience of the land resource base in different ecosystems;

2. to implement some of the recommendations of AGENDA 21 in the area of land resource planning and management, an assessment of current land use should be made at national and global levels.

This will require that:

1. existing long-term trend monitoring programmes, data collection and experiments are maintained and documented, and new ones supported in key agroecological zones in developing countries;

2. key species, biotic assemblages and processes contributing to soil resilience are identified;

3. quantitative indicators and threshold values of those attributes which determine soil resilience and sustainable land management are determined;
4. appropriate practices for different soils in agroecological zones are identified to ensure land management is conducted on a sustainable basis.

Any research conducted should:

- be developed in association with local communities and social scientists;
- build on past experience;
- use interdisciplinary teams;
- be targeted at specific agroecosystems

to obtain better information on land-use systems and changes in key soil properties related to soil resilience.

Soil resilience is a concept embracing many aspects discussed in this volume. A simplified definition covering the most important aspects is

The soil's ability to recover after disturbance.

International Advisory Committee:
Professor D.J. Greenland (CABI, Co-Chairman), Professor I. Szabolcs (HAS, Co-Chairman), Mr J.L. Nowland (CABI, Convenor), Professor I. Láng (Secretary General, HAS), Dr M.S. Swaminathan (CABI, Chairman, Governing Board), Professor R.S. Swift (Chairman, ISSS Commission II), Professor J.M. Lynch (Chairman, ISSS Commission III and Coordinator, OECD Agricultural Project Theme I), Dr G.S. Sekhon (Chairman, ISSS Commission VI), Professor G. Varallyay (HAS), Dr L.F. Elliott (USDA), Dr P.B. Tinker (NERC, UK) and Professor J.K. Syers (IBSRAM, Thailand).

Programme Committee:
D.J. Greenland (Chairman), I. Szabolcs, J.L. Nowland, R.S. Swift, J.M. Lynch, P.B. Tinker and J.K. Syers.

Organising Committee:
I. Szabolcs (Chairman), J.L. Nowland, I. Láng, D.J. Greenland, J.M. Lynch, G. Varallyay, R.S. Swift, P.B. Tinker and J.K. Syers.

Sponsors:
The Hungarian Academy of Sciences (HAS), CAB International (CABI) and The International Society of Soil Science (ISSS)
with additional support from:
The United Nations Environment Programme (UNEP), The European Environment Research Organisation (EERO), The Organisation for Economic Cooperation and Development (OECD), The Ford Foundation (FF), The Commission on the Application of Science to Agriculture, Forestry and Aquaculture (CASAFA) of the International Council of Scientific Unions (ICSU) and The British Council (BC).

Foreword

Land-based crop and animal husbandry is the major anchor of the global food security system. Fish catches are already reaching the limits of sustainable levels in most oceans. It is in this context one has to view with concern the continuing soil degradation and diversion of prime farm land for non-farm purposes. The problem is particularly acute in South and Southeast Asia, where in view of the growing pressure of population on land and water, there is no option except to achieve higher productivity per unit of land, water, energy and time.

China, for example, has set an annual food grain production target of 500 million metric tonnes by the year 2000, as compared to the present production level of 440 million tonnes per year. The National Report prepared by China in August 1991 for the UN Conference on Environment and Development states, 'The ecosystem of China's agriculture is faced with pressures greater than those of any other country'. The increased use of mineral fertilizer has been responsible for about half of the 100 million tonne increase in grain production between 1978 and 1988, as well as for much of the increase in commercial and high value crops. Nitrate pollution of ground-water is already an environmental problem. The productivity of soils has also begun to decline in some farming areas because of the use of continuous monocropping instead of crop rotations that sustain nutrient levels and limit pests and diseases.

The Punjab region of India and the other green revolution areas face similar problems. Population pressure also leads to the diversion of forest land for crop husbandry and to the destruction of habitats rich in biodiversity. This publication on Soil Resilience and Sustainable Land Use is thus a timely contribution. The papers included in this volume provide a wealth of information on scientific land management. The suggestions contained in the different papers will have to be converted into national level action plans. In this context, I would like to cite two examples from the work being done in India.

The first relates to scientific land use planning. India has nearly 100 million operational holdings, a majority of which are below 2 ha in size. Under such conditions farmers are being encouraged to form their own *Land Use Associations,* where the available arable land in an area is classified into the following three groups:

1. *Conservation areas,* such as protected areas, erodible land etc.
2. *Restoration areas,* where an eco-redevelopment programme will have to be initiated based on a malady—remedy analysis. India alone has over 75 million ha, classified as wastelands, due to a variety of abiotic stresses.
3. *Sustainable intensification areas.* Such land should not be diverted for non-farm uses and should be subjected to an effective soil health monitoring system.

A second area of action relates to the setting up of a *Biological Software Centre for Sustainable Soil Health Management.* Such centres are designed for assembling products and processes which can help to maintain/enhance soil health and productivity. The software could include items which can help improve the chemical, physical and microbiological aspects of soil fertility maintenance.

Some examples of such software are:

1. Earthworms and vermiculture.
2. Nitrogen-fixing trees and shrubs including stem nodulating species.
3. Rhizobial cultures, Azolla, blue green algae.
4. Tree species like Neem whose seed cake promotes slow release of mineral nitrogen from applied urea, and
5. Plants which help to control nematodes and soil pathogens.

A 'Sustainable Soil Health Software Library' could provide the most appropriate material to users, depending on the nature of the soil and farming system.

This book indicates how through appropriate packages of technology, services and public policies we can arrest further soil degradation and promote sustainable land use. We owe a deep debt of gratitude to Professor D.J. Greenland and Professor I. Szabolcs and to CAB International and the Hungarian Academy of Sciences for their invaluable contributions to organizing this Workshop and publishing the proceedings speedily.

M.S. Swaminathan, FRS
Centre for Research on Sustainable Agricultural
and Rural Development, Madras

Part I

Sustainable Agriculture and Soil Resilience

Chapter 1
The Ecological Foundations of Sustainable Land Use: Hungarian Agriculture and the Way to Sustainability

I. Láng

Hungarian Academy of Sciences, Roosevelt tér 9, H-1051 Budapest, Hungary

The future of the world depends on the way the interrelations among population growth, energy resource utilization and environment protection will be harmonized. The rapid growth of world population is a fact which determines political, economic, cultural, and scientific activities alike. From statistical data and forecasts the following trend emerges:

1950 – 2.5 billion
1960 – 3.1 billion
1970 – 3.7 billion
1980 – 4.4 billion
1990 – 5.2 billion
2000 – 6.2 billion
2025 – 8.4 billion

That is, there is an average growth of about one billion persons every ten years, which means ten times the population of Hungary every year.

At the same time, the percentage shares of the two main groups of regions, namely, the more and the less developed ones, is also changing and the proportion of people living in developing countries is significantly increasing (Table 1.1).

This shift in population ratios also means that the role and political weight of the developing countries will increase in international forums. The possibility to satisfy the basic human needs of this growing population belongs among the primary human rights. Hence, the provision of satisfactory conditions for this is a global interest and a global duty. There cannot be lasting peace and security on Earth as long as certain regions suffer from serious food shortages while others have problems of overproduction and excessive consumption. This is especially true for a world where everybody can be informed about events elsewhere in a short time

Table 1.1. Percentage share of world population.

	1950	1975	1990	2000	2025
More developed regions	33.1	26.9	23.6	20.2	16.0
Less developed regions	66.9	73.1	76.4	79.8	84.0

Source: *World Resources* 1990–1991. Washington DC

due to a developed information technology and where the global spread of nuclear and other weapons of mass-destruction could not be limited and controlled in a satisfactory way.

The concept of 'sustainable development' was first formulated in the early 1980s when Lester R. Brown's book *Building a Sustainable Society* was published in 1981. This summarized views suggesting that it would be possible to harmonize the material needs of society, the growth of population and the rational utilization of natural resources so that environmental pollution would also be minimized. This new economic development model is called 'sustainable development'. In the years that followed, scientists have dealt increasingly with the various aspects of sustainable development.

In December 1983 the UN General Assembly established the World Commission on Environment and Development, also known as the Brundtland Commission, as it was chaired by the Norwegian politician Mrs Gro Harlem Brundtland.

In February 1987, the Commission reported on its work in the book *Our Common Future* which the UN General Assembly later discussed and expressed its general acceptance of it. The report focused on the concept of sustainable development. It did not suggest, as did the report of the Club of Rome, *Limits to Growth,* that economic growth should be restrained. *Our Common Future* concluded instead that people in the Third World would only be able to satisfy their basic needs and eventually a higher level if economic development continued in their countries. Without such development, the Commission argued, poverty would be conserved on a world scale for the sake of environmental protection, and such a situation would be unjust and unacceptable.

The Report, *Our Common Future,* provided the hope that it would be possible to satisfy basic human needs for the entire human population and to manage the natural resources and the environment so that future generations will not be any worse off than the present one.

The concept of sustainable development, as presented in the Brundtland Commission's report, integrated economic and environmental policies, so that in cases of conflict between the two, ecological interests are given preference. However, fundamental changes are needed in society before sustainable development can be achieved. Changes need to be made in the relations between society as a whole and certain groups within it as well as the state organs responsible for environmental protection. Representatives of the various social groups need to be involved in the preparation of environmental decisions, in certain cases also in decision-making and in all cases in controlling the execution. Information on environmental issues should be available to the public. Above all, a democratic and self-governing society is a precondition for sustainable development.

Sustainable development will also require a modification in the lifestyle of the developed countries and should include a voluntary decrease of wasteful consumption, the development of a new value system, and a new relation to nature.

The technological aspects of sustainable development include radical decreases in energy and material use, and increases of energy efficiency, recycling, and the production of environmentally sound products.

The report, *Our Common Future,* stressed the need for the regulation of world population, but in such a way that does not conflict with cultural traditions and religious faith. The report also discussed the protection of internationally shared resources, 'the global commons'. These include the oceans and seas, the atmosphere, and outer space. The Commission also attached great significance to the protection of tropical rainforests, which have an important role in climate regulation, as well as in preventing desertification.

Economic prerequisites for sustainable development, as identified by the report, were: debt reduction for some countries and a decrease in military spending. The large debts of some countries have hindered economic progress, while military spending does not have much of a useful purpose. Reductions from the latter could well be used for environmental protection. Easing military tension and relieving poverty could, in an indirect way, effectively promote global environmental protection.

According to the definition of the Brundtland Report, 'Sustainable development is development that meets the needs of the present without compromising the ability of future generations to meet their own needs'.

The strategic imperatives for sustainable development are as follows:

- Reviving growth
- Changing the quality of growth
- Meeting essential human needs
- Ensuring a sustainable level of population
- Conserving and enhancing the resource base
- Re-orienting technology and managing risks
- Merging environment and economics in decision-making.

The report, *Our Common Future,* has been praised as well as criticized. On the whole, however, it has had a positive impact as it has attracted the World's attention to the connection between environment and development, opened up new ways of thinking about society, and mobilized institutions and people to act for sustainable development. The principle of sustainable development is now accepted by most governments and environmental movements.

On the initiation of the report, *Our Common Future,* the UN General Assembly decided in December 1989 to convene the UN Conference on Environment and Development which was held in Rio de Janeiro, Brazil on 3–14 June 1992.

The following documents were approved at the Conference in Rio:

- Rio Declaration on the Environment and Development
- Framework Convention on Climate Change
- Convention on Biological Diversity
- Declaration on the principles relating to forests
- AGENDA 21

The agricultural issues can be found in each document in one form or another.

The Rio Declaration includes 27 principles. These are general principles to be applied to different kinds of economic activity. Consequently, agriculture is not named separately, nor are industry, transport and communication. Let me quote Principles 3 and 4 as examples:

> The right to development must be fulfilled so as to equitably meet developmental and environmental needs of present and future generations.
>
> In order to achieve sustainable development, environmental protection shall constitute an integral part of the development process and cannot be considered in isolation from it.

The Framework Convention on Climate Change takes into account the carbon-absorbing capacity of newly planted forests when it deals with the calculation of the reduction of CO_2 emissions. This is a new development.

The agreement on biodiversity upgrades the economic significance of biological resources. This might greatly affect later plant improvement and animal husbandry.

The document on the 'Declaration on the principles relating to forests' particularly urges the protection and rational utilization of natural forests. However, it also deals with the issues of planted forests as follows:

> The role of planted forests and permanent agricultural crops as sustainable and environmentally sound sources of renewable energy and industrial raw material should be recognized, enhanced and promoted. Their contribution to the maintenance of ecological processes, to offsetting pressure on primary/old-growth forest and to providing regional employment and development with the adequate involvement of local inhabitants should be recognized and enhanced.

The document AGENDA 21 contains recommendations for international organizations, national governments and various social groups. It consists of 40 chapters. Among them, several are directly related to agricultural production. They are as follows:

- Integrated approach to the planning and management of land resources
- Promoting sustainable agriculture and rural development
- Strengthening the role of farmers.

The chapter on land resources in the document AGENDA 21 enlists the following objectives:

1. To review and develop policies to support the best possible use of land and the sustainable management of land resources.
2. To improve and strengthen planning, management and evaluation systems for land and land resources.
3. To strengthen institutions and coordinating mechanisms for land and land resources.
4. To create mechanisms to facilitate the active involvement and participation of all concerned, particularly communities and people at the local level, in decision-making on land use and management.

The following recommended programme areas can be found in the chapter on sustainable agriculture:

1. Agricultural policy review, planning and integrated programming in the light of the multifunctional aspects of agriculture, particularly with regard to food security and sustainable development.
2. Ensuring people's participation and promoting human resource development for sustainable agriculture.
3. Improving farm production and farming systems through diversification of farm and non-farm employment and infrastructure development.
4. Land-resource planning, information and education for agriculture.
5. Land conservation and rehabilitation.
6. Water for sustainable food production and sustainable rural development.
7. Conservation and sustainable utilization of plant genetic resources for food and sustainable agriculture.
8. Conservation and sustainable utilization of animal genetic resources for sustainable development.
9. Integrated pest management and control in agriculture.
10. Sustainable plant nutrition to increase food production.
11. Rural energy transition to enhance productivity.
12. Evaluation of the effects of ultraviolet radiation on plants and animals caused by the depletion of the stratospheric ozone layer.

In the chapter entitled 'Strengthening the role of farmers' of the document AGENDA 21, the following objectives can be found:

1. To encourage a decentralized decision-making process through the creation and strengthening of local and village organizations that would delegate power and responsibility to primary users of natural resources.
2. To support and enhance the legal capacity of women and vulnerable groups with regard to access, use, and tenure of land.
3. To promote and encourage sustainable farming practices and technologies.
4. To introduce or strengthen policies that would encourage self-sufficiency in low-input and low-energy technologies, including indigenous practices and pricing mechanisms that internalize environmental costs.
5. To develop a policy framework that provides incentives and motivation among farmers for sustainable and efficient farming practices.
6. To enhance the participation of farmers, men and women, in the design and implementation of policies directed towards these needs, through their representative organizations.

These recommendations will surely have an impact on the agrarian policy of many countries, Hungary among them, as well as on the research programmes.

Following this global overview, I am going to deal with the situation of Hungarian agriculture and with the outlining of the possible ways leading to sustainable development.

Hungarian Agriculture

Hungarian agriculture is in a specific situation. In the course of the democratic parliamentary elections held in 1990, the majority of the population voted for parliamentary democracy, involving a multi-party system, overwhelming private property, market-oriented economy, a constitutional state and the guaranteeing of civil rights. All this entails fundamental changes in agriculture, both in ownership and production relations. This process coincides with the demand for change – also existing in the developed countries – leading to sustainable development. That is, these two processes historically coincide in Hungary, hence, it is especially important to shape them in a coordinated and positively interacting way.

According to the definition formulated by the IUCN, UNEP and WWF, sustainable development is: 'improving the quality of human life while living within the carrying capacity of supporting ecosystems'.

Sustainable Nutrition Security was defined by Swaminathan (1987) as 'providing physical and economic access to balanced diets and safe drinking water to all people at all times'.

Hungary's natural resources have been studied thoroughly in the last one hundred years. We have deep knowledge of the soil, water and biological resources of the country, of the changes in climate, of the effectiveness of various management and production systems. It is a fact that the population of the country – which is 10.3 million at present – as well as the numerous tourists who visit the country can be supplied with basic food products. Moreover, food is also exported from Hungary. The value of food exports was US$2.5 billion in 1991.

The dynamics of *grain production* can be best characterized with the following figures (in million tonnes):

1938 – 7.1
1950 – 5.4
1960 – 6.9
1970 – 7.6
1980 – 14.0
1990 – 12.5
1991 – 15.6

The value of the indicator 'tonnes of cereal per capita' significantly exceeds one, which makes it possible to intensively breed poultry and pigs, or to export grains, or to do both.

The dynamics of meat production (in million tonnes) was as follows:

1938 – 0.751
1950 – 0.838
1960 – 1.070
1970 – 1.343
1980 – 2.018
1990 – 2.247
1991 – 1.983

Consequently the sustainable development of agriculture does not mean a guarantee of food security for the Hungarian population, as the conditions for this are given in the country's natural resources. However, economic access to a balanced diet is not simply a production issue; it is a complex social process, interrelated with the value of the GDP and the income-earning capacity of the various social groups. In this respect, the situation is not as simple as in the case of per capita agricultural production in which Hungary is among the leaders on the international 'top list'. These interrelations will be returned to later.

One of the key issues of Hungarian agricultural development is to decide what ratios agricultural and food products account for in the imports and exports of the country. The strategy used for the realization of such a decision and whether it succeeds under the competitive conditions of a market, with agricultural production being supported in many places, can fundamentally influence sustainable development in agriculture.

The ratio of agricultural and food products within Hungarian imports was 7.5% in 1987 and 6% in 1991. The corresponding figures in exports were 19% in 1987 and 25% in 1991. This indicates that Hungarian agriculture and food industry greatly contribute to the stability of the country's economy. Within exports, livestock and animal products account for 45%, garden products 26%, grains and fodder 17%, and other plant products 12% in value.

It is a strategic issue whether agricultural and food exports can be decreased without impairing the objectives of national development. If they can, less production would suffice, which in turn would help the conservation of natural resources. If they cannot, because the related export incomes are needed, the question is how goods of such a value can be produced with less damage to the environment. These issues are the key challenges for Hungarian agriculture.

I, personally, think that agricultural and food exports will be needed in the future too; however, it is not quantity but quality that should be raised and the marketing of more valuable products should be improved.

Land resources

Hungarian indicators relating to land resources are relatively good, both in respect of arable land per capita (0.47 ha) and in that of its utilization. As a result of research carried out for several decades, we have precise scientific knowledge of the properties of the country's soils.

It is advantageous for Hungary that the ratio of utilizable agricultural area compared to the total area of the country is relatively high (69.4%). If, besides arable land, we also take into account forests, reedy areas and fishponds, i.e. all regions yielding utilizable biological products, this figure will be as high as 88.4%.

In the last half century, the area of arable land has significantly decreased, while that of forests has increased (Table 1.2).

The expansion of settlements, industrial plants, the development of roads and the infrastructure have taken away large areas from agriculture in

Table 1.2. Changes in the area of arable and forest land.

Year	Arable land (000 ha)	Forest (000 ha)
1938	5618	1106
1950	5518	1165
1960	5310	1306
1970	5046	1470
1980	4735	1610
1990	4712	1695

Hungary, as has happened in other countries of the world. In April 1992 the Hungarian government decided on the afforestation of an additional 150,000 ha by the end of the century. This will also serve the purpose of stabilizing the CO_2 content of the air.

In respect of the utilization of land resources, we have to differentiate between the owner and the user of the land. Before the Second World War, most of the land in Hungary was privately owned, a smaller part of it was in the possession of communities and in state ownership. Hungarian agriculture was characterized by the predominant large estates, by the smaller peasant farms and by the agricultural workers who did not own any land. Following the post-war democratic land reform – when the large estates were distributed among the peasants – a great number of so-called dwarf holdings were formed. Some parts of the large estates were transformed into state farms. In the early 1950s a process started in the course of which private farms were forced – 'from above' – into production-type cooperatives, at the end of which the majority of the land became the property of cooperatives.

The distribution of land according to *groups of utilization and ownership* was as shown in Table 1.3 (in May 1989).

Following the democratic elections in the spring of 1990, the Parliament declared the principle and legally regulated the restoration of private ownership. In the case of land, this resulted in the earlier land owners putting in a claim for compensation, so that they might become land owners again. Thus the lands owned by the state and cooperatives will become privately owned – except those which remain government property. Land will be given to the new owners in the autumn of 1992 and the spring of 1993, and thus this year and next will be critical in respect of the development of agriculture.

Cooperative farms will also be transformed, separated or terminated according to the decision of the members among whom the property of the

Table 1.3. Distribution of land in Hungary in May 1989.

	Area (ha)	%
State farms and enterprises	2,667,554	28.68
Farms belonging to councils and other communities	278,142	2.99
Various individual small farms	678,148	7.28
Cooperative farms	5,679,191	61.05
In cooperative ownership	3,471,311	–
Privately owned	1,991,734	–
In state ownership	216,146	–
Total	9,303,035	100.00

cooperative will be distributed. They can decide then whether to leave the cooperative and continue their activity as independent private farmers, or – now as real owners – to form a new type of cooperative and to continue farming within that organization.

On the basis of experiences gained so far it is probable that in about two-thirds of the 1300 earlier cooperatives the new-type cooperative will be chosen, because the financial and technical conditions for successful separation from the cooperatives on a mass scale and the development of independent farming are not provided at present. Also the existing machinery, equipment and infrastructure can only be utilized in an efficient way within the frame of cooperatives.

Of the 134 state farms 23 will remain the property of the government, at least in part, 25–90%, (depending on the given farm) remaining government property; the rest will be privatized.

It is my personal opinion that this transformation in land ownership coinciding with marketing and financial crises and with drought will most probably lead to a temporary decrease in agricultural production, which might last for a few years. This decrease, however, is not expected to be so great as to endanger the food supply of the country.

In the last 40 years several adverse effects have had an impact on the soil resources of Hungary. They are:

- Soil erosion caused by water and wind;
- Physical degradation of the soil, and soil compaction;
- Acidification of the soil;
- Secondary salinization and alkalization.

Of these kinds of soil degradation, erosion is the most widespread and can be found on 40% of the country's arable land.

The earlier economic policy urging ever greater quantitative growth and large-scale farming has led to the formation of overlarge fields by abolishing roads and rows of trees, often resulting in merging soils of different quality in one field. The configurations of the terrain have often been also disregarded.

Big and heavy tractors and machines which have been used for intensive soil cultivation have caused soil compaction. More reasonable farm and field sizes will make it possible to reduce these kinds of damage and to promote sustainable development.

Soil amelioration has unfortunately diminished in Hungarian agriculture. It was done on 78,000 hectares in 1990 and only on 40,000 hectares in 1991. In practice this meant the liming of acidic soils. The amelioration of salt-affected and sandy soils has practically stopped. On the whole, it is a mere 1% of the arable land on which amelioration is still carried out. This is due to the radical decrease in financial support by the state and to the lack of capital on the part of the farms. Without larger-scale amelioration programmes, the sustainable development of agriculture is not possible.

Following the changes in land ownership, trading in land will start and, consequently, land will again have a price and value. Hungarian peasants have not been used to this in the last 40 years, so they have to learn it again. This will provide an opportunity to appreciate and protect natural resources more properly. However, this does not come automatically. There are warning examples of soil degradation also in countries with traditional parliamentary democracy and market-oriented economy. The organization of consulting services helping the farmers' work and the realization of government interventions and financial support in the interests of promoting the protection of the natural resources will be of special importance in the case of Hungary. Land, however, is not only the property of the individual, it is at the same time a national treasure; so the state should also take part in its protection.

Water resources

Hungarian agriculture is dependent on the quantity of natural precipitation and its distribution over the year. The national average of precipitation varied between 414 and 660 mm between 1983 and 1991. Within this, the distribution among regions shows the values of 400–700 mm. The number of days with precipitation is 110–130.

In 1990, 204,000 hectares were irrigated and 148,000 hectares in 1991. About 3–5% of the area of fields and gardens where plants are produced are irrigated. A higher ratio is not expected in the future. Irrigation is expensive, and is only profitable where valuable plants, like fruit, vegetables, flowers, or high-quality seeds are produced, and in gardens around houses.

Hence, Hungarian agriculture will primarily be characterized by *dry farming* in the long run. This should be harmonized with sustainable development. The soil stores the fallen precipitation in a natural way. Hence soil and plant cultivation and the applied technology should be chosen so as to make the soil suitable for this function.

The possibility of *global warming* is a great challenge for Hungarian agriculture. If such a change of climate also occurs in Europe, it will have an unfavourable effect on Hungary. The country's agriculture is exposed to drought. In four out of the last ten years (1982–1991), the national average precipitation was less than 500 mm, implicating a greater frequency of drought than was experienced in the last 50 years. Unfortunately, the drought of 1990 was followed by a very hot summer in 1992. It is disputable whether this is due to global warming, but it is a fact that drought has occurred in Hungary more often in recent years than in earlier decades.

From all this, we can draw the conclusion that an adaptive strategy should be worked out for Hungarian agriculture which seriously takes into account the decreasing quantity of precipitation and the increasing average temperature. In this respect, stress should be laid on water-saving farming, which is a complex system, involving the improvement of infiltration into the soil, mulching, the use of smooth rollers, more widespread application of irrigation, the acquisition and use of up-to-date machinery necessary for soil cultivation, and sowing to be done at the proper time, the stock-piling of fodder and seeds, the breeding of drought-resistant varieties etc. All this agrees with the concept of sustainable development.

Energy

Total energy consumption in Hungary was 1316 PJ (petajoule) in 1989, which decreased to 1175 PJ in 1991. This decrease was due to the fact that the government terminated a great part of the energy- and material-intensive production in the course of transforming the industrial structure. As in the whole East-European region, in Hungary too, energy efficiency greatly lags behind that of Western Europe. According to *World Resources 1992–93*, energy consumption in relation to GNP in constant US$ was 46 MJ in Hungary, only 7 MJ in Austria and 9 MJ in Finland in 1989. In the same year, energy consumption per capita in gigajoules was as follows:

Austria – 117
Finland – 169
Hungary – 107

That is, energy consumption per capita is hardly less in Hungary than in Austria, whereas five times as much energy is needed in relation to the production of the GDP. In Finland per capita energy consumption is 50% greater than in Hungary, but this is due to the colder climate.

Agriculture consumed 73 PJ at the end of the 1980s. From this, diesel fuel and petrol represented 61%, coal 4.4%, and electric energy 9.1%. Within the total energy balance, agriculture consumes 6–7%, together with forestry and water management 9–10%. These data refer to direct energy consumption. If we also calculate the energy necessary for the production of all materials and machinery applied in agriculture, this value will double.

According to the principle of sustainable development, much attention should be paid to the saving of energy in every field. In this respect, there also are significant reserves in agriculture. Soil cultivation, transportation, fodder drying, and the so-called closed animal keeping systems are all great consumers of energy. The use of fuel-saving tractors and trucks, for example, would lead to significant energy savings. Part of the imported materials could be substituted by agricultural by-products. For instance, in the heating of settlements, coal could be substituted by wood, by bio-briquet; organic manure can be used to produce bio-gas; a part of the nitrogen-based mineral fertilizers can be replaced by nitrogen produced through biological fixation; some wrapping materials can be substituted by materials produced from local by-products.

Solar energy can directly be utilized primarily in drying and partly in the heating of dwelling houses, sheds and stables. Significant energy-savings can be achieved through working procedures in which energy optimization is the primary objective.

The government has drastically reduced the budgetary support for energy consumption in the past two years. As a result, prices have gone up.

Coal briquet	1989 – Ft. 113 100 kg^{-1}
	1992 – Ft. 699 100 kg^{-1}
Natural gas	1989 – Ft. 3.24 m^{-3}
	1992 – Ft. 7.80 m^{-3}
Petrol (super)	1989 – Ft. 24 $litre^{-1}$
	1992 – Ft. 68 $litre^{-1}$

The high prices have decreased the profitability of agricultural production, but have definitely enhanced energy savings.

Biological diversity

There are 415 plant species and 619 animal species under protection in Hungary at present: 6% of the territory of the country is protected. Also protected are all the (approximately 2500) caves. In connection with the privatization of land, the settlement of the right of ownership and the compensation to be paid to the earlier owners of protected areas arise as new problems.

Forests occupy 18% of the country's territory. In the long run, this ratio will grow as a result of the afforestation of areas which are less suitable for farming as well as through the planting of trees on smaller areas, along roads and ditches. With new plantings, foresters prefer to use poplars, which grow quickly, acacia, Scots fir and black pine trees. Ecologists dispute the correctness of this practice, because the species are not native to Hungary and their growing in monoculture changes the original ecosystem. However, it is also a fact that acacia was planted in sandy and alkaline soils on Hungarian territory as early as the 18th century.

As regards the primary functions of forests, 80% serve economic, 16% protective and 4% recreational purposes.

According to the Hungarian 'Red Book', published in 1990, 2% of the one million plant species of the Earth and 1% of the estimated 45,000 vertebrate species can be found in Hungary. The Red Book classifies 110 vertebrates, 210 invertebrates and 730 plants as potentially endangered species.

Large-scale agricultural production has impoverished the genetic stock of plants and animals raised, although it has definitely contributed to obtaining higher yields. In the future, biodiversity should be increased, and the development of sudden stress-bearing populations as well as well-adaptive and disease-resistant varieties should be given preference. It is the task of scientific research to help develop this tendency with due consideration for the future. There is also a change needed in the system of qualification of varieties, giving preference to the ecologically adaptive new species, instead of applying the one-sided principle based on quantitative growth.

The former WEFSA dealt with the issue of biodiversity, and the lectures have been published in *The Biodiversity of Microorganisms and Invertebrates; Its Role in Sustainable Agriculture* (ed. D.L. Hawksworth, published by CAB International).

Use of chemicals

The spectacular growth achieved in Hungarian agriculture in the 1970s was primarily the result of the greatly increased use of *mineral fertilizers and pesticides*. The dynamics of the use of mineral fertilizers are shown in Table 1.4. There is no exact figure for 1991 yet, but it is estimated to be less than 100 kg ha^{-1}.

The reduction in the use of mineral fertilizers has economic reasons: money is short, the prices of industrial products have gone up more than those of the agricultural ones, and consequently the farms bought less fertilizer to save money. This is possible now without causing much damage because, as a result of the earlier intensive use of mineral fertilizers, the soil is still saturated with plant nutrients.

The intensive use of mineral fertilizers had negative environmental effects: soils poor in calcium became acidified. The heavy use of such fertilizers has also contributed to the increase of the nitrate content of underground waters in many parts of the country.

Table 1.4. Use of fertilizers in Hungary between 1960 and 1990.

Year	NPK active agent kg ha^{-1} agricultural area
1960	24
1970	122
1980	260
1985	305
1988	290
1989	275
1990	104

I think that in order to achieve the social objective of being self-sufficient in food and also of being able to export agricultural products, the rational and professional application of mineral fertilizers is not dispensable. Organic production methods are necessary, of course, but they cannot be applied to the whole territory of the country. According to my estimates, a national average of 150 kg ha^{-1} mineral fertilizer (active agent) will be necessary also in the future.

The use of *pesticides* also grew notably in the 1970s. In plant-growing, chemical protection was applied to twice as large a territory as the area of arable land (that is, the plants were treated twice on the average), whereas in orchards and vineyards this value was 5 to 6 times as high. The negative effects were apparent in the damage caused to the natural ecosystems, whereas pests and pathogens have become more resistant and, as a consequence, ever more effective pesticides have had to be applied. In the future, *integrated plant protection methods* will have to be used, including the greater use of resistant varieties, the application of biological protection, reduction of the quantity of chemicals used, and the application of such superselective agents and agrotechnological methods which will reduce only the population of undesired insects and fungi.

The spreading of professional knowledge and the establishment of advisory services for farmers have great significance in this respect.

Hobby gardening

Hobby gardening has been spreading all over the world and has become an especially popular activity in the industrially developed countries. According to the article 'The Gift of Gardening' hobby gardening is practised in 80% of the 9.3 million households in the United States. In Hungary this ratio is lower; we know about 830,000 hobby gardens, covering an area of about 48,000 ha. Gardening is, in the first place, a way of life. It provides recreation, relaxation and helps develop a feeling of closeness to nature.

The economic value produced in these small gardens should also not be underestimated. It contributes to the income of these families. As a result of the restructuring of Hungarian economy, serious social problems have arisen: unemployment rate is about 10% and it might reach 17–20%; more than 20% of the population live at or below the minimum subsistence level. So the reduction of expenditures and self-sufficiency in cheap food are social issues which also belong to sustainable development.

It is important that in the course of privatizing the land, local governments should allow opportunities for the development of hobby gardens in the territories close to the settlements. This would help the population to spend its spare time in a healthy way, and at the same time make it possible for these people to earn a modest but not insignificant extra income.

I would like to mention at this point that the number of lacto-vegetarians, semi-vegetarians and vegetarians in Hungary has been growing. These tendencies should be helped by various means, for example, by supporting sanitary propaganda and by granting special tax exemptions for products replacing meat. On the whole, the propagation of general knowledge of healthy nourishment should be improved in Hungary.

Postharvest systems

The way to sustainable development in agriculture cannot be restricted to the energy- and material-saving production technologies of basic goods. The processes following harvesting are also important, and thinking in the complex system of total biological production, as well as acting accordingly, is also a constituent of sustainable development.

According to a survey made in the 1980s, calculated in yearly average, 85.5% of the Hungarian plant biomass was formed in agriculture and 14.5% in forestry. The ratio of primary agricultural products is 40.7%, that of agricultural by-products 44.8%. Primary products in forestry represent 12.3%, whereas by-products in forestry 2.2% of the total biomass production.

About 64.7% of the primary agricultural products are used for forage and will only become utilizable animal products, i.e., meat, milk, eggs, etc. after transformations with great losses. It is the natural function of manure, a by-product of animal-keeping, to be returned to the soil to improve its quality. With concentrated animal-keeping, however, liquid manure can cause serious environmental damage. Hence, the decentralization of production, the formation of smaller stocks and a re-introduction of animal-keeping in litters would all promote the protection of the environment.

With postharvest systems, in the case of primary products, the development of local storage capacity and the establishment of decentralized primary processing factories might be those objectives of the national agrarian policy which at the same time help the realization of sustainable development. Decentralized processing provides job opportunities and simultaneously reduces the stress on the environment.

In the case of by-products and possibly also wastes, the aim is to utilize them locally and completely. By-products and wastes can be used to produce fodder, industrial raw materials, energy and manure. The utilization of by-products and wastes is an environmental demand; however, it also helps to achieve profitability in farming.

Conclusions

The goals identified by the US National Research Council in its Report on

Alternative Agriculture (1989) are also very informative for Hungarian agriculture:

1. Incorporating natural processes such as nutrient cycles, nitrogen fixation, and pest–predator relationship into the agricultural production process.
2. Reducing the use of off-farm inputs with the greatest potential to harm the environment and the health of farmers and consumers.
3. Making greater use of the biological and genetic potential of plant and animal species.
4. Improving the match between cropping patterns and the productive potential and physical limitations of agricultural lands to ensure long-term sustainability of current production levels.
5. Emphasizing improved farm management and conservation of soil, water, energy and biological resources.

These recommendations also apply to Hungarian agriculture. However, due to its unique historical background, these should be complemented with some more points enabling the achievement of sustainability:

- The counterbalancing and/or elimination of the deficiencies and damages resulting from the one-sided large-scale system and quantitative development that characterized the earlier decades;
- Consciously influencing the new ownership relations and farm systems in the interests of sustainable development;
- Preventing the adverse developments which usually appear in the early stages of an evolving market economy (wasting or causing damage to free natural resources, deforestation, exhaustion of the soil, etc.);
- Rapid dissemination of the new trends, methods and results of research.

Finally, I would like to express my thanks for the help of my colleagues who have contributed to the realization of this chapter.

References

Swaminathan, M.S. (1987) Building national and global food security systems. In: *Global Aspects of Food Production,* Oxford University Press, Oxford, pp. 417–449.
The gift of gardening. (1992) *National Geographic,* No. 2.
US National Research Council (1989) *Report on Alternative Agriculture.*

Chapter 2
Soil Resilience and Sustainable Land Management in the Context of AGENDA 21

H. Eswaran

USDA-Soil Conservation Service, PO Box 2890, Washington DC 20013, USA

Introduction

With only seven years to go before the end of the century, the global society is awaiting a paradigm shift in addressing global problems. The 'Earth Summit', which was recently convened in Rio de Janeiro, was, among other things, an admission by world leaders that major adjustments to current trends in land use were needed. This political impetus comes from environmental and not from agricultural concerns though the latter are equally disturbing and the cause of some of the environmental problems. Irrespective of the motivating reasons of the political community, the result will probably be the paradigm shift we have awaited and which will necessarily force us to look at land use and land management in a new light and require us to address new problems, develop new methods, and in general make radical changes in our research and development agenda.

The two driving forces in global concerns are sustainable agriculture and global climate change, and both are major propellants of research and development due to the political awareness and support they have received. No other efforts in recent history have received the kinds or magnitude of support from the global community. The soil science purist will argue that as soils are the basis for both these concerns, a major endeavour must be based on soils. Despite the validity of this, recent trends show less emphasis on soil research by national decision-makers and international donors. Consequently, only a small fraction of funding is going into soil research.

The thrust towards environmental sustainability was triggered by the report of the Bruntland Commission (WCED, 1987) which for the first time made a careful analysis of the finite limits to global resources and alerted the general public to the fragile nature of the living planet.

The concept of sustainable agriculture becomes pertinent and takes on a new dimension when viewed in the context of limits to resource availability and use (Dumanski *et al.* 1992). If such limits do not exist or are perceived not to exist, exploitation with disregard to consequences is the result, as has happened in the past. Sustainability hinges on the premise that arable agricultural land is finite. It also assumes that much of the land that is suitable for agriculture is already in use except for a few countries in Africa and South America. Of the 13.4 billion hectares of ice-free land on the world's surface, only about 25% is potentially arable. Of this potentially arable land, about 40% is suitable for productive agriculture and the remaining areas, largely in the intertropical regions, require high inputs and good management for sustained productivity.

The emerging crises are primarily related to the rising population competing for the limited land resource. Perhaps for the first time in the history of mankind, there will be insufficient land for agricultural expansion. Mankind is faced with the challenge of not only increasing productivity but also preserving the resource base to achieve intergenerational equity.

The Issues

The Food and Agriculture Organization (FAO, 1989) has reported that in the tropics cultivated land per person fell from 0.28 ha in 1971 to 0.22 ha in 1986. This is evidence of the great pressure on available land. In Southeast Asia and the Near East the amount of land cultivated is already close to the amount of potential land suitable for cultivation. With geometric increases in population, availability of land for agriculture becomes a critical resource issue in most of these countries.

Major issues which directly and indirectly require attention are:

- *Limits of land resource.* Many countries have reached or will be reaching the limits of their arable land resource base by the end of this century. Increased food and fibre production must result from increased productivity of land rather than increased land for production.
- *Rural poverty and unequal equity distribution.* Continued migration of people from rural to urban areas results partly from inadequate opportunities and poor support of agro-based industries in rural areas.
- *Degradation of arable land.* Degradation, and in some instances urbanization, of arable land reduces land available for agriculture at accelerating rates. This results in increasing use of marginal lands and fragile ecosystems which are more difficult to maintain under sustainable production.
- *Weak institutional framework in NARS.* There is a continuing lack of infrastructural and suitably qualified personnel in developing countries to coordinate and conduct research which reduces the ability of

National Agricultural Research Systems (NARS) to address problems and respond to changing needs.

- *Research emphasis on past concerns rather than future problems.* With the increasing pressure on land there is a need for a greater emphasis on resource-focused as opposed to commodity-focused research, and the need to provide an ecosystem emphasis in agronomic research and transfer. Additionally a holistic systems approach to agronomic research is required to address globally relevant problems through regional agro-ecological networks.
- *Global climate change.* Future uncertainties about anticipated climate change, specifically as these changes affect sustainability of production, risks associated with production, and the impact of climate change on the quality of the resource base, must be adequately presented if they are to influence decisions and actions.
- *Policy options for decision-makers.* Because there are insufficient policies based on meaningful research results, the role of scientists is clearly diminished. In addition there is frequently little effort by responsible and knowledgeable scientists to link national policies to global environmental concerns which are crucial to sustainable land management.

All lands have constraints, however minor they may be. However, there are some major constraints which are geographically widespread or are shared by many countries. Some of these are:

1. Moisture stress
2. Soil acidity
3. Nutrient limitations
4. Compaction of soils
5. Problem soils (such as acid sulphate soils and peats)
6. Low organic matter content
7. Susceptibility to flooding
8. Susceptibility to land degradation
9. Lack of soil conservation policies

In many countries, misuse of land has or is resulting in stressed lands which are degraded and whose production potential is considerably reduced. Stressed ecosystems are those that are degraded or have reached a stage of degradation at which they cannot support their original biotic communities or cannot support agriculture in the absence of relatively high inputs. The stresses may be biotic and specifically, anthropically induced or abiotic.

From a land resource point of view, examples of stressed ecosystems are:

- acid soils
- degraded lands of the humid tropics

- steep lands
- heavy clay soils
- wetlands
- lands of the semiarid tropics
- multiple stressed lands

For agricultural purposes, some ecosystems are naturally stressed. In the context of sustainable agriculture and technology to mitigate the effects of anticipated global climate change, there is an urgent need to understand better such ecosystems, develop methods to assess and monitor them, and evaluate technological options for their management. The subject must be addressed in a multidisciplinary manner and involve environmentalists, ecologists, agriculturists, soil scientists, and others.

Some Concepts

Soil systems, like most natural systems, are in dynamic equilibrium. Most changes are slow and imperceptible particularly when viewed in the time-frame of human lifespan. However, catastrophic events such as high intensity storms can accelerate erosion processes resulting in measurable changes. The changes are mainly in the structure and composition of the material and such changes are referred to as 'structural changes'. Changes are measurable directly or indirectly or may be inferred from behaviour of the system. Many of the changes are related to uses of the soil. These 'performance'-related changes are more important as they can be quantified, particularly in economic value terms.

In recent years, due to population pressures resulting in greater intensity of land use, land degradation has emerged as an important issue. This has prompted discussions on sustainability of systems. The concept of sustainable land management (Eswaran, 1992) is 'a system of technologies that aims to integrate ecological and socioeconomic principles in the management of land for agriculture and other uses to achieve inter-generational equity'. Sustainability deals with performance at certain acceptable levels over given time frames.

Many soil systems have been subject to degradation either naturally or induced by man. Sustainable systems may be established on these soils. However, an unanswered question is the ability of the soil to be restored to its previous performance level. Basically, this is the concept of 'resilience' which in this chapter is defined as: 'the ability of the system to revert to its original or near original *performance or state* that existed before the impressed forces altered it'.

The concept of soil resilience relates to either performance or to the state or structure of the system. The functional component, performance, is measured by yield or bearing capacity, or some other attribute that is use related. The structural component, relates to the pedological composition of the material such as thickness and arrangement of horizons, proportions of components such as organic versus mineral matter, etc. More attention is normally given to performance resilience of the system as this has immediate socioeconomic impact. Structural resilience has been largely academic as changes take place over much longer time frames, but become very important in the context of global climate change or where drastic modifications have been made to the soil. However, with recent concerns about global climate change, there is greater attention being paid to structural resilience.

In this concept, resilience is the ability of the system to recover. The term resilience has, however, been used in the sense of 'resistance to change' (Brinkman, 1990), which is an opposite concept. Brinkman (1990) refers to resilience against climate change, a notion of resistance to the change or an ability not to respond to impressed forces. For example he states, 'soils most resilient against the effects of such increasing aridity and rainfall variability would have high structural stability and a strongly heterogeneous system of macropores, hence a rapid infiltration rate, as well as a large available water capacity . . .'. This deals with the stability of the system and may be related to the resilience of the system.

In this chapter, the term resilience (noun) is used in the context of systems that have changed, with resilience being a measure of the ability of the system to return to near original state. The term 'resilient (adjective) soil' is employed *sensu strictu* Brinkman, to connote the buffering capacity of the soil or its tenacity to resist changes.

Any form of agriculture disturbs the natural equilibrium of the ecosystem. If this concept is accepted, we can invoke soil resilience to connote the ability of technology to sustain a performance level on the soil. On some soils under suitable environments, technology has enhanced the performance of the soil for some uses. The concept of resilience is to evaluate the ability of the soil to return to this performance level, if degradation or lowering of performance had resulted. In such evaluations, there is both a time dimension and a value dimension.

Degradation is not soil specific nor is it at the same rates in all soils. The converse, that resilience is not soil specific and that not all soils rebound at the same rate, is also true. Quantification of both degradation and resilience is necessary to apply the concepts. Figure 2.1, is a hypothetical representation illustrating the rates of degradation. The Oxisols, particularly those with initial high organic matter contents, can degrade very quickly whereas others do so more slowly. It is equally important to appreciate that the degradation pattern, as a function of time, is also different as shown in Fig. 2.1. The rates and patterns of degradation serve as indicators to predict degradation and consequently contribute to reduce rates of degradation.

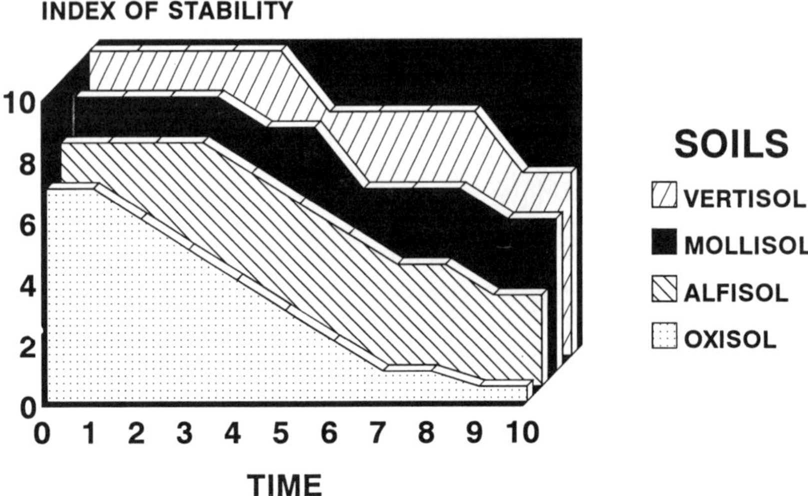

Fig. 2.1. Rates and patterns of soil degradation as a function of soils.

Resilience results from efforts in rehabilitation. Figure 2.2a suggests that to regenerate the original structure of a soil requires geological periods. It would be better to allow a badly degraded soil to revert to its natural habitat, through conservation reserve programmes, than to attempt to modify it. Costs associated with the latter are prohibitive. If the soil has not been badly degraded, then Figure 2.2b suggests that its resilience attributes may be managed to attempt to restore it to its original performance. Again, as illustrated in Figure 2.2b, each soil will respond differently and so a knowledge of the resilience characteristics of the soil is needed and management must be in this context.

Climate is a major control of desertification and consequently, resilience of productivity. A regional assessment of vulnerability to desertification can be made by an analysis of the prevailing climate. Figure 2.3 illustrates the potential, from a climatic point of view, of the soils in Africa to suffer desertification. As shown in the map, some regions are extremely vulnerable, whereas others are resilient. A more critical analysis can be made if the composition of the soils is also considered. Such a holistic analysis needs to be done and will be the challenge of the next decade.

The purpose of this volume is to consider aspects of monitoring and evaluating soil resilience in the framework of these considerations. In the past few decades, attention was focused on soil degradation and technologies to mitigate this process. This concern still remains. In addition, the new concept of resilience requires the evaluation of the ability of the system to recover. This emphasis stems from the large areas of degraded soils around the world. Consequently, the challenge of sustainable land management is twofold: first to maintain or enhance productivity and quality of the resource base for the good and prime lands of the world, and second to manage the degraded lands to bring them to an environmentally acceptable quality and an economically tolerable productivity. These are the basic precepts of AGENDA 21.

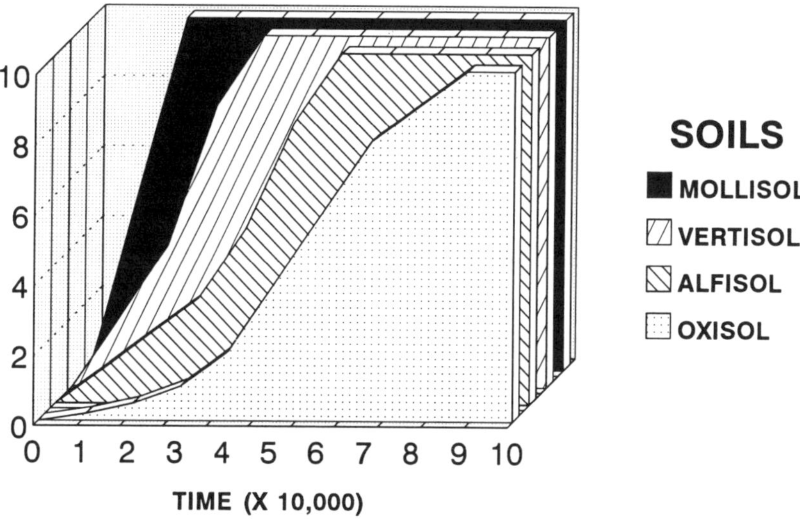

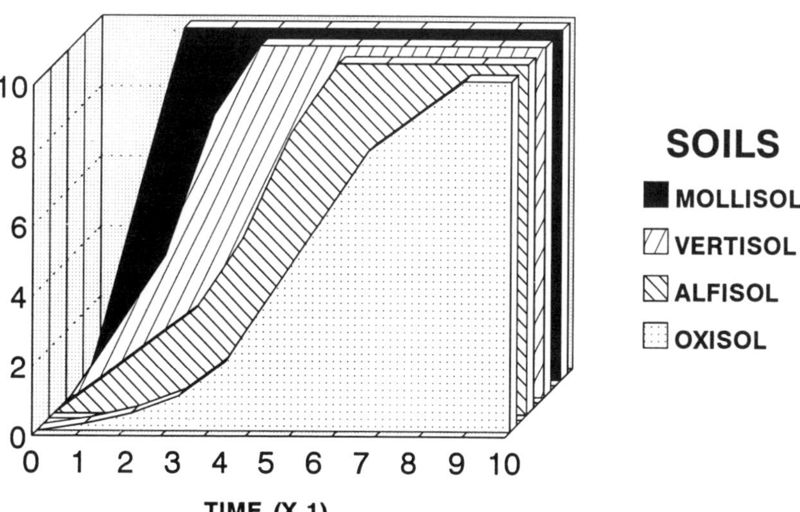

Fig. 2.2. Resilience of soils with respect to (a) profile development and (b) performance.

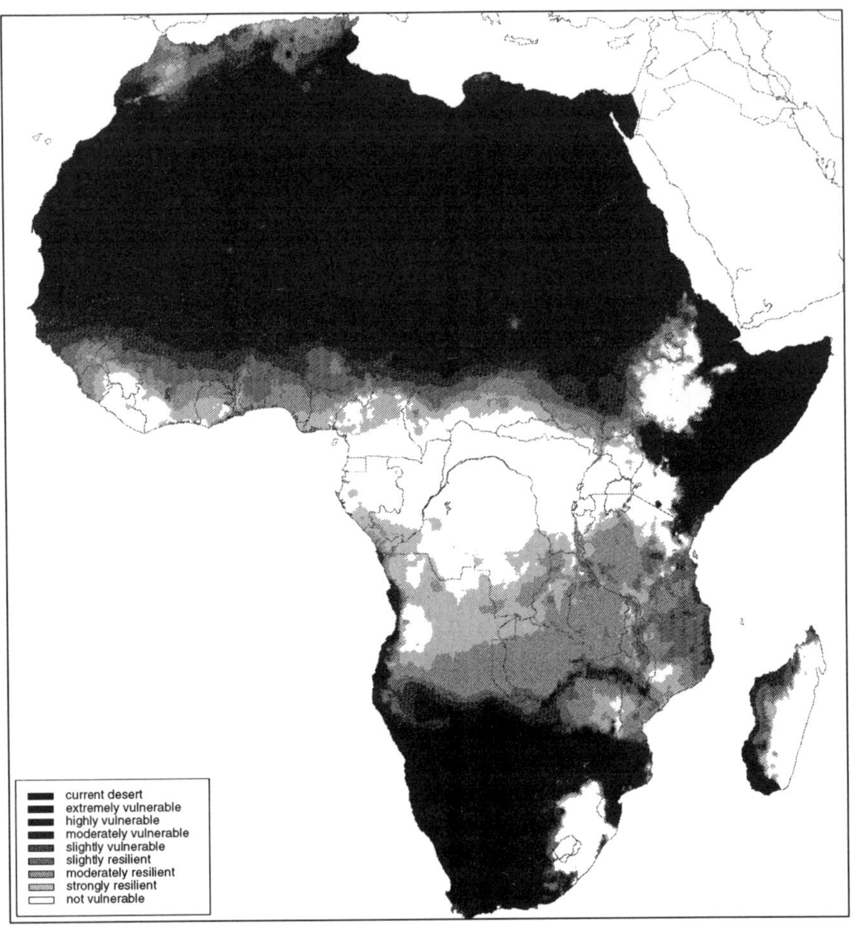

Fig. 2.3. A climatic assessment of vulnerability to desertification.

AGENDA 21

In June 1992, nations from around the world met at Rio de Janeiro, Brazil, and recognized that the achievement of sustainable development and enhancement of the environment are urgent and priority issues which must be addressed immediately to ensure the economic growth and well-being of people in the years to come. This United Nations Conference on Environment and Development (UNCED) culminated in AGENDA 21 which provides a comprehensive programme of action and a blueprint for activities in areas of environment in the context of sustainable development, to be adopted by Governments. The AGENDA calls for new imperatives, radical changes in approach to research and development, and massive global investments to attain the common goal of not only well-being of the human race but also a more rational management of the environment for equitable growth. This is a recognition that humans are part of the environment and that their survival is a function of the management of the other components of the environment. This is also a call for a more-efficient use of the resources and an assurance for intergenerational equity of the life-support systems of this planet.

AGENDA 21 is a voluminous document and so only the aspects relevant to this chapter are considered here. The Rio declaration enunciated 27 principles which provide the basis and intent of the declaration. Some relevant principles are:

- The right to development must be fulfilled to meet equitably present and future needs;
- The special situation and needs of developing countries require that they be given special priority;
- National authorities should endeavour to promote internalization of environmental costs;
- Environmental impact assessment as a national instrument must be incorporated in developmental programmes;
- Peace, development, and environmental protection are interdependent and indivisible.

The message in AGENDA 21 may be summarized as:

1. There is a clear emphasis on appropriate management of land, water, and biological resources, as a basis of cropping, grazing, and forestry. Socioeconomic constraints and drivers receive special attention and are to be considered in the context of biophysical factors.
2. Research and development must proceed holistically. Decision support systems with biophysical and socioeconomic databases are obvious priority needs for the sustainable use of land.
3. Problems of biodegradation and the need to recycle wastes, and degradation of the resource base must be given immediate attention; the resilience of the natural resource base has to be used to restore ecosystems.
4. Improved information flow, particularly to the local level is the key to sustainable development. National capacity building and institutional strengthening are integral components of this approach.
5. Sustainability is controlled by socioeconomics. Progress on resolution of income equity, alleviation of poverty, debt reduction or trading and other economic imperatives will impinge on natural resource research and development. There is an urgent need to integrate biophysical and socieconomic data and information systems so that a holistic approach is ensured in research and development activities.

Sustainable land management and specifically strengthening the scientific basis, is high on the priority list of AGENDA 21. Research is called for in these areas and specifically to:

1. develop, apply, and institute necessary tools for sustainable land management;
2. develop quality of life indicators in the areas of health, social welfare, state of the environment, and economy;
3. rationalize economic approaches to environmentally sound development and basically institute better resource management;
4. assist in long-term policy formulation, risk management, and technology assessment. This specifically calls for:

(a) interaction between science and decision-making;
(b) generation and application of knowledge, particularly indigenous knowledge;
(c) improving cooperation between scientists and enhanced collaboration with scientists in the less endowed parts of the world;
(d) participation of people in setting priorities not only in the development areas but also in research.

Direct and indirect human resource development is repeated throughout the document. Research networks involving National Agricultural Research Systems is an effective mode. However, the AGENDA is cognizant of the fact that to have the necessary impact there is also a parallel need for effective resource assessment and monitoring. This can be done through integrated information systems incorporating environmental monitoring, accounting, and risk assessment. The emphasis on monitoring is interesting as few have paid attention to this in the past. AGENDA 21 values the monitoring component of resource assessment to provide the dynamic inventory of resource conditions to assist in the policy decisions that impact resource use and management.

Finally, AGENDA 21 makes a strong plea for promoting and supporting research tailored to local environments for sustainable land management. It specifically requires consideration of:

1. Assessment of land potential and ecosystem functions.
2. Interactions between land resources and social, economic, and environmental systems.
3. Indicators for sustainable land management with emphasis on economic, social, environmental, and political factors.
4. Testing research findings through networks and pilot projects.

AGENDA 21 does not mention soil resilience because this is a new concept. However, the thrust of the activities proposed by AGENDA 21 clearly emphasizes the need to look at degraded soils from the point of view of their rehabilitation. As indicated previously, this adds a new dimension to research and development activities in the area of sustainable land management.

Challenges and Opportunities

The new concept of soil resilience provides an opportunity to address soil degradation in a positive and proactive manner and specifically from the point of view of sustainability. AGENDA 21 provides the framework to initiate a programme of international collaboration to address this and related challenges which are summarized as:

1. Identification and monitoring of parameters (including indicators) that control the enhancement of the quality of the land resource, and quantification of their impact in different agroenvironments.
2. Development of methodologies to evaluate the resilience of stressed ecosystems and implementation of preventive measures to retard degradation.
3. Identification of prime lands and application of modern technology to enhance and sustain their production and ensure their preservation for such uses.
4. Development of a conceptual, integrated framework for sustainable land management and its validation for different land use scenarios.
5. Development and utilization of systems-based analysis and strategy tools for assessing the impact of agricultural practices on the land resource and for developing environmentally sound land use options.
6. Development of methods and approaches to evaluate how land use changes affect global climate and the converse, which is the impact of global climate change on soil resources.
7. Development of new approaches to soil resource inventory and monitoring, utilizing recent computer technology, to make farm level resource information readily available to clients, particularly in developing countries.
8. Translation of soil resource information and soil research findings into implementable policy options.

As the century draws to a close, soil scientists throughout the world can help catalyse the new global environmental paradigm. Feeding and clothing an expanding population and striving for zero-degree degradation of land resources and manipulating the resilience attributes of land are still the most viable and exciting options for the 21st century. The knowledge of soil scientists is extremely relevant to the changing conditions of our world; but it will be insignificant if we fail to satisfy the critical needs of our governments and the people they represent. We can draw on the strength of our ancestors, revisit our capabilities as professionals and see beyond the obvious.

References

Brinkman, R. (1990) Resilience against climate change. In: Scharpenseel, H.W., Shoemaker, M. and Ayoub, A. (eds) *Soils on a Warmer Earth*, Elsevier, Amsterdam, pp. 51–60.

Dumanski, J., Eswaran, H. and Latham, M. (1992) A proposal for an international framework for evaluating sustainable land management. In: Dumanski, J., Pushparajah, E., Latham, M. and Meyers, R. (eds) *Evaluation for sustainable land management in the developing world.* Vol. 2, IBSRAM, Bangkok, Thailand, pp. 25–45.

Eswaran, H. (1992) Role of soil information in meeting the challenges of sustainable land management. 18th Dr R.V. Tamhane Memorial Lecture. *Journal of the Indian Society for Soil Science* 40, 6–24.

FAO (1989) Sustainable agricultural production: implications for international agricultural research. Compiled by the Technical Advisory Committee of the CGIAR. FAO Research and Technical Papers, no. 4. Rome, 131 pp.

WCED (World Commission on Environment and Development) (1987) *Our Common Future.* Oxford University Press, Oxford.

Chapter 3
The Concept of Soil Resilience

I. Szabolcs

Research Institute for Soil Science and Agricultural Chemistry of the Hungarian Academy of Sciences, PO Box 35, H-1525 Budapest, Hungary

Introduction

We come across more and more often the term 'soil resilience' without having a precise definition for this expression. When the environmental aspects, or agri-, horti- or sylvicultural value of a soil is discussed soil resilience is frequently mentioned; in most cases only as a general idea or as rhetorical rubric.

The ever-growing importance of the soil as a substantial part of the environment and irreplaceable basis for the production of food, fodder, and raw materials is evident and more and more appreciated. The formation of an up-to-date concept of soil resilience is necessary so that everybody concerned can be in a position to agree or disagree on its definition and apply it accordingly both in science and technology whenever the soil and its productivity are discussed.

The Role of the Soil in Nature

The role of soils in nature is complex, many-sided and includes biospheric, hydrospheric, atmospheric and lithospheric functions. Their interaction is illustrated in Fig. 3.1.

The productivity of soils, with regard to their utilization, is always determined by the properties and attributes of the whole soil body (Arnold *et al.*, 1990). Evidently, topsoil conditions have a special role in the nutrient and water supply of the soil biota (including organisms both above and below the soil surface).

The most important functions of the pedosphere include (but are not limited to) the following:

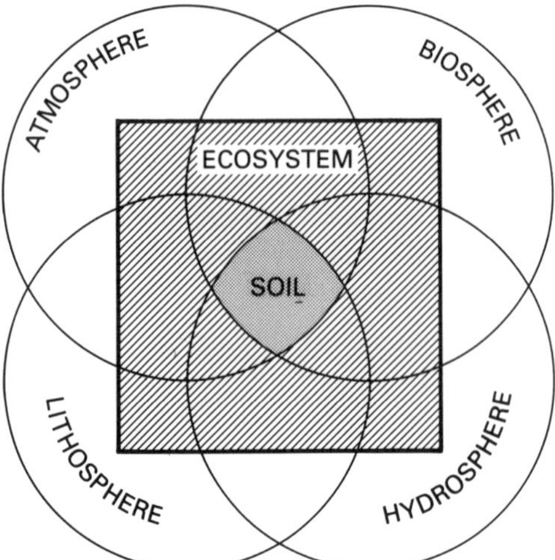

Fig. 3.1. Schematic diagram of the interaction of lithosphere, atmosphere, biosphere, hydrosphere, ecosystems and soils (Szabolcs, 1989).

1. The pedosphere regulates many biotic processes, including the supply of plants with mineral nutrients and water from the soil to build up their biomass which will then become another source of nutrition underlying the food chain. This role determines the biodiversity of the soil as well as its productivity. The function of the soil as a larder of the biomass is rather remarkable because, as shown in Table 3.1, in several ecosystems the bulk of organic material is in the soil and only a smaller part can be found in other biological elements of the environment.

2. The pedosphere constitutes an interface between the biosphere and the geosphere sustaining, regulating, and controlling several turnovers and fluxes of substances, particularly the cycle of so-called bio-elements, the elements of mineral nutrition which, through a number of intermediate stages, are again returned into the soil.

3. The soil, as a porous system, contributes to the chemistry of the water and heat balance of the atmosphere. The pedosphere regulates the exchange of various gases between the atmosphere and the gaseous phases of soil by absorbing oxygen, exuding carbon dioxide and other gases, such as methane, hydrogen, hydrogen sulphide, nitrogen oxides, and ammonia, etc. Evaporation of water from soil influences atmospheric water content.

4. The pedosphere redistributes water in the various hydrological fluxes, transforms precipitation into infiltration, surface runoff, subsurface intersoil runoff, and ground water runoff, and also alters the chemical composition of the precipitation. The top layer of the pedosphere is often washed away by the runoff of water in the course of erosion.

Table 3.1. Organic matter content in different ecosystems.

Type	Overground phytomass (t ha^{-1})	Humus (t ha^{-1})
Coastal marshes and terrestrial swamps	10–20	128
Tundra	3–10	320
Taiga	270	100
Grasslands	16	355

5. The pedosphere has important lithospheric functions as a specific and stratified mantle of the earth's surface protecting it from destructive impacts. It buffers and regulates destructive processes, e.g. horizontally and vertically acting exogenic processes.

Soil Properties and Soil Attributes

These two terms have often been applied as synonyms and there is no exact definition indicating the differences between them. So, somewhat arbitrarily, the word 'properties' will be applied in this chapter to those characteristics of the soil which can be measured directly and unambiguously, and expressed in exact weights, measures, and terms. There is no dispute about how to express for example the specific weight, moisture content, particle size distribution or the pH of a soil, and there are plenty of conventional methods for establishing these data. On the other hand there are important soil characteristics which cannot be measured or expressed so unambiguously like soil fertility or soil productivity. To this second group belongs the term 'soil resilience', which exists beyond doubt but its interpretation requires some circumscription if we want to characterize this soil attribute in a given place and time, and for a given purpose.

When we try to characterize soil resilience as a significant attribute of the soil, it is necessary to define what resilience is in general. The latter is not a very difficult task. To produce a definition which makes it possible to measure the resilience of a given soil in a given place in respect of its correlation to soil forming and environmental processes, and of the targets of soil utilization and land use, including both scientific and practical aspects, is quite another story.

The Resilience of Soils

Resilience has been defined from various points of view for various purposes. In general it means the tolerance against stress, or that proportion of the total work of deformation which the body can give back following the removal of the deforming forces. In engineering and technics 'resilience' is the elastic limit of a body.

These definitions are correct. However they give but little information in this general form to enable us to analyse soil resilience as a soil attribute. The first step on the road to characterize soil resilience correctly is to pose the question of 'Against what?' as soil's have no resilience in general (or in absolute terms). What they have is a capability of behaving in this or that way in the face of certain more or less well-defined agents, forces, or effects. The precondition of characterizing the resilience of a soil requires the definition of these agents, effects and forces.

In order to answer the above question we have to go back to the mass and energy fluxes dominating soil processes and resulting in soil behaviour in general. Soil resilience is partly a result and partly a counterpart of such fluxes. In this respect three kinds of mass and energy fluxes can be enumerated.

1. Atmospheric, including radiation, heat, and precipitation from and through the atmosphere.
2. Endogenic soil fluxes including weathering, soil formation, the synthesis and decomposition of organic matter, accumulation, leaching, and redistribution of compounds, erosion, oxidation–reduction, release and fixation of nutrients, etc.
3. Exogenic (anthropogenic) mass and energy fluxes, closely related to and overlapping endogenic fluxes, including the effects of agricultural operations (e.g. tillage, the application of chemicals, irrigation, drainage, terracing, removal of crop yields), as well as contamination with industrial output, radioactive fission products, etc.

Soil resilience includes all the processes that enable soils to counteract stress and alterations. Consequently soil resilience includes such important properties as the buffering capacity of a soil in respect of chemical, physical, and biological impacts. It is not necessary to give a detailed definition of these three main types of soil buffering capacity but it should be noted that all of them have important theoretical and practical significance, e.g. physical buffering capacity plays an important role as far as the hazard of erosion is concerned, whereas chemical buffering capacity comes into play in the case of soil acidification or alkalization. It is somewhat more difficult to define the biological buffering capacity of a soil which concerns several biological processes as well as the soil biota *per se.*

It follows from this that a soil may have high buffering capacity against chemical agents (e.g. a heavy textured soil or a calcareous soil against acidification) and, at the same time, it may have very low buffering capacity against biological stress (e.g. in relation to the loss of vulnerable biota from the given soil).

Soil resilience also includes the important capability of the soil known as renewability. It is very important that the soil as a natural resource is capable of conditional renewability, i.e. it can regenerate itself after several kinds of deterioration and degradation within reasonable limits. Soil renewability is also important in practical terms because following stress it can regain its original or nearly original state through the action of natural and anthropogenic soil-forming processes.

In connection with the phenomena described above the transforming ability of soils should also be mentioned. The detoxication, decomposition, and transformation of many substances in the soil contribute to its renewability, e.g. decomposition of toxic organic matter, some kinds of adsorption, etc.

Perhaps it is not necessary to quote many other examples in order to argue for the statement that soil resilience should always be determined concretely, because high or low resilience has no absolute meaning as regards soil value in respect of practical purposes. Rocks in a desert may be very resilient against chemical, physical or biological agents, but do not have much value for production. On the other hand a soil, very vulnerable to such agents, can be very valuable for agricultural purposes (e.g. a humous, loamy soil) provided that, in the course of utilization, the limits of its resilience are not surpassed.

When speaking of soil resilience the aspects of entropy should also be mentioned. The definition of entropy is closely related to the second law of thermodynamics (i.e. the principle of entropy increase) which provides a criterion for the determination of the direction of natural processes. They proceed in the direction through which $\mathrm{d}S_{sys}$ increases. It also provides a criterion for equilibrium: they cease when $\int \mathrm{d}S_{sys}$ attains a maximum, which is zero at equilibrium and positive for spontaneous processes. The heat transferred to a closed system divided by T, the temperature, is equal to or less than the entropy increase for any possible processes. The total entropy change in the closed system $\mathrm{d}S_{sys}$ is the sum of the changes inside the system $\mathrm{d}S_{int}$ and the entropy transferred to the system from its surroundings $\mathrm{d}S_{sur}$.

$$\mathrm{d}S_{sys} = \mathrm{d}S_{int} + \mathrm{d}S_{sur} \tag{3.1}$$

$$\mathrm{d}S_{sur} = \frac{\mathrm{d}q}{T} \tag{3.2}$$

where: q = heat transferred to the system from its surroundings.

For reversible processes of the equilibrium state of the system

$$\mathrm{d}S_{\mathrm{int}} = 0 \tag{3.3}$$

and for a spontaneous or natural process in the system

$$\mathrm{d}S_{\mathrm{int}} > 0 \tag{3.4}$$

The term 'entropy' has been applied in soil science and biology based on the above regularities in a somewhat adjusted and simplified interpretation, characterizing the loss of internal energy of the system caused, for example, by decomposition of organic matter. In biology the ability of plants to transform solar energy into chemical energy by photosynthesis has been considered as diminishing the entropy on the earth's surface. With a certain exaggeration it can be said that biological processes act against entropy.

It is accepted that soil-forming processes, which are inseparable from biological processes, also reduce the entropy of the system in comparison with an environmental system without soil formation (e.g. rocks).

In other words, the resilience of soils is in reverse proportion with the entropy in the soil.

This concept of soil resilience can be described in a simple equation.

$$SR = BC_{\mathrm{pH}} + BC_{\mathrm{ch}} + BC_{\mathrm{b}} + \int_{t_1}^{t_2} \frac{\mathrm{d}PSF}{\mathrm{d}t} + \int_{t_1}^{t_2} \frac{\mathrm{d}AF}{\mathrm{d}t} \tag{3.5}$$

where: SR = soil resilience
BC_{ph} = physical buffering
BC_{ch} = chemical buffering
BC_{b} = biological buffering
PSF = pedological soil fluxes
AF = anthropological soil fluxes

Such an approach can contribute not only to the interpretation of soil resilience but also, in cases of further studies, to its modelling, estimation and even measurement.

Conclusions

An advanced study of soil behaviour as well as a more up-to-date interpretation of soil properties and attributes are necessary to achieve sustainable land use. Practically it is soil behaviour that decides whether our attempts at effective land utilization will fail or will be successful.

Soil resilience which cannot be measured by simple methods but can and should be estimated in given circumstances, is directly correlated with the sustainability of land use.

Soil-forming processes must be better understood, measured and monitored if an acceptable determination of soil resilience in a given place and under given circumstances is to be achieved.

References

Arnold, R.W., Szabolcs, I. and Targulian, V.O. (eds) (1990) *Global Soil Change* International Institute for Applied Systems Analysis, Laxenburg, Austria.

Szabolcs, I. (1980) Mass and Energy Flow in Soil Formation. *Acta Geologica Academiae Scientiarum Hungaricae* 23, 251–261.

Szabolcs, I. (ed) (1989) Ecological Impact of Acidification. *Proceedings of the Joint Symposium 'Environmental Threats to Forest and Other Natural Ecosystems'*, University of Oulu, Finland, 1–4 November 1988, Budapest.

Chapter 4
Sustainable Land Use Systems and Soil Resilience

R. Lal

Department of Agronomy, The Ohio State University, 2021 Coffey Road, Columbus, Ohio 43210, USA

Introduction

The origin of all living things on earth can be directly or indirectly linked to soil. Soil is the substance of things hoped for, and the evidence of things not seen (Bear, 1986). The mere fabric of human life and civilization is woven on earthen looms (Bradley, 1935). The importance of the most basic of all resources is further enhanced by the realization that soil resources of the world are finite in geographic extent. Only 11% of the earth's total land area of 13.4×10^9 ha is currently being cultivated. The remaining potentially cultivable land is marginal for agricultural use because most of it is either inaccessible or is severely constrained by unfavourably steep terrain, shallow rooting depth, extremes of moisture and/or temperature regimes, lack of availability of essential inputs, or is located in ecologically sensitive regions. The per capita arable land area of 0.3 ha in 1990 is expected to decrease to 0.25 ha in the year 2000, 0.15 ha in the year 2050 and about 0.10 ha in the year 2150. Human needs, with per capita arable land area of 0.10 ha, can only be met by science-based and innovative technology. Consequently, there is an overwhelming interest in sustainable land use systems that prevent or minimize soil degradation and restore productive capacity and life support processes of degraded lands.

'Agricultural sustainability' and 'soil degradation' are complex concepts, have subjective values, and conceptual understanding depends on users' immediate and long-term concerns and needs. The scientific community needs to standardize and operationalize these concepts and create objectivity in developing a rational and science-based strategy for resource management. The objective is to use these concepts in practical and operational terms to alleviate the pressing problems of modern agriculture, e.g.

increase per capita productivity and decrease risks of soil and environmental degradation. In that context, agricultural sustainability implies 'an increasing trend in per capita productivity per unit consumption of the non-renewable or the limiting resource, or per unit degradation of soil and environmental characteristics or per unit reduction in soil's life support processes'. The concern is not only to increase per capita productivity but also to maximize production per unit soil loss, per unit energy input, per unit reduction in soil organic carbon, per unit efflux of radiatively active gases, per unit consumption of ground water, per unit increase in concentration of nitrate, phosphate or other pollutants in natural waters. Therefore, an index of sustainability depends on productivity and changes in soil and environment (eqn 4.1)

$$I_S = f(P_i \cdot S_p \cdot W_r \cdot C_f)_t \tag{4.1}$$

where: I_S is the index of agricultural sustainability
P_i the per capita productivity per unit input of the limited resources
S_p is the change in soil properties or the life-support processes
W_r is the depletion of water resources or change in water quality
C_f is the alteration in climatic factor with possible long-term effects.

Agricultural productivity, therefore, is intimately linked to the soil as a limited and a non-renewable source.

Soil Resilience

Soil is a dynamic entity, and dynamism is life. Because early civilizations originated and flourished on fertile soils, most ancient cultures developed reverence for the soil on which they depended. Many cultures (Indo-Aryans, Hebrews, Greeks) believed that humans originated from land, and that land is a living entity worthy of worship. For example, the Maori of New Zealand believe that 'The land is the mother that never dies'. The Indo-Aryans portrayed the eternal divine power as the one whose 'rivers are the veins, the trees are hairs, the air the breath, and passing ages the movement' (Bhagwad Gita, 500–900 BC). These beliefs are similar to the modern 'Gaia Hypothesis' that proposes that 'earth is a living organism', and 'is a complex entity involving the biosphere, atmosphere, oceans and soil; the totality constituting a feedback or cybernetic system which seeks an optimal physical and chemical environment for life' (Lovelock, 1979).

Soil dynamism is evidenced by continuous formative and evolutionary changes until a dynamic equilibrium is achieved. Anthropogenic perturbations alter this state of equilibrium, and may change the rate and direction of principle processes (Rozanov, 1990). With the advent of modern technology, human action is becoming progressively drastic due to development of bigger and faster machines, and concentrated and highly active chemicals

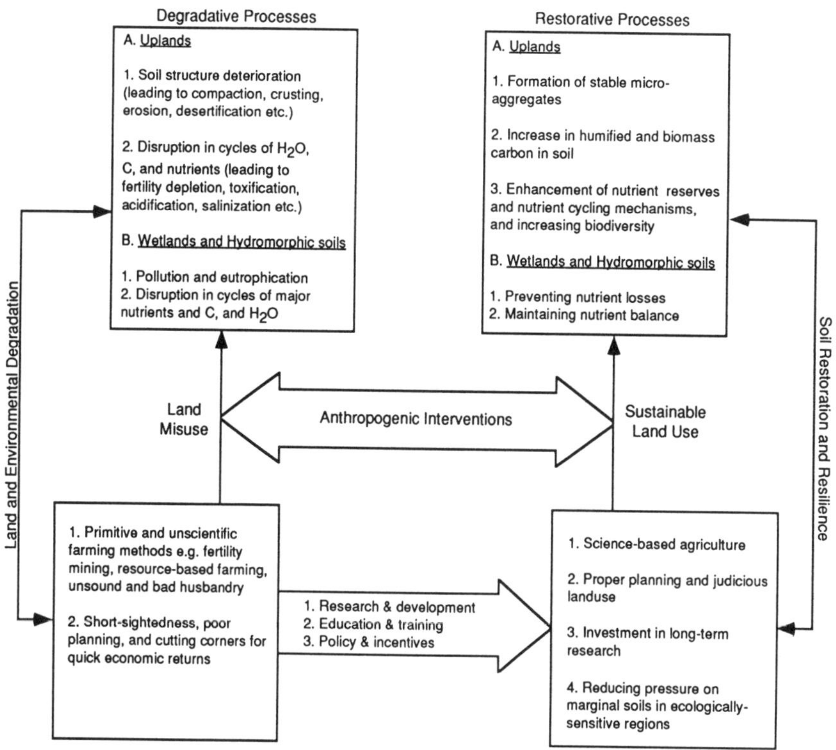

Fig. 4.1. Soil degradative and restorative processes and anthropogenic interventions.

designed to enhance soil fertility and accentuate plant growth. The greater are human demands the more potent are the technologies likely to have drastic effects on soil processes.

The term 'soil resilience' implies ability to bounce or spring back into shape or position after being stressed. Soil, being a dynamic and an organic entity, has built-in ability to restore its life-support processes provided that the disturbance created by human intervention is not too drastic and sufficient time is allowed for the life-support processes of soils to restore themselves. However, soil can undergo irreversible degradation if it is drastically disturbed and its life-support processes are severely jeopardized. Through constant misuse and mismanagement 'man may sap the vitality of Gaia by reducing productivity and by deleting key species in her life-support system' (Lovelock, 1979). The 'Magic of Mother Earth' is her ability to heal herself. Soil resilience depends on the balance between soil restorative and soil degradative processes (Fig. 4.1). It depends on the harmonious and symbiotic action of all living organisms in creating favourable environments for life to thrive. While humans value crops and livestock, other species are not necessarily pests, weeds or vermin. They play an important role in the self-regulatory mechanisms of Mother Earth.

Soil degradative processes are set in motion by deterioration of soil structure and disruption in cycles of carbon, depletion of nutrient reserves within the soil, and weakening of nutrient recycling mechanisms. Soil restorative or self-regulatory processes are strengthened by science-based agriculture, discriminate and judicious land use, reduction of pressure on marginal lands and fragile ecosystems, and adoption of improved technology. The degree of soil resilience depends on the magnitude of alterations in the life support processes of the soil due to anthropogenic perturbations (eqn 4.2).

$$S_r = S_a + \int_{t_i}^{t_f} (S_n - S_d + I_m) dt \qquad (4.2)$$

where: S_r is soils resilience
S_a is the antecedent soil condition
S_n the rate of new soil formation
S_d is the rate of soil degradation/depletion
I_m is the management inputs from outside the ecosystem.

The magnitude and sign of the term $(S_n - S_d + I_m)$ is critical in determining soil resilience. In addition to S_n and S_d, human intervention involving resource management is critical to soil resilience.

Rate of new soil formation (S_n)

Soil is a renewable resource as long as the term $(S_n - S_d)$ is positive; otherwise it is a non-renewable entity, at least within a human lifetime. The time frame is an important consideration in this analysis (Darvoll, 1973; Watkins *et al.*, 1975; Bear, 1986). Accurate estimates of the rate of soil formation are not known (Johnson, 1987). Most available data are based on informed opinion and indirect evaluation. Chamberlin (1909) 'guesstimated' the rate of soil formation at about 2.5 cm (1 inch) in a thousand years. Bennett (1939) observed that rate of soil renewal for Cecil sandy clay loam was 0.005 tonne ha^{-1} $year^{-1}$ under forest cover and 0.03 tonne ha^{-1} $year^{-1}$ under grass. He calculated that it takes nature from 300 to 1000 years or more to build 2.5 cm of topsoil. Hudson (1976) estimated the rate of new soil formation to be about 2.5 cm in 30 years. Some progress has been made in revising the early estimates, and measured rates of soil formation for some soils are shown in Table 4.1.

Friend (1992) estimated that worldwide the rate of soil formation is about 2.5 cm per 150 years. It would seem therefore, that most soils are non-renewable within the human life span. The major exception to this rule may be the formation of alluvial flood plains by the rivers carrying a heavy sediment load, e.g. the Yellow River, the Ganges etc.

Table 4.1. Experimentally determined rates of soil formation in the tropics.

Country	Region	Rate (cm year^{-1})	Soil	Reference
Soils of volcanic origin				
Indonesia	Humid tropics	0.73	Andisol	Van Baren (1931)
New Guinea	Humid tropics	0.06	Andisol	Ruxton (1966)
Trinidad	Humid tropics	0.46–0.50	Andisol	May (1960)
Kenya	Humid tropics	0.14–0.24	Andisol	Dunne *et al.* (1978)
Hawaii	Humid tropics	0.85	Andisol	Mohr and Van Baren (1954)
Residual soils				
Africa	Humid tropics	0.0013	Oxisol	Aubert (1960)
Cameroon	Humid tropics	0.007	Alfisol	Boulad *et al.* (1977)
Ivory Coast	Humid tropics	0.013–0.045	Ultisol	Leneuf and Aubert (1961)
Senegal	Semiarid tropics	0.0013–0.0017	Alfisol	Nahon and Lappartient (1977)
Senegal	Semiarid tropics	0.42	Aridisols	Aubert and Maignier (1949)
Zimbabwe	Subtropic	0.0011	–	Owens and Watson (1979)
Zimbabwe	Subtropic	0.0041	–	Owens and Watson (1979)
Kenya	Semiarid tropics	<0.1	Alfisol	Dunne *et al.* (1978)
Alluvial soils				
Pakistan	Semiarid tropics	0.013	Inceptisols	Hall *et al.* (1982)

Soil degradation (S_d) rate

Soil degradation is the decline in the capacity of the soil to produce goods of value to humans. It has plagued the earth ever since human exploitation of land began. Many ancient civilizations thrived on 'good soil' and vanished as soils degraded due to misuse. Typical examples are the Riparian and Harappan Kalibangan cultures in the Indus valley, the Mesopotamian and Lydian Kingdoms in the Mediterranean region, and the Mayan civilization in Central America. These great civilizations vanished along with the depletion of the original mantle of fertile topsoil. The problem of soil degradation has been drastically accentuated by changes in land use since the 18th century (Richards, 1991; Williams, 1991). The data in Table 4.2, and those of Oldeman (Chapter 7, this volume) show that of the total land area of the world of 13.4×10^9 ha, about 2.0×10^9 ha are degraded to some degree. Asia and Africa combined account for a total of 1.24×10^9 ha of degraded land.

There are several processes of soil degradation. Most prominent among these are wind and water erosion, chemical degradation and fertility depletion, and physical degradation resulting from decline in soil structure (Table 4.3). Estimates of soil degradation in dry climates are equally alarming (Table 4.4). Such statistics play an important role in creating

Table 4.2. Human-induced soil degradation from 1945 to 1990 (World Resources Institute, 1992–93).

Region	Total degraded area (10^6 ha)	Degraded area as % of vegetated land
World		
Moderate, severe, extreme	1215.3	10.5
Light	749.0	6.5
Total	1964.4	17.0
Europe		
Moderate, severe, extreme	158.3	16.7
Light	60.6	6.4
Total	218.9	23.1
Africa		
Moderate, severe, extreme	320.6	14.3
Light	173.6	7.8
Total	494.2	22.1
Asia		
Moderate, severe, extreme	452.5	12.0
Light	294.5	7.8
Total	747.0	19.8
Oceania		
Moderate, severe, extreme	6.3	0.8
Light	96.6	12.3
Total	102.9	13.1
North America		
Moderate, severe, extreme	78.7	4.4
Light	16.8	0.9
Total	95.5	5.3
Central America and Mexico		
Moderate, severe, extreme	60.9	24.1
Light	1.9	0.7
Total	62.8	24.8
South America		
Moderate, severe, extreme	138.6	8.0
Light	104.8	6.0
Total	243.4	14.0

awareness about the magnitude and severity of the problem and in formulating a global strategy in addressing it. Indeed, if these statistics on soil degradation are correct, the challenge they present to the human race is one of the greatest because soil resources are finite and essentially non-renewable. It should be a matter of the greatest urgency for decision-makers and opinion shapers to prevent and control soil degradation. However, careful analyses of these data on land/soil degradation may reveal several problems.

The term soil degradation is vague and estimates of its extent are highly subjective. It is therefore important that soil degradation by different processes be defined quantitatively. To do so, is to delineate threshold values or critical limits of soil properties beyond which life support processes of the soil are severely jeopardized. Important soil properties whose critical limits need to be defined are rooting depth, plant-available water capacity, soil organic matter content, soil structural attributes, and capacity and intensity factors and limiting levels of principal nutrients. These limits differ for different soils, climatic conditions, farming systems, land use, and managerial inputs. The GLASOD project (ISRIC/UNEP, 1990) and FAO (1991) map of soil resources are steps in the right direction, and the information provided can lead to development of cause–effect relationships between soil degradation and agronomic productivity, but much still needs to be done.

Inputs and improved technology (I_m)

The majority of soils in the regions outlined in Tables 4.2–4.4 have undergone drastic changes in their properties due to intensive cultivation, land misuse, and other anthropogenic perturbations. However, the impact of these changes on productive potential has not been determined. Productivity of degraded soils can be enhanced by improved management. The schematic in Fig. 4.2 highlights the importance of management to soil degradation and productivity. Soil degradation must be determined scientifically. Improved management and judicious land use can set in motion the restorative processes even on drastically disturbed lands (Thomas and Jansen, 1985). Results from a case study depicting effects of erosion on yields of several crops for different levels of inputs are shown in Table 4.5. With low input of amendments able to reverse the degradation process, only 26% of the maximum yield potential was realized for maize. With high input, however, yields of most crops were almost doubled. Bunding and water management and growing rice resulted in realization of full productive potential even with low inputs.

The impact of improved technology in reducing soil erosion and degradation on crop land in USA was measured by Lee (1990), who found that sheet and rill erosion from crop land declined by 11.6% between 1982 and 1987. The conservation Reserve Program authorized in the Food Security Act of 1985 is believed to have reduced soil erosion in USA by about 0.5 billion tons, or roughly one-third, in its first five years from 1986 to 1991. The projected reduction for a 10-year period from 1986 to 1995 is estimated at about 60% (Table 4.6). This dramatic reduction is not only a major success story but also an example of how improved technology can assist natural soil resilience to restore soil productivity and prevent further soil degradation. Wolman (1967) developed a schematic of the sequence of land use

Table 4.3. Types and extent of soil degradation. (Recalculated from World Resources Institute, 1992–93.)

Region	Total degraded area (10^6 ha)	Degraded area as % of vegetated land
World		
Water erosion	1100.0	56
Wind erosion	550.0	28
Chemical degradation	235.8	12
Physical degradation	78.6	4
Total	1964.4	100
Europe		
Water erosion	113.9	52
Wind erosion	41.6	19
Chemical degradation	26.2	12
Physical degradation	37.2	17
Total	218.9	100
Africa		
Water erosion	227.3	46
Wind erosion	187.8	38
Chemical degradation	59.3	12
Physical degradation	19.8	4
Total	494.2	100
Asia		
Water erosion	433.2	58
Wind erosion	224.1	30
Chemical degradation	74.7	10
Physical degradation	15.0	2
Total	747.0	100
Oceania		
Water erosion	83.4	81
Wind erosion	16.4	16
Chemical degradation	1.1	1
Physical degradation	2.0	8
Total	102.9	100
North America		
Water erosion	60.1	63
Wind erosion	34.4	36
Chemical degradation	0.0	0
Physical degradation	1.0	–
Total	95.5	100
Central America and Mexico		
Water erosion	46.5	74
Wind erosion	4.4	7
Chemical degradation	6.9	11
Physical degradation	5.0	8
Total	62.8	100

Table 4.3. *(Cont'd)*

Region	Total degraded area (10^6 ha)	Degraded area as % of vegetated land
South America		
Water erosion	124.1	51
Wind erosion	41.4	17
Chemical degradation	70.6	29
Physical degradation	7.3	3
Total	234.4	100

Table 4.4. Areas within arid, semiarid and subhumid zones of the world affected by desertification (UNEP, 1991)

		Affected by desertification	
Land use	Total area (10^6 ha)	Area (10^6 ha)	%
Range land	3700	3100	80
Rainfed crop land	570	335	60
Irrigated land	131	40	30
Total	4409	3475	70

changes in USA and sediment yield over a period of two centuries from 1800 to 2000. As the forests were cleared and cultivated, the grasslands were ploughed, and construction activity increased as did the sediment yield. However, adoption of conservation practices reduced sediment yield for the same land use system. The pattern of sediment yield in relation to land use and conservation technology was similar for all soils, but the amplitude of change differed depending on soil properties, terrain characteristics and climate. In addition to benefits in crop yield, scientific management can also set in motion soil restorative processes. Vast areas of land that were once the 'dust bowl' of the south-central United States have been transformed into highly productive lands by science-based agriculture. Similarly, vast areas of salt-crusted barren lands in Haryana and Punjab provinces in India have been transformed into highly productive lands by science-based improved land use systems.

Agriculture in North America, Western Europe and other developed economies has undergone several phases of development (Mengel, 1990). Per capita productivity has drastically increased with each phase beginning with the replacement of human power by horse power, and going on to improvements in farm implements, use of motorized equipment and chemi-

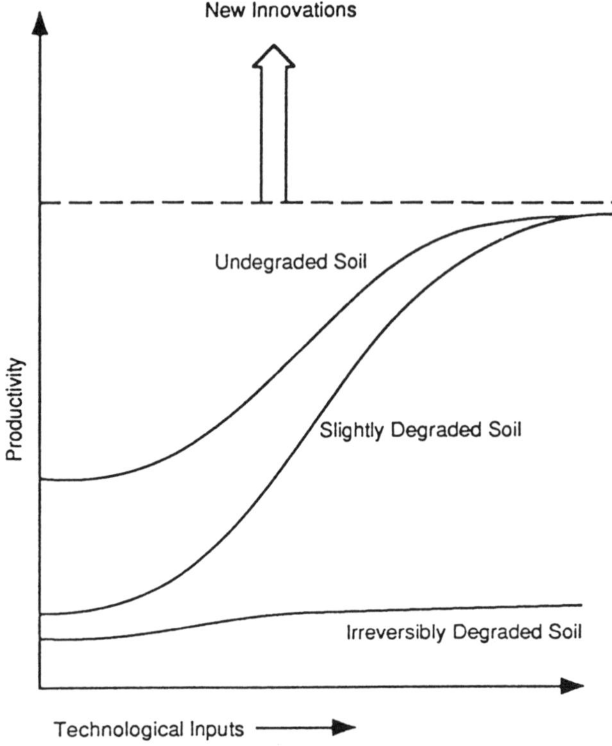

Fig. 4.2. Soil degradation and productivity in relation to technological innovations.

cal fertilizers and pesticides, and use of improved and input responsive varieties. These technological innovations have been used most successfully on undegraded naturally resilient prime agricultural land. Similar innovations need to be developed for soils degraded by different processes.

Time scale (t)

Soil may be a renewable resource over the geological time scale, but not over the human life span. With increasing demographic pressure, however, issues of productivity and sustainability must be addressed over the human life span, and the even shorter common economic planning time span of 5 to 10 years or less. Therefore, it is appropriate to test the impact of innovative technologies on soil restoration or resilience over a time span of 20–25 years. With the exception of volcanic soils and some aeolian and alluvial deposits, the amount of new soil formation (S_n) over such a time span may be small or negligible in real terms.

Table 4.5. Input effects on yield of different crops on an eroded soil in Kenya (% of yield obtained on uneroded soil). (Recalculated from Ford, 1986.)

	% Production loss		
Crop	Low input	Intermediate input	High input
Maize	26	39	48
Beans	32	44	51
Cassava	36	48	59
Bunded rice	92	94	92
Irish potato	27	45	49

Table 4.6. Changes in actual and projected crop land soil erosion in the USA from 1985 to 1995 (Brown, 1991).

Soil loss	Mg(10^6)
Excessive soil loss in 1985	1620
Reduction in 1986–1990	−550
Projected reduction in 1991–1995	−400
Projected excessive soil loss in 1995	670
Reduction in soil loss for the decade ending in 1995	−950 (59%)

Factors Affecting Soil Resilience

The interacting factors that affect soil resilience can be grouped into two categories: (a) endogenous; and (b) exogenous (Table 4.7). Endogenous factors are related to inherent soil properties and micro- and meso-climate. Factors that enhance soil resilience are sufficient rooting depth, loam to clay loam texture, structurally active soils containing high activity clay minerals, a high proportion of stable micro-aggregates, gentle to rolling terrain, good internal drainage, and favourable microclimate. High soil organic matter content and other characteristics of fertile soil depend on these conditions. There are several examples of such highly productive soils, e.g. alluvial soils along major rivers, Mollisols, Inceptisols, Andisols and soils formed on highly basic parent materials. Because of their high inherent nutrient levels and favourable soil physical properties, these soils are remarkably resilient even with poor or lax management. Life support processes of such soils can

Table 4.7. Factors affecting soil resilience.

Endogenous factors	Exogenous factors
Rooting depth	Land use and farming system
Soil texture and clay mineralogy	Technological innovations
Parent material	Inputs and management
Landscape position and terrain	
Moisture regime	
Micro- and mesoclimate	
Soil biodiversity, e.g. flora and fauna	

withstand some degree of mistreatment and neglect without being seriously harmed or suffering lasting deleterious effects.

Land use, soil, crop and livestock management are important exogenous factors. Choice of appropriate land use and use of science-based improved technology can enhance and sustain high productivity and accentuate resilience of even fragile soils in ecologically sensitive regions. Using improved scientific technology on prime agricultural lands has a synergistic effect on soil resilience. Adopting an appropriate land use and science-based improved technology are crucial to sustainable use and to foster the resilience of soils which have either been degraded due to past misuse or have inherent constraints, e.g. shallow depth, steep terrain, high erodibility. Choosing appropriate land use and adopting science-based technology play a major role in restoring soil productivity.

The importance of science-based inputs, judicious land use, and adoption of modern technology on soil resilience is evident from comparison of soil erosion data from USA in the 1930s and 1980s (Table 4.8). Survey techniques and terminology used in the 1930s were different from those used in the 1980s. Nonetheless, the comparison highlights the significance of land use in soil resilience. Estimates of soil erosion by Bennett (1939) showed that half of the US land area had been affected by erosion before and during the 1930s. In addition to 40 million ha of essentially ruined crop land, the process of soil erosion had already damaged or was threatening to damage more than one-third of the arable land. Approximately 75% of the US crop land in the 1930s was susceptible to some degree of erosion. Bennett (1939) considered that lack of increase in yields of corn and cotton for the 60-year period from 1871 to 1930 was due to the impact of severe soil erosion. However, estimates of soil erosion 50 years later showed that the extent and magnitude of soil erosion were drastically less in 1989 than in 1939. Technological advances and use of energy inputs apparently resulted in high production without excessively stressing these resources. Miller *et al.* (1985) observed that much of the 'ruined' land described by Bennett in 1930s was used for producing timber and forages in 1980s and 1990s. The soils of the

Table 4.8. Estimated extent of erosion damage in USA.

Estimated soil erosion in 1930s (Bennett, 1939)		Estimated soil erosion in 1980s (USDA, 1989)		
Erosion condition	Area (10^6 ha)	Land use	Total area (10^6 ha)	Area eroding >T (10^6 ha)
Total US land area (48 states)	771	Crop land	170	70[a]
Degraded land of all types (includes crop land)		Pasture land		
		Range land	163	29
Essentially ruined	24	Forest land	160	10
Severely eroded	90	Minor land	24	4
Moderately eroded in the beginning	314	Total non-federal rural land	571	118
Damaged crop land (includes harvested, idle, failure, fallow)				
Essentially ruined for cultivation	20			
Severely damaged	20			
One half to all top soil gone	40			
Moderately eroded, erosion beginning	40			
Land not now damaged (land in forest, swamp or marsh, etc.)	284			
Land on which damage is undefined (desert, badlands, western mountains)	59			

[a]Includes wind and water erosion.

'ruined' lands of the 1930s were sufficiently resilient that they had been restored, at least partially, to be agronomically productive and environmentally safe and contributed less sediments to the oceans, waterways, lakes, and reservoirs.

Land Use, Soil Management and Sustainability

Evaluating land capacity, adopting appropriate land use, and employing farming systems and soil and crop management practices that realize the potential of the land and restore its productivity are important considerations in long range planning for effective development and sustainable use of land resources. Scientific management of any land use is at least as much as, if not more, important than the type of land use. In the context of soil degradation and restoration, 'how' to use the land is as important as 'what' to use it for. Furthermore, the amount and quality of biomass produced depends on the magnitude and flux of: (a) nutrients, (b) water, and (c) energy through the soil–plant management system. It is not only the efficiency but also the total flux of nutrients, energy and water through the system that govern productivity and sustainability.

Soil structure and sustainability

Soil structure management is crucial to soil resilience and agricultural sustainability. Although a complex attribute which is difficult to characterize and quantify, soil structure regulates the fluvial and transport pathways, and circulatory and respiratory processes in soils. It regulates fluid retention and movement, nutrient transformations and translocation, faunal activity and species diversity, and the strength and rigidity of the rooting media (Fig. 4.3). For most soils soil structure maintenance and enhancement

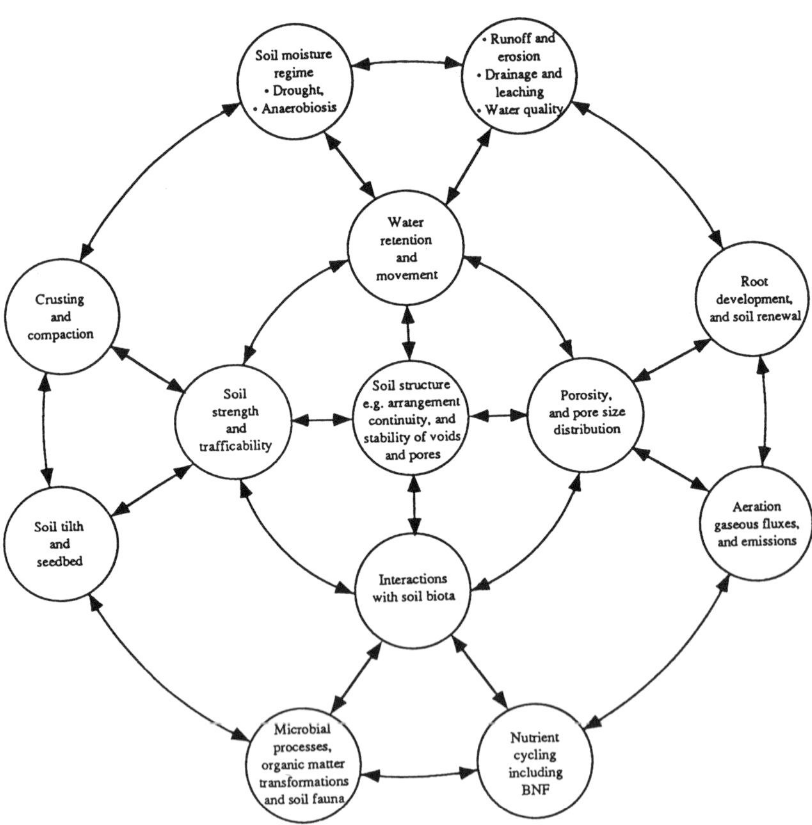

Fig. 4.3. Soil structure and its impact on soil processes and agricultural sustainability.

Table 4.9. Effects of land use and management on physical and chemical properties of soil at Sanborn plots in University of Missouri (Modified from Gantzer *et al.*, 1990).

Cropping system	Management system	Soil organic matter content (%)	Clay content (%)	Soil erodibility factor K (mg ha^{-1} $year^{-1}$)
Corn	Unfertilized	1.2	26.8	0.52
	Fertilized	2.4	30.0	0.40
6-year rotation	Unfertilized	1.6	14.6	0.62
	Manured	2.5	18.4	0.50
	Fertilized	2.2	17.6	0.52
Timothy	Unfertilized	2.2	16.7	0.53
	Manured	3.5	16.7	0.44

depend on land use. Soil structure enhancing land use systems are those that optimize, which for most soils means increase, soil organic matter content, and the activity and species diversity of macro- and meso-fauna. These systems include mulch farming, conservation tillage, frequent use of planted fallows and cover crops, appropriate rotations and crop sequences and combinations, etc. Nutrient management and soil fertility improvement, through judicious use of organic manures and inorganic fertilizers, also play important roles in improving soil structure. Results of long-term experiments conducted in India (IARI, 1984; Nambiar and Abrol, 1989; Sekhon, 1989) showed improvements in water stable aggregates and moisture retention at field capacity due to the use of inorganic fertilizers and organic manures. The data in Table 4.9, from the 100-year-old experiment conducted at Sanborn field at University of Missouri, show significant improvements in soil organic matter, clay content, and top soil depth by the use of organic manure and chemical fertilizers. Both soil physical and nutritional properties are enhanced when nutrients depleted due to crop removal and other losses are judiciously replaced.

Soil erosion management

Soil erosion control is crucial to sustainable management of soil resources and to soil resilience. In some cases, taking the pressures off marginal lands can probably be the best solution for controlling accelerated erosion. In other cases, intensive cultivation of prime agricultural land, creating off-farm employment and developing income generating capabilities in non-agricultural sectors can be important options. The choice of appropriate erosion control measures must be carefully made with due consideration of soil types, land form and terrain characteristics, rainfall regime and hydrology, cropping/farming system, and the socioeconomic factors.

Table 4.10. Management effects on runoff and soil loss from sloping lands in Thailand (Anecksamphant *et al.*, 1990).

	Ustic Kandihumult[a] (Chiang Rai)			Ultic Haplustalf[b] (Chiang Mai)		
Treatment	Runoff (cm)	Erosion (t ha^{-1})	Rice grain yield (t ha^{-1})	Runoff (cm)	Erosion (t ha^{-1})	Corn grain yield (t ha^{-1})
Traditional	15	120	1.10	3.2	1.9	2.6
Alley cropping	8	69	1.15	4.5	1.5	2.6
Grass strips and perennial crops	7.5	65	0.97	3.0	2.0	3.0
Hillside ditches with grass strips	6.0	65	0.83	2.0	0.6	3.1
Agroforestry	10.5	95	1.13	6.0	4.5	3.0
Bare fallow	21	230	–	15.0	10.0	–
Forest	5	10	–	3.0	1.0	–
LSD (0.05)			0.19			NS

[a]Slope 20–50%, rainfall 1794 mm $year^{-1}$
[b]Slope 18–40%, rainfall 1434 mm $year^{-1}$

1. Traditional = ploughing up and down the slope
2. Alley cropping = with *Leucaena* hedges
3. Agroforestry with coffee and grain crops

Farming systems and erosion control

Conservation effective measures are those that enhance soil structure, decrease raindrop impact, improve infiltration capacity, and decrease runoff rate and amount. These techniques include practices such as no-till or conservation-tillage, mulch farming through cover crops and planted fallows, multiple cropping, and creation of a multistorey canopy e.g. through agro-forestry. The data in Table 4.10 from Thailand show the impact of management systems on runoff, erosion and crop yield. Conservation effectiveness of management systems differed among two soils and ecoregions. Forestry was apparently the best land use for Chiang Rai because soil erosion remained severe even with alley cropping, grass strips and agroforestry systems. Soil erosion was, however, effectively controlled by these practices adopted on an Alfisol at Chiang Mai.

Several long-term soil restorative experiments were conducted at the International Institute of Tropical Agriculture (IITA), Ibadan, Nigeria. The data from these experiments are shown in Tables 4.11–4.13. It seems that growing deep-rooted shrubs or woody perennials is crucial to improving infiltration rates and restoring structure of a severely degraded soil with a compacted sub-soil. Table 4.11 shows that improvements in infiltration rate

Table 4.11. Restorative effects of crop rotations and agroforestry systems on infiltration rate (cm h^{-1}) of an eroded Alfisol in western Nigeria (IITA, Block D).

	Plough-till from 1973–1979			No-till from 1973–1979[a]		
Rotation system	1979	1984	% Change	1979	1984	% Change
Crop rotation:						
Maize–maize	10.5	7.1	−32.3	10.2	7.8	−23.5
Maize–pigeon pea	10.2	4.7	−53.9	15.6	4.2	−73.0
Agroforestry system:						
Maize–mucuna	9.0	4.6	−48.0	19.2	7.3	−61.9
Maize–leucaena	4.2	6.4	+52.3	16.2	16.6	+2.5
Natural fallow	7.2	6.5	−9.7	6.0	7.8	+30.0

[a]No-till system was adopted from 1973 to 1984 in all treatments.

were brought about only by growing *Leucaena leucocephala*. In this severely eroded soil, a natural fallow for about five years was not effective in restoring soil structure. Bush fallowing can, however, improve infiltration rate and soil structure, if the soil is not severely degraded. For example, the data in Table 4.12 show gradual improvement in infiltration rate with time after the land reverted to bush fallow. The data in Table 4.13 show that an increase in the organic carbon content of the surface layer was achieved by growing annual or seasonal cover crops that produce large amounts of biomass, e.g. *Mucuna utilis*. Woody shrubs or deep-rooted perennials were not as effective in producing biomass as aggressively growing cover crops.

Residue management and erosion control

In most soils under cultivation a regular and sizeable addition of organic material is essential to maintain a favourable level of soil organic matter and to stimulate biotic activity of soil fauna, e.g. earthworms and termites. Structural collapse of soils with predominantly low-activity clays can be avoided by maintaining a content of about 2% soil organic carbon and by enhancing the activity of soil fauna. Crop residue mulch is an important ingredient of any improved farming/cropping system. Frequent applications of 4–6 t ha^{-1} of residue mulch applied to the soil surface is beneficial for soil and water conservation, regulating soil moisture and temperature regimes, improving soil structure, enhancing biological activity of soil fauna, and protecting soils from high intensity rains and ultradesiccation. Mulching also suppresses weed growth.

Although the beneficial effects of mulching are widely recognized, economic procurement of mulch material in sufficient quantity is a serious

Table 4.12. Restorative effects of natural fallowing on infiltration rate (cm h^{-1}) of an Alfisol (Lal, 1992).

Treatment	Pre-clearing 1978	Cultivation duration (years) 1	2	3	4	5	6	7	8
Continuous cropping (watershed 4)	146.0	30.6	28.2	12.6	12.6	20.4	13.8	10.2	10.8
Traditional cropping (watershed 7)	156.0	46.8	37.8	19.8[a]	1.92	24.0	43.2	115.8	193.0

[a]Reverted to natural fallow.

practical problem. Management of crop residues as a source of mulch is, therefore, closely linked with the cropping system, tillage methods, and planted fallows. A range of cultural practices are available to procure adequate amounts of residue mulch for soil protection and fertility enhancement, e.g. cover crops, conservation tillage, sod seeding, agroforestry, etc. Live mulch, alley cropping, ley farming, planted fallows, use of industrial by-products are some of the cultural practices specifically adopted to procure mulch. Suitability of a practice depends on the specific biophysical and socioeconomic environment.

Crop management and erosion control

A continuous ground cover is necessary to provide a buffer against sudden fluctuations in micro- and mesoclimate, and prevent degradative effects of

Table 4.13. Restorative effects of crop rotations on soil organic carbon content (%) for 0–10 cm depth of an eroded Alfisol in western Nigeria (Block D, IITA).

	Plough-till from 1973–1979		No-till from 1973–1979	
Rotation treatment	1982	% Increase over 1979	1982	% Increase over 1979
Maize–maize	1.10	7.8	1.30	0.0
Maize–pigeon pea	1.50	47.1	1.41	8.5
Maize–mucuna	1.40	37.3	1.48	13.8
Maize–leucaena	1.22	19.6	1.34	3.0
Natural fallow	1.30	27.5	1.46	12.3

Initial level of organic carbon in 1979: Plough-till = 1.02% No-till = 1.30%

raindrop impact or high velocity winds. Timely planting, use of viable seed at optimum rates, use of improved cultivars and cropping systems, correct fertilizer use, and appropriate pest management are all important aspects of good farming practices. Benefits of timely planting include provision of a buffer against uncertain rains, unfavourable soil temperature regimes, pest infestation, and unfavourable markets. Planted fallows, of either legume or grass covers, are generally more effective in restoring soil fertility and improving soil physical properties than natural fallows. Soil organic matter can often be increased and soil structure improved over a period of 2 to 3 years.

There are several methods of managing cover crops. The live mulch system involves growing grain/food crops through a low-canopy cover crop. Mixed cropping creates diversity and decreases soil erosion risks. Agroforestry and mixed farming are important techniques to create diversity. Mixed farming can be a stable system for small land holders provided that pastures are lightly grazed, stocking rate is low, and animal waste is applied to the land to replenish soil fertility.

Energy management

The overall energy input and output must be regulated to enhance the output:input ratio, and attain a positive thermodynamic energy balance with respect to biological yield vs. production inputs. Both energy efficiency and flux must be increased. A low output subsistence system can be highly efficient in terms of energy use, but is economically unsustainable because of low productivity. The energy efficiency of a high-input system can be improved by reducing nutrient losses due to leaching and erosion, and

Table 4.14. Effects of organic and conventional farming for 40 years on wheat yield at Spokane, Washington, USA. (Recalculated from Reganold *et al.*, 1987.)

Year	Organic farm ($Mg\ ha^{-1}$)	Conventional farm[a] ($Mg\ ha^{-1}$)	Ratio of organic:conventional farm
1982	4.79	4.45	1.08
1983	4.25	4.42	0.96
1984	4.25	4.67	0.91
1985	4.38	4.35	1.00
1986	4.85	4.35	1.11
Average	4.50	4.45	1.01

[a]Average of two farms using inorganic fertilizer.

Table 4.15. Some indices of sustainability for plant, soil and water resources.

Plant	Soil	Water
1. Biomass production	1. Soil structure	1. Water balance
2. Productivity	2. Net rate of soil renewal	2. Water quality
3. Nutritive quality	3. Nutrient fluxes and cycling	3. Water: air ratio in the root zone
	4. Soil organic matter content	
	5. Rooting depth	
	6. Gaseous fluxes	

enhancing nutrient capital through judicious inputs of chemical fertilizers and organic or other amendments (liming).

The source of nutrients, organic or inorganic, is not crucial as long as nutrients are supplied in adequate quantity and in readily available form. The data in Table 4.14 show no significant difference between wheat yields from 40 years of cultivation using organic manures or inorganic fertilizers. However, hoping to increase and sustain agricultural production by adding chemicals alone, without improved and efficient systems of soil and crop management, is bound to cause frustrations and disappointments. It is the judicious combination of both management and inputs that is crucial to a sustainable use of natural resources.

Technological Options for Sustainable Land Use

A truly sustainable system meets the following criteria: (a) it enhances soil resilience through long-term maintenance of the biological and ecological integrity of natural resources; (b) it sustains a desirable level of support to the social, political and economic well being of a farm, community or region; and (c) it enhances the quality of life. In the context of agriculture, sustainability of a land use system can be assessed by evaluating the characteristics of plant, soil and water resources (Table 4.15). The system is sustainable if: (a) a favourable level of soil structure is maintained, (b) soil erosion is controlled, (c) nutrients are effectively recycled, (d) soil organic matter content is optimized, and (e) water and energy regimes are maintained favourable to the ecological integrity of the system. The basic principles of sustainable use of soil and water resources are the same for all ecoregions. Technological options, however, may be soil and site specific. Some basic principles and underlying technological options for sustainable use of soil and water resources are outlined in Tables 4.16 and 4.17 for the humid and subhumid/semiarid tropical regions, respectively. Judicious use of these options is likely to sustain productivity and the life support systems of soil and water resources.

Table 4.16. Best management practices for sustainable land use in humid tropics.

Arable land use	Pasture development
• Use prime land, and avoid steep and shallow soils • Forest removal by manual methods, or by shearblade • Use cover crop and mulch farming techniques for soil and water conservation • Frequent use of planted fallows is necessary • Wherever feasible, integrate woody perennials and livestock with food crop annuals	• Use prime land, and avoid steep/shallow soils • Proper clearing methods are essential, e.g. manual, slash and burn, etc. Tree defoliants can also be used in regions with low tree density. Dead trees can be left standing • Seeding with suitable and ecologically adapted mixture of grass and legumes is necessary • Maintain soil fertility as per soil test values. Balanced fertilization is important • Stocking rate should be low, and controlled and rotational grazing should be adopted • Use live fences and vegetative hedges
Agroforestry	**Forest plantations**
• Use prime land of high inherent fertility • Land clearing should be done by manual methods or shearblade techniques • Choice of native tree species which do not aggressively compete with annuals is critical • Proper tree management is crucial • Choice of appropriate crops and cropping sequences is critical to success • Soil fertility management should be done in relation to cropping intensity and soil test values	• Use prime land with no severe limitations • Clear existing vegetation by manual methods of slash and burn or by use of shearblade. Some roots and stumps can be left intact • Seed a leguminous cover crop immediately • Establish tree seedlings through the leguminous cover by suppressing it through chemical or mechanical means. Cover crop management is crucial to tree establishment • Use balanced fertilizer based on soil test value and tree requirement • Use effective soil and water conservation techniques

Conclusions

Soil is the essence of all life on earth. It is a non-renewable and a finite resource. The rise and fall of ancient civilizations depended on the ability of people to maintain and enhance the life support processes in their soil resources. Judicious use through discriminate and objective management of soil, water and climate is as crucial to human survival and welfare now as it was for the extinct civilizations who could not manage these most basic of all resources. Sustainability of soil resources must be assessed in terms of trends in per capita productivity with reference to changes in soil properties, water characteristics, and climatic factors.

Soil resilience depends on antecedent soil conditions, temporal changes in soil properties, land use and management systems. The rate of new soil formation is rather slow and ranges from 0.01 to 0.003 cm year^{-1} for most soils developed *in situ*. For a soil to be renewable, the rate of soil depletion

Table 4.17. Best management practices for sustainable land use in subhumid and semiarid tropics.

Water conservation • Rough seedbed, deep ploughing at the end of rains • Tied ridges • Microcatchments • Use broadbed and furrow system in Vertisols • Supplementary irrigation where needed • Use wind breaks of leguminous shrubs and trees • Reduce weed competition **Soil fertility management** • Apply farmyard manure, organic wastes, and nitrogeneous fertilizers • Apply micronutrients where needed • Turn under as much crop residue and biomass as possible • Increase fertilizer use efficiency through water conservation, split dose application, and placement in the vicinity of row zone • Use cover crops and green manures	**Crop management** • Improved crops and cultivars developed/adapted for harsh environment • Multiple cropping based on inter-cropping, rotation cropping, relay cropping and mixed cropping systems • Use agroforestry systems based on native tree species where appropriate • Use ley farming based on growing appropriate forages, light/controlled grazing, and relevant cropping sequences • Improve crop stand through use of good seed, high seed rate, appropriate seeding equipment • Seed in furrow to minimize sand blasting and decrease the adverse effects of high soil temperature

should be less than or equal to the rate of soil formation. The rate of soil depletion and degradation of soil productivity depend on land use and soil management, and may be independent of profile depth for soils with root-restrictive subsoil characteristics. Choice of appropriate land use and use of science-based improved technology can enhance and sustain soil productivity and accentuate resilience of even fragile soils in ecologically sensitive ecoregions.

Science-based land use, 'how' rather than 'what', is important to sustainability and resilience. Processes, practices and policy issues involved in sustainable land use are outlined in Fig. 4.4. The principal processes involved in soil resilience are: (a) control of soil organic matter content; (b) improvement in soil structure; (c) increase in soil biodiversity; (d) reduction in soil degradation and erosion rates below the soil formation rate; and (e) increase in nutrient capital and recycling mechanisms. Basic principles of technological options to achieve these are known, and include practices of good farming, e.g. conservation tillage, mulch farming, crop rotations involving cover crops and planted fallows, mixed farming with controlled grazing and low stocking rate, agroforestry, and use of inorganic fertilizers and organic manures. Above all, it is important to use improved crops and cultivars, and introduce new crops and species with potential to develop agro-based industries. Although basic principles are the same, these techniques should be validated, adapted and fine-tuned under local edaphic conditions and socioeconomic environments. Once developed,

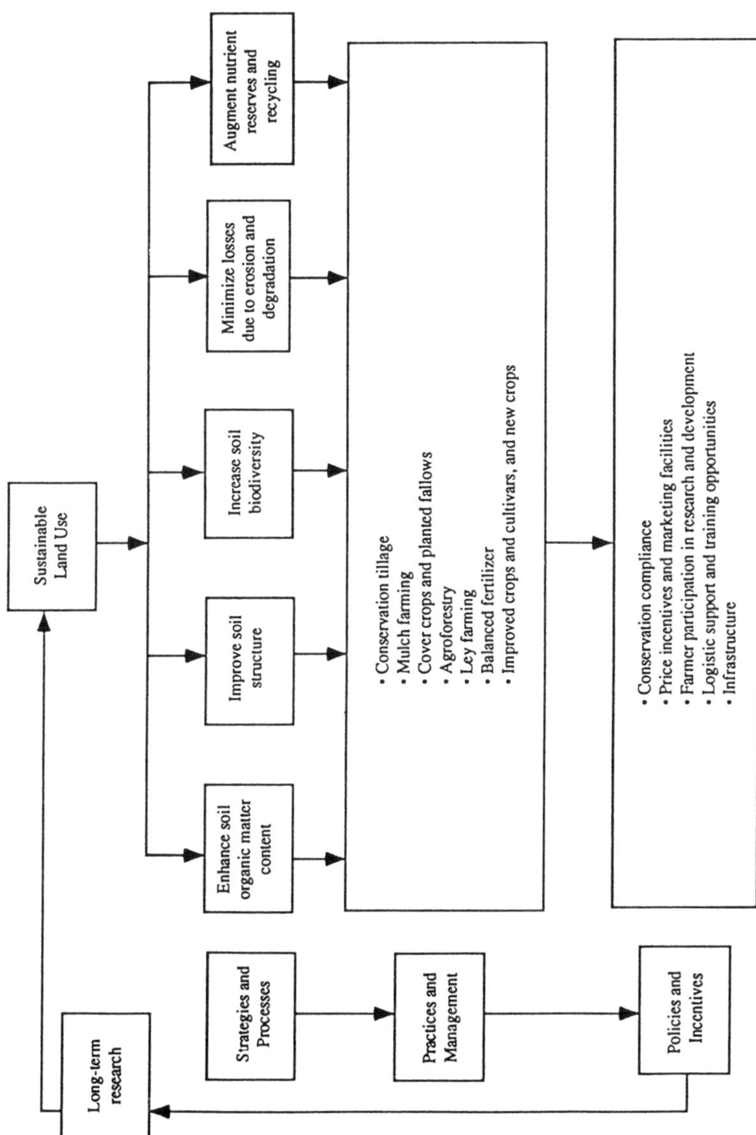

Fig. 4.4. Processes, practices and policies involved in land use and soil resilience, and the need for long-term research.

widespread adoption of these technologies depend on policies and incentives, and availability of credit and marketing facilities. Participation of the farming community is crucial at all stages of research, development and testing of improved technology. Above all, long-term and system-oriented research is needed to address the sustainability issue. Land use, agricultural sustainability, and soil resilience are long-term issues. Properly designed, adequately equipped, and well-managed long-term land use and soil management experiments are needed for different soils and ecoregions to address these issues. Specifically, a long-term research programme is needed with the following objectives.

1. To standardize criteria, delineate threshold values of soil properties, and develop quantitative and reliable methods of assessment of soil degradation by different processes at the local, regional and global scales. There is a need to improve the reliability and accuracy of statistics on soil degradation. Baseline data are needed on extent, degree and type of soil degradation in relation to geology, terrain, vegetation, climate, current land use, and soil properties, with data banks for different agroecological regions.
2. To develop objective and quantifiable criteria of soil resilience, and assess resilience of major soils in relation to parent material, soil properties, land use, and management systems.
3. To develop appropriate land uses and restorative measures for enhancing the economic productivity of degraded soils. Assessing accurately the extent of soil degradation, establishing the cause–effect relationship, choosing appropriate land use, and developing restorative measures are important research and development priorities. It is important to identify what land is restorable at what cost and for what use.
4. To identify the reasons for non-acceptability of available research information by farmers. Although innovative and research-proven concepts, and subsystems exist, several of these are not being accepted by the farming community. There is a need to evaluate these concepts in the context of socioeconomic environments of the farming community.
5. To evaluate soil dynamics and evolution under different land uses and farming systems for principal soils and ecoregions. The agronomic productivity should be assessed in terms of land use impact on soil properties.
6. To increase per capita productivity per unit input of non-renewable resource and per unit decline in important soil properties, and
7. To develop an index of agricultural sustainability. To be functional, the sustainability concept must be based on quantitative indices, e.g. trends in per capita productivity, soil and water conservation, productivity restoration, water quality in relation to dissolved and suspended load, use of non-renewable resources, greenhouse gas emissions, etc.

References

Anecksamphant, C., Boonchee, S. and Sajjapongse, A. (1990) Management of sloping lands for sustainable agriculture in northern Thailand. *14th Congress ISSS*, Kyoto, Japan, VI, 198–202.

Aubert, G. (1960) Influences de la vegetation sur le Sol en Zone Tropicale Humide et Semihumide. Rapports du Sol et de la Vegetation. *Collog. Soc. Nat. Fr.* 11–12.

Aubert, G. and Maignier, R. (1949) L'erosion Aeolinée dans le Nord du Senegal et du Soudan Francais. *Bull. Agr. Cong. Belgique* 40, 1309–1316.

Bear, F. (1986) *Earth, the Stuff of Life.* University of Oklahoma Press, Norman.

Bennett, H.H. (1939) *Soil Conservation.* McGraw-Hill, New York.

Boulad, A.O., Muller, J.P. and Bocquire, G. (1977) Determination of the age and rate of alterations of ferrallitic soil from Cameroon. *Science and Geology Bulletin* 30, 175–178.

Bradley, J.H. (1935) *Autobiography of Earth.* Coward-McCann, New York.

Brown, L. (1991) The global competition for land. *Journal of Soil and Water Conservation* 46, 394–397.

Chamberlin, T.C. (1909) Soil wastage. In: Proceedings of a conference of Governors in the White House, Washington, DC 1908. US Congress 60th, 2nd Session, House Document 1425, Washington DC.

Darvoll, J. (1973) Resources, renewable and nonrenewable. In: *Resources and Population.* Academic Press, New York.

Dunne, T. (1979) Sediment yield and land use in tropical catchments. *Journal of Hydrology* 42, 281–300.

FAO (1991) World soil resources. *World Soil Resources Report 66*, FAO, Rome, Italy.

Ford, R. (1986) *Land, People and Resources in Kenya.* World Resources Institute, Washington DC, p. R-15.

Friend, J.A. (1992) Achieving soil sustainability. *Journal of Soil and Water Conservation* 47, 156–157.

Gantzer, C.J., Anderson, S.H., Thompson, A.L. and Brown, J.R. (1990) Estimating soil erosion after 100 years of cropping on Sanborn field. *Journal of Soil and Water Conservation*, 45, 641–644.

Hall, G.F., Daniels, R.B. and Foss, J.E. (1982) Rate of soil formation and renewal in the USA. In: *Determinants of Soil Loss Tolerance.* American Society of Agronomy, Madison, WI.

Hudson, N.W. (1976) *Soil Conservation.* B.T. Batsford Ltd, UK.

IARI (1984) Highlights of research of a longterm fertilizer experiment in India (1971–82). *IARI-LTFE Res. Bull.* 1, 100 pp., New Delhi, India.

ISRIC/UNEP (1990) *World Map of the Status of Human-induced Soil Degradation* (GLASOD), Wageningen, The Netherlands.

Johnson, L.C. (1987) Soil loss tolerance: fact or myth. *Journal of Soil and Water Conservation* 42, 155–160.

Lal, R. (1992) Tropical agricultural hydrology and sustainability of agricultural systems: a ten year watershed management project in southwestern Nigeria. *Technical Bulletin*, The Ohio State University/IITA, 303 pp.

Lee, L.K. (1990) The dynamics of declining soil erosion. *Journal of Soil and Water Conservation* 45, 622–624.

Leneuf, N. and Aubert, G. (1960) Attempts to measure the rate of ferralization. *Transactions 7th International Congress of Soil Science,* Paris, VI, 225–228.

Lovelock, J.E. (1979) *Gaia: A New Look at Life on Earth.* Oxford University Press, Oxford.

May, R.C. (1960) Rate of clay formation and mineral alteration in a 400-year old volcanic ash soil at St Vincent, B.W.I. *American Journal of Science* 258, 354–368.

Mengel, K. (1990) Impacts of intensive plant nutrient management on crop production and environment. *14th Congress ISSS,* 12–18 August 1990, Kyoto, Japan, Plenary Papers, pp. 42–52.

Miller, F.P., Rasmussen, W.D. and Mayer, L.D. (1985) Historical perspective of soil erosion in the United States. In: Follett, R.F. and Stewart, B.A. (eds) *Soil Erosion and Crop Productivity.* ASA-CSSA-SSSA, Madison, WI, pp. 23–48.

Mohr, E. and van Baren, F. (1954) *Tropical Soils.* Interscience Publ., New York.

Nahon, D. and Lappartient, J.R. (1977) Time factor and geochemistry in iron crusts genesis. *Catena* 4, 249–254.

Nambiar, K.K.M. and Abrol, I.P. (1989) Longterm fertilizer experiments in India: an overview. *Fertilizer News. Special Issue* 34 (4), 11–20.

Owens, L.B. and Watson, J.P. (1979) Rates of weathering and soil formation on granite in Rhodesia. *Soil Science Society of America Journal* 43, 160–166.

Reganold, J.P., Elliott, L.F. and Unger, Y.L. (1987) Longterm effects of organic and conventional farming on soil erosion. *Nature* 26(330), 370–372.

Richards, J.F. (1991) Land transformation. In: Turner, B.L., Clark, W.C., Kates, R.W., Richards, J.F., Mathews, J.T. and Mayer, W.B. (eds) *The Earth as Transformed by Human Action: Global and Regional Changes in Biosphere over the Past 300 Years.* Cambridge University Press, New York, pp. 163–178.

Rozanov, B.G. (1990) Human impacts on evolution of soils under various ecological conditions of the world. *14th International Congress of Soil Science,* 12–18 August 1990, Kyoto, Japan, Plenary Papers, pp. 53–62.

Ruxton, B.P. (1966) *The Measurement of Denudation Rates.* Institute Australian Geography, Sydney, Australia.

Sekhon, G.S. (1989) Nutrient balances in the soils and crop productivity. *Fertilizer News: Special Issue* 34 (4), 27–32.

Thomas, D. and Jansen, I. (1985) Soil development in coal mine spoils. *Journal of Soil and Water Conservation* 40, 439–442.

UNEP (1991) *Status of Desertification and Implementation of the United Nations Plan of Action to Combat Desertification.* UNEP, Nairobi, Kenya, 77 pp.

US Department of Agriculture (1909) *Bureau of Soils.* Bulletin no. 55, Washington, DC.

USDA (1989) The second RCA appraisal. Soil, water and related resources on non-federal land in the United States, Washington, DC.

Van Baren, F. (1931) Quoted by D. Brusden (1979) 'Weathering'. In: Embleton, C. and Thornes, J. (eds) *Processes in Geomorphology,* Edward Arnold, London, pp. 73–129.

Watkins, J.S., Bottino, M.L. and Morisawa, M. (1975) *Our Geological Environment.* W.B. Saunders, Philadelphia.

Williams, M. (1991) Forests. In: Turner, B.L., Clark, W.C., Kates, R.W., Richards, J.F., Mathews, J.T. and Mayer, W.B. (eds) *The Earth as Transformed by Human Action:*

Global and Regional Changes in Biosphere over the Past 300 Years. Cambridge University Press, Cambridge.

Wolman, M.G. (1967) A cycle of sedimentation and erosion in Urban river channels. *Geography Annual* 49A, 385–395.

World Resources Institute (1992–1993) *Towards Sustainable Development. A Guide to the Global Environment.* World Resources Institute Washington, DC, 385 pp.

Chapter 5

The Biological Dimension of Soil Resilience: the Impact of Molecular Biology

J.M. Lynch

Microbiology & Crop Protection Department, Horticulture Research International, Littlehampton, West Sussex, BN17 6LP, UK. Present address: School of Biological Sciences, University of Surrey, Guildford, Surrey, GU2 5XH

Introduction

Biological communities in soil are complex, containing microorganisms (viruses, bacteria, yeasts, filamentous fungi and protozoa) and animals, yet the composition of the communities can be quite stable. However, this has often been difficult to investigate experimentally. Modern environmental issues such as the need to investigate the impact of genetically modified microorganisms (GMMOs) on the soil environment have been a stimulus to question how best to investigate and determine the baseline communities of soil and their stability. This is eased experimentally by combining some of the techniques coming from molecular biology with traditional techniques in microbial ecology such as the use of selective media. Moreover, the techniques being developed for these purposes are equally relevant to determining the impact of other agents, such as pollutants, on the soil biota and hence determining the biological dimension of soil resilience.

In this chapter some of the background principles which govern the existence, activity and analysis of the soil biota will be outlined. The emphasis will be on cropped soils, but many of the ideas will be relevant to soils bearing natural vegetation or forests. The exploitation of these ideas in the study of the environmental impact of GMMOs will then be considered.

Substrates

The development of the soil biota is dependent on the substrate input. The major source of carbon and energy for the biota in the surface 5 cm of soils growing arable or vegetable crops is crop residues such as straw, with root-derived carbon (rhizodeposition) (Whipps and Lynch, 1985) providing only about 10% of the carbon (Lynch, 1988). However, if the total soil profile is considered, rhizodeposition provides a similar amount of carbon to the crop residue. Rhizodeposition products are more soluble and readily available for biodegradation than crop residues, which are usually mainly lignocellulosic.

The utilization of substrates leads to the formation of microbial biomass. In terms of saprophytic ability of organisms utilizing the crop residue substrates, a soil biomass level is formed which is fairly constant during the year and which in temporal terms can be called resilient. However, as the plant root grows in the soil and soil samples are taken from between crop rows, there is an increase in the biomass which approximately reflects the increase in substrate availability being provided by the roots (Lynch and Panting, 1980). A crucial factor in the development of the rhizosphere biomass is that because the qualitative nature of the substrate being provided to the biota is different from native soil organic matter, the species composition of the biomass, or the biodiversity, becomes different from that in the bulk soil. In this respect therefore it can be argued that the rhizosphere biomass is only resilient while the substrate input is being provided. Similarly, the biomass associated with crop residue decomposition will only be resilient while the substrate is available.

Process Values

The saprophytes utilizing crop residues and the rhizodeposition are both involved in metabolic processes which can be of significance to the plant and soil but obviously the events taking place closest to the plant, namely in the rhizosphere, are those which are likely to be of greatest significance to it.

The process values of rhizosphere organisms vary between the beneficial effects, such as N_2-fixation, soil stabilization and biocontrol, and the harmful which include pathogenesis and phytotoxicity (Lynch, 1990). There is also a group of processes which we can term neutral or variable, such as nutrient flux. It is recognized now that many of the processes are under genetic control including the recognition processes of the biomass to the root. As such, it is now possible to deploy the modern techniques available from molecular biology to investigate the populations themselves and the processes in which they are involved.

Probably the greatest area of study in soil microbiology this century has been N_2-fixation. This has led to the characterization of the *nif*-gene complex which encodes the nitrogenase enzyme responsible for N_2-fixation.

As a consequence of this it is possible to take sequences from that gene and to use them as DNA probes which can be hybridized to *nif*-genes in colonies of bacteria isolated from nature. In a series of cooperative experiments between the National Institute of Agro-environmental Sciences and HRI (Kimura *et al.*, 1993), the *nif*-H and *nif*-DK sequences from *Klebsiella pneumoniae* were used to hybridize to colonies of bacteria which were isolated from soil in which rice or wheat was grown. The advantage of these two sequences is that they are highly conserved and present in the *nif*-genes of most N_2-fixing organisms. By the classical enumeration method for N_2-fixing bacteria using N-free medium, total counts as colony forming units (cfu) were 6×10^5–13×10^5 cfu g^{-1} dry soil. About 40% of total anaerobic bacteria detected on the medium were determined to be *nif*-gene carriers. The *nif* gene-carrying bacteria which grew on the complete medium were estimated to be 2.5×10^6–5×10^6 cfu g^{-1} dry soil, accounting for 7–11% total bacteria on the medium. Consequently, therefore, the DNA probe method revealed that in soil and in the rhizosphere there are a significant number of *nif* gene-carrying anaerobic bacteria which cannot be detected by the classical method where N-free medium is used. There were five times as many such *nif* gene-carrying bacteria that could grow only on the complete medium as total bacteria detected on the N-free medium. The implication of these results is that the *nif*-gene appears to be extremely pervasive in soil. It should be noted, however, that the *nif*-gene probe is only recognizing the genetic potential for a process to occur and indicates nothing about the expression of that gene. Indeed, it would seem to be clear from these observations that only a small proportion of the *nif*-gene complex present in soil is expressed. This might perhaps explain why attempts to introduce N_2-fixing organisms into soils *per se*, as opposed to the inoculation of legumes, seldom gives an increase in soil N_2-fixation. Barriers to expression of that gene are both genetic, such as there being an inadequate promoter sequence alongside the structural gene sequence, and biochemical, such as there being inadequate ATP coming from carbohydrate utilization, for the nitrogenase enzyme to function. It is therefore clear to me that in looking at the biological dimension of soil resilience, it is not only the population biology of the organisms that should be considered, but also the population biology of their genes and the capacity to express those genes. In this respect, there is a range of other DNA probes already available to study soil and rhizosphere processes. For example, the pervasiveness of the chitinase gene from *Streptomyces* sp., an enzyme putatively involved in biocontrol of fungal pathogens which carry chitin in their cell walls (Kimura *et al.*, 1993) was also investigated. The cellulase gene complex has also been extensively studied (Teeri *et al.*, 1990) and it seems reasonable that DNA probes could be deployed from these studies to investigate soil decomposition processes.

Potential for Gene Transfer

One of the environmental concerns is that genes in one organism might be transferred to other organisms. The greatest attention has been focused on the plasmids which can carry many genes encoding functions which could influence plants and soils. It is generally considered that plasmids can be involved in conjugation processes. In order to investigate this, we have used populations of *Enterobacter cloacae* or *Pseudomonas cepacia* in model experimental systems composed of glass columns packed with vermiculite or soil (Sun *et al.*, 1993). Both of these common rhizosphere organisms were engineered to carry the transmissible recombinant plasmid R388::Tn1721 carrying resistance to the antibiotics tetracycline and trimethoprim as gene donors. The recipients were the same two species with nalidixic acid resistance carried on the chromosome. Therefore, donor, recipient and transconjugant cells could be recovered on selective isolation media containing the respective antibiotics. Only transconjugants will grow on a medium containing all three antibiotics. The *E. cloacae* proved to be more stable and reached a steady state with effluent populations of the donor and transconjugant being up to 3×10^{10} and 3×10^{5} respectively. By contrast in columns of soil, transconjugants could only just be detected and then only for the first two days after inoculation. However, the fact that such conjugation steps can occur at all in soil indicates that the soil biota is likely to be in a state of continuous evolution and we can expect the number of effective strains of an organism and hence the total biodiversity of soil to be continuously increasing as a consequence of these gene exchange processes. The important exercise is to endeavour to describe quantitatively the gene exchange processes occurring in soil.

Knudson *et al.* (1988) proposed a simple mass action model for conjugative plasmid transfer and obtained experimental evidence to support its predictions. The model was posed in terms of a set of difference equations and, unfortunately, neither it nor their experimental work took into account leaching and diffusion. Thus the model represents a homogeneous batch system and does not represent these two important characteristics of soil. Lynch and Bazin (1990) have used the underlying assumptions of this model to formulate a new one for gene transfer in a well-mixed thermodynamically open ecosystem. For this we used a set of differential equations and included Monod substrate-dependent growth terms for the three populations of bacteria, the gene donors (D), the recipients (R) and the transconjugants (T). The equations of balance are over time (t):

$$\frac{dD}{dt} = \frac{\mu_d DS}{K_d + S} - \frac{D}{\theta} \tag{5.1}$$

$$\frac{dR}{dt} = \frac{\mu_r RS}{K_r + S} - \frac{R}{\theta} \tau \left[(D + T)r\right] \tag{5.2}$$

$$\frac{dT}{dt} = \frac{\mu_t RS}{K_t + S} - \frac{T}{\theta}\tau\,[(D + T)R] \tag{5.3}$$

$$\frac{dS}{dt} = \frac{S_r - S}{\tau} - \frac{\mu_d DS}{(K_d + S)Y_d} - \frac{\mu_r RS}{(K_r + S)Y_r} - \frac{\mu_t TS}{(K_d + S)Y_t} \tag{5.4}$$

where: θ = retention time of the system
τ = conjugational rate transfer constant
S = substrate concentration
S_r = input nutrient concentration
Y_d, Y_r, Y_t = yield constants for each of the bacterial populations
μ_d, μ_r, μ_t = maximum specific growth rates
K_d, K_r, K_t = saturation constants

The basis of the model is on the lines of the layer models developed by Addiscott and Wagenet (1985) but Dr Elizabeth Scott at King's College London modified these models and integrated them into the LEACHM model of Wagenet and Hutson (1989). The combined models can be applied to any microbial interaction, be it gene exchange, predator/prey or biocontrol in the rhizosphere. An exciting development from this has been that Dr Scott has shown that the predator/prey process is chaotic but in finding a degree of order (strange attractors), some elements of prediction of how populations will develop with time is possible.

Environmental Impact

A great stimulus to the study of the soil biota has come from the need to check the level of hazard in releasing genetically-engineered organisms into the environment. The investigative processes and the methodology which are being developed are equally relevant to any environmental impact study, such as the consequence of the addition of a pollutant.

The UK Department of the Environment is funding a project on the release of GMMOs into the rhizosphere at Littlehampton in association with the NERC Institute of Virology and Environmental Microbiology in Oxford, the University of Aberdeen and King's College London. Our strategy in the first year of study has been to investigate the baseline ecology of wheat rhizosphere growing at Littlehampton on a silt loam soil and the sugar beet growing at Oxford on a heavy Oxford clay, giving contrasting crops, soils and meteorology. The experiments are done in the field and also in a contained glasshouse using intact monoliths of soil (60 cm deep × 15 cm diameter) extracted from the field.

Using a variety of isolation techniques for bacteria, filamentous fungi, yeast and protozoans including selective agar media and identification of the bacteria to species level using fatty acid methyl ester profiling, we have found by analysis that there is much greater similarity between the populations in the field and glasshouse than was found in a similar study on the leaves of wheat (phyllosphere). Both in Oxford and Littlehampton we have found *Pseudomonas aureofaciens* to be a constant component of the biota associated with leaves and roots. The ubiquity of this organism led us to select it for genetic marking. Dual markers were inserted on the chromosome to minimize the potential for gene exchange with other microorganisms. The first functional gene used was *lacZY* which codes for lactose utilization and can be detected on a medium containing X-gal, the positive colonies turning blue. The second gene is *xylE* which codes for catechol utilization, turning it yellow; this latter gene was associated with another carrying kanamycin resistance. It is thus possible to determine the recovery on selective media of the GMMO or any recipient of either of its marker genes. The presence of the *lacZY* gene combined with the use of X-gal enables a most probable number technique to be used which enables single cells to be detected in a soil suspension.

The results of all these studies show that it is possible to determine the bacteria, filamentous fungi and yeasts. For some fungi which have ubiquitous spores in the environment, such as *Penicillium* spp., and therefore lead to a large number of colony-forming units, the populations (but not biomass) can also be determined with confidence. By contrast, for fungi such as *Pythium* and *Trichoderma,* biometrically sound data can only be collected using selective isolation media but sometimes chemical markers can be used. Indeed, we have demonstrated that the number of colony-forming units can be inversely related to the soil biomass using chitin or ergosterol as markers (Lumsden *et al.,* 1990). Our preliminary studies indicate that, as expected, colonization of all organisms is dominantly in the rhizosphere. However, a spore-former like *Penicillium* is picked up preferentially in soil, although this may or may not reflect the major location of the vegetative state. A critical appraisal of our results for 1991/92 will be published when our biometric analyses are complete.

Needs for Study: Monitoring/Technology

Clearly these are early days in the study of the biodiversity of the rhizosphere and its resilience. It does point, however, to the utility of the emerging monitoring technology.

To develop the application of the monitoring technology, it is also necessary to develop profiles of the microorganisms (including genetic construction, ecology and toxicology), crops (including associated insects and wildlife), environment (including soil and meteorology) (Lynch, 1991). It should then be possible to determine strategies for environmental impact by following perturbations on the baseline ecology. A perturbation which is recovered to its baseline rapidly would be the index of a resilient soil.

Acknowledgements

The application of molecular ecology to soil science necessitates inputs from large multidisciplinary teams and in this respect I appreciate inputs from my collaborators, at HRI (John Whipps, Frans de Leij and John Fenlon); King's College, London (Michael Bazin, Elizabeth Scott and Sun Li); the NERC Institute of Virology and Environmental Microbiology, Oxford (Mark Bailey); University of Aberdeen (Ken Killham, Jim Prosser and Anne Glover) and the National Institute of Agro-Environmental Sciences, Tsukuba, Japan (Ryosuke Kimura and Kiyotaka Miyashita). The work referred to here was supported by the Agricultural and Food Research Council, the Department of the Environment and the Organization for Economic Cooperation and Development.

References

Addiscott, T.M. and Wagenet, R.J. (1985) Concepts of solute leaching in soils: a review of modelling approaches. *Journal of Soil Science* 36, 411–424.

Kimura, R., Lynch, J.M., Katoh, K. and Miyashita, K. (1993) Enumeration of specifically functional soil bacteria. *Proceedings 6th International Symposium on Microbial Ecology*, Barcelona, 1992 (in press).

Knudson, G.R., Walter, M.V., Porteous, L.A., Prince, V.J., Armstrong, J.L. and Seidler, K.J. (1988) Predictive model of plasmid transfer in the rhizosphere and phyllosphere. *Applied and Environmental Microbiology* 49, 416–422.

Lumsden, R.D., Carter, J.P., Whipps, J.M. and Lynch, J.M. (1990) Comparison of biomass and viable propagule measurements in the antagonism of *Trichoderma harzianum* and *Pythium ultimum*. *Soil Biology and Biochemistry* 22, 187–194.

Lynch, J.M. (1988) The terrestrial environment. In: Lynch, J.M. and Hobbie, J.B. (eds) *Micro-organisms in Action – Concepts and Applications in Microbial Ecology*. Blackwell Scientific Publications, Oxford, pp. 103–131.

Lynch, J.M. (ed.) (1990) *The Rhizosphere*. John Wiley, Chichester.

Lynch, J.M. (1991) Developments in rhizosphere microbiology. *Australian Microbiologist* 13, 99–100.

Lynch, J.M. and Bazin, M.J. (1990) The need to quantify rhizosphere population and community dynamics. *Transactions XIV Congress of the International Society of Soil Science* 3, 4–9.

Lynch, J.M. and Panting, L.M. (1980) Cultivation and the soil biomass. *Soil Biology and Biochemistry*, 12, 29–33.

Sun, L., Bazin, M.J. and Lynch, J.M. (1993) Plasmid exchange dynamics in a model soil column. *Molecular Ecology* 2, 9–15.

Teeri, T.T., Jones, A., Kraulis, P., Ronvinen, J., Penttilä, M., Harkki, A., Nevalainen, H., Vanhanen, S., Saloheimo, M. and Knowles, J.C. (1990) Engineering *Trichoderma* and its cellulases. In: Kubicek, C.P., Eveleigh, D.E., Esterbauer, H., Steiner, W. and Kubicek-Pranz, E.M. (eds) *Trichoderma reesei Cellulases: Biochemistry, Genetics, Physiology and Applications.* Royal Society of Chemistry, Cambridge, pp. 156–167.

Wagenet, R.J. and Hutson, J.L. (1989) LEACHM: a process-based model of water and solute movement transformations, plant uptake and chemical reactions in the unsaturated zone. Version 2.0. Continuum 2. New York State Water Resources Institute, Cornell University, Ithaca, New York.

Whipps, J.M. and Lynch, J.M. (1985) Energy losses by the plant in rhizodeposition. *Annual Proceedings of the Phytochemical Society of Europe* 26, 59–71.

Chapter 6

Ecological–Economic Assessment of Soil Management Practices for Sustainable Land Use in Tropical Countries

A-M.N. Izac

Inland Valleys Program, International Institute of Tropical Agriculture, PO Box 5320, Oyo Road, Ibadan, Nigeria.
Present address: International Centre for Research in Agroforestry, PO Box 30677, Nairobi, Kenya

Introduction

Soil resilience breaks down, leading to irreversible soil degradation and unsustainability, whenever either soil erosion, soil pollution or soil nutrient depletion reach threshold levels. In the agricultural systems found in industrialized countries, soil erosion and soil pollution are the two main causes of soil degradation as nutrient depletion is counteracted by the use of inorganic fertilizers (which often contributes to pollution). In tropical agricultural systems, and in Africa in particular, soil erosion and nutrient depletion are the principal processes which jeopardize soil resilience. Up to the present time, few chemicals have been used by farmers in these systems, so that levels of pollution due to soil amendments are still far from threshold levels.

Soil and biological scientists have been investigating erosion and pollution processes in industrialized countries for a number of years, and have devised soil management techniques to palliate these problems. Since these techniques were not adopted voluntarily by farmers to the extent originally hoped, methods of economic analyses of soil erosion and pollution were developed and governmental policies implemented to provide additional incentives to encourage adoption by farmers. In spite of this relatively long research and policy history, there are many parameters which are still unknown in current biological and economic analyses of erosion and pollution. Some of the conclusions reached during a recent European symposium on soil management in the European Community, for instance, point to very basic gaps:

> A number of questions about erosion remain unanswered. Firstly, it is necessary to determine the scale of the problem. There have been

 Soil Resilience and Sustainable Land Use (eds D.J. Greenland and I. Szabolcs)

> several reports of serious erosion within the European Community but so far there is no informed overview of the extent of the problem. This leaves policy makers without an adequate basis from which to judge the severity of the problem.
> ... there seem to be no clear answers yet to the following questions: Is there a productivity loss associated with erosion? If so, at what stage in the erosion process does a decline in productivity become apparent? To what extent is the increase in soil erosion related to current farming practice? At what stage does the situation become irretrievable?
> ... As with soil erosion, there is a need to monitor the levels of various pollutants in the soils of the European Community.
> (Bullock, 1987, pp. 584–585)

Given this dearth of empirical evidence it is not surprising that the reasons for the relative success of some policies and the relative lack of success of others are controversial. Analysts concur only on two points, namely, the complexity of the issues and the need for policies to be adapted to the particular socioeconomic circumstances of a region or country. One of them summarized the debate thus: 'Making soil and water conservation work has proven difficult for many reasons – some technical, some socio-economic, some political' (Pierce, 1987, p. 34).

Erosion and nutrient depletion processes have been studied in tropical agriculture and soil management practices have been devised to remedy these problems. As in the case of temperate agriculture, some basic questions remain unanswered. However, in comparison with what has happened for temperate agriculture, economic and policy analyses have been extremely rare (Anderson and Thampapillai, 1990). Consequently, the reasons for the low adoption levels of strategies promoting on-farm soil resilience are not clear (see for instance Whittome *et al.*, 1992, concerning the adoption of alley cropping in West Africa).

Before the best method of promoting soil resilience in tropical countries can be determined there needs to be an understanding of the economic processes at play in farmers' soil management decisions. In particular, we need to address the question: Are these processes such that individual farmers will voluntarily adopt levels of soil management which society considers adequate, or are these processes such that farmers' interests and society's interests in sustainable land use are divergent? If individual and societal interests coincide, the cornerstone of the promotion of soil resilience in tropical countries will be the dissemination of information to farmers and systematic extension efforts. If individual and societal interests diverge, however, the cornerstone of the promotion of soil resilience will be governmental policies providing incentives for farmers to adopt socially desirable soil management techniques.

The objectives in this chapter are to discuss: (i) whether farmers' and society's interests regarding sustainable soil management converge or diverge; and (ii) the principal avenues for action open to researchers and policy makers for promoting soil resilience in tropical agricultural systems. A brief discussion of agricultural systems hierarchy serves to set the analysis of economic processes in soil management within an appropriate framework. Economic processes are then examined at two different levels in this hierarchy, the farming system and watershed/regional levels. The policy implications of this analysis are drawn, and finally recommendations are made concerning research and policy priorities in soil management in tropical countries.

Hierarchy of Agricultural Systems as a Background to the Assessment of Soil Management Techniques

Fig. 6.1 represents a spatial hierarchy of agricultural systems with supraregional systems (which can transcend national boundaries) at the highest level and soil systems at the lowest level. The latter focuses on plant–soil interactions within crop fields which is the level at which specific biological processes such as nutrient uptake may be investigated. The highest level in the hierarchy occupies the largest land area and the lowest level covers the smallest spatial unit. Systems at level n are constrained and controlled by systems at level $n + 1$, and in turn they constrain systems at level $n - 1$ (Allen and Starr, 1982). Individual farmers (farming systems) for example, have to take the environment provided by the village/catchment level as a constraint in their decisions concerning their farming systems. Likewise, the dynamics and behaviour of catchment systems are controlled by, and have to operate within, the framework of regional river basin systems.

Interactions link systems together at different levels. Fig. 6.2 shows some of the interactions between the farming system, catchment, regional and supraregional levels. Fig. 6.2 also serves to illustrate how farmers integrate a wide range of ecological and economic parameters belonging to levels higher than the farming system, in their decisions to manage their lands in a given way.

It also shows how decisions made at the farming system scale have repercussions at the same scale, as well as at lower and higher scales in the hierarchy. These are mediated through various economic and biological processes such as nutrient cycling and the market mechanism. Because these processes transcend farm boundaries, it is helpful to establish a (rather academic) distinction between the economic processes which occur at the farming system scale and those which are manifest at the watershed, regional or national scale, in the following discussion of soil management.

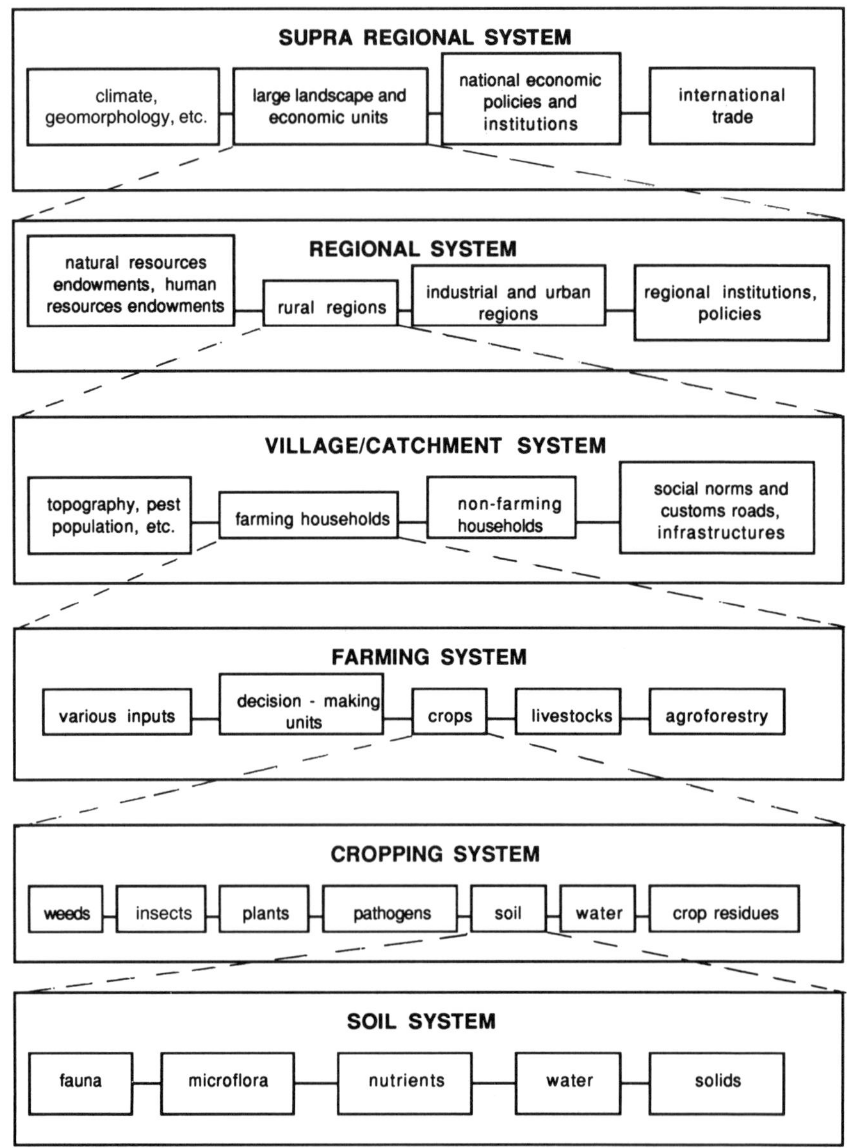

Fig. 6.1. Hierarchy of agricultural systems.

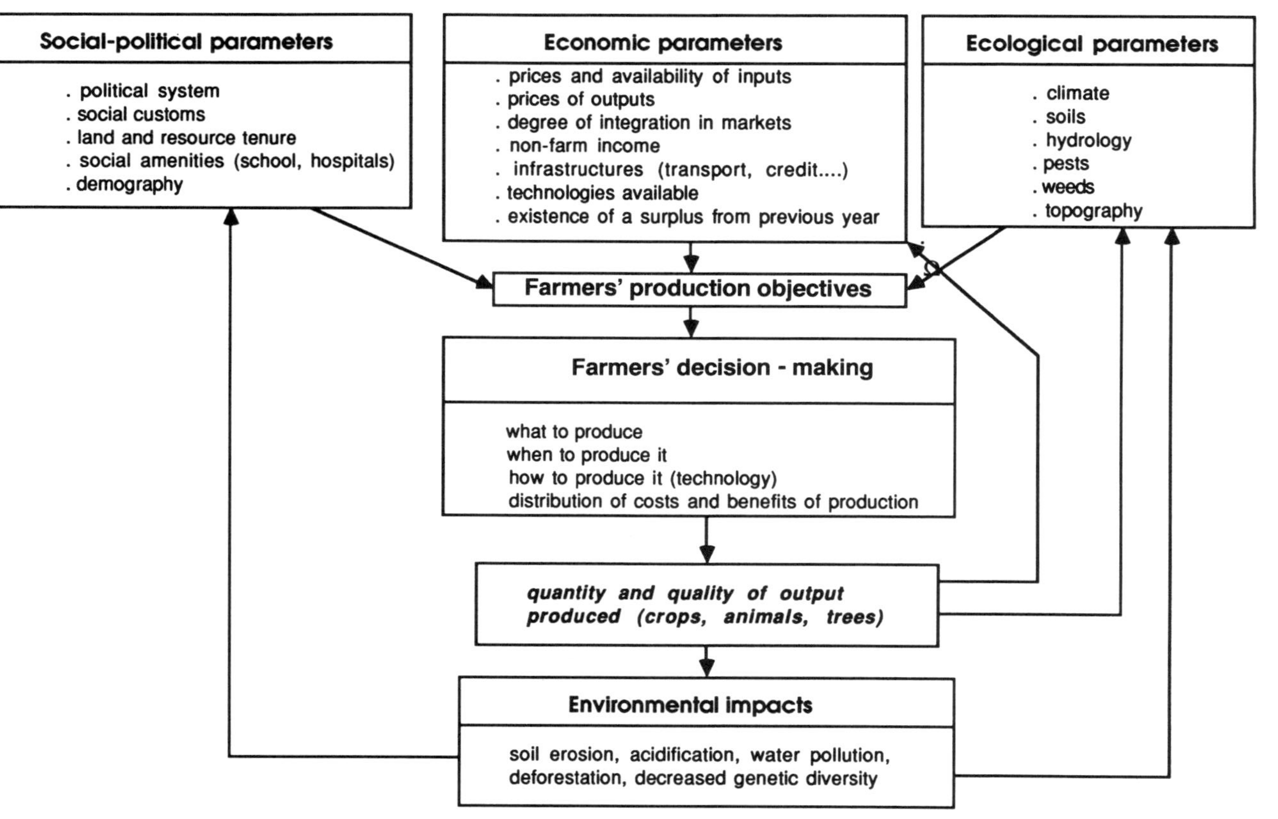

Fig. 6.2. Interactions between different scales.

Economic Processes at the Farming System Scale

The decision by an individual farmer to adopt a given soil management technology is made on the basis of a number of factors (Fig. 6.2). In spite of their complexity, such decisions have two fundamental aspects: (i) farmer's production objectives; and (ii) the consequences of alternative courses of action, as they are perceived by the farmer (see Keeney, 1982 for a discussion of decision analysis).[1]

Farmers' production objectives will of course vary among individuals but can be broadly stated as follows: to increase yields (revenues) and/or yield stability, now or in the future, in such a way that the overall advantages of these increases outweigh their overall disadvantages. The consequences of alternative soil management practices will be assessed by farmers on the basis of this objective, and in terms of their advantages and disadvantages. These advantages and disadvantages are all the monetary and non-monetary benefits and costs of soil management practices, as perceived by the farmers.

The management practices which have been devised for promoting soil resilience in tropical countries address the problems of soil erosion and nutrient depletion, as mentioned above. They consist of the building of physical structures on farmers fields and/or of improved soil fertility management by organic (grown) inputs. The use of inorganic fertilizer is not considered in this chapter. Relevant economic processes are similar to those concerning any other purchased inputs, and are thus well known. Moreover, the environmental effects of fertilizer use in tropical systems (e.g. soil acidification, Kang, 1989) imply that inorganic inputs are not a key feature of soil resilience and sustainable land uses in such systems. The soil management practices considered in this chapter are summarised in Table 6.1.

Each of these options entails various monetary and non-monetary costs. The monetary (or out-of-pocket) costs are straightforward. They are the costs farmers must incur to hire additional labour (e.g. for building physical structures, planting trees or green manures) and to purchase necessary materials (e.g. seeds, pipes for drainage). The non-monetary or opportunity costs are not actual disbursements for the farmer, but are nevertheless very real costs. They consist of all the opportunities for generating benefits which a farmer foregoes when choosing a given management option. For example, if green manures are planted using family labour, one opportunity cost of this management option will be the benefits which family labour would

[1]This assumes that farmers are rational, in the sense that they seek to follow courses of action which, in their judgement, are congruent with their production objectives. The evidence available concerning farmers' behaviour in developing countries (e.g. Gladwin, 1980; Celis *et al.*, 1991) suggests that this is a realistic assumption.

Table 6.1. On-farm soil management strategies.

Construction of physical structures	Organic matter management
Contouring, ridging, terracing Windbreaks Improved drainage	Management of quantity of residues (i) *in situ* use (e.g. crop residues) (ii) transfer within the farm (iii) planted (e.g. green manures, alley cropping, agroforestry) (iv) bought outside the farm (v) composting of residues Management of placement of residues (i) on soil surface (including low tillage) (ii) incorporated Management of timing of application (i) preplanting (ii) post-planting

have generated if the time spent planting green manures had been spent planting cash crops, or engaging in social activities (depending upon the production objectives of the farming household).[2] These various costs (Table 6.2) are all individual, since they are borne uniquely by individual farmers, and recur on a yearly basis, starting with the first year of implementation. They are thus all incurred at the farming system scale (Figs 6.1 and 6.2).

The benefits generated by these soil management options (Table 6.3) are much more difficult to assess than their costs. These benefits are made up of all the beneficial functions of soil processes which are enhanced through the management options. All these options result in increasing intrinsic soil fertility in an ecologically sustainable manner. There are two kinds of benefits which are associated with such an increase. First, from a quantitative perspective, yields will increase. Second, from a qualitative perspective, soil resilience and sustainability (including biodiversity) will be increased by enhancement of the long-term fertility status. As shown in Table 6.3, yield increases are a monetary benefit, since crop yields are evaluated by markets. Enhanced soil resilience, ecological sustainability and biodiversity are non-monetary benefits (there are no markets for resilience, sustainability and biodiversity). Increased sustainability benefits consist more specifically in the benefits of decreased risks of yield fluctuations (and crop failure) on a seasonal basis, the benefits of enhancing the quality of the soil resource base and the benefits of enhancing the capacity of the system to adjust to

[2]The available empirical evidence shows that small-scale farmers are very aware of the opportunity costs (in addition to the more obvious monetary costs) of their management decisions (Izac and Tucker, 1992).

Table 6.2. Individual costs of soil management practices.

Management options	Costs to an individual farmer
• Management of quantity and quality of residues (i) *in situ* use (e.g. crop residues) (ii) transfer within the farm (iii) planted (e.g. green manures, trees) (iv) bought outside the farm (v) composting of residues • Contouring, ridging terracing • Windbreaks • Improved drainage	• Opportunity costs of not using residues in another way (e.g. fodder, soil amendments on other fields) for (i), (ii) Not using the land in another way (e.g. crops) for (iii) and for windbreaks, drainage, ridges • Opportunity costs of family labour used in (ii), (iii), (iv), (v) and in building all physical structures (e.g. foregone participation in other farming activities, in social activities) • Monetary costs of additional hired labour for (ii), (iii) (planting, pruning), (iv), (v) and building all physical structures • Monetary cost of purchase and transport for (iv), windbreaks and improved drainage
• Management of placement of residues (i) on soil surface (ii) incorporated • Management of timing of application (i) preplanting (ii) post-planting	• Opportunity cost of possible losses in germination for (i) • Opportunity cost of family labour for (i) and (ii) • Monetary cost of hired labour for (i) and (ii) • Opportunity cost of family labour • Monetary cost of hired labour

Table 6.3. Individual and social benefits of soil management practices.

	Time			
Benefits	Year 1	Years 2–5	Years 6–10	Years 11–50
Monetary and individual				
Increased yields through increased soil fertility	+	++	++	+++
Non-monetary, individual and social				
Increased sustainability of system through:				
Decreased risks of yield fluctuations with usual climatic variability	0	+	++	+++
Enhanced soil resource base	0	0	+	++
Enhanced capacity of system to adjust to exogenous changes without generating increased flows of pollutants	0	0	+	++
Increased biodiversity of soil biota	0	+	++	+++
Possible increased biodiversity of fauna and flora	0	0	+	++

0: no measurable benefit.
+, ++, +++: benefit is measurable and its intensity ranges from low (+) to high (+++).

changes without releasing increased flows of detrimental chemicals (e.g. nitrate, pesticides) and materials (e.g. weed seeds) in the environment.

These qualitative and non-monetary benefits are much more difficult to quantify than the monetary benefits of recommended soil management practices. Biological scientists have not reached a consensus about how to assess these various qualitative functions of soil biological processes; it would thus be highly unrealistic to assume that farmers will be aware of all these functions. Furthermore, individual farmers will probably be concerned only by those benefits which accrue directly to them. In all likelihood, therefore, these farmers will acknowledge only the monetary benefits of increased yields and the non-monetary benefit of decreased risks of yield fluctuations. They will balance these benefits against all the monetary and opportunity costs represented in Table 6.2. These benefits, however, tend to be cumulative over time; they reach a greater amplitude after the first 5 or 10 years of a practice being implemented whereas the costs occur regularly on a yearly basis. An example is provided by agroforestry adoption in Java. It takes a minimum of 5 years for the cumulative benefits to exceed the initial land preparation and planting costs in these agroecosystems (Barbier, 1990, p. 202).

There are then two basic differences between the costs and the benefits of sustainable soil management. The costs have to be borne by the farmer implementing these practices on his/her land and they occur on a yearly basis, starting with the first year of adoption. The benefits are not all received by the farmer adopting the practices, are cumulative over time and are in most cases low during the first years following adoption. This raises the issue of the time frame of relevance to farmers.

It is well-known in economics that individuals in a market system tend to use relatively short 'planning horizons' (e.g. 2–3 years) to make decisions. It is indeed a prerequisite to stay in business in the short term in order to be able to survive in the long term. This is all the more so for subsistence-oriented small-scale farmers who are always uncertain about their 'survival' in farming from one year to the next. It is therefore likely that even if they were aware of the medium- and long-term non-monetary benefits of soil management, most farmers would not give a high priority or value to these benefits because they occur over the long term, a period of time of little relevance to their immediate needs. The soil management options which have the highest likelihood of being adopted by farmers in developing countries are thus those which result in sufficiently significant short-term increased yields and decreased risks to compensate for the yearly costs of implementation. This is summarized in Fig. 6.3 where these costs of implementation are represented as a decreasing function over time. The corresponding total benefit curve however, is an exponential function of time, for the reasons mentioned above. Two scenarios are represented in Fig. 6.3. In scenario A, total benefits (TB_a) increase relatively rapidly over time, so that

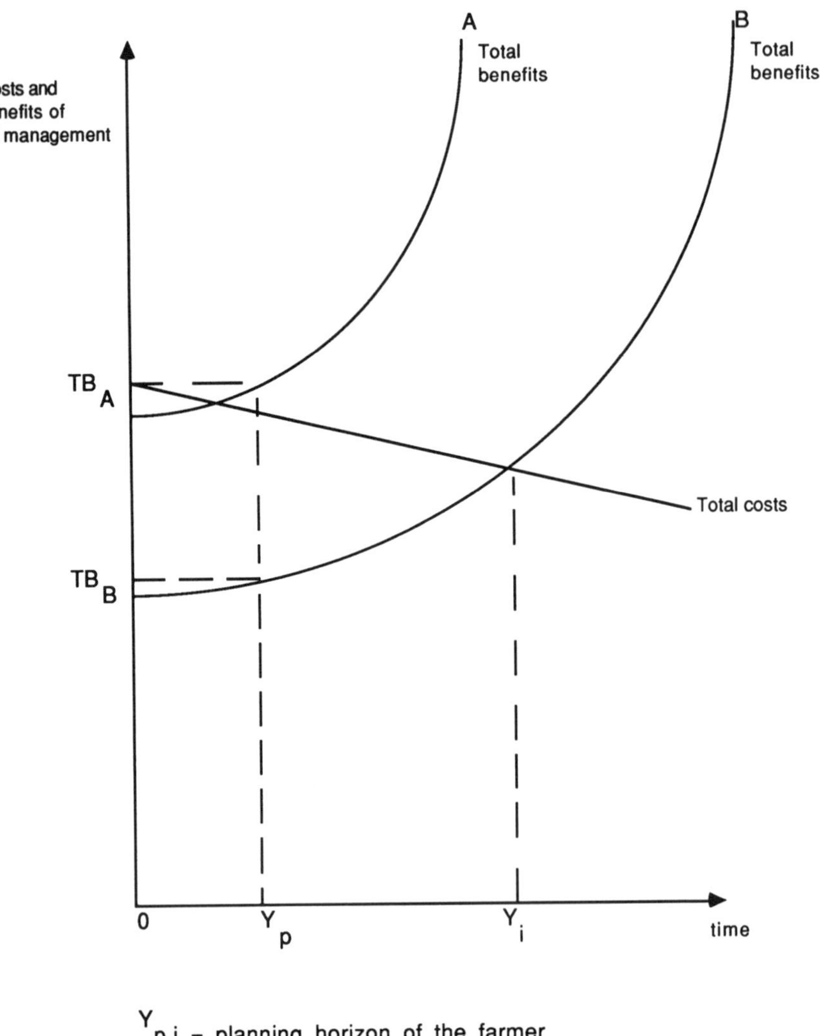

Fig. 6.3. Trade-offs between the individual costs and benefits of soil management over time (farming system scale).

they become greater than total costs within the planning horizon of the farmer (Y_p). Adoption of some improved soil management practices would logically ensue.

In scenario B, total benefits (TB_B) increase relatively slowly and remain inferior to total costs over the farmer's planning horizon. Adoption is then unfeasible. An example is provided by recent evidence from the USA which indicates that farmers generally 'fail to act' against the onsite and individual yield losses caused by erosion (Miranda, 1992). This balancing of monetary and non-monetary, present and future, individual costs and benefits over the duration of farmers' planning horizons constitutes the fundamental economic process of soil management at the farming system scale.

There are very few empirical studies documenting the extent of the benefits and costs of soil management strategies for farmers in developing countries. One such study (Huszar and Cochrane, 1990) shows that adoption of conservation measures such as bench terracing and grass strips by upland farmers in Indonesia initially led to a decrease in farmers' profits (total benefits minus total costs). Few farmers in the area had sufficient cash available to withstand these initial years of lower profitability relative to farmers who did not adopt any conservation measures (Huszar and Cochrane, 1990).

Economic Processes at the Watershed and Regional Scales

The non-monetary and social benefits of soil management which are related to resilience, sustainability and biodiversity occur at the farming system, watershed system and regional system scales (e.g. agroecosystem sustainability, decreased water pollution, increased food security). These benefits are likely to be valued more highly by society than by individual farmers because society generally has a longer planning horizon than individuals and because many of these benefits are off-farm benefits (e.g. decreased downstream water pollution).

As already seen, individual farmers will not take into consideration most off-farm benefits in their soil management decisions since these benefits are received by other individuals in society. Empirical evidence concerning the extent of on-farm and off-farm benefits of soil management practices in developing countries is not readily available (Anderson and Thampapillai, 1990). However, the evidence available from developed countries indicates that the monetary off-farm benefits of soil conservation alone (i.e. the monetary gains of controlling erosion for society) far outweigh on-farm benefits. The benefits of soil conservation can be measured in terms of the actual costs of erosion which would be avoided if conservation measures were adopted. In the USA the on-farm costs of erosion (yield losses, increased fertilizer costs) have been estimated to be of the order of 500 million dollars per year (Crosson, 1986), whereas the off-farm monetary costs of erosion (e.g. water sedimentation and turbidity, increased flooding) have been evaluated at about 5 billion dollars per year (Clark *et al.*, 1985). Similar differences have also been reported in Australia (Russell *et al.*, 1990) and Canada (Van Kooten *et al.*, 1989). The anecdotal information available concerning on-farm and off-farm costs of erosion in developing countries is

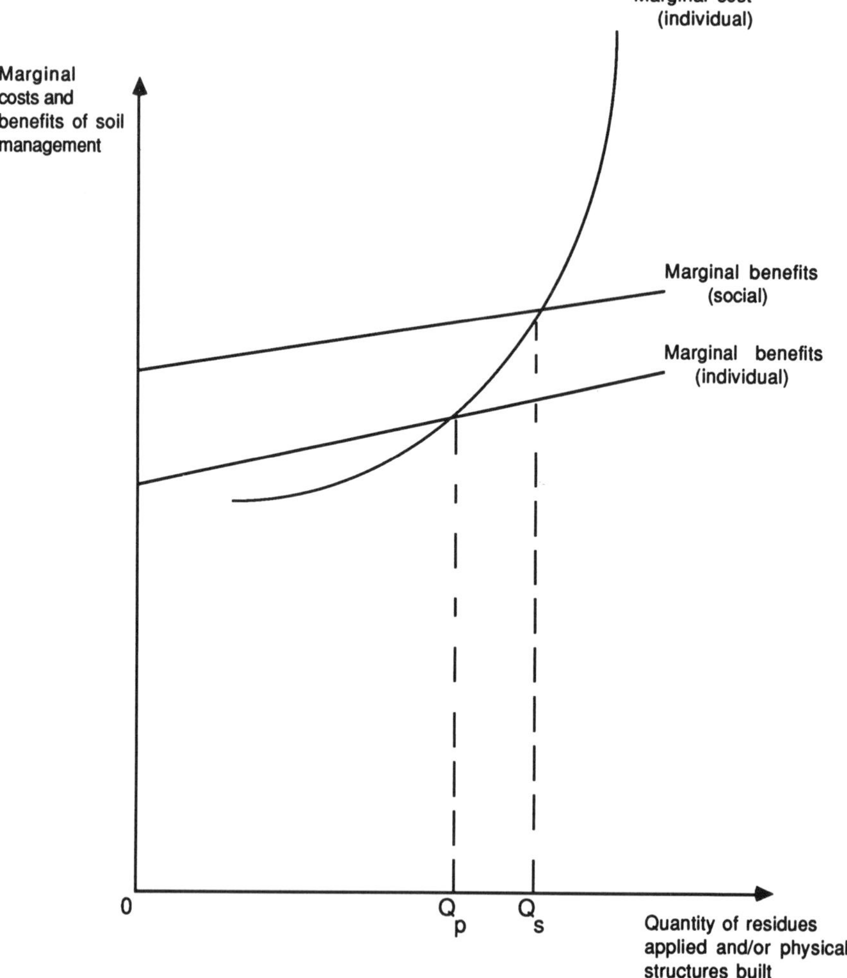

Fig. 6.4. Trade-offs between the individual and social costs and benefits of soil management at a given point in time (regional scale).

in agreement with the evidence from these countries (see Crosson, 1987 for details).

Such significant differences between the private and social monetary benefits of soil management indicate that an even greater discrepancy exists between the private benefits and social monetary and non-monetary (sustainability) benefits of soil management. This situation is represented in Fig. 6.4.

Individual farmers who decide to adopt soil management strategies (scenario A in Fig. 6.3) have to bear all the costs (monetary and non-monetary) of these practices. Therefore, at a given point in time, a farmer thus faces a marginal cost curve[3] which is made up of all these costs, as shown in Fig. 6.4. This farmer is also confronted with a marginal benefit curve[4] representing the monetary benefits and whatever non-monetary benefits of soil management are valued by this farmer. For the reasons just discussed, society has a marginal benefit curve which is higher than that of the farmers'. The farmer will decide to apply quantity Q_p of residues (it is the quantity which maximizes net individual returns). However, the optimal quantity which should be applied from society's perspective, as it maximizes net social returns, is Q_s. The amount of soil management practices voluntarily adopted by farmers (at the farming system scale) will thus be suboptimal from society's viewpoint because it will not take into account the non-monetary benefits of soil management at the watershed and regional scales. Agroecosystem sustainability is then unlikely to be attained. These trade-offs between individual costs and social benefits of soil management (represented in Fig. 6.4) are a fundamental economic process at the watershed and regional scales of analysis.

The foregoing analysis differs from some economic models of soil erosion in developed countries proposed in the literature (e.g. McConnell, 1983). The conclusions derived from such models, that soils can be depleted by farmers in a 'socially optimal' way, contrast with the conclusions reached here. This is because these models are built on assumptions which restrict their applicability to situations where: (i) all financial and real-asset markets function smoothly and efficiently; and (ii) there are no off-site costs of erosion (and thereby no social benefit of erosion control) (McConnell, 1983). Such situations are improbable in developing countries where the imperfect working of markets has been documented (Roitman, 1990). Moreover, whatever the acceptability of the ethical implications of the assumption regarding off-site costs of erosion in developed countries, this assumption is particularly unwarranted in the context of sustainable agricultural development in tropical countries. The analysis presented here is not based on similarly restrictive assumptions. It should thus have a greater operational relevance for developing countries.

---

[3]Marginal cost is the increase in total cost associated with the application of one additional unit of residues.

[4]Marginal benefit is the increase in total benefit associated with the application of one additional unit of residues.

Policy Implications

The economic processes discussed above indicate that private and social interests concerning soil management are highly likely to diverge. Consequently, the extent of soil management practices voluntarily adopted by farmers (that which is in their own best interest) will almost certainly be inferior to that which is socially optimal.

In such circumstances, providing farmers with information and advice on soil management practices will not be an effective means of increasing adoption levels. Incentives will have to be offered to farmers to compensate for the fact that they bear all the costs of implementation of soil management whereas society receives most of the benefits of this management. Positive steps thus need to be taken to render soil management more attractive to farmers in order to promote soil resilience and to ensure agricultural sustainability. Various policy instruments are available for accomplishing this. Without attempting to identify 'the best' policy for promoting a socially optimal level of adoption of soil management in developing countries, it is possible to discuss the relevance of generic categories of policies to soil management. Such a discussion will serve to devise more specific recommendations for researchers and decision-makers.

Three principal kinds of policies have been used in developed countries (with variable results) to increase adoption of soil management practices: regulations, taxes/subsidies and price incentives. Regulations, which legally force farmers to adopt various strategies, and taxes levied on farmers who degrade their soils, would be inappropriate, at least in developing countries. Both types of policies require farmers to bear the entire financial burden of adoption costs and are thus not equitable (especially so when most farmers are resource-poor). In addition, their successful implementation (e.g. monitoring of compliance, system of fines) necessitates a complex and costly institutional infrastructure which most tropical countries do not possess.

Subsidies provided to farmers who adopt recommended soil management practices would be more equitable since they would decrease the financial burden borne by individual farmers. However, they would be costly for society to implement in the absence of the necessary institutional framework (e.g. for monitoring adoption on individual farmer's fields). Subsidy types of policies are therefore generally not appropriate to the institutional environment characterizing tropical countries.

The manipulation of relevant prices to provide additional financial incentives for farmers to adopt soil management practices would however be a possible course of action in these countries. For example, it has been argued that increases in prices of imported fruits in Java through the use of import taxes would encourage the adoption of agroforestry based on fruit production, as this measure would increase the profitability of local fruit production (Barbier, 1990). A policy based on this type of price manipulation would need to be implemented over the initial period of years during which the on-farm benefits of soil management are likely to be low relative to the corresponding costs. The use of price manipulation policies must be based on realistic macroeconomic models to permit the evaluation of the series of changes which will be triggered by the manipulation of one or two prices in the economy. This is necessary to ensure that these manipulations will not have counterproductive consequences for farmers or other groups in society through the chain of economic interactions characterizing any economy. Price manipulation policies are equitable and inexpensive for a government to implement, as long as reliable and sufficiently detailed macroeconomic models are available to analyse all the consequences of the policies and fine tune them.

Finally, a type of policy which has until now rarely been implemented in developed countries but which appears to have a good potential for tropical countries is the support by governments of local communities to manage soil resources in a community-based fashion. For example, funds can be made available at the village level to groups of farmers who wish to implement soil management strategies (e.g. contour banking, reafforestation). The same could be done with seeds (e.g. for green manures) and seedlings (e.g. for agroforestry) and whatever materials are needed by farmers (e.g. to construct windbreaks). Such direct support, if coupled with a land tenure system which gives local communities legal title to farming land, has the potential to increase adoption at the regional and national scale to a socially optimal level. Cases where such a policy has been implemented on an experimental basis with some success are discussed by Moorehead (1989) and Cernea (1987), and formation of Land Use Associations is urged by Dr M.S. Swaminathan in the Foreword to this book.

Such a policy would have relatively low implementation costs as it would not require as complex an institutional and administrative infrastructure as subsidy policies do. It would be equitable in that society would share some of the costs of adoption of soil management practices. It would have three additional advantages in the case of tropical Africa.

First, community-based management of soil resources would be congruent with the current institutional environment of farmers in the region, in which land, labour and water are already managed partly individually and partly communally at the village level (Izac and Swift, 1992). Second, such soil management would enable farmers to decrease their individual costs of production (over and above the funds and/or materials received) by pooling their managerial skills and some of their inputs, such as labour, thereby realizing economies of scale. Third, such management would be better adapted to ecological constraints in the zone since a number of soil degradation problems (e.g. erosion) occur at the watershed or village level in tropical Africa.

Such a community-based management is obviously not a panacea.

Transaction costs for individual farmers may be high in some communities and much discussion and conflict can arise from collective soil management. However, given the long experience of local communities in tropical Africa with communal management and conflict resolution, it appears to be a feasible alternative to policies of regulations, taxes and subsidies.

Research Implications

Research priorities are implicitly contained in the previous arguments. There are four principal research directions which need to be pursued to increase the relevance of soil science and resource economics to soil degradation problems in tropical countries.

First, the beneficial effects of soil management (Table 6.3) must be quantified by soil scientists. Both quantitative effects (yield increases) and qualitative effects (resilience, sustainability, biodiversity) need to be evaluated as functions of different levels of soil management practices. That is, the response curve of crops to these practices must be assessed, for different representative crops in different representative agroecological zones, and the functional form of the corresponding total benefit curves in Fig. 6.3 must be identified. It is indeed important to quantify these beneficial effects across different agroecological zones and for representative farming systems within these zones so as to facilitate the extrapolation of the data obtained to clearly delineated geographical zones (e.g. through the use of a geographical information system).

Second, the production objectives of farmers and their planning horizons must be identified by economists at representative on-farm research sites, throughout these zones.

Third, the cost and benefit functions of soil management at the farming system, watershed and regional scales have to be evaluated by economists. As already mentioned, cost functions are relatively easy to assess. Benefit functions, however, can only be evaluated once soil scientists have measured the magnitude of the beneficial effects of sustainable soil management. This evaluation of cost and benefit functions in representative farming systems will permit the analysis of the economic processes represented on Figs 6.3 and 6.4. There are economic techniques available for assessing the value of non-monetary benefits such as decreased water pollution or increased biodiversity (e.g., Ahmad *et al.*, 1989; Vincent *et al.*, 1991). These techniques (sometimes called natural resources accounting) have controversial dimensions. Researchers will need to develop feasible techniques of evaluation of non-monetary benefits which are appropriate to the specific socioeconomic constraints of tropical countries.

Such evaluations of the costs and benefits of soil management are not only needed for the development of appropriate policies, but are also required to provide a strong incentive for policy-makers to actually implement these policies.

Fourth, research into the means of promoting the adoption of soil management should follow two paths. Initially, policy analysts need to devise price manipulation policies and community-based support policies which are appropriate for representative specific socioeconomic and ecological environments. They can then assess the equity, economic and ecological effects of these policies using the analytical tools provided by resource and environmental economics for evaluating environmental and resource policies (see for example Dubgaard and Nielsen, 1989). Recommendations concerning the policy options which are likely to lead to a socially optimal level of adoption of soil management practices in specific representative areas could then be provided to appropriate policy-makers.

Then, available soil management techniques, developed principally by soil and biological scientists, should be further adapted to the ecological and socioeconomic conditions which characterize tropical countries. There remain a number of technical issues which need to be resolved before all of these soil management techniques are technically operational in the zone. For example, no-tillage may become a viable soil management technique in Africa when herbicides are developed which effectively control the mixed vegetation which grows in the tropical humid forest zone (Akobundu and Kang, 1992). Beyond the resolution of these 'technical' problems, the relevance of research on soil management practices in tropical countries could also be increased through a greater integration of the contributions from appropriate disciplines. Interdisciplinary teams of scientists, including social scientists, working together in what the French call 'milieu réel', (which is more apt than the English 'on-farm') with farmers, would be more likely to elaborate soil management strategies which lend themselves to a community-based implementation. There is nothing new in such a call for greater interdisciplinarity and greater consideration of farmers' real world constraints in technology development research. However, given the paucity of true interdisciplinary research in the field, it may be necessary to repeat this call over a period of time until more research organizations become aware of it and start motivating their scientists to undertake such research.

Conclusion

It has been argued in this chapter that two fundamental economic processes characterize soil management practices in tropical countries. One process, which occurs at the farming-system scale, is the trading-off by farmers of the monetary and non-monetary costs of these practices with their associated monetary and non-monetary benefits. Farmers bear all the monetary and opportunity costs of soil management practices, starting from the initial year of implementation. The benefits they receive, however, are cumulative over time and may not match costs during the initial years following adoption. Given the relatively short length of the planning horizon of farmers, it will often be in their best individual interest not to adopt soil management practices. The other process, which takes place at the watershed system and regional system scales, is the balancing of individual costs and social benefits of soil management.

These social benefits are likely to be higher than private benefits; it follows that for those farmers who decide to adopt soil management practices, the actual quantity of soil management undertaken will very likely be inferior to that which is socially optimal.

Economic processes characterizing soil management practices at the farming system and regional scales are thus such that it is probable that voluntary adoption by individual farmers will be socially suboptimal and that some form of government intervention will be needed to promote sustainable land uses in tropical countries. For a variety of reasons concerning equity and costs of implementation for a government, the types of intervention which are the most likely to succeed in these countries are price manipulation policies and policies encouraging community-based implementation of soil-management practices.

Research priorities for soil and biological scientists and economists have been derived from this analysis. Indeed, much remains to be done to promote effectively the adoption of sustainable soil management by farmers in tropical countries.

References

Ahmad, Y.J., El Serafy, S. and Lutz, E. (eds) (1989) *Environmental Accounting for Sustainable Development.* The World Bank, Washington DC, 100 pp.

Akobundu, I.O. and Kang, B.T. (1992) No-till crop production in the tropics. Manuscript prepared for IITA 25th Anniversary Book, IITA, Ibadan (in press).

Allen, T.F.H. and Starr, T.B. (1982) *Hierarchy Perspectives for Ecological Complexity.* University of Chicago Press, Chicago, 310 pp.

Anderson, J.R. and Thampapillai, J. (1990) *Soil Conservation in Developing Countries. Project and Policy Intervention.* Policy and Research Series, World Bank, Washington DC, 45 pp.

Barbier, E.B. (1990) The farm-level economics of soil conservation: the uplands of Java. *Land Economics* 66 (2), 199–211.

Bullock, P. (1987) Report on session I: Soil protection. A need for a European Programme? In: Barth, H. and L'Hermite, P. (eds) *Scientific Basis for Soil Protection in the European Community.* Elsevier Applied Science, London, pp. 581–588.

Celis, R., Milimo, J.T. and Wanmali, S. (eds) (1991) *Adopting Improved Farm Tech-*

nology: a Study of Smallholder Farmers in Eastern Province, Zambia. IFPRI, Washington, DC, 409 pp.

Cernea, M.M. (1987) Farmers organizations and institution building for sustainable development. In: Davis, T.J. and Schirmer, I.A. (eds) *Sustainability Issues in Agricultural Development.* World Bank, Washington DC, pp. 116–136.

Clark, E.H., Haverkamp, A. and Chapman, W. (1985) *Eroding Soils: the Off-Farm Impacts.* The Conservation Foundation, Washington DC.

Crosson, P.R. (1986) Soil erosion and policy issues. In: Phipps, T.T., Crosson, P.R. and Price, K.A. (eds) *Agricultural and the Environment* Resource for the Future, Washington DC, pp. 35–73.

Crosson, P. (1987) Soil conservation and small watershed development. In: Davis, T.J. and Schirmer, I.A. (eds) *Sustainability Issues in Agricultural Development* World Bank, Washington DC, pp. 182–187.

Dubgaard, A. and Nielsen, A.H. (eds) (1989) *Economic Aspects of Environmental Regulations in Agriculture.* Wissenschaftsverlag Vauk, Kiel, 329 pp.

Gladwin, C. (1980) Cognitive strategies and adoption decisions: study of non-adoption of an agronomic recommendation. In: Brokensha, D., Warren, D.M. and Werner, O. (eds) *Indigenous Knowledge Systems and Development.* University Press of America, New York, pp. 9–28.

Huszar, P.C. and Cochrane, H.C. (1990) Constraints to conservation farming in Java's uplands. *Journal of Soil and Water Conservation* 45, 420–423.

Izac, A-M.N. and Swift, M.J. (1992) An ecological–economic framework for developing sustainable agricultural technologies for sub-Saharan Africa. Submitted to *Agricultural Economics.*

Izac, A-M.N. and Tucker, E. (1992) Analyse systémique des agrosystémes de bas-fonds d'Afrique Occidentele. In: Raunet, M. (ed.) *Bas-Fonds et Riziculture.* CIRAD, Montpellier, pp. 203–212.

Kang, B.T. (1989) Nutrient management for sustained crop production in the humid and sub-humid tropics. In: Van der Heide, J. (ed.) *Nutrient Management for Food Crop Production in Tropical Farming Systems.* Institute for Soil Fertility, Haren, The Netherlands, pp. 3–28.

Keeney, R.L. (1982) Decision analysis: an overview. *Operations Research* 30, 803–838.

McConnell, K.E. (1983) An economic model of soil conservation. *American Journal of Agricultural Economics* 65, 83–89.

Miranda, M.L. (1992) Landowner incorporation of onsite soil erosion costs: an application to the Conservation Reserve Program. *American Journal of Agricultural Economics* 74, 434–443.

Moorehead, R. (1989) Changes taking place in common-property resource management in the inland Niger delta of Mali. In: Berkes, F. (ed.) *Common-Property Resources.* Belhaven Press, London, pp. 256–272.

Pierce, F.J. (1987) Complexity of the landscape. In: Halbach, D.W., Runge, C.F. and Larson, W.E. (eds) *Making Soil and Water Conservation Work: Scientific and Policy Perspectives.* Soil Conservation Society of America, Iowa, pp. 15–36.

Roitman, J.L. (1990) The politics of informal markets in sub-Saharan Africa. *Journal of Modern African Studies* 28, 671–696.

Russell, I.W., Izac, A-M.N. and Cramb, R.A. (1990) *Towards an Evaluation of the Off-Site Costs of Soil Erosion in Queensland.* Discussion paper 2/90, Department of Agricultural, University of Queensland, 119 pp.

Van Kooten, G.C., Weisensel, W.P. and de Jong, E. (1989) Estimating the costs of soil erosion in Saskatchewan. *Canadian Journal of Agricultural Economics* 37, 63–75.

Vincent, J.R., Crawford, E.W. and Hoehn, J.P. (eds) (1991) *Valuing Environmental Benefits in Developing Countries.* Michigan State University Special Report 29, East Lansing.

Whittome, M., Spencer, D.S.C. and Bayliss-Smith, T. (1992) IITA and ILCA on-farm alley cropping research; history, current status and lessons for extensionists. Paper presented at the International Conference on Alley Farming, 14–18 September 1992. IITA, Ibadan.

Part II

The Extent of Soil Degradation

Chapter 7
The Global Extent of Soil Degradation

L.R. Oldeman

International Soil Reference and Information Centre (ISRIC), PO Box 353, 6700 AJ Wageningen, The Netherlands

Introduction

Past and present human intervention in the utilization and manipulation of environmental resources are having unanticipated consequences. The Earth's soils are being washed away, rendered sterile or contaminated with toxic materials at a rate that cannot be sustained. As stated by Brundtland *et al.* (1987): 'There is a growing realization in national and multilateral institutions that not only many forms of economic development erode the environmental resources upon which they are based, but that at the same time environmental degradation can undermine economic development'.

Although soil degradation is recognized as a serious and widespread problem, its geographical distribution and total areas affected are only very roughly known. Dregne (1986) states that sweeping statements on the fact that soil erosion is undermining the future prosperity of mankind do not help planners, who need to know where the problem is serious and where it is not. This feeling was also expressed by the United Nations Environment Programme (UNEP). This organization indicated the need to produce, on a basis of incomplete knowledge, a scientifically credible global assessment of soil degradation in the shortest possible time. 'Politically it is important to have an assessment of good quality now instead of having an assessment of very good quality in 15 or 20 years' (ISSS, 1987).

UNEP's project 'Global Assessment of Soil Degradation (GLASOD)' was implemented and coordinated by the International Soil Reference and Information Centre (ISRIC) in the Netherlands. Three years later the World Map of the Status of Human-Induced Soil Degradation was published (Oldeman *et al.*, 1990). Since the map sheets did not provide information on the global extent of the soil degradation, mainly because of the projection of the topographic base map with scale distortion at higher latitudes, the map units were digitized and linked to a Geographic Information System, capable of converting the Mercator Projection into an 'equal area' projection. It is now for the first time possible to make a regional and global quantification of the extent and seriousness of the various soil degradation processes.

Methodologies for the Assessment of Global Soil Degradation

The preparation of a World Map of Soil Degradation within a three year period was accomplished by a cooperative effort with a large group of soil and environmental scientists worldwide, who were asked to give their expert opinion on the types, degrees, areal coverage and human-induced causes of soil degradation in their regions. A team of 21 regional correlators was provided with a simplified topographic base map. In order to ensure uniformity of interpretation they were asked to use general guidelines for the assessment of the present status of human-induced soil degradation (ISRIC, 1988). These guidelines gave definitions of the various types of soil degradation, of the degree to which the soil was degraded, and of the various human-induced interventions that had caused the soil to deteriorate to its present situation.

The methodology called for a stepwise approach. The first step involved the delineation of mapping units on the topographic base map based on physiographic characteristics. These units should show maximum homogeneity of topography, soils, climate, vegetation and land use.

The second step involved an evaluation of the various types of soil degradation that may occur in the mapping unit. For each recognized type of soil degradation an indication was given of the degree of soil degradation, the relative extent of the area within the mapping unit being affected, the recent past rate of the degradation process and the causative factor(s). Additionally, information was provided on the relative extent of the wasteland and stable terrain portions of the mapped unit.

The third step was the compilation of the regional information into a global map and the development of a legend for the map. Each map unit in its final version contained no more than two types of soil degradation. Although in total 12 different types of soil degradation are included in the legend, only four major groups of soil degradation are represented on the map by different colours. The severity of the major type of soil degradation is indicated on the map by a different shading of the colour. The severity index is a combination of the estimated degree of the soil degradation process and the relative extent of the area affected. Although it is recognized that this compilation results in a simplification of information provided, the final GLASOD map itself provides sufficiently detailed information for the user group for which it was intended: policy-makers and decision-makers. The major objective of the map is to strengthen awareness of the dangers resulting from inappropriate land and soil management. However, each map unit is provided with a symbol giving more detailed description of the two dominant types of soil degradation.

Characteristics of the GLASOD Map

Soil degradation types

The information on soil degradation processes does not relate to the relative fragility of the ecosystem. It describes situations where the balance between the natural resistance of the soils (and their vegetative cover) and the climatic aggressivity has been disturbed by human intervention. Soil degradation is defined as a process which lowers the current and/or future capacity of the soils to produce goods or services.

Two categories of soil degradation processes are recognized. The first group relates to displacement of soil material. The two major soil degradation types in this category are soil erosion by water forces or by wind forces. The second group deals with soil deterioration *in situ*. This can either be a chemical or physical soil degradation process.

Water erosion (W) – bluish green on the GLASOD map

The displacement of soil material by water can have several negative consequences. The removal of part of the usually fertile topsoil reduces the productive capacity of the soil, whereas in extreme cases the rooting depth can become restricted for agricultural crops. Although measurements of crop yield reductions caused by soil erosion are difficult, mainly because over time farmers may substitute increasing amounts of fertilizers to compensate for the loss of the natural fertility of the soil, studies in the USA have established relationships between soil erosion and reduced crop yield (Batie, 1983). In fragile soils with a low structural stability run-off water may lead to rapid incision of gullies, eating away valuable soils and making the terrain eventually unsuitable for farming. In the GLASOD approach these two forms of water erosion are distinguished:

1. Loss of topsoil (Wt). This form of water erosion is generally known as surface wash or sheet erosion. It occurs almost everywhere, under a great variety of climatic and soil conditions and land uses. Loss of topsoil is often preceded by compaction and/or crusting resulting in a decrease in the infiltration capacity of the soil.
2. Terrain deformation (Wd). Although the total area affected by rills and

gullies is far less compared to loss of topsoil its effects are more spectacular and more easy to observe in the field. Control of active gullies is difficult and restoration is almost impossible.

Although the displacement of soil material by water may lead to off-site effects such as sedimentation of reservoirs, lakes and harbours; excessive siltation of the basin land; destruction of coral reefs and shellfish beds; river bed filling, flooding and riverbank erosion, these phenomena could not be mapped at the scale of the GLASOD map.

Wind erosion (E) – yellowish brown on the GLASOD map

The displacement of soil material by wind is nearly always caused by a decrease of the vegetative cover of the soil, either due to overgrazing or to the removal of vegetation for domestic use or agricultural purposes. It is a widespread phenomenon in arid and semiarid climates. In general, coarse-textured soils are more vulnerable to wind erosion than fine-textured soils. Three types of wind erosion are recognized by GLASOD:

1. Loss of topsoil by wind erosion (Et), defined as a uniform displacement of the topsoil.
2. Terrain deformation (Ed), defined as an uneven displacement of soil material, leading to deflation hollows and dunes. Although mapped separately, this type of wind erosion may be considered as an extreme form of loss of topsoil.
3. Overblowing (Eo) is defined as coverage of the land surface by wind-carried particles. In contrast to off-site effects of water erosion, this off-site effect of wind erosion occurs on relatively large areas and is mappable. Overblowing may seriously influence infrastructure (road, railroads), buildings and waterways, and may cause damage to crops.

Chemical degradation (C) – red on the GLASOD map

Chemical degradation of the soil does not refer to cyclic fluctuations of the soil chemical conditions of relatively stable agricultural systems, in which the soil is actively managed to maintain its productivity, nor to gradual changes in the chemical composition as a result of soil forming processes. Chemical degradation processes are very different.

1. Loss of nutrients and/or organic matter (Cn). Loss of nutrients is a common phenomenon in countries with low-input agriculture. It occurs if agriculture is practiced on poor or moderately fertile soils, without sufficient application of manure or chemical fertilizers. The rapid loss of organic matter of the topsoil after clearing of the natural vegetation is also included.
2. Salinization (Cs) is defined as a change in the salinity status of the soil. It

can be caused by improper management of irrigation schemes, mainly in the arid and semiarid regions covering small areas. Salinization may also occur if seawater or fossil saline groundwater intrudes groundwater reserves of good quality, usually in coastal regions or in closed basins with aquifers of different salt content when there is an excessive use of groundwater. Finally, salinization takes place where human activities lead to increased evapotranspiration in soils on salt-containing parent material or with saline groundwaters.

3. Acidification (Ca) may occur in coastal regions upon drainage/oxidation of pyrite-containing soils. Acidification is also caused by over-application of acidifying fertilizers. In both cases the agricultural potential of the land is reduced.

4. Pollution (Cp). Many types of pollution can be recognized, such as industrial or waste accumulation, excessive use of pesticides, acidification by airborne pollutants, excessive manuring, oil spills, etc. This form of soil degradation is generally restricted to heavily industrialized nations with high population densities, although the effect of acidification by airborne pollutants may lead to deposits at considerable distance from their source.

Physical degradation (P) – pink on the GLASOD map

Three types are identified:

1. Compaction, crusting and sealing (Pc). Whereas compaction of the soil is usually caused by the use of heavy machinery, sealing and crusting of the topsoil occurs if the soil cover is not sufficiently protected against the impact of raindrops. Particularly soils low in organic matter with poorly sorted sand fractions and appreciable amounts of silt are vulnerable.

2. Waterlogging (Pw). Human intervention in natural drainage systems may lead to flooding by river water and submergence by rainwater. It should be noted that the construction of paddy fields is not included as it is considered to be improvement of the terrain for wetland rice cultivation.

3. Subsidence of organic soils (Ps). This phenomenon is caused by drainage and/or oxidation of organic soils. It is only identified on the map if the agricultural potential is negatively affected.

Miscellaneous terrain types

Although human-induced soil degradation is widespread throughout the world, large areas are not affected by human intervention, or stabilized by human intervention. Some of the land is either unsuitable for agricultural activities (climatic, topographic and soil constraints) or poorly accessible. On the GLASOD map, two categories of land are recognized.

Stable terrain (S) – light grey on the GLASOD map

The stable terrain is subdivided into three categories depending on the type of human intervention or its absence.

1. Stable terrain under natural conditions (SN). Absence of any kind of human intervention because the type of land, the climatic conditions or the accessibility is not suitable for living or for agricultural activities. Large portions in northern Canada, northern Europe and the former USSR are included, but also large parts of the Himalaya and Andes mountains, as well as certain portions of the rainforests in South America and South Africa.

2. Stable terrain with a permanent agricultural land use (SA). If agricultural land is well managed, no soil degradation will occur and productivity levels will not decrease.

3. Terrain stabilized by human intervention (SH). With the growing awareness of the dangers of soil degradation, efforts to conserve this precious natural resource are growing. Examples of conservation practices are reforestation, terracing, gully control, improved water management.

Wastelands – dark grey on the GLASOD map

Historic or recent natural processes have rendered some areas into non-used wastelands. There is no appreciable vegetative cover or agricultural potential. The GLASOD map recognizes in this group active dunes (D), deserts (A), salt flats (Z), rock outcrops (R), arid mountain regions (M), and ice caps (I).

These miscellaneous terrain types are only identified on the GLASOD map, if the mapping unit does not include any form of human-induced soil degradation. This implies that even if a mapping unit has an infrequent occurrence of a certain type of soil degradation, the whole mapping unit is coloured according to that type of soil degradation, although the vast majority of the land in that mapping unit may be stable.

The degree of soil degradation

Soil degradation processes by definition result in a loss in soil productivity, although the ways in which this happens differ greatly with the various soil degradation processes. In the GLASOD approach, the degree to which the soil is presently degraded is related in a qualitative manner to the agricultural suitability of the soil, to its reduced productivity, to its possibilities for restoration to full productivity and in relation to its original biotic functions.

The degree of soil degradation should also be related to a time scale. The GLASOD approach does not provide information on human-induced soil degradation occurring in the past, although the original guidelines specified three historic periods: early civilizations occurring up to 250 years ago; the period of European expansion in the Americas, Australia, Asia and Africa, 50 to 250 years ago; the post-Second World War period, related to the explosive increase in human population and standards of living respectively in the developing countries and the industrialized western society.

The human-induced soil degradation type, its degree and areal coverage only relates to the change in conditions over this last post-Second World War period. The following four degrees of soil degradation are specified:

1. Light. The terrain has a somewhat reduced agricultural suitability, but is suitable in local farming systems. Restoration to full productivity is possible by modifications of the management. Original biotic functions are largely intact.
2. Moderate. The terrain has a greatly reduced productivity, but is still suitable for use in local farming systems. Major improvements are required to restore the terrain to full productivity, which are beyond the means of local farmers in developing countries. Original biotic functions are partially destroyed.
3. Strong. The terrain has virtually lost its productive capacity and is not suitable for use in local farming systems. Major investments and/or engineering works are required to rehabilitate the terrain, which are often beyond the means of national governments in developing countries. Original biotic functions are largely destroyed.
4. Extreme. The terrain is unreclaimable and beyond restoration. It has become human-induced wasteland. Original biotic functions are fully destroyed.

It should be realized that these general definitions are qualitative and that judgements made by the experts in the field are subjective, although the guidelines provide some quantitative tools to support these estimates.

The relative extent of soil degradation

At the scale of the GLASOD map it is obviously not possible to indicate individual events of soil degradation. Expert estimates are, however, provided to indicate the relative frequency of occurrence of each soil degradation event within each mapping unit. Five classes were recognized:

1. Infrequent: up to 5% of the unit is affected
2. Common: 6–10% of the unit is affected
3. Frequent: 11–25% of the unit is affected
4. Very frequent: 26–50% of the unit is affected
5. Dominant: over 50% of the unit is affected

Causative factors

The concept of human-induced soil degradation implies by definition a social problem. No person will intentionally destroy this precious natural resource. But increasing pressure on the land, the increased desire for better living conditions, and higher standards of living, the search for land to survive, have resulted in some kinds of intervention that have caused the soil to degrade. The GLASOD approach distinguishes the following types of causative factors:

1. Deforestation or removal of the natural vegetation (f). Clearing of the land for agricultural purposes, large scale commercial forestry, road construction, urbanization.
2. Overgrazing (g). Actual overgrazing of the vegetation may not only lead to vegetation degradation, but can cause soil compaction, wind and water erosion.
3. Agricultural activities (a). This includes a wide variety of agricultural practices, such as insufficient or excessive use of fertilizers, use of poor quality irrigation water, improperly timed use of heavy machinery, absence of anti-erosion measures on land susceptible to water and wind erosion, etc.
4. Overexploitation of the vegetation for domestic use (e), e.g. for fuel needs, fencing, etc. There is not a complete removal of the vegetation but the remaining vegetation does not provide sufficient protection against soil erosion or sealing and crusting of the topsoil.
5. Bioindustrial and industrial activities (i). This causative factor is directly related to the soil degradation type 'soil pollution'.

Global Extent of Soil Degradation

The global extent of soil degradation is calculated in two steps. First the surface area for each individual GLASOD map unit is calculated. For each map unit, the relative extent of the two dominant soil degradation types, its degree and causative factor are known. This allows for an estimation of the area affected by each degradation type per GLASOD unit. The portion of the GLASOD unit not affected by soil degradation is either stable or non-used wasteland or other land used for non-agricultural purposes.

The GLASOD map covers the earth surface between 72° N and 57° S. Summation of the surface area of all GLASOD units leads to a total land area of 13,103 mha (million hectares), which compares favourably with FAO's estimate of 13,069 mha (FAO, 1990).

The GLASOD map and accompanying statistics do not allow assessment of soil degradation on a country by country basis. However, statistics are provided for seven continental regions (Fig. 7.1). Human-induced soil

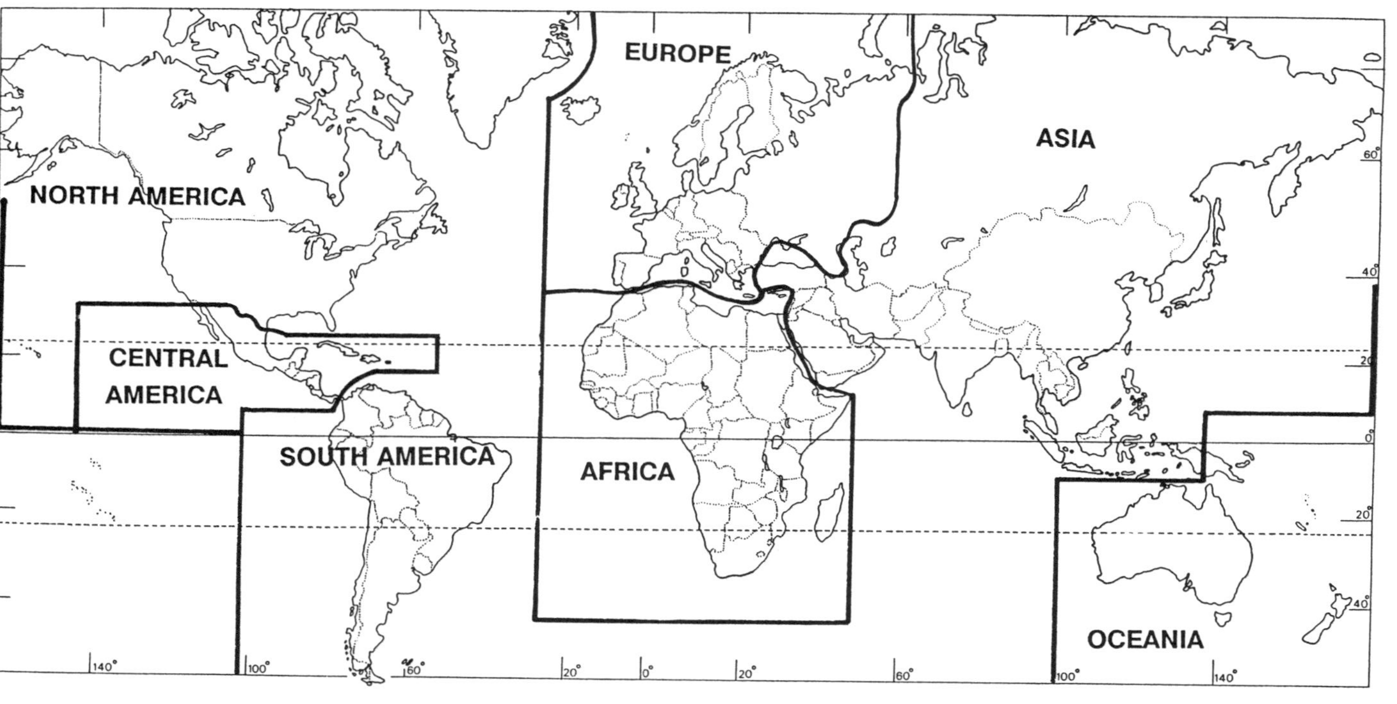

Fig. 7.1. World region boundaries.

degradation worldwide has affected 1966 mha or 15% of the total land area. These percentages range from 5% in North America, 12% in Oceania, 14% in South America, 17% in Africa, 18% in Asia, 21% in Central America, to 23% in Europe (including the European part of the former USSR).

Human-induced soil degradation has affected 24% of the inhabited land area. These values range from 12% in North America, 18% in South America, 19% in Oceania, 26% in Europe, 27% in Africa and Central America, and 31% in Asia.

Global extent of soil degradation types

Water erosion

Water erosion is by far the most important type of soil degradation, affecting about 1100 mha worldwide (56% of the total area affected by human-induced soil degradation). The most widespread form of water erosion is the loss of the topsoil (on 920 mha), whereas terrain deformation (rills and gullies) occurs on 175 mha. Water erosion occurs on all continents and under all climatic conditions, although somewhat more in humid climates (Table 7.1). Almost 80% of the terrain affected by water erosion has a light to moderate degree of degradation. This implies that around 225 mha are degraded by water erosion to such an extent that they are no longer suitable for agricultural land use. The most important causative factor is defores-

Table 7.1. Global and continental extent of water erosion (mha).

<table>
<tr><th></th><th>Light</th><th>Moderate</th><th>Strong + Extreme</th><th>Total</th><th>Degraded soils (%)</th><th>Dryland zone[a]</th><th>Humid zone[a]</th></tr>
<tr><td>Africa</td><td>58</td><td>67</td><td>102</td><td>227</td><td>46</td><td>122</td><td>105</td></tr>
<tr><td>Asia</td><td>124</td><td>242</td><td>73</td><td>441</td><td>59</td><td>165</td><td>276</td></tr>
<tr><td>S. America</td><td>46</td><td>65</td><td>12</td><td>123</td><td>51</td><td>35</td><td>88</td></tr>
<tr><td>C. America</td><td>1</td><td>22</td><td>23</td><td>46</td><td>74</td><td rowspan="2">38</td><td rowspan="2">68[b]</td></tr>
<tr><td>N. America</td><td>14</td><td>46</td><td>–</td><td>60</td><td>63</td></tr>
<tr><td>Europe</td><td>21</td><td>81</td><td>12</td><td>114</td><td>52</td><td>48</td><td>66</td></tr>
<tr><td>Oceania</td><td>79</td><td>3</td><td>+</td><td>83</td><td>81</td><td>70</td><td>13</td></tr>
<tr><td>World</td><td>343</td><td>526</td><td>223</td><td>1094</td><td>56</td><td>478</td><td>615</td></tr>
</table>

[a]Dryland zone is defined as the climatic region with an annual precipitation/evapotranspiration ratio of 0.65 or less (UNEP, 1992a). The humid zone has a ratio of more than 0.65.
[b]North + Central America.

tation (43%), whereas overgrazing (29%) and agricultural mismanagement (24%) are also important human interventions.

Wind erosion

Wind erosion is a widespread phenomenon in arid and semiarid zones, usually on coarse textured soils with a limited vegetative cover. The global extent of soils affected by wind erosion is around 550 mha (28% of the total terrain affected by soil degradation). Loss of topsoil by wind forces is by far the most important type of wind erosion (on 455 mha), whereas terrain deformation (82 mha) and overblowing (12 mha) are relatively less important in extent, although these types of wind erosion may have very serious effects on human activities. The degree of wind erosion is mainly light or moderate. Wind erosion is mainly caused by overgrazing (60%), followed by agricultural mismanagement (16%) and exploitation of the vegetative cover for domestic use (16%). Deforestation (8%) is the fourth causative factor leading to wind erosion. Table 7.2 indicates the continental distribution of wind erosion affected terrain.

Chemical soil degradation

A total of almost 240 mha is affected by chemical degradation worldwide, which is about 12% of the total area affected by human-induced soil degradation (Table 7.3). In South America, almost 30% of soil degradation is due

Table 7.2. Global and continental extent of wind erosion (mha).

	Light	Moderate	Strong	Total	Degraded soils (%)	Dryland zone[a]	Humid zone[a]
Africa	88	89	9	186	38	186	1
Asia	132	75	15	222	30	206	16
S. America	26	16	–	42	17	28	14
C. America	+	4	1	5	7	38[b]	1[b]
N. America	3	31	1	35	36		
Europe	3	38	1	42	19	39	3
Oceania	16	–	+	16	16	16	+
World	269	254	26	548	28	513	36

[a]Dryland zone is defined as the climatic region with an annual precipitation/evapotranspiration ratio of 0.65 or less (UNEP, 1992a). The humid zone has a ratio of more than 0.65.
[b]North + Central America.

Table 7.3. Continental and global extent of chemical soil degradation (mha).

	Loss of nutrients	Salini-zation	Pollution	Acidifi-cation	Total	Degraded soils (%)	Dryland zone[a]	Humid zone[a]
Africa	45	15	+	1	62	12	33	29
Asia	15	53	2	4	74	10	54	20
S. America	68	2	–	–	70	29	17	53
C. America	4	2	+	–	7	11	2[b]	5[b]
N. America	–	+	+	+	+	+		
Europe	3	4	19	+	26	12	4	22
Oceania	+	1	–	–	1	1	1	+
World	136	77	21	6	240	12	111	130

[a]Dryland zone is defined as the climatic region with an annual precipitation/evapotranspiration ratio of 0.65 or less (UNEP, 1992a). The humid zone has a ratio of more than 0.65.
[b]North + Central America.

to chemical processes, nearly all as a result of loss of nutrients and/or loss of organic matter. In North America and Oceania, chemical soil deterioration is of minor importance in comparison to other forms of human-induced soil degradation. In Africa, Asia, Central America and Europe, about 12% of the degraded soils are chemically affected. Loss of nutrients is the major type in Africa and South America, whereas salinization is the most important form of chemical degradation in Asia. Pollution is the dominant type of human-induced soil degradation in Europe. Acidification plays a role on 6 mha, of which 4 mha are located in Asia.

Chemical soil degradation is mainly caused by agricultural mismanagement (56%) and deforestation (28%) but industrial and bioindustrial activities are responsible for pollution.

Physical soil degradation

This type of human-induced soil degradation is recognized on 'only' 83 mha or around 4% of the total area worldwide affected by soil degradation (Table 7.4). The major form of physical soil degradation is compaction, sealing and crusting (68 mha). Nearly 33 mha are in Europe, where the main causative factor relates to the use of heavy machinery. Sealing and crusting, mainly as a result of cattle trampling and an insufficient cover by natural vegetation or crops, is found in Africa (18 mha) and Asia (10 mha). Waterlogging, mainly caused by human intervention in natural drainage systems, is the major form of physical soil degradation in Central America, but is also identified in South America. Subsidence of organic soils is mainly found in the coastal

Table 7.4. Continental and global extent of physical soil degradation (mha).

	Compaction, sealing and crusting	Water-logging	Subsidence of organic soils	Total	Degraded soils (%)	Dryland zone[a]	Humid zone[a]
Africa	18	1	–	19	4	14	5
Asia	10	+	2	12	2	10	2
S. America	4	4	–	8	3	+	8
C. America	+	5	–	5	8	1[b]	5[b]
N. America	1	–	–	1	1		
Europe	33	1	2	36	17	9	27
Oceania	2	–	–	2	2	1	1
World	68	11	4	83	4	35	48

[a]Dryland zone is defined as the climatic region with an annual precipitation/evapotranspiration ratio of 0.65 or less (UNEP, 1992a). The humid zone has a ratio of more than 0.65.
[b]North + Central America.

swamps of Southeast Asia, but also affects the soil productivity in some parts of the European section of the former USSR. Agricultural mismanagement (80%) and overgrazing (16%) are the two major causative factors of human-induced physical soil degradation.

Global extent of the severity of soil degradation

The assessment of yield reductions as a result of soil erosion is difficult, because over time, farmers may apply increasing amounts of fertilizers both to substitute nutrients lost by topsoil erosion and to enhance the natural fertility of the soil. They may change tillage practices, use different kinds of machinery, grow improved varieties of plants. Despite these complicating factors, studies in the USA have shown a relationship between soil erosion and reduced yields on many soils (Batie, 1983). Technological innovations continue to increase productivity of soils, but the average rate of change has declined from 2.2% annually during the 1950–1965 period to 1.8% annually during 1965–1979 in the USA.

Table 7.5 shows that over 300 mha are strongly degraded on a world scale. For this area – about the size of India – restoration to original productivity can only be achieved through major investments and engineering works. The terrain has virtually lost its productive capacity. More than 40% of this strongly degraded terrain is in Africa and 36% is located in Asia.

A much larger portion of the earth surface – more than 900 mha – has a moderate degree of human-induced soil degradation. By definition this

Table 7.5. Global extent (mha) of the degree of soil degradation for the major types of soil degradation.

	Light	Moderate	Strong + extreme	Total
Water	343	527	224	1094
Wind	269	254	26	549
Chemical degradation	93	103	43	239
Loss of nutrients	52	63	20	135
Salinization	35	20	21	76
Pollution	4	17	1	22
Acidification	2	3	1	6
Physical degradation	44	27	12	83
Total	749	911	305	1965

terrain is still suitable for use in local farming systems but has a serious decline in productivity. However soil resilience is such that it can be restored to its original productive capacity. This category needs the full attention of national decision-makers. If no efforts are undertaken to rehabilitate this moderately degraded land, one may fear that a major portion of this land will further deteriorate in the future and may become unreclaimable. Water erosion is again the main type of soil degradation, occurring on more than 500 mha, followed by wind erosion on some 250 mha. Loss of nutrients and/or rapid loss of organic matter in the topsoil occupies an area of over 60 mha. About one-third of this moderately degraded terrain is found in Asia, about 20% is located in Africa, and 12% in South America. Deforestation is the major cause of soil degradation in this moderately degraded portion of the earth's surface (38%), followed by agricultural mismanagement (28%) and overgrazing (25%).

An important portion – 750 mha – is now lightly degraded, characterized by somewhat decreased productivity, but farmers can restore the terrain to its full productivity by some modification of their management. Water and wind erosion are again the major types of soil degradation, respectively 343 mha and 269 mha. This light degree of soil degradation occupies 295 mha in Asia, 174 mha in Africa and 105 mha in South America. In Oceania, more than 95% of the degraded soils are in the 'lightly degraded' category.

Soil degradation and causative factors

The GLASOD map recognizes five different types of human intervention that have caused soil to degrade to its present status: deforestation, overgrazing, agricultural practices, overexploitation of the vegetative cover and

Table 7.6. Causative factors of human-induced soil degradation (mha).

	Deforestation	Overexploitation	Overgrazing	Agricultural activities	(Bio)industrial activities
Africa	67	63	243	121	+
Asia	298	46	197	204	1
S. America	100	12	68	64	–
C. America	14	11	9	28	+
N. America	4	–	29	63	+
Europe	84	1	50	64	21
Oceania	12	–	83	8	+
World	579	133	679	552	23

bioindustrial and industrial activities. Table 7.6 illustrates the global and continental extent of each causative factor, and Table 7.7 shows how each causative factor is related to the type of soil degradation.

More than 50% of the soil degradation caused by deforestation is in Asia, while it is the dominant causative factor of soil degradation in South America. In Africa, the major causative factor of soil degradation is overgrazing. Agricultural mismanagement is an important causative factor in Asia and Africa, and is the most important cause of soil degradation in North America. In Oceania, overgrazing is the dominant cause of soil degradation. Although (bio)industrial activities are a relatively minor problem compared to other causative factors, it is important to note that out of 23 mha worldwide 21 mha are located in Europe.

Table 7.7. Type of soil degradation and causative factors worldwide (mha).

	Causative factor				
Type of soil degradation	Deforestation of natural vegetation	Overexploitation of natural vegetation	Overgrazing	Agricultural activities	(Bio)industrial activities
Water erosion	471	36	320	266	–
Wind erosion	44	85	332	87	–
Chemical degradation	62	10	14	133	23
Physical degradation	1	+	14	66	–
World	579	133	679	552	23

Deforestation, overgrazing and agricultural activities are all main contributors to water erosion, mainly as a result of exposing the soil to the aggressive forces of rainfall or as a result of cultivating on steep slopes without proper antierosion measures. Overgrazing is the major cause of wind erosion, by exposing soil to the aggressive force of wind. Agricultural activities such as improper water management in irrigation schemes and insufficient application of fertilizers to restore nutrients removed by crops are major causes of chemical degradation. Also deforestation, leading to a rapid decline of organic matter in the topsoil, is an important cause of chemical degradation. Obviously (bio)industrial activities are the causative factors of soil pollution. Physical degradation is mainly caused by agricultural activities (heavy machinery) and to a certain extent by overgrazing (trampling by cattle).

Soil degradation and land use

Although the statistical information derived from the GLASOD map does not give direct information about the relationship between land use and human-induced soil degradation, the types of human intervention that have caused the soils to degrade to their present status are all related to land use activities (except the (bio)industrial activities). Therefore the human-induced soil degradation figures are related to agricultural or pastoral use or to exploitation of forest and woodland.

Table 7.8. Total land area and 'other land' area (mha) according to FAO (1989) and GLASOD.

	Total land area		'Other land'	
	FAO	GLASOD	FAO	GLASOD
Africa	2964	2966	1301	1303
Asia[a]	4376	4254	1597	1475
S. America	1753	1769	237	253
C. America	300	306	102	108
N. America	1832	1885	701	754
Europe[a]	1002	951	197	171
Oceania	842	882	198	238
World	13069	13013	4333	4302

[a]Asia includes the Asian part of the former USSR and Europe includes its European part. The FAO figures have been adapted (with information from Dokuchaev Institute of Soil Science) since the USSR is separately listed in the FAO statistics.

The statistics reported in FAO's production yearbooks give information on five categories of land (FAO, 1990):

- Arable land, which refers to land under temporary crops, temporary meadows, land under market and kitchen gardens and land temporarily fallow. Soil degradation as a result of mismanagement of agricultural practices directly relates to arable land.
- Land under permanent crops, which refers to land cultivated with crops that occupy the land for long periods. It excludes land under trees grown for wood or timber. Also in these areas agricultural activities may lead to soil degradation.
- Permanent pasture, which refers to land used permanently for herbaceous forage crops, either cultivated or growing wild. Overgrazing of these areas may lead to soil degradation.
- Forest and woodland, which refers to land under natural or planted stands of trees, whether productive or not, and includes land from which forests have been cleared. Deforestation and overexploitation of the vegetative cover are the main human-induced factors that cause soil degradation in this category.
- 'Other land' includes unused but potentially productive land, wasteland, urban land, barren land, parks, etc. This land is not affected by human-induced soil degradation.

The GLASOD map has identified separately physiographic units that are for 100% wasteland. Since the total area degraded by human action is known as well as the total area under agricultural occupation, pastures and forests from the FAO statistics it is possible to calculate this 'other land' portion by combining figures from the FAO and GLASOD statistics (Table 7.8).

Soil degradation on agricultural land

The total agricultural land includes land under permanent crops and arable land. According to FAO this area occupies 1475 mha (Table 7.9). The total area affected by human-induced soil degradation as a result of agricultural mismanagement of the land occupies 552 mha. Adding the estimated 10 mha of agricultural land in Europe affected as a result of (bio)industrial activities, the total area affected by human-induced soil degradation on agricultural land is 562 mha. In other words, worldwide 38% of the agricultural land is affected by human-induced soil degradation. It is also known that of this 560 mha, about 285 mha is moderately degraded. This implies that around 20% of the agricultural land worldwide is moderately degraded (around 6% is strongly degraded). In Central America almost 75% of the agricultural land suffers from soil degradation, and 65% of the agricultural land in Africa is affected.

Table 7.9. Global and continental extent (mha) for agricultural land, permanent pastures, forest and woodland, and the portion of these areas affected by human-induced soil degradation.

	Agricultural land			Permanent pasture			Forest and woodland		
	Total[a]	Degraded	%	Total[a]	Degraded	%	Total[a]	Degraded	%
Africa	187	121	65	793	243	31	683	130	19
Asia	536	206	38	978	197	20	1273	344	27
S. America	142	64	45	478	68	14	896	112	13
C. America	38	28	74	94	10	11	66	25	38
N. America	236	63	26	274	29	11	621	4	1
Europe	287	72	25	156	54	35	353	92	26
Oceania	49	8	16	439	84	19	156	12	8
World	1475	562	38	3212	685	21	4048	719	18

[a]Source: FAO, 1990.

Soil degradation on pasture land

The total area under permanent pastures worldwide accounts for 3212 mha according to FAO (1990). Although a major portion of the rangeland area is affected by vegetation degradation, the GLASOD data refer only to human-induced soil degradation. Overgrazing is the obvious causative factor leading to soil degradation, although it may be assumed that in Europe around 20% of the area affected by human-induced (bio)industrial activities (4 mha) has affected the pasture land in Europe. This implies that worldwide, around 685 mha of permanent pasture land are affected by human-induced soil degradation or around 21%, of which almost 40% (271 mha) are moderately degraded. A relatively large proportion of the permanent pasture land is affected in Europe (36%) and Africa (31%), whereas on the other continents between 10% and 20% is affected by soil degradation. Dregne *et al.* (1991) estimated that 73% of the dryland rangelands worldwide are desertified, which implies soil and vegetation degradation in the dryland zone.

Soil degradation on forest and woodland

Deforestation and overexploitation of the natural vegetative cover for domestic use are the major causative factors of human-induced soil degradation in forest and woodland. FAO (1990) estimates the total area worldwide of forest and woodland in 1989 to be around 4049 mha. The total area affected by soil degradation in the forest and woodland is around 719 mha or around 18%. Of this almost 50% (348 mha) is moderately degraded. Table 7.9 shows for each continent the percentage of forest and woodland affected by soil degradation. Soil degradation caused by deforestation and overexploitation of the vegetative cover in South America is observed on 'only' 13% of the forest and woodland, whereas it has affected 25% of forest and woodlands in Europe and Asia.

Concluding Remarks

Biswas and Biswas (1974) estimated that already some 10% of the world's arable land was despoiled by human activities. The GLASOD figures indicate that almost 40% of agricultural land has been affected by human-induced soil degradation, and that more than 6% is degraded to such a degree that restoration to its original productivity is only possible through major capital investments. Rehabilitation of the degraded soils to their original productivity is only possible on the lightly and moderately affected soils. GLASOD does not provide information about the essential elements for soil resilience. It may however help to indicate where soil rehabilitation is possible. In setting priorities for rehabilitation in differently affected degraded areas, convincing arguments may exist for intervention in strongly degraded areas if human populations have to be supported for humanitarian reasons. However, for sustainable economic and agricultural development, stabilizing inputs would have to be primarily in moderately degraded areas (UNEP, 1992b).

The GLASOD map does not indicate areas where degraded land has been rehabilitated, nor does it show areas with sustainable management systems. A programme was initiated in 1991 by the World Association of Soil and Water Conservation in a worldwide network of about 200 collaborating experts to prepare a World Overview of Conservation Activities and Techniques (WOCAT).

Rehabilitation of degraded land can only be done effectively if the building blocks of the land are well defined and quantified. There is a pressing need for a system which can store detailed information on natural resources of all kinds in such a way that this information can be assessed and combined in order to analyse each combination of land, water, vegetation and population which exists within a country or a region from the point of view of potential use, in relation to food requirements, socioeconomic factors, environmental impact or conservation. The International Society of Soil Science, in a joint effort with ISRIC, UNEP and FAO, has developed an international concept for a World Soils and Terrain Digital Database – SOTER. Once created, SOTER would provide the necessary ingredients for a wide variety of soil and land development related issues. As pointed out by

Higgins (personal communication, 1990): 'Only through development and application of such techniques can we catalyze the required breakthrough in land resource use, which is essential to halt and reverse current soil degradation in developing countries'.

Although the main objective of GLASOD was to increase the awareness of policy-makers and decision-makers of the dangers resulting from inappropriate land and soil management, a computerized land resource information system is a prerequisite for policy formation, development planning at all levels, efficient use of both internal and external resources, and for the implementation of a programme to assess soil resilience and its importance to sustainable land use.

References

Batie, S.S. (1983) *Soil Erosion: Crisis in America's Croplands?* The Conservation Foundation, Washington.

Biswas, A.K. and Biswas, M.R. (1974) *Environmental Considerations for Increasing World Food Production.* UNEP, Nairobi.

Brundtland, G.H., Khalid, M. *et al.* (1987) *Our Common Future.* Report of World Commission on Environment and Development presented to the chairman of Intergovernmental Intersessional Preparatory Committee, UNEP Governing Council. Oxford University Press, Oxford.

Dregne, H.E. (1986) Soil and Water Conservation: A Global Perspective. *Interciencia* 2, (4).

Dregne, H.E., Kassas, M. and Rozanov, B. (1991) A New Assessment of the World Status of Desertification. *Desertification Bulletin* 20, 6–18. UNEP, Nairobi.

FAO (1990) FAO Yearbook 1989. Production. FAO Statistical Series no. 94. Vol. 43, FAO, Rome.

ISRIC (1988) Guidelines for general assessment of the status of human-induced soil degradation. In: Oldeman, L.R. (ed.) *Working Paper & Preprint 88/4,* ISRIC, Wageningen (in English and French).

ISSS (1987) In: Van de Weg, R.F. (ed.) Proceedings of the Second International Workshop on a Global Soils and Terrain Digital Database (18–22 May 1987, UNEP, Nairobi). SOTER Report 2. ISSS, Wageningen.

Oldeman, L.R., Hakkeling, R.T.A. and Sombroek, W.G. (1990) *World Map of the Status of Human-induced Soil Degradation: an Explanatory Note.* Wageningen, International Soil Reference and Information Centre. UNEP, Nairobi, 27pp. + 3 maps.

UNEP (1992a) *World Atlas of Desertification.* Middleton, N.J. and Thomas, D.S.G. (eds) Edward Arnold, London.

UNEP (1992b) *Proceedings of the Ad-hoc Expert Group Meeting to Discuss Global Soil Databases and Appraisal of GLASOD/SOTER,* 24–28 Feb. 1992, UNEP, Nairobi.

Chapter 8
Soil Degradation in Hungary

P. Stefanovits

University of Agricultural Sciences, Godollo, POB 303, H-2103, Hungary

Using the GLASOD (Global Soil Degradation) Assessment (Chapter 7) and Hungarian cartographical work a comparison can be made of the ratios of different types and degrees of soil degradation for the World, Europe and Hungary. The basic data for Hungary are derived from cartographical work (Stefanovits 1963, 1964; Lang, 1980; Baranyai *et al.*, 1987, Pécsi *et al.*, 1989). Detailed information was available for agricultural lands. For forest areas estimates were based on aerial photographs and soil survey experience.

The GLASOD data for Europe are given in Table 8.1 and estimates for Hungary are presented in Table 8.2. The tables indicate that the ratios of types and degrees of soil degradation are different in Europe and Hungary. These data can be understood if we take into consideration that the geomorphology of Europe is more dissected in Hungary and there are also significant differences in geological structure. In addition, e.g. with respect to acidification and to subsidence, different criteria have been used.

Soil degradation caused by **water erosion** will be considered first. Water erosion occurs on one-third of the total area of the country in the form of surface creep and linear erosion.

Figures 8.1–8.3 showing the erosion map of Hungary, the distribution of the degrees of erosion and the clay mineral associations of Hungary, respectively, demonstrate that the ratio of these two types of erosion is related to the quantity and quality of clay minerals. The ratio of different degrees of surface erosion is also dependent on the quantity and quality of clay minerals which can be seen by comparing Figs 8.2 and 8.3. In the areas of high clay content where the dominant clay minerals are swelling clays, the proportion of strong erosion is high and the number of erosion channels km^{-2} is also high.

Estimating the erosion condition of agricultural land it was found that 48% was damaged by water erosion. Because of the improper use of soil the

 Soil Resilience and Sustainable Land Use (eds D.J. Greenland and I. Szabolcs)

Table 8.1. Human-induced soil degradation in Europe (mha).

Type	Light	Moderate	Strong	Extreme	Total
Loss of topsoil	18.9	64.7	9.2		92.8
Terrain deformation	2.5	16.3	0.6	2.4	21.8
Water	21.4	81.0	9.8	2.4	114.5 (52.3%)
Loss of topsoil	3.2	38.2	–	0.7	42.2
Terrain deformation	–	–	–	–	–
Overblowing	–	–	–	–	–
Wind	3.2	38.2	–	0.7	42.2 (19.3%)
Loss of nutrients	2.9	0.3	–	–	3.2
Salinization	1.0	2.3	0.5	–	3.8
Pollution	4.1	14.3	0.1	–	18.6
Acidification	0.1	0.1	–	–	0.2
Chemical	8.1	17.1	0.6	–	25.8 (11.8%)
Compaction	24.8	7.8	0.4	–	33.0
Waterlogging	0.5	0.3	–	–	0.8
Subsidence organic soils	2.6	–	–	–	2.6
Physical	27.9	8.1	0.4	–	36.4 (16.6%)
Total	60.6 (27.7%)	144.4 (66.0%)	10.7 (4.9%)	3.1 (1.4%)	218.9 (100%)
Total degraded surface	218.9		Deforestation:		83.8 (38.3%)
Total waste land	0.9		Overgrazing:		50.0 (22.8%)
Total other terrain	730.7		Overexploitation:		0.5 (0.2%)
			Agric. activities:		63.9 (29.2%)
Total	950.5		(bio)Industrial:		20.6 (9.4%)

extent of eroded areas and the ratio of moderate and strong erosion are increasing. The intensification of erosion is the result of the extension of maize and sunflower production over sloping areas for economic reasons.

The role and value of soil conservation differ from place to place. Soil protection is usually taken into consideration when land use and the direction and methods of cultivation are decided upon. However, technical soil protection (like terracing) is not considered. Another example is that strip cultivation and contour farming has never been practised in Hungary.

The collection of detailed data about soil degradation started during the 1950s. The content of the maps compiled enabled decision-makers to understand the need for soil conservation. The result was a fruitful decade which began by planning the work of soil conservation with government support and the organization of the Soil Conservation Company. This period lasted

Table 8.2. Human-induced soil degradation for Hungary (thousand ha).

Type	Light	Moderate	Strong + Extreme	Total
Loss of topsoil	1210	1335	850	3395
Terrain deformation	420	500	350	1270
Water	1630	1835	1200	4665 (25.3%)
Loss of topsoil	900	230	360	1490
Terrain deformation	–	–	–	–
Overblowing	350	–	–	350
Wind	1250	230	360	1840 (9.9%)
Loss of nutrients	–	–	–	–
Salinization	20	–	–	20
Pollution	50	–	–	50
Acidification	2000	2000	700	4700
Chemical	2070	2000	700	4770 (25.7%)
Compaction	2500	1500	–	4000
Waterlogging	800	600	350	1750
Subsidence organic soils	1000	500	–	1500
Physical	4300	2600	350	7250 (39.1%)
Total	9250	6665	2610	18,525

only a decade after which, the supports and subsidies were discontinued and the soil conservation organizations were disbanded. At the same time economic interest in maize production increased, causing a decrease in the area of soil protective plant production.

The importance of **wind erosion** was affected by the changes in land ownership. When the fields were owned by smallholders there were wooded pathways across the fields and trees around the farms. After the Second World War, with the introduction of large-scale farming most of the wooded pathways and small farms were replaced by vast fields (100–200 ha) suited to machine cultivation, serial fertilization and chemical plant protection. The lack of trees increased wind erosion and the regular use of heavy machines caused degradation of soil structure. As a result not only sandy soils, but also the degraded powdery top layer of other soils formed on loess, became victims of wind erosion. These losses could be quantified from the data on wind erosion damage recorded by the State Insurance Company (Petrasovits and Karacsony, 1990).

Data are not available for nutrient loss as one of the categories of **chemical soil degradation**, because fertilization conceals the amount of nutrients lost by leaching and water erosion. Indirect data might be obtained by

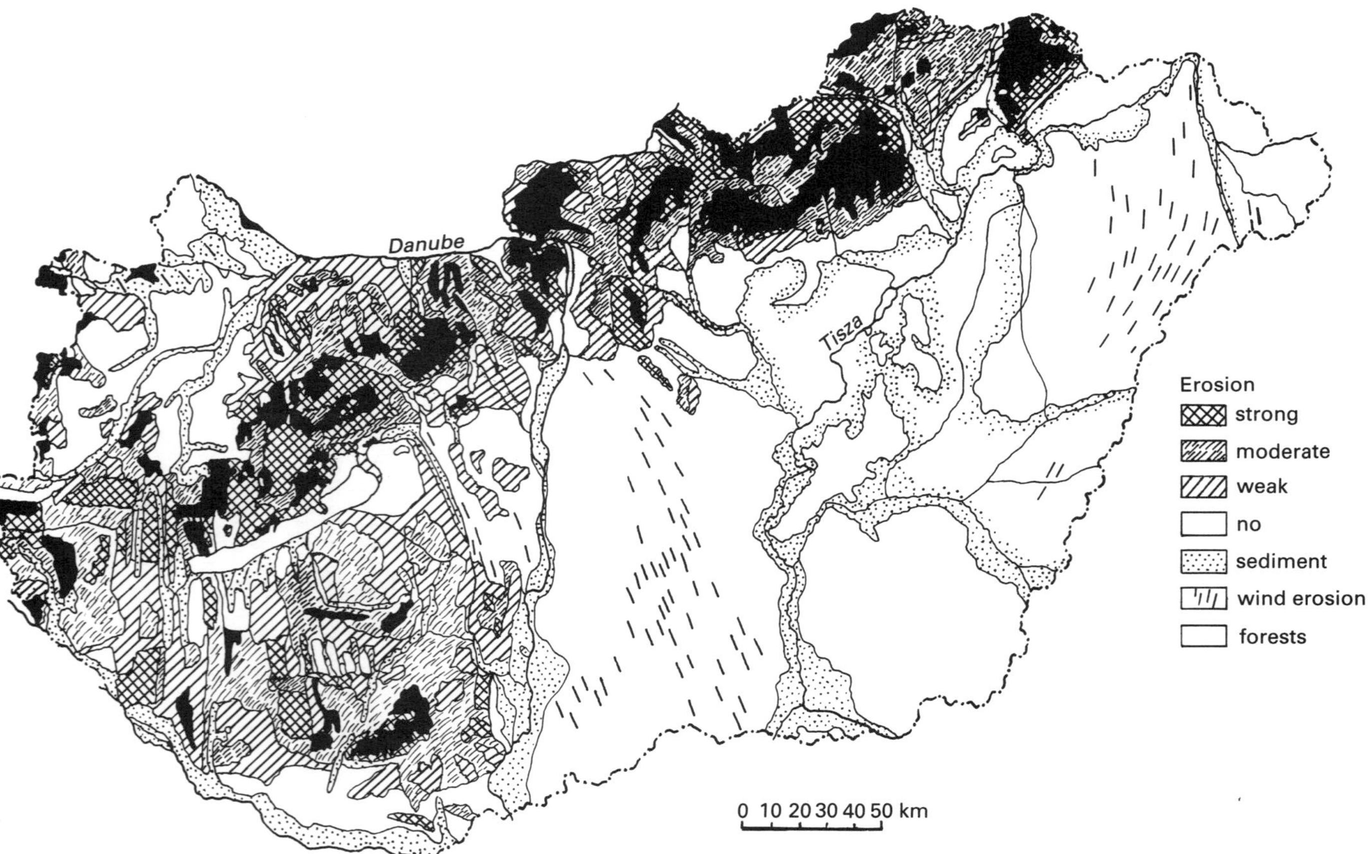

Fig. 8.1. Map of soil erosion of Hungary.

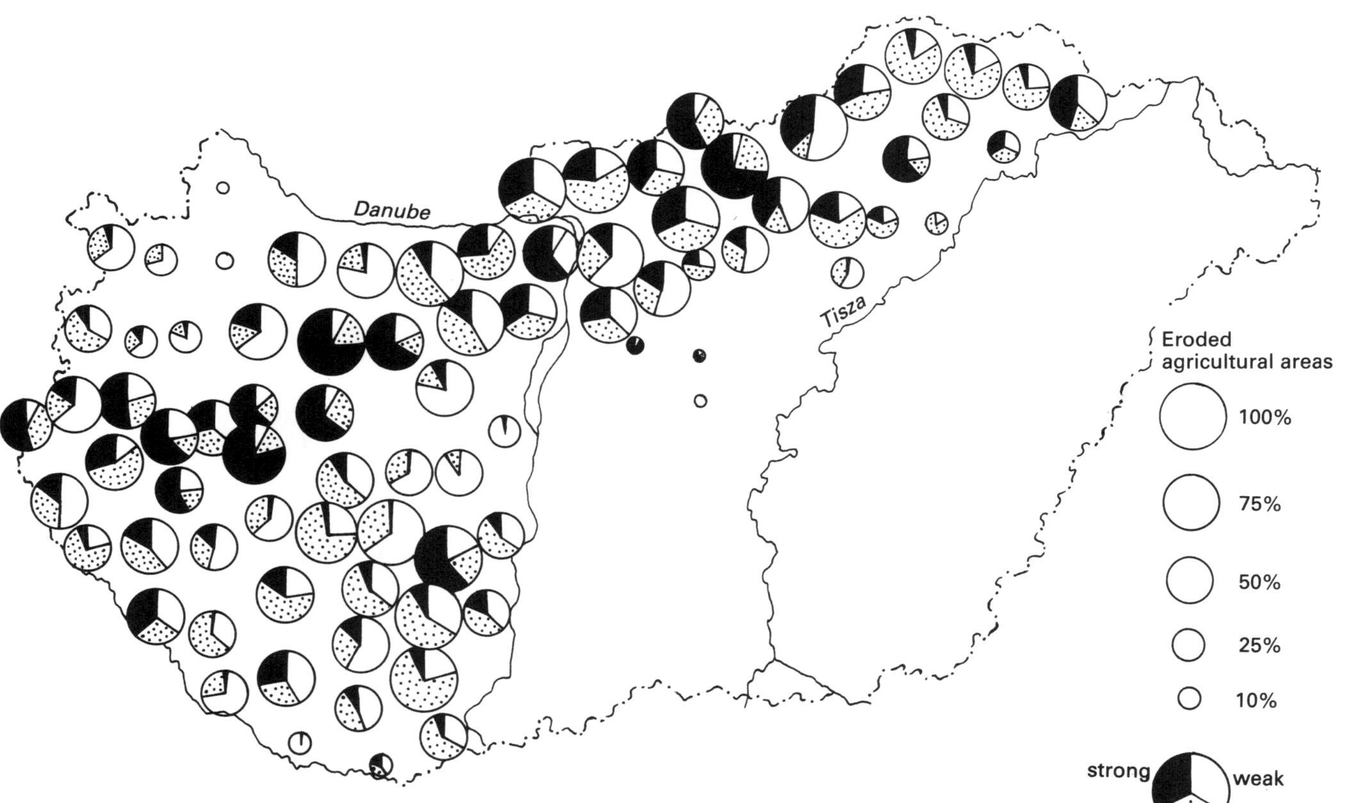

Fig. 8.2. Distribution of the degree of soil erosion of Hungary.

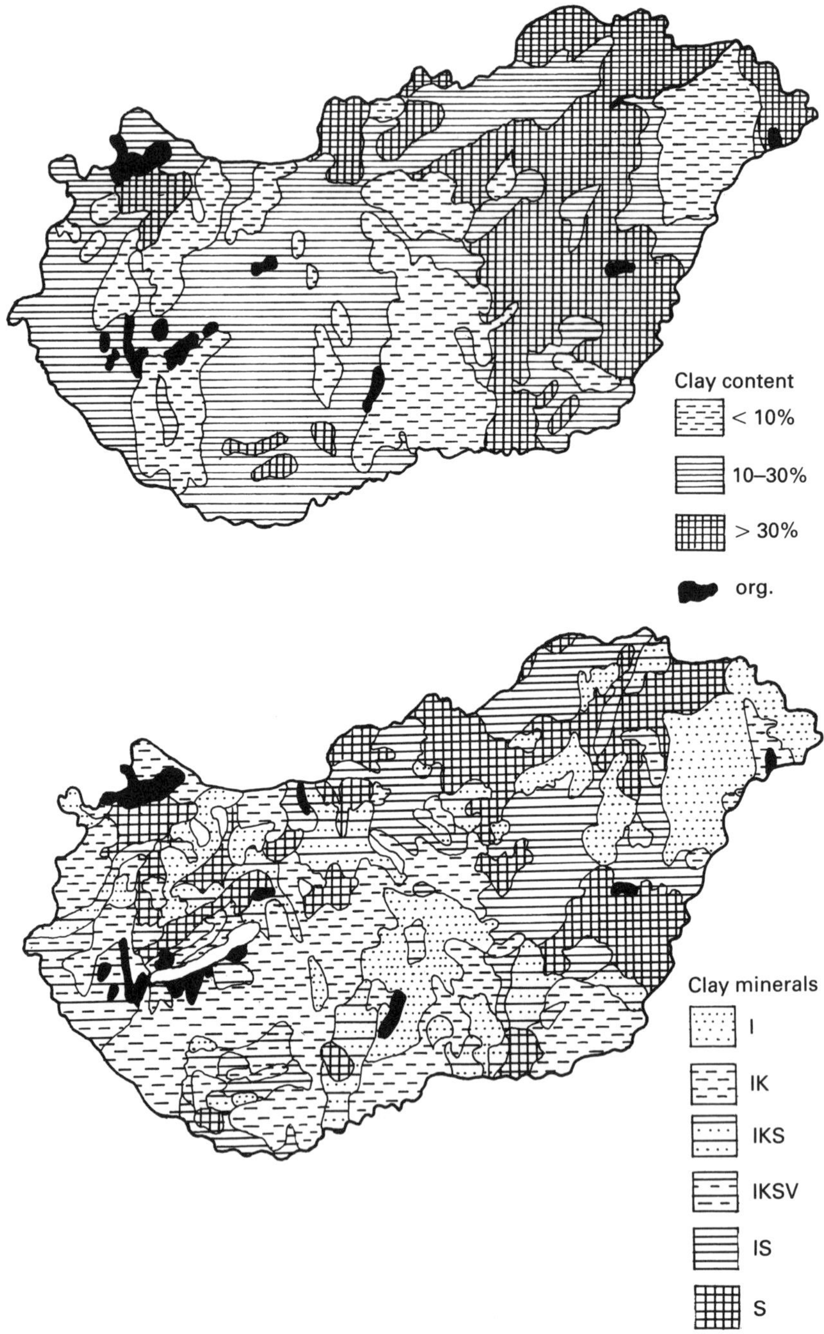

Fig. 8.3. Map of clay content and clay mineral associations in the soils of Hungary. I = illite, K = kaolinite, S = smectite, V = vermiculite.

measuring the eutrophication of waters but in this case the place of origin could not be determined within the watershed. The other problem is that leached nutrients cannot be separated or distinguished from the material arriving from sewage. That is why these factors are neglected.

According to the data obtained through soil tests conducted every three years by the former Soil Conservation Service the decrease of soil organic matter is significant in large areas. The loss of organic matter averages 0.1–0.2%. As most forest soils in Hungary contain only 2% organic matter this loss is very significant.

The changes in organic matter have been determined in terms of quantitative losses (Micheli *et al.*, 1991). Long-term intensive cultivation may also affect the quality (composition) of soil organic matter. In fact the increasing proportion of organic matter characterized by small molecular weight and the decrease of larger, more stable fractions may be more important than the changes in quantity.

Salinization, and secondary salinization, caused by human activity, occurs mainly adjacent to water reservoirs and canals where the increase of saline groundwater level is not prevented. This form of soil degradation can be detected mainly in the area of Lake Tisza and along canals crossing salt-affected areas.

Acidification of soils affects the largest area. It can be detected both in forest and meadow soils. The degree of acidification varies greatly, because the effects depend on the parent material and the natural soil-forming processes. The acidification of soils is described in several studies (Baranyai *et al.*, 1987; Stefanovits, 1986). The data show that acidification is increasing and its effects on other soil properties and on the environment are also increasing. The effects can be divided into two classes: air pollution (acid deposition) and acidification caused by fertilization. Their proportion changes according to the quantity and quality of acidic deposition and rainfall, the amount and type of fertilizers used, and plants grown. Whereas soil acidification caused by air pollution is higher in industrial centres, acidification of areas far from cities and industrial centres is mainly due to fertilizers.

Liming is used to remedy soil acidification. Lime application should follow the increases in amounts of fertilizers used and growing air pollution. However, the enormous increase in fertilizer use from 1967 has been followed by a decrease in liming. By the end of the 1970s the amount of applied fertilizers reached 280 kg ha^{-1} $year^{-1}$ (in active ingredients), but lime doses fell to half. This was due to the decrease in state subsidies for lime. An upward trend has occurred only in the last few years. Due to economic reasons and changes in land ownership the rate of fertilizer application decreased to less than 70 kg ha^{-1} in 1991 but the acidity of soils remained or increased further in the wake of previous fertilization and growing air pollution.

The most significant process of **physical soil degradation** is compaction, both in terms of extent and in the effects on production. It occurs in large areas of mainly chernozem soils because frequent and improper cultivation leads to structural degradation of the ploughed layer and compaction on wheel tracks. The topsoil becomes dusty and structureless while the compacted base of the ploughed layer may have a bulk density higher than 1.5 g cm^{-3}. Compacted layers reduce the permeability of soils and lead to stagnant water in flat areas, and surface erosion on sloping areas.

In the GLASOD classification the subsidence of organic soils appears under the heading of physical degradation. However biological and chemical processes may also contribute although the effect appears to be due mainly to the physical characteristics.

Conclusions

Soil degradation on agricultural land is common in Hungary. To avoid further degradation and ameliorate that which has already occurred the following tasks have to be completed:

- On sloping areas appropriate land use and cultivation techniques have to be used, and where necessary forest areas should be extended.
- In cultivated fields the use of soil protection measures, including use of soil protection cover crops, should be increased.
- Wooded pathways are to be maintained or developed to avoid or at least decrease the damage caused by wind erosion.
- Fertilizer use should be rationalized and lime applied in the areas where acidification may occur.
- Erosion control measures should be supported by economic decisions and interests.
- An effective soil conservation service and extension service should be developed and supported so that it can function effectively.
- Research work on degradation processes and control methods should be supported.
- Most importantly, at every level of education the importance of soil degradation and control should be emphasized.

References

Baranyai, F., Fekete, A. and Kovács, I. (1987) *A Magyarországi Talajtápanyag-vizsgálatok Eredményei.* Mezögazdasági Kiadó, Budapest.

Láng, I. (ed.) (1980) A mezögazdaság agroökológiai potenciálja az ezredfordulón. Budapest.

Micheli, E., Barabas, E. and Stefanovits, P. (1991) The effect of long-term fertilization on the quantity and quality of soil organic matter. *Proceedings 10th Humus*, Planta, Prague.

Pécsi, M. *et al.* (eds) (1989) *National Atlas of Hungary*. Kartográfiai Vállalat, Cartographia, Budapest.

Petrasovits, I. and Karácsony, J. (1990) A hazai erózió kutatás jelene és jövöje. *Agrár-világ* 1, 17–22.

Stefanovits, P. (1963) *Magyarország Talajai* 2nd edn. Akadémiai Kiadó, Budapest.

Stefanovits, P. (1964) *Talajpusztulás Magyarországon* (*Magyarázatok Magyarország eróziós térképéhez*). OMMI Kiadvány, Budapest.

Stefanovits, P. (1986) A talajok savasodásának néhány újabb adata. *Magyar Tudomány* 31, 339–341.

Chapter 9

Degraded Lands and Their Rehabilitation in India

I.P. Abrol[1] and J.L. Sehgal[2]

[1]*Deputy Director General, Indian Council of Agricultural Research, Krishi Bhawan, Dr Rajender Prasad Road, New Delhi, 110 001 India:* [2]*Director, National Bureau of Soil Survey and Land Use Planning, Nagpur, India*

Introduction

A large proportion of the land area of India shows clear evidence of advanced and continuing degradation seriously affecting the country's productive resource base. This, in turn, is threatening to undermine India's capacity to increase food production and alleviate rural poverty. On-going degradation of the resource base has adversely affected the supply of fuel-wood, fodder and timber which are far below the domestic needs. The socio-economic and ecological consequences of land degradation are far-reaching, cutting across many sectors and affecting most geographical areas of the country. Hardest hit by the problems associated with land degradation are the tribal and rural poor and landless people who rely heavily on the productive capacity of communal and public lands to support livestock grazing and to provide fuelwood.

Major physical manifestations of the processes causing degradation are the removal of topsoil by water and wind erosion, reduced capacity of soils to hold moisture, increased proneness to runoff and a gradual increase in the concentration of soluble salts in the root zone of soils.

Increasing pressure on land resources caused by the rapidly increasing population is further accelerating the processes of degradation. Overgrazing by livestock on private, communal and state-owned lands is perhaps the single leading cause of removal of vegetative cover. As the human and live-stock populations increase, so too will the pressure to overexploit already marginal grazing lands. Other causes of degradation are deforestation resulting from commercial and industrial exploitation, gathering of fuel-wood, foraging, inappropriate land use and cultivation practices, the lack of appropriate conservation measures, the land tenure system and related

 Soil Resilience and Sustainable Land Use (eds D.J. Greenland and I. Szabolcs)

socioeconomic factors. Other factors that promote degradation include activities like road construction, the spread of canal irrigation, uncontrolled urbanization, surface mining etc. This chapter describes the nature and extent of land degradation problems, their impact and the current efforts to rehabilitate and conserve the resource base.

Geographic Setting

India has a geographical area of 329 million hectares (mha) and lies north of the equator between the latitudes 8° 04′ and 37° 06′ N and longitudes 68° 07′ and 97° 25′ E. The climate in India is characterized by three distinct seasons: cool and mainly dry from October to February; hot and dry from March to June, and rainy from mid-June to September. Rainfall, the most important climatic parameter, is received during two distinct periods. The dominant south-west monsoon is active during June to September and the north-east monsoon during the winter months. Generally more than 80% of the total annual precipitation is received during the June–September rainy season, much of it in several high-intensity storms.

On the basis of average annual rainfall, India can be divided into three major zones; those with low, medium and high volumes of precipitation. Low rainfall regions receive less than 750 mm rainfall annually; these are often further subdivided into arid (less than 500 mm annually) and semiarid (500–750 mm annually). Medium rainfall (or subhumid) areas receive 750–1150 mm of annual precipitation and high rainfall (or humid) regions receive more than 1150 mm of rainfall annually (Table 9.1).

Status of Land Degradation

Based on existing sources of information a map showing the land degradation status of India using the methodology of GLASOD has been prepared. Two categories of human-induced soil degradation processes are recognized. The first deals with degradation through the displacement of soil material principally by water and wind. The second deals with internal soil deterioration resulting from the accumulation of chemical substances, e.g. salts, loss of nutrients, or through physical processes, including waterlogging.

Table 9.2 gives the approximate size of areas affected by the major degradation processes. About 200 mha, representing almost 60% of the total geographical area, is affected by various land degradation problems.

Among the degradation processes, erosion by water resulting in loss of topsoil (Wt) or in terrain deformation is by far the major factor causing loss in

Table 9.1. Total and area sown according to rainfall.

Normal rainfall (mm)	Reporting area (mha)	Net sown (mha)	Net irrigated (mha)	Rainfed (mha)
< 750	85.732	47.923	12.961	34.962
750–1150	96.625	51.225	15.966	35.259
> 1150	119.863	43.856	9.139	34.717
	302.220	143.008	38.066	104.948

Source: Directorate of Economics and Statistics (1990).

soil productivity and covers 155 mha. Erosion by wind is the dominant factor in western India affecting some 13.0 mha. Of these the loss of topsoil (Et) has affected the productivity of nearly 8.0 mha whereas terrain deformation (Ed) and overblowing account for about 5.0 mha. Soil deterioration due to chemical processes is estimated to affect nearly 15.0 mha, of which the loss or depletion of nutrients accounts for 3.9 mha whereas the accumulation of excess salts, soil salinization, is the primary factor in reducing the produc-

Table 9.2. Status of soil degradation in India.

Degradation type	Area (mha)		% of total area	
Water Erosion		155.2		47.2
Loss of topsoil (Wt)	142.4		43.3	
Terrain deformation (Wd)	12.8		3.9	
Wind Erosion		13.0		4.2
Loss of topsoil (Et)	8.0		2.5	
Terrain deformation (Ed)	3.5		1.2	
Overblowing (Eo)	1.5		0.5	
Chemical Deterioration		14.9		4.5
Loss of nutrient (Cn)	3.9		1.2	
Salinization (Cs)	10.9		3.3	
Physical Deterioration		16.3		4.9
Waterlogging (Pw)				9.6
Stable terrain under natural conditions (Sn)		31.8		
Land not fit for agriculture (including salt flats, rock outcrops, ice caps etc.)		16.8		5.1
Balance area of mapping units 1 to 7		80.9		24.6
Total		328.9		100.0

tivity of an estimated 10.9 mha. On about 7 mha the accumulation of excess water for a significant part of the year, waterlogging, is the main factor reducing the productivity of otherwise productive soils. The area of stable terrain, where human-induced degradation problems are relatively insignificant due to the presence of dense forest, is estimated at 32 mha. The degree of degradation may vary depending on the inherent soil properties, slope, topographic position, amount and intensity of rainfall, wind speed etc.

Erosion by water

This is the most serious cause of degradation in the Indian context. Narayana and Ram Babu (1983) analysed the existing soil loss data and concluded, as a first approximation, that soil erosion was taking place at an average rate of 16.35 tonnes ha^{-1} $year^{-1}$ totalling for all India 5334 m tonnes $year^{-1}$. Nearly 29% of the total eroded soil was permanently lost to the sea, and nearly 10% was deposited in reservoirs, resulting in the reduction of their storage capacity by 1–2% annually. The remaining 61% of the eroded soil was transferred from one place to another (Table 9.3). More recently Singh *et al.* (1992) presented an iso-erosion rate map of India based on 21 observed and 64 estimated soil loss data points spread over different land resource regions of the country. Iso-erosion rate lines were drawn by superimposing the available maps of soil, rainfall erosivity, slope, land use, forest vegetation and irrigation. The number identifying each iso-erosion rate line is the numerical value of the erosion rate in tonnes ha^{-1} $year^{-1}$ along this line. The annual water erosion rate values ranged from less than 5 tonnes ha^{-1} $year^{-1}$ for dense forests, snow-clad cold deserts, and the arid regions of western Rajasthan to more than 80 tonnes ha^{-1} $year^{-1}$ in the Shiwalik hills. Ravines along the banks of the rivers Yamuna, Chambal, Mahi, Tapti and Krishna, and the shifting cultivation regions of Orissa and the north-eastern states revealed soil losses exceeding 40 tonnes ha^{-1} $year^{-1}$. The annual erosion rate in the Western Ghats coastal regions varied from 20 to 30 tonnes ha^{-1} $year^{-1}$.

Black soils (vertisols and vertic subgroups) occupying about 71 mha in Peninsular India in the states of Maharashtra, Madhya Pradesh and Gujarat and the northern parts of Karnataka and Andhra Pradesh are highly erodible. Annual precipitation in these areas ranges from 500 to 1500 mm. Runoff can be 40% or more depending on rainfall volume and intensity, and on slope. A large part of these soils, about 25 mha, is farmed under rain-fed conditions, some 12 mha are cultivable fallows and the remainder is mostly under degraded forests and forest. Soil depth varies from less than 300 mm to more than 900 mm. The soils are subject to severe sheet and rill erosion with an annual soil loss of about 20 tonnes ha^{-1} $year^{-1}$. Even soils under normally cultivated crops, like sorghum and cotton, on slopes of 1–3%, have been observed to erode annually at the rate of 16 tonnes ha^{-1} $year^{-1}$.

Table 9.3. Soil erosion and sediment estimates.

Particulars	Amount (m tonnes)
Total soil erosion	5334
Sediment load of major, medium, and minor rivers	2052
Sediment deposition in reservoirs	480

Source: Narayana and Ram Babu (1983).

Red soils (Alfisols, Inceptisols, Ultisols), spread over about 90 mha, are another major soil group subject to high erosion by rain water. These soils are extensive in eastern Madhya Pradesh, on the Bihar Plateau, in Orissa and in parts of West Bengal, Andhra Pradesh, Karnataka, Kerala and Tamil Nadu. Rainfall in these areas ranges from 750 to 2000 mm year^{-1}. Most red soils, being shallow and of limited water retention capacity, allow rapid surface runoff and erosion. Major soil losses are due to sheet, gully and hill-side erosion. Sheet erosion is a serious constraint in nearly 75% of the red soils. Soil loss in these areas varies from 4 to 10 tonnes ha^{-1} year^{-1}. The lateritic soils occurring in rolling and undulating land suffer from severe soil erosion because they are located in the higher rainfall zone. Owing to high-intensity rains, these soils have been reported to lose about 40 tonnes ha^{-1} year^{-1} of soil in the absence of soil conservation measures.

The most spectacular erosion is observed in the form of gullies and ravine formation. Nearly 4 mha of land along the banks of major rivers are damaged in this way.

In the north-eastern states of India shifting cultivation is practised widely. With increasing population pressure 'fallow' periods which lasted 20–30 years 3 to 4 decades ago have been reduced to 3 to 6 years, further aggravating erosion and degradation problems. Soil erosion from hill slopes of 60–70% during the first, second and third years of cultivation was reported to be 145, 170 and 30 tonnes ha^{-1} year^{-1} respectively.

Wind erosion

Soil degradation resulting from wind erosion is a serious problem in the arid and semiarid regions in the states of Haryana, Gujarat, Punjab and Rajasthan. Destruction of the natural vegetative cover resulting from excessive grazing and the extension of agriculture to marginal areas is the chief human-induced factor leading to accelerated erosion. Wind erosion is also prevalent in the coastal areas where sandy soils dominate, and in the cold desert regions.

Soil salinization

The expansion of irrigation has been one of the key strategies in achieving self-sufficiency in food production. The net irrigated area in India has increased from about 20 mha in 1950 to about 45 mha at present. A large expansion in the irrigated area has been achieved through transported canal water. In almost all cases the groundwater table, which was several metres deep prior to the introduction of irrigation, rises when irrigation begins. Where the groundwater table is within about 2 m of the surface, the groundwater table contributes significantly to evaporation from the soil surface and causes soil salinization. In most canal-irrigated areas the problems of soil deterioration through accumulation of salts have reached serious dimensions. According to one estimate nearly 50% of canal-irrigated areas are affected by salt problems due to a lack of or inadequate artificial and/or restricted natural drainage, inefficient use of irrigation water and sociopolitical reasons. Salt problems have also increased where saline groundwaters have been used for irrigation in the absence of good quality irrigation water.

In many coastal regions excessive exploitation of groundwater has caused intrusion of sea water thus worsening the salinity problem. Salinity is a major threat to the maintenance of sustainable production in the irrigated areas and ameliorative measures are needed to arrest the process.

Table 9.4 gives an approximate geographic distribution of salt-affected soils. In India there are broadly two kinds of salt-affected soils: (i) saline; and (ii) alkali (or sodic). Whereas saline soils owe their low productivity to

Table 9.4. Geographic distribution of salt-affected soils.

Category	States in which the soils occur	Approximate area (mha)
Salt-affected soils of the arid and semiarid regions	Gujarat, Haryana, Punjab, Rajasthan, Uttar Pradesh	1.00
Salt-affected soils of the Indo-Gangetic plains	Bihar, Haryana, Madhya Pradesh, Punjab, Rajasthan, Uttar Pradesh	2.50
Salt-affected soils of the medium and deep black soil regions	Andhra Pradesh, Gujarat, Karnataka, Madhya Pradesh, Maharashtra	1.42
Coastal salt-affected soils		
Coastal salt-affected soils of the arid regions	Gujarat	0.714
Deltaic coastal salt-affected soils of the humid regions	Andhra Pradesh, Orissa Tamil Nadu and West Bengal	1.394
Acid salt-affected soils	Kerala	0.016
Total		7.044

Table 9.5. Trend in flood affected area and population.

Decade	Annual average area affected by floods (mha)		Annual average flood affected population (million)	Annual average total flood damages (Rs. million)
	Total	Cropped		
1950s	6.86	2.08	17.50	623
1960s	5.86	2.47	15.45	1041
1970s	11.19	5.55	43.35	6741
1980s	16.57	6.91	53.01	15,904

Source: Centre for Science and Environment (1991).

an excess of neutral soluble salts, e.g. chlorides and sulphates of sodium, calcium and magnesium, alkali soils are distinguished by the presence of measurable to appreciable quantities of sodium carbonate.

Although reliable estimates are not available, the accumulation of toxic substances of industrial and urban origin are increasingly contributing to land degradation. In some intensively cultivated areas, in the Indo-Gangetic plains and elsewhere, farmers use large quantities of fertilizers (400 kg ha^{-1} or more of nutrients). There are a few reports indicating that the use of excessive amounts of nutrients results in groundwaters being polluted by enrichment with nitrates.

Physical degradation

Problems of physical soil degradation are generally related to the reduction of soil organic matter, which makes the soil more prone to crusting, increased runoff etc. There are more and more reports of subsoil compaction restricting root growth and crop yields in the intensively cultivated areas. By far the most serious physical degradation problem is the development of areas where surface water stagnates due to changes in hydrology, landscape, development activities, silting up of river beds etc.

Problems of flooding have been increasing rapidly over the years (Table 9.5). This is attributed primarily to large-scale deforestation of the catchment areas, changes in land use and the destruction of surface vegetation. The problems are aggravated by increased sedimentation and the reduced capacity of drainage systems.

Impact of Land Degradation

The importance of land degradation in India has been increasingly recognized over the past thirty to forty years. The initial concerns were largely

Table 9.6. Siltation rate of selected reservoirs.

	Year of		Sedimentation rate mha 100 km^{-2} year^{-1}	
Reservoir	Impounding	Observation	Assumed	Observed
Beas	1974	1981	4.29	23.59
Chambal	1960	1976	3.61	5.29
Maithan	1956	1979	1.62	12.15
Nizamsagar	1931	1973	0.29	6.34
Panchet	1956	1974	2.47	9.92
Ramganga	1974	1974	4.29	17.30
Tungabhadra	1953	1972	4.29	6.11

Source: Narayana and Ram Babu (1983).

related to siltation of large multipurpose water reservoirs constructed at enormous cost. The capacity of several reservoirs has decreased at a much faster rate than envisaged at the planning stage (Table 9.6.) This adversely affects the capacity to sustain the gains in productivity achieved over the past decades. Table 9.5, which gives the annual average of areas affected by floods, provides a clear indication of the continuing degradation resulting in increased runoff from river catchments, affecting life and property.

Direct experimental evidence of the effect of degradation on productivity is limited. Among the earlier studies, a survey of the soils of Sholapur district of Maharashtra in 1870 and again in 1945 showed that in a period of 75 years nearly 17% of the land with a medium depth of soil (more than 450 mm) was converted into shallow soils of depth less than 450 mm (Basu *et al.*, 1960).

Vittal *et al.* (1990) evaluated the effect of topsoil depth on the yield of rain-fed crops grown in an Alfisol over a period of 4–5 years. Yield responses were up to 2.5 times greater in soils with deeper topsoils when rainfall in the critical period exceeded evapotranspiration than under drier conditions. Yield variations were explained by topsoil depth in cereals and by interaction between rainfall during the critical period and topsoil depth in cereals and castor bean ($R^2 > 0.72$). Yield losses expected to result from erosion, based on 56 years of rainfall data were 138, 84 and 51 kg ha^{-1} cm^{-1} for sorghum, pearlmillet and castor bean respectively. Continuing land degradation through the loss of topsoil is one of the major factors of low and unstable crop yields in the rain-fed semiarid tropics of India.

There are vast areas of degraded common grazing lands, uncultivable wastelands and degraded forests that pose a serious threat to adjoining productive cropland. Their degradation continues primarily because of the huge populations of livestock and of uncontrolled grazing and foraging far in excess of the sustainable capacity. Common property resources (CPR)

Table 9.7. Some indicators of the physical degradation of common property resources (CPR) in a span of 30 years.

Indicator	State		
	Andhra Pradesh	Madhya Pradesh	Rajasthan
Products collected by villagers			
In past	32	46	29
At present	9	22	8
No. of watering points in grazing CPRs			
In past	17	16	48
At present	4	3	11

Source: Jodha (1990).

have, in the past contributed a significant, indeed a major part, of such supplies as fuelwood, animal pastures and employment for the poorer sections of the society. Jodha (1990) showed that there was a rapid decline in common property resources and their productivity over the past few decades. The study covered 82 villages in different parts of the country and showed that over about 30 years there was a reduction in CPR by 31–55%. The physical degradation of CPR had resulted from the disappearance of a number of plant and tree species which the villagers used to gather from the common lands. Similarly CPR traditionally used for grazing more productive milch cattle, working bullocks etc. are now grazed by sheep and goat. The number of watering points have also declined significantly (Table 9.7).

The impact of soil degradation in irrigated areas due to soil salinization and waterlogging have not been evaluated adequately. India has, over successive 5-year plan periods, made large investments in creating an irrigation potential of nearly 45 mha. Canal irrigation accounts for 16.36 mha, tank irrigation for 3.3 mha and the remaining 25.5 mha is irrigated with water from wells and other sources. Problems of salinization and alkalization are most serious in the areas that have received canal irrigation. A few studies are indicative.

The left bank canal of the Tungabhadra irrigation project in Karnataka state was commissioned in 1953. A study made 30 years later (Bowonder and Ravi, 1984) showed that about 33,000 ha had been seriously affected by waterlogging and salinity. It was further estimated that this area was expanding at the rate of 6000 ha annually. The yield of about 20,000 ha had already fallen to zero forcing the cultivators to abandon their lands. In the initial years of soil degradation, cultivators tend to switch to more tolerant crops like rice, only to find that even the production of tolerant crops is substantially reduced, and their production accelerates the degradation of additional areas due to inefficient water use.

In the Nagarajunsagar project command area, nearly 25,000 ha of the 140,000 ha under irrigation have been affected by salinity and waterlogging in a period of 14 years.

Based on a survey of 110 farming households in the Gauriganj Block of Sultanpur district, UP, within the command of Sharda Sahayak irrigation project, Joshi and Jha (1991) observed that during the 12–13 years following the commissioning of the canal system more than 87% of the farms were affected by salt-related problems. On the affected farms nearly 29% of the farm area remained uncultivated. Even where the salt-affected soils were cultivated land use intensity was low despite the availability of irrigation, and cropping options were also curtailed on the degraded soils. It was further observed that the yields of paddy and wheat were 41–56% lower in the degraded soils. Net incomes were reduced by 87–92% on the salt-affected soils. Paddy was the only option on waterlogged soils though the net incomes were reduced by 54–55% compared with paddy grown on normal soils. For a paddy unit costs rose by about 60% and for wheat by 85% when cultivation was extended to the salt-affected soils. These results, though based on a small sample size, suggest that the economic trade-offs between creating additional irrigation capacity and improving drainage and other components of existing irrigation systems could be highly competitive.

Technologies for the Rehabilitation of Degraded Soils

India has, over the years, developed research and education institutions to service agricultural development including rehabilitation of degraded lands. The organizations involved include Central Research Institutions and the State Agricultural Universities. In recent years, the State Agricultural Universities have been strengthened through a National Agricultural Research Project (NARP), supported by the World Bank, whereby the capacity to undertake region-specific research has been strengthened by creating research institutes and stations to cater to the needs of all major agroclimatic regions in a state. The Central Research Institutes which have a major mandate to evolve technologies for the rehabilitation of degraded soils include:

- Central Soil and Water Conservation Research and Training Institute, Dehra Dun and its eight regional stations located in major agroclimatic regions.
- Central Research Institute for Dryland Agriculture, Hyderabad.
- Indian Grassland and Fodder Research Institute, Jhansi.
- Central Soil Salinity Research Institute, Karnal.
- Central Arid Zone Research Institute, Jodhpur.

Over the years, these institutes have evolved soil and water conservation technologies for checking soil degradation and for rehabilitating degraded soils. These efforts have led to the development of soil and water conservation technologies that are being promoted throughout the country. These include soil and water conservation engineering works such as bunds and graded bunds, terraces and drainage channels, gully stabilization with engineering and/or vegetative measures, collection of runoff in storage/percolation structures; various other soil and water conservation practices; cultivation of alternate crops/varieties; improved crop management, grazing land improvement and agroforestry development. Similarly technologies have been developed for the rehabilitation of soils degraded due to the accumulation of salts, particularly those widespread in the Indo-Gangetic plains. Implemented in differing combinations appropriate to the local conditions many of these practices have a place in improving soil and water conservation and in rehabilitating degraded lands.

Programmes and Progress in Rehabilitating Degraded Soils

In India efforts to rehabilitate degraded lands and/or adopt soil and water conservation measures have been promoted largely through central government-sponsored programmes implemented by the relevant state department/s and field agencies. Some of the important programmes taken up in the past are described here.

First is the soil conservation in the catchment areas of River Valley Projects. This programme was initiated in the third Five Year Plan period (1961–1966) following concern highlighted by several surveys revealing that the siltation rate of most reservoirs was much higher than the rates assumed at the time of planning (Table 9.6). Initially the programme was taken up in 13 catchments and later was extended to include 27 river valley project catchments involving a total area of about 69 mha. For the implementation of these programmes the Central Government provides 100% support; 50% as grant and 50% as loan to the state government concerned. The projects themselves are carried out by the states' departments responsible for agriculture, soil conservation and forests. In 1988–1989 nearly 2.27 mha were treated at an expense of 2,729 million rupees. Although the area treated each year has remained almost constant over the past decade, unit costs have increased sharply (Subaramaniyan and Samuel, 1990).

In addition to the specific programme for soil and water conservation, the government has sponsored several programmes for the benefit of the rural poor, particularly in areas with low and uncertain rainfall. Most of these programmes included a component relating to the conservation and improvement of the resource base. For example, in 1970 a Rural Works Programme was initiated in 54 districts in 13 states concentrated in the western and central zones of the country. Although the intent of the programme was to create permanent productive assets, almost from the inception of the scheme a very substantial share of the funds available was used to provide employment in the creation of a large number of works of low priority and little or no productive or developmental impact.

In order to correct the situation this programme was replaced in 1973 by the Drought Prone Area Program (DPAP), the intent being to change from a relief and employment oriented programme to one aimed at 'drought proofing'. Major attention was paid to restoration of the ecological balance, soil and moisture conservation, afforestation, restructuring cropping patterns, and the improvement of agronomic practices among other activities. During the period 1980–1985 various soil and water conservation treatments were provided covering 427,000 ha. However, doubts have frequently been expressed about the validity of achievements in terms of reported area coverage and the impact of soil conservation engineering works. This is also true for other rural development programmes, e.g. the Desert Development Programme, started in the seventh plan period (1985–1986) and covering 21 districts in five states over an area of 36.2 mha; the National Rural Employment Program, etc. According to available information, till the year 1987–1988, soil conservation measures had been taken on an area of about 33 mha. However, little information is available about what these measures accomplished in stabilizing the resource base.

Lessons from Past Efforts

In India soil and water conservation measures recommended and carried out at field level have been largely at the instigation of the central government and/or state governments. Employment generation and the mitigation of drought conditions were the key elements. The most common soil conservation technique used at field level was the construction of contour or graded earthen bunds at vertical intervals of 1–2 m (on slopes of 6–8% or less). Experience has shown that small, resource-poor farmers are often reluctant to realign their plot boundaries to conform to acceptable contour grades. Farmers are also reluctant to have these bunds within their field since these take a considerable part of their land out of cultivation. For these and other reasons technical considerations are often compromised. As a result, in the event of high rainfall the bunds are frequently breached and the purpose for which they were constructed is rarely effected. Erosion on these fields is often accelerated due to soil disturbance. Some have even questioned the technical appropriateness of the technology especially in the black soil regions with medium and high rainfall. By concentrating runoff and impeding its natural flow, earthen bunds, whether graded on the contour or along the plot boundaries, exacerbate the problem and they themselves become a cause of water stagnation and soil erosion.

In the past many soil conservation projects were carried out without involving the landowners in an effective manner. In fact there was a complete lack of a mechanism to interact with the farmers. They were left to stand aside and watch as silent spectators. Bunds, having steeply sloping sides, are themselves subject to erosion. Consequently earthen bunds need regular maintenance and periodic reconstruction, failing which, they disintegrate within 3–5 years. As they are not fully convinced of the benefits and because they have limited resources farmers take little interest in maintaining these structures.

Yet another concern in some of the past soil and water conservation programmes was the manner in which they were carried out in isolation from the crop production programmes. Thus whereas seeds and fertilizers are considered an input for improved productivity, soil and water conservation are not. For this reason farmers have not been enthusiastic about these programmes. A small and resource-poor farmer is interested in improving the production of his food grains and income and is not worried about long-term sustainability. Unless he is convinced that soil and water conservation measures will add to his production or in some other way help him make a better use of his investments, efforts for reducing soil erosion are not his priority. We consider that if more emphasis could be focused on *in-situ* water conservation, the benefits might be more visible and technologies more acceptable to the farmers. When a farmer sees improved water availability in his field he is encouraged to try other methods for improved production.

In most programmes soil and water conservation measures were only taken on private arable lands. These efforts are frequently paralleled though not complemented by the efforts of the departments of forestry, most of which have their soil conservation division undertaking the planting and tending of trees and implementing other soil conservation works on the state-owned lands.

It is apparent that ultimately many of the government-sponsored activities became instruments for providing temporary employment to the rural unemployed rather than developing cropland and raising its productivity.

The Watershed Approach

Notwithstanding past setbacks there have been welcome qualitative changes in approaches to the management of resources in the past decade or so. It has been increasingly realized that there is a need to adopt an integrated approach to the management and development of resources including soils, water, crop, range, forests and livestock based on the natural ecological unit of the watershed.

Since 1983–1984 the Indian Council of Agricultural Research, working through the Central Research Institute for Dryland Agriculture, Hyderabad and the Central Soil and Water Conservation Research and Training Institute, Dehra Dun, and collaborating closely with the State Agricultural Universities and line departments, have made efforts to improve the management of 47 experimental model watersheds in different agroclimatic zones selected for intensive development and for in-depth monitoring and evaluation. In 1986–1987, a National Watershed Development Programme for Rainfed Agriculture was initiated with the primary objective of stabilizing agricultural production in rain-fed areas by significantly increasing investment. The programme concerned areas in the rainfall zones 500–1125 mm and above 1125 mm per annum. The main objectives of the programme were:

- taking the watershed as a basis, to conserve and upgrade croplands and degraded lands;
- to develop and demonstrate location specific technologies for soil and moisture conservation and for the stabilization of crop production;
- to augment the fodder, fruit and fuel resources of village communities through appropriate alternate land use systems.

In recent years there has been a marked increase in both the level of interest in this approach and in the willingness to commit resources for implementation. However, relative to the size of the problem and the amount of resources committed to these activities, integrated watershed development programmes are still in their infancy. The experience gained from the implementation of experimental and other projects (Vaidyanathan, 1991), including those funded by the World Bank, highlights several issues that need to be dealt with to ensure the success of future programmes.

Data base for planning

Watershed planning calls for a great deal of information. At present there is no field agency to generate the required data rapidly enough for the preparation of development plans. It is necessary to evolve simple, rapid, and inexpensive techniques of obtaining the required information on soil depth, texture, slope etc. to form the basis of planning. It is also necessary to train local people to carry out these studies.

Integration of activities

A basic tenet of the watershed approach is the integration of various activities at field level. In the past this has rarely happened. As the Agricultural Finance Corporation Report (1988) described, in one of the schemes:

> the line departments separately prepare their part of the plan and the two are superficially combined and called a watershed management plan. There was no effective arrangement for collecting basic data needed for integrated planning. For this reason there was no proper coordination between agricultural and forest departments resulting in 'hillocks' and 'nallas' around agricultural lands remaining bare and denuded. Such a situation will need to change.

Technologies

Soil and water conservation techniques recommended and adopted have largely been developed at the experimental stations. However, quite often such methods have not proved suitable in various natural situations that require a multidisciplinary approach to developing visible solutions. For this reason, too, it is increasingly realized that researchers and planners have much to learn from the farmers whose knowledge is often based on time-tested observations and a strong analytical capability. What is needed is to study closely indigenous practices for soil and water conservation, the farmer's experience in adoption and the need for innovation based on local knowledge and resources, the farmer's perception of the problem and his immediate and long-term needs (Kerr, 1991).

The above considerations also call for building flexibility in the funding 'norms' and the use of funds at field level. Most projects to date have suffered from complicated financial procedures leading to undue delays. As we proceed to make corrections based on the lessons learned from past experience and adopt new approaches to technology generation there is reason to believe that there will be significant advances in the control of degradation in the decades to come.

References

Agricultural Finance Corporation (1988) Report on evaluation study of Soil Conservation in the River Valley Project of Nizamsagar, Bombay p. 108.

Basu, J.K., Kaith, D.C. and Rama Rao, M.S.V. (1960) Soil conservation in India. *Farm Bulletin 58*, Farm Information Unit, Directorate of Extension, Ministry of Food and Agriculture, New Delhi, 64 pp.

Bowonder, B. and Ravi, C. (1984) Waterlogging from irrigation projects: an environmental management problem. Centre for Energy, Environment and Technology, Administrative Staff College India, Hyderabad, 41 pp. (mimeographed).

Centre for Science and Environment (1991) *Floods, Flood Plains and Environment,* A Citizens' Report, New Delhi, 167 pp.

Directorate of Economics and Statistics, Ministry of Agriculture, Government of India (1990) *Indian Agriculture in Brief,* 23rd edn, 460 pp.

Jodha, N.S. (1990) *Rural Common Property Resources: Contribution and Crisis,* Foundation day lecture, Society for Promotion of Wasteland Development, New Delhi.

Joshi, P.K. and Jha, D.N. (1991) Farmlevel effects of soil degradation in Sharda Sahayak Irrigation Project, Working Paper on Future Growth in Indian Agriculture no. 1, International Food Policy Research Institute.

Kerr, J.M. (ed.) (1991) Farmer's practices and soil and water conservation programmes: summary proceedings of a workshop, 19–21 June 1991, ICRISAT Centre, India, Patancheru, AP 502324, India: International Crops Research Institute for the Semi-Arid Tropics (ICRISAT) and Winrock International.

Lal, R. and Stewart, B.A. (1990) Soil degradation – A global threat In: Lal, R. and Stewart, B.A. (eds) *Advances in Soil Science*, pp. 13–17.

Narayana, V.V.D. and Ram Babu (1983) Estimation of soil erosion in India. *Journal of Irrigation and Drainage Engineering* 109, 419–434.

Singh, Gurmel, Ram Babu, Pratap Narain, Bhushan, L.S. and Abrol, I.P. (1992) Soil erosion rates in India. *Journal of Soil and Water Conservation* 47, 97–99.

Subramaniyan, S. and Samwel, J.C. (1990) Soil and water conservation in the catchments of River Valley Projects in India. *International Symposium on Water Erosion, Sedimentation and Resource Conservation,* 9–13 October 1990, proceedings pp. 103–115.

Vaidyanathan, A. (1991) *Integrated Watershed Development: Some Major Issues* Foundation Day Lecture May, 1991, Society for Promotion of Wastelands Development, New Delhi, 20 pp.

Vittal, K.P.R., Vijayalakshmi, K. and Rao, U.M.B. (1990). The effect of cumulative erosion and rainfall on sorghum, pearlmillet and castor bean yields under dry farming conditions in Andhra Pradesh, India, *Experimental Agriculture* 26, 429–439.

Chapter 10

Constraints in Managing Soils for Sustainable Land Use in Drylands

B.G. Rozanov

(Moscow State University) United Nations Environment Programme, PO Box 30552, Nairobi, Kenya

Introduction

Two key words determine the major issues in this volume – sustainability and resilience – recently both getting more and more fashionable, particularly in mass media and international political or semitechnical documents, with the result that they are ill-defined and grossly misused in many cases.

There are some inherited uncertainties attached to both terms. As the concept of sustainability does not imply any change in itself, either regressive (degradation of natural resources, civil strife) or progressive (soil improvement, socioeconomic development), but rather presupposes a quasistationary equilibrium, the introduction of the combined term **sustainable development** without clearly defining ecological or socioeconomic implications of such a combination, led to further confusion.

Therefore, one has to be very careful when using either term. For instance, for many experts sustainability means 'the same forever'. The socioeconomic consequence of such an understanding includes vigorous promotion of so-called 'traditional technologies' in the developing countries at a time of unprecedented rates of population growth when such technologies cannot either satisfy basic human needs or protect the environment. They were very good and appropriate a hundred years ago, but are absolutely useless now leaving their users in poverty and misery forever. It is just pure misuse of terminology and attached concepts, although, naturally, certain elements of the traditional technologies can be utilized with success. At the same time, resilience is not the synonym of stability, although the term is too often used in this way.

Common sense often helps in such difficult situations. Therefore, it is better to agree from the very beginning on what we are talking about and proceed with the necessary work, while time may provide more accurate

 Soil Resilience and Sustainable Land Use (eds D.J. Greenland and I. Szabolcs)

definition. This is precisely what happened with the definition of desertification. In spite of a rather vague definition provided by the United Nations Conference on Desertification in 1977 (United Nations, 1977), the world was engaged in combating this menace, experiencing both success and failure for 14 years until a new, more precise and operational definition was adopted in 1990 (Odingo, 1990), then refined in 1991 (UNEP, 1991). Now we are going to witness intergovernmental negotiations on the global convention on desertification and the specific Chapter 12 of Agenda 21 adopted in June 1992 by the United Nations Conference on Environment and Development elaborates six problem areas within this concept, slightly redefined once more for the purpose (UNCED, 1992).

World Drylands

According to the latest concept adopted by UNEP for environmental assessment purposes, drylands are the temperate, warm or hot areas with a P/PET ratio less than 0.65 (UNEP, 1991, 1992). They are subdivided into hyperarid, arid, semiarid and dry subhumid territories (Table 10.1).

The data show that 64% of the global drylands and 97% of hyperarid deserts are concentrated in Africa and Asia, although Australia is the driest continent in respect of the percentage of its total land surface.

There are three closely interrelated but very different phenomena that determine specific problems of the drylands and their use: **aridity**, **drought** and **desertification/land degradation**. These parts are distinguished not only by their physical nature but also by the major human strategies available to cope with them.

Aridity is a permanent climatic characteristic of an area that is determined and maintained by general global atmospheric circulation with certain local specifics imposed by topography; this characteristic is mainly manifested in the lack of atmospheric precipitation to support biological productivity. The human strategy for coping with aridity is:

1. to adapt living standards, behaviour and technology to corresponding climatic characteristics in accordance with the degree of aridity; and
2. to provide fresh-water supply for both human and animal consumption as well as for technological needs by either tapping underground aquifers or water transfers from outside.

Drought is a deficit or too-late arrival of atmospheric precipitation. It is usually defined in relation to the degree of probability as regards the 'normal' or more or less usual amount/date for a given locality within this or that aridity zone. Drought affects negatively local biological productivity (as well as consumer's water supply or electricity production by hydrostations)

Table 10.1. Distribution of world drylands by continents and degrees of aridity (mha) (UNEP, 1991, 1992).

Type of dryland	P/PET	Africa	Asia	Australia	Europe	North America	South America	World Total	%
Hyperarid	< 0.05	672	277	0	0	3	26	978	16
Arid	0.05–0.20	504	626	303	11	82	45	1571	25
Semiarid	0.21–0.50	514	693	309	105	419	265	2305	38
Dry subhumid	0.51–0.65	269	353	51	184	232	207	1296	21
Total	< 0.65	1959	1949	663	300	736	543	6150	100
%	–	32	32	11	5	12	8	100	

for a specific year or a season when it occurs. The human strategy towards drought includes:

1. accurate and reliable forecasting in time and space;
2. utilization of flexible production systems that incorporate diversified land use and specific technologies to be applied in time of drought; and
3. maintaining insurance reserves at various levels of social organizations to cope with disastrous effects when drought does occur.

Desertification is land degradation in arid, semiarid and dry subhumid areas resulting mainly from adverse human impact (UNEP, 1991). The main human strategy to combat desertification is adoption of ecologically appropriate land use and general land care in a favorable socioeconomic, legislative and political environment.

The above three phenomena and related human strategies to survive and develop constitute a unity, of which none of the parts can be treated separately in any programme designed to amend or mitigate either the environmental or the socioeconomic conditions of the drylands. These three characteristics of the drylands can be regarded as natural constraints to managing soils to achieve sustainable land use in these areas.

Present Status of World Drylands

By definition, all 6.1 bha (billion hectares) of the world's drylands are naturally subjected to aridity, to different degrees. However, there is no drought or desertification in 0.9 bha of hyperarid deserts like the Sahara. On the other hand, 5.2 bha of potentially productive drylands, 84% of the world's drylands, are prone to drought and desertification, usually in proportion to the degree of aridity, although with many important

Table 10.2. Global status of desertification/land degradation in the drylands of the world, by continents (UNEP, 1991).

Continent	Irrigated lands			Rainfed croplands			Rangelands			Total agriculturally used drylands		
	Total (mha)	Degraded		Total (mha)	Degraded		Total (mha)	Degraded		Total (mha)	Degraded	
		mha	%		mha	%		mha	%		mha	%
Africa	10.42	1.90	18	79.82	48.86	61	1342.35	995.08	74	1432.59	1045.84	73.0
Asia	92.02	31.81	35	218.17	122.28	56	1571.24	1187.61	76	1881.43	1311.70	69.7
Australia	1.87	0.25	13	42.12	14.32	34	657.22	361.35	55	701.21	375.92	53.6
Europe	11.90	1.91	16	22.11	11.85	54	111.57	80.52	72	145.58	94.28	64.8
N. America	20.87	5.86	28	74.17	11.61	16	483.14	411.15	85	578.18	428.62	74.1
S. America	8.42	1.42	17	21.35	6.64	31	390.90	297.75	76	420.67	305.81	72.7
Total	145.50	43.15	30	457.74	215.56	47	4556.42	3333.46	73	5159.66	3562.17	69.0

exceptions. It is not the arid but rather the semiarid areas which have the highest rate of land degradation mostly due to the differences in population pressure and patterns of land use.

It is estimated (Table 10.2) that about 3.6 bha, or 70% of potentially productive drylands, are currently threatened by the various forms of land degradation known as desertification.

1. Degradation of vegetation and sometimes soil in 3.3 bha or about 73% of the total area of dry rangelands, including all uncultivated non-forest lands and bushlands whether used at present as rangeland or not.
2. Decline in fertility and soil structure leading gradually to soil loss in 216 mha of rain-fed croplands, or nearly half of their total area in the drylands.
3. Degradation by waterlogging, salinization and/or alkalinization of 43 mha of irrigated croplands amounting to nearly 30% of the total area in the drylands.

As for the current human-induced soil degradation, it affects about 1 bha within the area of productive drylands, that is almost one-fifth of their total area (UNEP, 1992a). This includes mainly soil erosion by wind and water, and salinization of irrigated croplands. Soil erosion affects severely both rain-fed croplands and rangelands which are over-exploited and often left without protective plant cover. The situation is particularly serious on marginal lands in semiarid areas where cultivators encroach on the rangelands during occasional rainy years, and leave them bare during dry ones thus accelerating soil erosion. In places where the soil cover is thin enough, a few such cycles will result in hard rock coming to the surface thus transforming semidesertic dry steppe into true stony desert (Hammada, Gobi).

Drylands in almost 90 developing countries represent the largest global area of extreme and persistent poverty of the rural population. The poverty of people in these areas grows with the advance of desertification accentuated by a fast growth of population or alternatively, in some cases by migration. Degradation and ultimate loss of natural resources lead often to civil strife as evidenced by the present situation in the Horn of Africa.

People in the drylands, struggling daily with almost universal poverty, have no means to maintain or to improve their lands and have only one choice – to degrade them further. The sustainability of traditional technologies has not been able to match the present rate of population growth. Moreover, the growth of consumption demands now requires much greater norms of production in comparison with the not so distant past. In case of prolonged drought in certain areas, global food security is already in a precarious state. It will become much worse if desertification continues unabated.

Concept of Soil Resilience in Drylands

The above background characteristics of the world drylands seem to give little hope for appreciable benefit from soil resilience in these areas. However, this is not true. The most striking fact is that in spite of unprecedented growth of human pressure throughout the world drylands, both in qualitative and quantitative terms, the comparison of soil maps composed now and several decades ago, at least at a small scale, does not reveal a substantial soil change even in those areas where soil degradation processes were reported to be particularly active. This is a paradox which requires special investigation.

The most probable explanation of this situation can be found in the spatial pattern of soil degradation, particularly soil erosion. In the latter case, apparently, the major process is that of redistribution of the material within the catchment area without much outflow either by wind or water. Some of the material, naturally, goes out of the area, but its major part is redistributed along the slope in the process of planation which could be active both in meso- and macroscales.

The present-day figure of soil loss in the world is estimated as 25.4 billion tonnes per year (UNEP, 1992b); this is the total removal of material from land surface to the ocean. Some time ago geologists estimated this at 24 billion tonnes, which is practically the same. On the basis of the total area of the global land surface, this is something like 1.9 t ha^{-1} $year^{-1}$ as an average. This removal is certainly matched by the process of soil formation.

On the other hand, this is an average value which cannot actually be found anywhere. The actual rate of soil erosion varies from zero to more than 1000 t ha^{-1} $year^{-1}$ depending on a complex of local characteristics, the pattern of land use being one of the important ones. In certain areas there is accumulation of new material brought from the surrounding territories.

The second explanation of the above paradoxical situation can be connected with the methodology of assessing soil degradation and resilience. In estimating the degree of soil degradation and resilience in the drylands, and maybe elsewhere, the most common procedure is to measure how many tonnes of soil have been lost per hectare per year. In drylands this index has no meaningful value or practical importance: the most important point is what remains as a result of the erosion. Losses of the top 20 cm of some Aridisols on 50 m thick loess and of the same top 20 cm of some Luvisols, or Alfisols in African dry savanna have very different ultimate consequences. In the first case there will be no appreciable change in land capability, whereas in the second no capability will be left to support plant life whatsoever. The general concept of so-called allowable soil loss is total nonsense if the remaining soil is not considered. What is allowable in one area, may be considered catastrophic in another.

Furthermore, while considering the remaining posterosional soil we may come across very different situations determined by the nature of the exposed material. One such situation may be where the exposed material will be potentially more fertile than the removed topsoil. Naturally, the opposite case also occurs, and probably more often.

In any case it is no less important to assess the result of soil degradation than the process when soil resilience is estimated.

The need to define resilience introduces a new area of fundamental and adaptive research for the near future, in view of the growing human impact on the environment including soil. Till now we have not developed a full concept of soil resilience. At least, we do not know how to measure it or how to compare two soils in respect of this characteristic. Just qualitatively or conceptually it might be said that dryland soils appear more resilient than was previously thought.

Taking into account the above considerations, there is a need to distinguish very clearly between the resilience of land and that of soil. This distinction is particularly important for the drylands where all natural phenomena are accentuated by aridity. The resilience and relatively quick recuperation of dryland ecosystems after severe droughts is too often considered by ecologists in terms of a two-part subsystem of plants and animals completely forgetting the third component – soil. If the soil is not destroyed during the drought, as by wind erosion for example, the ecosystem might recover very quickly indeed, particularly in case of a decrease in human and animal pressure as a result of emigration or massive cattle loss (Olsson and Rapp, 1991). Otherwise the result may be quite different.

Constraints on Sustainable Land Use in Drylands

The constraints on sustainable land use in drylands originate from two sources. First, they are connected with the natural conditions of these areas as described above. Second, they are closely related to socioeconomic and political conditions, which are usually not very favourable in the developing countries in general and in the areas affected by desertification in particular.

The development of irrigated agriculture is limited by the availability of fresh water and by the use of backward technologies leading eventually to soil degradation. With 30% of the irrigated land already affected by degradation processes to some degree, any new irrigation project will bring about additional soil salinization because lands of lower and lower quality are being developed for irrigation. Where an irrigation project is constructed with modern sophisticated technology, including adequate water quality control and artificial drainage, the cost becomes so high that the investment is hardly profitable. Governments are providing certain subsidies for the development of irrigated agriculture, but that is not always paying under the existing trade climate.

As regards rain-fed agriculture and the pastoral use of extensive rangelands, they both have rather limited potential for large-scale production and both are very vulnerable to recurrent drought. At the same time for many developing countries affected by desertification, nomadic or seminomadic pastoralism or a combination of pastoralism with rain-fed agriculture are the only way of life and constitute the backbone of their economies.

It is the economic situation which constitutes the main constraint to managing soils for sustainable land use because any additional investment in the protection or rehabilitation of drylands should come from outside, whereas the investments have no immediate return and have long periods of gestation. Thus they are not attractive for the holders of capital. At present the major emphasis is on foreign aid, both bilateral and multilateral. However, governments, both local and national, do not always attach sufficient priority to land problems in their long lists of needs, and the major part of foreign aid goes to other areas.

Conclusions

The constraints on managing soils for sustainable land use in drylands are different in nature. Some are connected with the specificity of the natural conditions including soils, others with the prevailing socioeconomic conditions. Appropriate management of soils is the prerequisite of sustainable land use in the drylands, but this requires substantial investment which is not available from local sources and can only be achieved through international cooperation. Certain problems still require extensive research, e.g. soil resilience, particularly in view of the diversity of dryland soils in different ecological conditions. Another area of research in the drylands is the relationship between the processes of soil degradation and results in different ecological situations; to answer the question whether soil erosion is always harmful or whether it may be beneficial in certain cases depending on the nature of the soil parent material and soil formation. The above problems concerning the soils of drylands are recommended as priorities for fundamental and adaptive research by the appropriate scientific institutions.

References

Odingo R.S. (ed.) (1990) *Desertification Revisited.* UNEP, Nairobi.

Olsson K. and Rapp A. (1991) Dryland degradation in central Sudan and conservation for survival. *Ambio* 20, 192–195.

UNCED (1992) *AGENDA-21*. United Nations, Rio de Janeiro.

UNEP (1991) *Status of Desertification and Implementation of the United Nations Plan of Action to Combat Desertification.* UNEP, Nairobi.

UNEP (1992a) *World Atlas of Desertification.* Edward Arnold, London.

UNEP (1992b) *Saving Our Planet: Challenges and Hopes.* UNEP, Nairobi.

United Nations (1977) *Desertification: Its Causes and Consequences.* Pergamon Press, Oxford.

Part III

Avoiding and Combating Soil Degradation

Chapter 11
Determinants of Resilience in Soil Nutrient Dynamics

H. Tiessen, J.W.B. Stewart and D.W. Anderson

College of Agriculture, University of Saskatchewan, Saskatoon, Canada S7N 0WO

Introduction

Resilience is the ability of a system to return to dynamic equilibrium after disturbance. Resilience differs from stability in that stable systems may be able to withstand disturbance under a particular set of conditions, but are not able to recover once one or more of the conditions are changed (Tiessen and Anderson, 1991). With respect to nutrient cycles, resilience is imparted by a balance between nutrient mobilization for biological uptake and nutrient stabilization in the soil or other system components to avoid leaching or erosion loss. Natural ecosystems have developed mechanisms to cycle nutrients effectively between the biotic and soil compartments, or to maintain nutrients within the biotic compartment with minimal loss (Vitousek and Reiners, 1975). Adaptations to specific natural disturbance regimes such as seasonal water deficits, flooding or fire frequently form part of natural ecosystems. Such adaptations may be a useful point of departure for the understanding and design of managed ecosystems.

Ecosystems managed for production are frequently highly disturbed through harvest, tillage, fire, stocking, destocking etc. The sustainability of a production system depends on the strategies employed to maintain system resilience in the face of such disturbances. With respect to nutrient cycles strategies must be implemented that minimize nutrient losses (other than harvest), while at the same time maintaining adequate levels of nutrient availability for the desired production levels. Managed systems with net nutrient exports also require adequate rates of nutrient replacement.

This chapter discusses nutrient cycling strategies in different temperate and tropical ecosystems, and explores the relevance of these strategies to artificial disturbance and management regimes. Phosphorus (P) is the main example of the important interrelated biogeochemical elements (C,N,P,S), since P has a wider range of interaction and reaction with both inorganic and organic soil colloids, and therefore can illustrate mechanisms of cycling appropriate for several other nutrients (Stewart *et al.*, 1983). Several studies of organic matter composition show that C:N, C:S and C:P ratios in organic matter have to be interpreted with caution and with knowledge of prevailing environmental factors in soils both along environmental gradients and even within profiles (Stewart and Cole, 1989). Phosphorus is also the main limiting nutrient in many tropical soils.

The Phosphorus Cycle

Phosphorus is one of the main limiting elements for crop production. In many natural ecosystems P availability limits overall ecosystem productivity through its effect on plant production and nitrogen fixation (Walker and Syers, 1976). Correlations between available P and soil carbon are therefore frequently observed as an expression of the overall P limitation on ecosystem productivity and soil organic matter accretion.

Plants take up P from the soil solution which contains only a minute fraction of total soil P, but which is constantly replenished by the hydrolysis of labile inorganic P (P_i) or by mineralization of organic P (P_o). Labile P_i fractions, in turn, are in exchange with more stable P_i compounds through slow reactions. Therefore, the P supply to vegetation depends on the

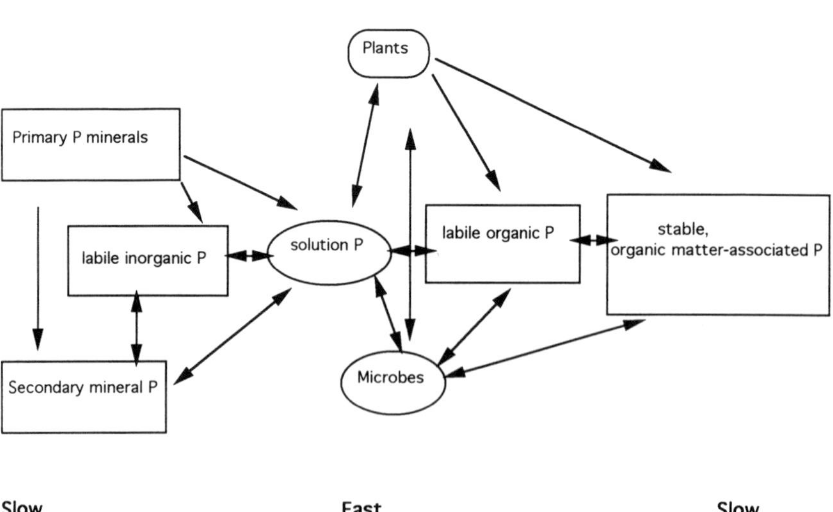

Fig. 11.1. A conceptual phosphorus cycle.

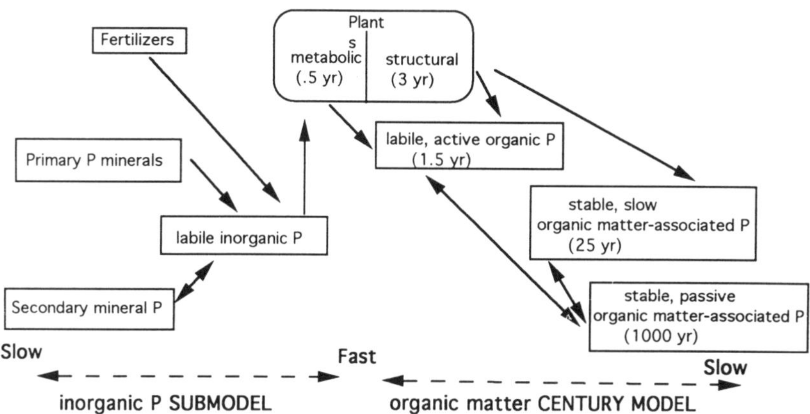

Fig. 11.2. Representation of the soil organic matter and associated phosphorus cycle based on the CENTURY model.

quantities of labile P_i, the rates of transformation between labile and more stable P_i, and the size and rates of transformation of the mineralizable P_o pool in the soil (Tiessen, 1991). Figure 11.1 presents a conceptual phosphorus cycle of a soil. It includes measurable inorganic and organic P pools, some of which are relatively inert and cycle slowly while others cycle much more rapidly (cf. reviews Cole *et al.*, 1977; McGill and Cole, 1981; Tate, 1984; Smeck, 1985; Stewart and Tiessen, 1987). Faster components, by implication, are more important for short-term supply to plants, but are also more susceptible to losses.

Simulation models have been developed on temperate soils that combine this conceptual P cycle with soil organic matter models (Fig. 11.2) and can simulate long-term C, N and P dynamics in soils across different climatic zones (Parton *et al.*, 1988, 1989). Applying such a model across limited environmental gradients has provided the opportunity to study the controls on soil organic matter (SOM) and nutrient cycling. At a microscale, variations in the P cycle may be caused by the influence of different plant litter types, different decomposing microorganisms, interactions of organic compounds and associated P in organo–mineral complexes which determine the persistence and build up of P_o, and changes in the rates and intensities of reaction of P_i with soil colloids (Coleman *et al.*, 1983). The net effects of this variation can be expressed at the ecosystem level in simplified relationships between model components and environmental controls during soil development and under different management, which may provide a framework to understanding nutrient dynamics and resilience in very different ecosystems.

Influence of soil formation and weathering

In most unweathered soils P occurs predominantly in the form of apatites, calcium phosphates containing carbonate, fluoride, sulphate, hydroxide, and a number of different cations. Less commonly, P occurs in primary minerals where it substitutes isomorphously for silicate in crystal structures. Weathering and leaching cause the dissolution of primary phosphates and result in the formation of secondary Fe and Al phosphates with low solubilities (Walker and Syers, 1976). The distinction between primary (acid-soluble) and secondary (alkali-soluble) phosphates is useful because the balance between primary Ca-associated and secondary Al+Fe P_i forms reflects the weathering stage of the soil (Tiessen *et al.*, 1984a). During weathering and leaching, total soil P contents are reduced, and at the same time, organic matter and associated P_o pools accumulate in the soil as a result of the establishment of vegetation and the recycling of plant litter (Walker and Syers, 1976).

Phosphorus 'fixation'

Phosphate is a highly charged anion which interacts tenaciously with Fe and Al compounds in soils. Phosphorus 'fixation' is, as any sorption process, reversible, and in many weathered soils, secondary Fe- and Al-phosphates are a potential source of P available to plants (Hughes, 1982; Hughes and LeMare, 1982), particularly if they are also associated with organic matter (LeMare, 1982). One of the problems in assessing the reversibility of sorption is that sorbed P undergoes further transformations with time (Mattingly, 1975; Parfitt *et al.*, 1989), possibly by recrystalization (Barrow, 1983), solid-state diffusion (Willett *et al.*, 1988), or exchange into P pools which are not in immediate exchange with the solution (Fardeau and Jappe, 1980; Salcedo *et al.*, 1991).

Isotopic dilution methods can be used to measure the size of the P pool which can act as a 'sink', but from which solution P can also be replenished at rates relevant to plant uptake in the field, i.e. at rates varying from days to years. Integrated studies of P sorption, Fe and Al activities, and P_i fractionations with the procedures of Kurmies (1972) or Hedley *et al.* (1982) have proved to be successful in elucidating P_i chemistry and transformations. These methods can follow the fate of P_i or fertilizer P with liming, incubations, exhaustive cropping or similar controlled manipulations. Correlative studies with other soil properties, along environmental gradients, with differences in cultivation or soil development have helped to develop the understanding of the P_i cycle (Fig. 11.1), and promise further successes in new ecosystems.

Organic phosphorus transformations

Studies of the availability, cycling and chemical composition of P_o suffer from greater methodological problems. P_o fractionations are included in Hedley *et al.*'s (1982) fractionation, but the biological functions of the chemically extractable fractions are not easily defined. Whereas some measurable P_o fractions changed predictably under manipulations such as incubations, fertilizations, manuring, exhaustive cropping, etc. an understanding of the role of P_o in the environmental P cycle must be based on an understanding of organic matter transformations (Tiessen, 1991). Separation of recent litter, floatable organic matter, clay-associated or mineral-stabilized organic matter (Feller *et al.*, 1983), and analysis of the associated P (Tiessen *et al.*, 1983) are methods that have shown some success. Short-term turnover of P_o is largely microbially mediated and microbial P measurements using one of the chloroform methods are useful in this context, although they may be inaccurate in highly P sorbing soils.

The turnover of organic matter constantly releases P into the solution from P_o mineralization. Through the mineralization of P_o, the pool size of 'total available' P is strongly time dependent, and 'plant available P' is a functional concept rather than a measurable quantity. Therefore available P must be defined relative to an ecosystem and its components, taking into consideration the P demand and the period over which P supply is required by a plant, plant community or crop. Over the period of P demand such as a cropping season, year, or growth cycle, considerable recycling can occur which needs to be taken into consideration. The element of time in the definition of plant available P has not received much attention in the past, since efforts were concentrated on annual crops under arable agriculture with well-defined uptake periods and little recycling. Complex management systems such as agroforestry, including alley cropping and shifting cultivation, and managed rotations including fallows, mulches and pastures, are gaining importance in 'sustainable' agriculture in which long-term nutrient cycles are managed. The nutrient availability in such systems is clearly governed by longer-term transformations beyond the 'immediately available' and easily measurable nutrient pool. Phosphorus may be limiting for the accumulation or subsequent mineralization of C and N, and this in turn may limit P availability through reduced P_o mineralization. Net mineralization phases with fertility depletion may be followed by the regeneration of fertility under fallow or rotational agriculture (Tiessen *et al.*, 1992). An understanding of resilience of the long-term nutrient dynamics must be based on the entire complex of P_i and P_o transformations including mineral interactions and the turnover of organic matter.

In ecosystems where biotic P cycling through closed recycling of plant litter is important, such as in nutrient-limited tropical rain forests, chemical P transformations in the soil are of limited relevance, and an apparently simpler analysis of critical ecosystem components such as leaf P, litter P, plant P, decomposition rates, root growth and root distribution and the role of soil fauna, bacteria and fungi must be evaluated. In the absence of a long-lived isotopic tracer for P, only the use of repeated elemental budgets in different components along seasons has provided insight into the P cycle in such ecosystems (Medina and Cuevas, 1989), and an understanding of P transformations invariably hinges on an understanding of the associated carbon cycles.

System linkages

The foregoing paragraphs show the importance of existing linkages in nutrient and SOM cycles of entire ecosystems. Although P was chosen as a specific example, SOM turnover, litter recycling, mineral interactions and environmental controls all need to be taken into account if a predictive capability of fertility is to be developed. Similar arguments hold for the cycles of S and N, and to some extent even for cations, particularly in weathered soils where mineral nutrient pools are small and SOM-associated CEC is important. This emphasizes the importance of SOM stabilization for the resilience of nutrient dynamics. Pools of SOM with different stabilities provide a spectrum of nutrient availabilities with different rates of release and response to different kinds of disturbances. Some processes involved in SOM stabilization in temperate regions (Duchaufour, 1976) are: (i) direct stabilization of chemically resistant materials such as lignins in the soil (this corresponds to the structural material in CENTURY; (ii) modification of organic matter by microbial decomposition and formation of SOM (probably corresponding to 'metabolic' material); but also (iii) direct stabilization of organic compounds by interaction with carbonates, sesquioxides or clays to form little-transformed 'humin' (suggesting a shunt to passive SOM). Other studies of SOM have yielded some clearly different fractions based on particle size fractions (Tiessen *et al.*, 1984b). Coarse material is recent litter that is still actively being transformed and readily available. Fine clay-sized material is largely of microbial origin and associated nutrients are highly available. Intermediate-size fractions contain highly stabilized SOM with radiocarbon ages around 1000 years, which contain nutrients of low (slow) availability. These concepts were confirmed in studies on N cycling by Ladd *et al.* (1977). Clearly, both the pools and the processes in the transformations of litter to SOM and the SOM cycle are much more complex than the simplified SOM model (Fig. 11.2) implies. Nevertheless, the model successfully predicted the levels of loss or gain of SOM under native or agricultural Gramineae on grassland and temperate forest soil with varying climate and management. It is a new challenge to expand this predictive ability to new ecosystems without getting side-tracked by complex mechanistic models, and at the same time develop such mechanistic models to advance the understanding of underlying processes.

Global change processes may interrupt successful mechanisms of nutrient cycling or shift agricultural production into regions where different climatic, social, economic and technological conditions prevail and where the evaluation and prediction of sustainability call for unprecedented interdisciplinary efforts. In view of the complexity of the processes involved, the assessment and prediction of resilience under complex and changing environmental conditions and management will require a logical framework for integrating process information and subsequent monitoring. Sensitivity analyses for process models that concentrate on critical soil and ecosystem properties, such as the CENTURY model for SOM and its extensions for nutrient cycles, will test the agreement and reliability for a wide range of temperate agricultural regions. In more weathered soils where part of the soil organic matter is strongly bound to sesquioxides, organic matter and associated nutrient cycling activities are restricted, and pool partitioning and turnover rates must be revised from those developed on temperate grasslands. Comparisons of different ecosystems will facilitate this process.

Ecosystem Examples

Temperate grasslands

In subhumid temperate grasslands, high soil organic matter (SOM) and underground biomass levels can build up because of limited leaching and slower decomposition. The nutrient reservoir associated with high SOM levels imparts considerable resistance to fertility loss resulting from natural or agricultural disturbance. Disturbance by drought, fire or grazing is an integral part the functioning of this ecosystem, and recovery is usually fast and complete.

North American grassland soils have been found to remain productive for nearly a century of continuous agricultural use. The nutrient stores associated with high SOM levels have imparted resilience to an agricultural system, which, when used with limited inputs has mined soil nutrients. Agricultural production has to a large extent utilized the mineralization of organic N from degrading SOM pools, and summer fallowing to speed up SOM mineralization has been successful in supplying N for crop production over several decades (Janzen, 1987). In addition, P_o mineralization can account for much of the crop P demand in such soils (Tiessen *et al.*, 1983). This has led to the recognition of the importance of SOM to nutrient cycling, and prompted extensive studies of SOM pools and their transformations.

Soil organic matter is not uniform but has components that turn over at different rates. A portion with rapid turnover affects nutrient cycling and availability in the short term, and is highly responsive to management. Portions with slower turnover rates are important for N, S and P supply in the long term, contain some of the soil's cation exchange capacity and affect soil physical properties such as aggregation, bulk density and water infiltration. Large amounts of organically bound nutrients therefore remain relatively unavailable in the short term. Soil management can supplement short-term nutrient availability through speeding up natural processes by fallowing and tillage, or through fertilization. Alternatively low acquisition/export rates such as those associated with extensive grazing operations may be supported by natural processes of nutrient supply.

The relative importance of organic matter for nutrient cycling depends on the stability of its components, but also on other soil properties. Subhumid temperate soils formed on fertile glacial deposits are relatively young, unleached and rich in primary minerals. Although SOM levels are important for nutrient supply to plants and crops, they are not essential for ecosystem survival.

After severe erosion during a dry period in the 1930s, prairie soils were found to return to a productive cycle after either abandonment to native grasses or being seeded down to cultivated grasses. Available nutrient reserves in these soils were replaced by nutrients from weatherable minerals. Glacial till material from the overburden of strip mines showed accretion rates of 28 g C m^{-2} $year^{-1}$ and 2.4 g N m^{-2} $year^{-1}$ indicating that typical SOM levels for the region might be reached after 250–350 years (Anderson, 1977). The soil parent materials thus impart resilience. The recovery from such drastic disturbance is clearly too slow to permit an agricultural production system to continue. One of the critical questions is whether rates of nutrient release from a disturbed system are adequate for the desired biological production. For instance Tiessen *et al.* (1983) showed that the total P exported by crops from a prairie soil was less than that released from mineralizing SOM over 60 years of cultivation. But at the later stages of the mineralization process only the more-resistant SOM fractions remained and release rates were too slow to fulfil crop demands. During the period of mineralization surplus P_i was precipitated as apatite-like compounds whose low solubility apparently does not satisfy crop acquisition rates. As a consequence fertilizer use in the area has greatly increased over the past 20 years. An assessment of productivity would have to determine the balance of rates of P_o turnover and solubilities of P_i with crop demands, which has only been done through an empirical estimation of fertilizer responses. *A priori* predictions of fertility based on the underlying biological and chemical processes, which would be required for the prediction of sustainability or resilience of nutrient supplies, have not been attempted.

On parent materials where production is sustained by biological adaptation, resilience is naturally limited. A moderately productive grassland on saline soils for example may be stable if not disturbed, but takes a very long time to recover from a disturbance such as cultivation, and cannot be considered to be resilient. Sandy soils stabilized by continuous vegetative cover may be regarded similarly. Such ecosystems are stable unless disturbed by fire, overgrazing or cultivation, but once a disturbance has occurred, recovery is difficult.

Temperate forest

Temperate forests have formed on soils in more humid environments than the grasslands. Greater leaching has mostly produced lower equilibrium SOM levels, lower base saturation and smaller nutrient stores. In intact forests, greater amounts of nutrients are stored in aboveground biomass than in the readily available pools in the soil, and a disturbance such as fire is a major disruption from which the ecosystem recovers only slowly or enters into a transition to grassland. The lower levels of interactions of leached forest soils with inorganic nutrients, and lower organic nutrient pools, mean that greater proportions of nutrients present are in available and leachable form than in natural grassland soils. Frossard *et al.* (1989) showed losses of 40% of total P from the solum of 10,000 year old subhumid forest soils. Such data indicate that P mobility may be greater than frequently assumed, and that long-term sustainability requires careful attention to the leaching potential of critical elements. When such soils are used for agriculture, nutrient and SOM losses are critical and must be minimized. Nutrients must be supplied at replacement rates for crop exports, and replacement must be carefully timed and dosed to avoid leaching. Traditional rotational agriculture systems have aided biological replacement of N, and maintained organic matter levels often above those of the original forest soil. Such SOM management increases the stability of nutrients against loss. For instance, the introduction of graminaceous grain crops mimic a transition to grassland upon disturbance with subsequent greater SOM retention than could be achieved with root crops or trees. The acidity of leached forest soils has been successfully controlled by liming, and this in combination with crop rotations has produced highly productive and sustained agriculture in many regions.

In some forest ecosystems in which decomposition is retarded nutrients accumulate in organic form. Reasons may be biological (cold temperature montane forests, or waterlogging) or chemical (SOM stabilization in minerals of Andisols, Rankers). In extreme cases nutrient tie-up is so extensive that forest is replaced by sedges or moss. Some such soils have been used for agricultural purposes, e.g. upon drainage, but nutrient imbalances predominate since SOM has not been biologically processed to fit uptake requirements. Such soils are used far from their equilibrium condition, offer practically no resilience, and production must be closely managed.

Semiarid tropical grasslands

The most serious limitation to production in semiarid tropical grasslands and savannahs is usually water. Many water-limited systems are hard to irrigate because their mineralogy is in balance with a dry regime; swelling clays, clay pans etc. will cause physical problems, and salinization frequently occurs due to the high evaporation rates under tropical conditions. Many tropical grasslands also occur on old land surfaces with highly weathered soils. Frequently the parent materials are preweathered from previous geological ages with wetter climates, and present conditions of low leaching intensities are no indication for present mineralogy. Low nutrient contents and high sesquioxide activities limit inorganic nutrient cycling, and low productivity, high decomposition rates and frequent fires limit SOM accumulation. The high sesquioxide activities limit P availability both by reducing P_i solubility, and by stabilizing P_o. In extreme cases much P in the soil can be fixed in lateritic nodules, which also act as efficient sinks for added fertilizer P (Tiessen *et al.*, 1990). Predicting P fertility therefore requires not only measurements of available P_i fractions but also of potential P_i sorption by the soil, and of the role of P_o. Soil fertility depends on a balance of both organic matter and inorganic nutrient cycles. This balance is easily destroyed by too frequent or too hot (late season) fires which eliminate the timed release of P from recent organic residues and expose P and other nutrients to fixation, sending soil productivity into a downward spiral with declining SOM levels and increasing erosion rates. The management of such fragile lands through rotations, mulches, selective burning etc. is difficult, and few adequate tests are available for nutrient availability and long-term management planning. It is important to recognize the strong P limitation on N fixation in such ecosystems. Given sufficient inputs in lime and fertilizers, semiarid savannahs can be successfully managed by the reduction of P fixation and the input of available nutrients, as the experience of the Brazilian Cerrados shows (Goedert, 1983). Without the inputs, the soils lack resilience.

Humid tropical forest

Except for those developed on highly basic parent materials, soils (e.g. Alfisols in West Africa) under humid tropical upland forest (precipitation zones above 1500–2000 mm $year^{-1}$) are mostly poor to very poor depending on the capacity of the soils for nutrient and organic matter retention. Soils with moderate retention capacity, i.e. moderate clay content and some proportion of high activity clays, can be managed through a combination of organic and mineral nutrition. Soils at Yurimaguas (upper Amazon basin, Peru) on relatively heavy sediments have been successfully maintained fertile through liming, fertilization and the judicious maintenance of organic matter levels and soil cover (Sanchez *et al.*, 1982). No measure of resilience or sustainability was found in these experiments, but an empirical optimization of the production system was followed with interventions that counteracted fertility decline, acidification, and erosion as these problems became apparent. The management system has also found practically no acceptance among farmers, pointing to socioeconomic constraints against its acceptance (Fearnside, 1987).

Oxisols are common under tropical forests and savannahs. These soils show extensive and durable microaggregation by which a portion of SOM is effectively locked up through interactions with clays and sesquioxides. Total SOM levels may therefore be much greater than active SOM levels, and this distinction is important in the evaluation of critical changes in SOM levels (Tiessen *et al.*, 1992).

Soils of low retention capacity, such as the sandier soils of the northern Amazon, many of which were formed from preweathered Guyana plateau sediments, cannot be successfully managed after complete forest clearing. Traditional slash–burn agriculture operates in small clearings for short cultivation periods so that the biological linkages for nutrient recycling of the intact forest are not broken. The abundant litter layer of the intact forest contains the majority of the active roots and symbiotic fungi, which recycle nutrients from litter decomposition directly to the vegetation. After slash–burn, fertility declines in parallel with the disappearance of the litter layer (Jordan, 1989). Regeneration of the forest system after abandonment depends on such factors as the size of the clearing, pointing to the importance of complex biotic linkages in the system's resilience. Soil analyses of inorganic nutrient availability, or even of the SOM would be unable to predict any ecosystem fertility, since fertility is entirely determined by biotic processes and pools.

Conclusions

Attempts to assess soil organic matter dynamics under complex and changing environmental conditions and management have been aided by the use of logical frameworks for integrating information, by predictive modelling and by comparing SOM dynamics across different ecosystems. Soil organic matter and associated nutrient dynamics are an important component of soil resilience. Many of the initial studies were based on temperate grassland and forest ecosystems, where SOM management has been traditionally practised through rotations and ley farming. On grassland soils formed from nutrient-rich parent materials soil organic matter is important to many desirable soil properties but nutrient cycling associated with organic matter is not essential for ecosystem survival. Modern high-input management systems have been able to maintain productivity in the face of declining SOM cycling and organic nutrient supplies. A sensitive analysis of resilience with limited input agriculture needs to develop the ability to balance supply and production demands at the critical junction between nutrient supply controlled by mineralization rates and that controlled by solubility/desorption processes. This transition point has been reached in many modern agricultural systems, but needs to be reevaluated for the study of lower input, 'organic', and long-term sustainable farming systems. The challenge to the development of modelling approaches to this assessment is the determination of the level of abstraction permissible in a model that can predict SOM and nutrient dynamics across systems and landscapes but may have to incorporate specific cycling events in order to predict nutrient supply rates at the juncture of biologically versus chemically limited supply.

The evaluation and management of soil resilience and nutrient cycling in tropical systems faces even greater challenges in the incorporation of mechanistic approaches to overall system understanding. Mechanistic events such as fixation play an important role in weathered soils which require mechanistic shunts in nutrient cycling models (such as adaptation of the CENTURY model to depict nutrient dynamics and thus resilience). In addition, the essential importance of maintaining biological linkages in some highly adapted tropical ecosystems must be recognized. This has not found its way into current efforts of the integration and modelling of agro-ecosystem understanding and remains a major challenge to effective prediction.

References

Anderson, D.W. (1977) Early stages of soil formation on glacial till mine spoils in semi-arid climate. *Geoderma* 19, 11–19.

Barrow, N.J. (1983) On the reversibility of phosphate sorption by soils. *Journal of Soil Science* 34, 751–758.

Cole, C.V., Innis, G.S. and Stewart, J.W.B. (1977) Simulation of phosphorus cycling in semi-arid grasslands. *Ecology* 58, 1–15.

Coleman, D.C., Reid, C.P.P. and Cole, C.V. (1983) Biological strategies of nutrient cycling in soil systems. In: MacFadyen, A. and Ford, E.D. (eds) *Advances in Ecological Research*. Academic Press, New York, Chapter 13, pp. 1–55.

Duchaufour, P. (1976) Dynamics of organic matter in soils of temperate regions: its action on pedogenesis. *Geoderma* 15, 31–40.

Fardeau, J.C. and Jappe, J. (1980) Choix de la fertilisation phosphorique des sols tropicaux: emploi du phosphore 32. *Agronomie Tropicale* 35, 225–231.

Fearnside, P.M. (1987) Rethinking continuous cultivation in Amazonia. *BioScience* 37(3), 209–214.

Feller, C., Bernhardt-Reversat, F., Garcia, J.L., Pantier, J.J., Roussos, S. and Van-Vliet-Lanoe, B. (1983) Étude de la matière organique de différentes fractions granulométriques d'un sol sableux tropical. Effet d'un amendement organique (compost). *Cahiers ORSTOM ser. Pedol.* 20(3), 223–238.

Frossard, E., Stewart, J.W.B. and St Arnaud, R.J. (1989) Distribution and mobility of phosphorus in grassland and forest soils of Saskatchewan. *Canadian Journal of Soil Science* 69, 401–416.

Goedert, W.J. (1983) Management of the Cerrado soils of Brazil. A review. *Journal of Soil Science* 34, 405–428.

Hedley, M.J., Stewart, J.W.B. and Chauhan, B.S. (1982) Changes in inorganic and organic soil phosphorus fractions induced by cultivation practices and by laboratory incubations. *Soil Science Society of America Journal* 46, 970–976.

Hughes, J.C. (1982) High gradient magnetic separation of some soil clays from Nigeria, Brazil and Colombia. I. The interrelationships of iron and aluminium extracted by acid ammonium oxalate and carbon. *Journal of Soil Science* 33, 509–519.

Hughes, J.C. and Le Mare, P.H. (1982) High gradient magnetic separation of some soil clays from Nigeria, Brazil and Colombia. II. Phosphate adsorption characteristics, the interrelationships with iron, aluminium and carbon, and comparison with whole soil data. *Journal of Soil Science* 33, 521–533.

Janzen, H.H. (1987) Soil organic matter characteristics after long-term cropping to various spring wheat rotations. *Canadian Journal of Soil Science* 67, 845–856.

Jordan, C.F. (1989) An Amazonian rain forest: the structure and function of a nutrient stressed ecosystem and the impact of slash and burn agriculture. *Man and the Biosphere Series*: vol. 2. UNESCO, Paris, 176 pp.

Kurmies, B. (1972) Zur fraktionierung der Bodenphosphate. *Die Phosphorsäure* 29, 118–151.

Ladd, J.N., Parsons, J.W. and Amato, M. (1977) Studies of N immobilization and mineralization in calcareaous soils II Mineralization of immobilized nitrogen from soil fractions of different particle size and density. *Soil Biology and Biochemistry* 9, 319–325.

Le Mare, P.H. (1982) Sorption of isotopically exchangeable and non-exchangeable phosphate by some soils of Colombia and Brazil, and comparisons with soils of southern Nigeria. *Journal of Soil Science* 33, 691–707.

Mattingly, G.E.G. (1975) Labile phosphate in soils. *Soil Science* 119, 369–375.

McGill, W.B. and Cole, C.V. (1981) Comparative aspects of C, N, S and P cycling through soil organic matter during pedogenesis. *Geoderma* 26, 267–286.

Medina, E. and Cuevas, E. (1989) Patterns of nutrient accumulation and release in Amazonian forests of the upper Rio Negro Basin. In: Proctor, J. (ed.) *Mineral Nutrients in Tropical Forest and Savanna Scosystems.* Blackwell Scientific Publications, Oxford, UK.

Parfitt, R.L., Hume, L.J. and Sparling, G.P. (1989) Loss of availability of phosphate in New Zealand soils. *Journal of Soil Science* 40, 371–382.

Parton, W.J., Stewart, J.W.B. and Cole, C.V. (1988) Dynamics of C,N,P and S in grassland soils: a model. *Biogeochemistry* 5, 109–131.

Parton, W.J., Sanford, R.L., Sanchez, P.A. and Stewart, J.W.B. (1989) Modelling soil organic matter dynamics in tropical soils. In: Coleman, D.C. *et al.* (eds) *Dynamics of Soil Organic Matter in Tropical Ecosystems.* NIFTAL, University of Hawaii, pp. 153–171.

Salcedo, I.H., Bertino, F. and Sampaio, E.V.S.B. (1991) Reactivity of fertilizer P in tropical soils by isotopic exchange kinetics of ^{32}P. *Soil Science Society of America Journal* 55, 140–145.

Sanchez, P.A., Bandy, D.E., Villachica, J.H. and Nicholaides, J.J. (1982) Amazon basin soils: management for continuous crop production. *Science* 216, 821–827.

Smeck, N.E. (1985) Phosphorus dynamics in soils and landscapes. *Geoderma* 36, 185–199.

Stewart, J.W.B. and Cole, C.V. (1989) Influences of elemental interactions and pedogenic processes in organic matter dynamics. *Plant and Soil* 115, 199–209.

Stewart, J.W.B. and Tiessen, H. (1987) Dynamics of soil organic phosphorus. *Biogeochemistry* 4, 41–60.

Stewart, J.W.B., Cole, C.V. and Maynard, D.G. (1983) Interactions of biogeochemcial cycles in grassland ecosystems. In: Bolin, B. and Cook, R.B. (eds), *The Major Biogeochemical Cycles and Their Interactions.* Scientific Committee on Problems of the Environment, Paris, pp. 247–268.

Tate, K.R. (1984) The biological transformation of P in soil. *Plant and Soil* 76, 245–256.

Tiessen, H. (1991) Characterization of soil phosphorus and its availability in different ecosystems. In: *Trends in Soil Science.* Council for Scientific Research Integration, Trivandrum, India, pp. 83–99.

Tiessen, H. and Anderson, D.A. (1991) Concepts and methodologies for measuring the sustainability of managed ecosystems. In: *Evaluation for Sustainable Land Management in the Developing World.* Vol 2: Technical papers. International Board for Soil Research and Management, Bangkok, Thailand. IBSRAM Proceedings no. 12(2), 525–544.

Tiessen, H., Stewart, J.W.B. and Moir, J.O. (1983) Changes in organic and inorganic phosphorus composition of two grassland soils and their particle size fractions during 60–70 years of cultivation. *Journal of Soil Science* 34, 815–823.

Tiessen, H., Stewart, J.W.B. and Cole, C.V. (1984a) Pathways of phosphorus transformations in soils of differing pedogenesis. *Soil Science Society of America Journal* 48, 853–858.

Tiessen, H., Stewart, J.W.B. and Hunt, H.W. (1984b) Concepts of soil organic matter transformations in relation to organo-mineral particle size fractions. *Plant and Soil* 70, 287–295.

Tiessen, H., Frossard, E., Mermut, A.R. and Nyamekye, A.L. (1990) Phosphorus sorption, and properties of ferruginous nodules from semi-arid soils from Ghana and Brazil. *Geoderma* 48, 373–390.

Tiessen, H., Salcedo, I.H. and Sampaio, E.V.S.B. (1992) Nutrient and soil organic matter dynamics under shifting cultivation in semi-arid Northeastern Brazil. *Agriculture Ecosystems and Environment* 39, 139–151.

Vitousek, P.M. and Reiners, W.A. (1975) Ecosystem succession and nutrient retention: a hypothesis. *BioScience* 25, 376–381.

Walker, T.W. and Syers, J.K. (1976) The fate of phosphorus during pedogenesis. *Geoderma* 15, 1–19.

Willett, I.R., Chartres, C.J. and Nguyen, T.T. (1988) Migration of phosphate into aggregated particles of ferrihydrite. *Journal of Soil Science* 39, 275–282.

Chapter 12
Maintaining Nutrient Status of Soils: Macronutrients

P. Stangel[1], C. Pieri[2] and U. Mokwunye[3]

[1]*President, Stangel and Associates Agricultural Consultants, P.O. Box 830, Florence AL 35630, USA;* [2]*Natural Resources Ecologist, The World Bank;* [3]*Director, Africa Division, International Fertilizer Development Center (IFDC)*

Introduction

The purpose of this chapter is to address the strategies that farmers and increasingly communities and national planners use to maintain and/or build the nutrient status of soils. These strategies are influenced by a range of external pressures and are likely to vary significantly under the influence of these factors in a particular agroclimatic zone.

External Factors Affecting Maintenance of Nutrient Status

There are at least seven external factors that affect the strategies farmers use to maintain the nutrient supply in a given soil. These are: (i) population pressures; (ii) migration of people; (iii) level of income; (iv) degree of government intervention; (v) the desire of society to preserve the nation's soil resource base; (vi) growing public concern about the impact of agriculture on the environment; and (vii) the agroecological environment in which the soils exist. The relation of population growth to grain yield and use of fertilizers is illustrated in Fig. 12.1.

Population pressures

The world's population, currently (1992) 5.4 billion, is expected to reach 6.1 billion by the year 2000 and will nearly double (8.5 billion) by the year 2025 (PRB, 1992). Overall growth is expected to stabilize near the end of the 21st century at which time the planet will be inhabited by about 10.5 billion

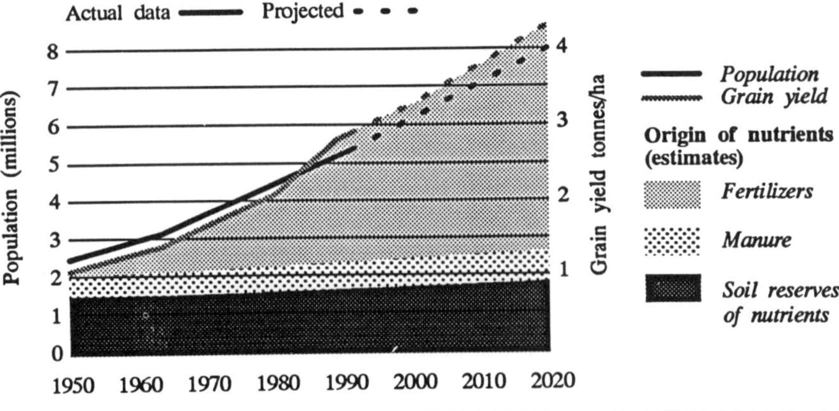

Source: Stapel (1982) with latest data added.

Fig. 12.1. Global trends in population growth, grain yield, and origin of plant nutrients.

people. Although current growth rates and absolute numbers projected for the end of the century and early in the 21st century are much lower than the forecasts made in the early 1980s, the disproportionate population distribution relative to available land and water resources is still a major cause for concern (Table 12.1).

Growth centres

Approximately 77% of the world's population now lives in developing countries. This population, growing at an annual rate of 2.0% in 1992, will double by 2026 (Table 12.1). In contrast, population growth over this period will average less than 0.5% per year for many developed countries, particularly in Europe where annual population growth rate is now 0.2% and rapidly approaching zero; a few countries (Germany, Hungary, and Bulgaria) will actually have a net loss in population. This means over 95% of the world population growth between 1992 and 2025 and 84% of the world's total population by 2025 will occur and reside on land located 23.5° latitude north and south of the equator. Within this zone lies only 38% of the world's land mass.

Africa and South America account for the greater part of the tropical zone, with 43% and 28%, respectively. Asia has 20%, Central America 4%, and Australia 5% of the remaining tropical area. This means that the vast majority of the population pressures will be exerted on the soils located in the tropics.

This population distribution also has major implications for the food

Table 12.1. World population estimates for 1992 and 2025 and other indicators.

Region/country	Population mid-1992 (10^6)	Growth rate (%)	Population mid-2025 (10^6)	Doubling time	% Urban	Per capita income (US $)	Fertilizer use ha^{-1} (kg of nutrients)
Africa	654	3.0	1540	23	30	630	19.5
North Africa	147	2.6	274	27	43	1070	60.4
Western Africa	182	3.0	449	23	23	410	8.7
Eastern Africa	206	3.2	528	22	19	230	8.7
Middle Africa	72	3.0	182	23	38	460	8.7
Southern Africa	47	2.7	106	26	52	2390	59.2
Asia	3207	1.8	4998	39	31	1680	124.9
Excluding China	2042	2.0	3407	34	34	2520	
Western Asia	139	2.8	313	24	62	–	74.1
Southern Asia	1231	2.2	2115	31	26	440	70.9
Southeast Asia	451	1.9	696	36	29	–	–
East Asia	1386	1.2	1839	57	34	2910	211.6
Latin America	453	2.1	729	34	70	2170	42.9
Central America	118	2.5	204	28	64	2170	72.9
Caribbean	35	1.8	49	38	58	–	–
South America	300	1.9	476	36	74	2180	34.9
Europe	510	0.2	516	338	75	12,990	186.2
Northern Europe	93	0.3	97	242	83	17,930	–
Western Europe	178	0.2	174	179	82	–	–
Eastern Europe	96	0.2	103	369	63	2080	–
Southern Europe	144	0.2	141	344	68	12,860	–
Former USSR	284	0.7	362	104	66	–	93.1
North America	283	0.8	363	89	75	21,580	86.9
Oceania	28	1.2	39	57	71	13,190	32.3
Developed countries	1224	0.5	1392	148	73	17,900	107.5
Developing countries	4196	2.0	7153	34	34	810	81.4
Developing countries (excluding China)	3031	2.3	5562	30	37	1000	–
World	5420	1.7	8545	41	43	3790	93.3

Source: PRB (1992), IFDC (1992).

surplus countries of the temperate zone. Using the assumption that most developing countries will not be able to finance their additional food requirements through imports but instead will meet these from increased domestic production, and considering the current cost to maintain food surpluses, many developed countries will have to modify downward their current rates of crop production. This, coupled with the growing concern over the impact agriculture has on the environment, will almost certainly result in a different strategy for maintaining the nutrient status of soils in temperate zones from that which farmers will follow in the tropics.

Asia

Asia accounted for approximately 60% of the world's population in 1992 (Table 12.1). This proportion is not likely to change by the end of the century. Middle East and South Asia (particularly Bangladesh, India, Iran, Pakistan, and Sri Lanka) will account for nearly 50% of the population increase between 1992 and 2000. This region already accounts for 22% of the world's population and with a 1992 population growth rate of 2.2% will add 200 million additional people by 2000 and will double its present number by 2023.

East Asia, including China is another pressure centre. This subregion, which accounted for 26% of the world's population in 1992, has seen its population growth slow down in recent years and is currently nearing the world's average of 1.7% per year. The main problems will occur in east and southeast China, where population densities are already among the highest in the world and opportunities for increasing the area of arable land are minimal.

Pressures also exist in other parts of Asia, but these are not as acute as in those areas just described. Indonesia may be one exception; 75% of its estimated 185 million people are centred on the island of Java, which already has a population density of nearly 800 people km^{-2}, one of the highest in the world. Indonesia is engaged in a massive transmigration programme to move people to the Outer Islands in an effort to prevent further increases in population densities; however, there is some doubt that such moves will alleviate the present problems on Java (Arndt, 1983).

Africa

African countries accounted for 12% of the world's population in 1992. This portion is expected to climb to 14% by 2000. Most African countries have population growth rates ranging from 2.6% to 4.1%. The subregions of North, West, and East Africa as a group average increases of 3.0% per year (Table 12.1). If this rate of growth continues, their populations will double by 2015. The rates of growth for some of the larger countries in this region are even higher. Ghana, Zimbabwe, and Tanzania have annual population growth rates of about 3.3%. Kenya, with an annual growth rate of 4.1%, will double its population by 2010. Thus, population pressures will increase sharply by 2000, particularly in West and East Africa because of the large population base and growth rate of the countries involved.

Effects on land use

Although substantial quantities of new land will undoubtedly be brought into production by the turn of the century, an increasing number of scientists believe that the estimates by FAO (1981) are overly optimistic. For example, Arndt (1983), in studying the potential of new land development as a factor in determining the success of transmigration schemes in Indonesia, notes that most of the good agricultural land has already been developed. New land, if it is to be brought into production, will require inordinately large amounts of capital and technical know-how, both of which are currently very limited. He concludes that as a result of increasingly high costs of development, countries like Indonesia and China will look primarily to existing lands for producing the needed food. A similar conclusion was drawn for developing countries in general in *The Global 2000 Report to the President* (CEQ, 1980).

FAO (1981, 1992) has repeatedly estimated that the major portion of the additional production required to feed the population of the developing countries must come from existing land. They estimated that 60% of that increase will come from higher yields and 14% from increased cropping intensity.

In summary, population pressures will clearly have a major impact on strategies that farmers adopt to maintain the nutrient status of their soils. Pressure will be greatest in the developing countries, particularly those in South and Southwest Asia, Central America, and East Africa. Other likely pressure points are Java (Indonesia), Eastern China, Northeast Brazil, and some countries in West Africa. In contrast, population growth over the next two to three decades will place little pressure on agriculture in most developed countries and, in particular, those already having surplus production of agricultural products. Furthermore, this situation may become even less critical as the agriculture in Eastern Europe and the Newly Independent States (NIS) of the former Soviet Union (FSU) becomes more productive if and when these republics complete their shifts from centrally planned to market-driven economies.

Migration of people

For various reasons, people the world over migrate from one region to another and from rural to urban areas. This has a major impact on the need to transport the food and fibre necessary to feed and clothe these people. It is equally important to recognize that with this food and fibre goes the transport of substantial quantities of plant nutrients. For example, an average person living in Asia may consume annually upwards of 300 kg of rice and smaller quantities of vegetables and fish. Collectively these contain about 2–6 kg of nutrient per person per year. This represents a net flow of roughly 6–19 million tonnes of nutrients annually from the field to the home, and if the home is off the farm, it represents a net flow away from the farm to the village and increasingly to the cities.

Roughly 75% of the populations of developed nations live in cities. In many of these countries 5% or less of the total population is classified as rural. In contrast, developing countries have between 30% and 80% of their population living in rural areas (Table 12.1). However, depending upon the country, urban populations are growing at rates two to three times that for the nation as a whole. In certain countries like Brazil, Argentina, Venezuela, and Mexico, the flow of people from rural areas to the cities is so great that net rural populations are actually declining even though the region may show a birth rate surpassing deaths by 2%–3% per year. This increasing trend in migration, which results in the transporting of essential plant nutrients from rural to urban areas, must be taken into account in striving to: (i) maintain nutrient balance in soils; and (ii) develop a clean environment including soil, groundwater, and neighbouring coastal waters. Research to develop the strategy to deal with such situations is badly needed.

Desire for improved income and a more diversified agriculture

As individual incomes rise, there is a growing tendency first to increase consumption of basic food crops such as rice, maize, and wheat and then shift away from these cereals to a more diversified diet of fruits, vegetables, fish, poultry, and red meats. This creates a demand for a broader range of crops and pressures farm groups either to diversify their agricultural activities and/or specialize in specific areas in which they have a comparative advantage in production. It leads to the development of large-scale poultry, swine, and cattle industries which produce large amounts of nutrient-laden manures that are not easy to handle. This loop in the nutrient cycle of these industries is weak and often leads to large nutrient losses.

Nutrient levels maintained in soils must reflect these options and along with other inputs and infrastructure establish an environment that will allow the potential of the enterprise to be realized but at the same time take into account the magnitude of losses, where these occur, and their net impact on the environment (either at the rural and/or urban level).

The role of government

Governments can and do play a very important role in determining how plant nutrients are used at the farm level and transported to off-farm locations. This includes such economic factors as price control, subsidies, credit, and non-price factors; policies such as building of infrastructure (farm-to-market roads, ports, rail and road terminals, and irrigation and sanitation systems); and establishment of efficient research and development facilities as well as technology transfer systems to get results to the user. By doing this, the farmer and/or a village is provided with an increased number of options to realize the agronomic potential but within specific economic and socioeconomic constraints. The specifics of these points were confirmed by Desai and Stone (1987) and Stone and Desai (1989) in their review of the reasons for the success of China and India in increasing both nutrient use and crop production and by Reardon and Islam (1990) in their review of the issues facing agriculture in West Africa. The policy lessons learned and applicable for other developing countries are as follows:

- There is usually a considerable unexploited potential for increased nutrient use (including fertilizer) in the existing agronomic environment of agricultural systems in most developing countries. The speedy removal of the most binding constraints (such as inadequate infrastructure and institutional capacity to supply inputs and market outputs) is the quickest means to convert this agronomic potential into real production through increased use of nutrients. (Price incentives are no substitute for this process.)
- Once the key constraints are removed, governments tend to underestimate and/or overcontrol the flow of nutrient supplies to the farm. Setting policy that will encourage the fertilizer sector to maintain adequate stocks of nutrients in strategic locations to allow the response to market demand is critical to exploitation of the agronomic potential. A flexible and sensitive system to move nutrients quickly from regional stocking areas to localities, in response to market forces is required.
- In choosing among policies or initiatives to resolve existing constraints, the habit of employing particular short-term expedients must be examined to understand their accumulated long-term implications for maintaining the health and efficiency of the dynamic processes.
- The paths and rates of growth in nutrient use on farms with no previous history of use of external sources of fertilizer differ greatly from those on farms that have already experienced the benefits of fertilizer but current use may be low or nil due to external constraints.
- Ethnic makeup and associated customs of users of fertilizer have a major impact on the speed with which farmers accept new changes in agricultural practices.

Governments can be a significant factor in determining the capabilities farmers will have and use in maintaining the nutrient status of soils needed to meet specific production goals.

Pressure to preserve the land resource base

There is a growing awareness by the general public and increasingly by national policy makers that land is a finite resource and must be preserved. Legislation is now in place in a number of developed countries that clearly regulates soil and land use. The creation of such institutions as the International Board for Soil Research Management (IBSRAM) and the International Fertilizer Development Center (IFDC), and the abilities of these institutions to establish linkages with other International Agricultural Research Centers (IARCs) and national institutions, are indicators of the importance the global community and national leaders of an increasing number of governments are placing on the importance of preserving the world's soil resource base. Further, the close linkage between these two institutions and other interested agencies is evidence of the strong relationship that exists between crop production, nutrient management, and effective soil and land use management.

As will be shown in a later section, much of the agriculture practised in the developing world takes the form of effectively mining nutrients from the soil. It is clear that to move from 'soil mining' agriculture to 'soil building' agriculture or to use Freudenberger's (1988) term to move to 'regenerative' agriculture, nutrient inputs will be needed that are far beyond the financial means of the average farmer, who is now living close to subsistence. But what choices or alternatives are open to the farmer? Subsistence farmers do not deliberately set out to destroy the land resource base. Circumstances force them to take options that result in land degradation. In the long term, it may be considerably more economical for individual nations to establish policies and put into effect programmes that limit agriculture as much as possible to the existing (cleared) land, and build soil productivity in these areas rather than continue to eke out a meagre living on marginal soils, whether these are savannahs, steep lands, or areas now covered by forests.

More and more governments recognize the importance of preserving the land base and accept the fact that to do so requires building the necessary infrastructure and establishing key policies that deal with the relevant technical, economic, and social issues. These moves will greatly impact strategies to maintain the nutrient status of soils.

Increased concern over the environment

The recent global conference on the environment and development (UNCED, 1992) highlighted the importance that the world's leaders placed on improving the quality of the environment surrounding the planet. Strategies were outlined and agreements reached that were designed to curb emissions of greenhouse gases, slow the decline in biodiversity, and reduce the pollution of rivers and marine waters. Effective nutrient management is a key element in achieving the goals outlined in these agreements. In an increasing number of developed countries, governments have passed legislation that sets limits on the amounts of nitrate nitrogen permissible in drinking water and on the levels of nitrogen and phosphates that are acceptable in lakes and streams. In addition, strict attention is directed toward controlling the use of phosphorus fertilizers which contain heavy metals including lead, cadmium, and mercury and their effects on soils and the crops grown on these soils.

While contaminants of soil and water may come from many sources, government officials and environmental activists alike are scrutinizing the important role played by various sources of nutrients (including chemical fertilizer as well as urban and industrial wastes) as potential sources of contaminants. As a result nutrient management and the levels maintained in a given soil will become affected not only by the goals of the farmers themselves but increasingly by the regulations imposed by the government regarding clean air, soil, and water. Increasingly, scientists and general environmentalists are suggesting that nutrient balance sheets be established for fields, farms, and watersheds so that the fate of nutrients may be followed and their impact on the environment assessed. This will be discussed more fully later in this chapter.

Agroclimatic zone

The challenge of maintaining the nutrient status of soils necessary to support land management systems that are sustainable can best be discussed within the framework of major agroecological zones. It follows that practices and resource endowment related to nutrient management will vary significantly between zones. While the Technical Advisory Committee (TAC) of the Consultative Group for International Agriculture Research (CGIAR) has identified 18 such agroecosystems, others, most notably UNEP (1991), have reduced these to six or less. Although there are merits for a more detailed approach, the authors have chosen the most recent classification developed at the UNEP-sponsored FAO/Netherlands Conference on Agriculture and the Environment (UNEP, 1991). This conference led to *The Den Dasch Declaration and Agenda for Action on Sustainable Agriculture and Rural Development.* Five major agroecological zones were identified, namely: (i) drylands and other areas of uncertain rainfall; (ii) irrigated farm lands; (iii) humid and perhumid areas; (iv) mountainous and hilly regions; and (v) coastal zones and small islands. (We have modified item (v) to include megacities near coastal zones, small islands, and navigable rivers.)

A brief description of the area, the countries involved, and the key issues relative to sustaining plant nutrients in soils is discussed in the remainder of this section.

Drylands and other areas of uncertain rainfall

This area represents roughly one-third of the earth's land surface and perhaps 15% of its population (840 million in 1992), most of whom live in the tropics at or below subsistence standards.

Areas within this agroecosystem that need increased focus on nutrient management and either maintaining, restoring, or building the plant nutrient status of the soil are: Sudano-Sahelian zone of sub-Saharan Africa (from Cape Verde to Somalia); southern Africa (the Greater Maghreb); western Asia; southern Pakistan (Baluchistan, Sind); Rajasthan in India; northern China (Great Wall and North) and Mongolia; northeastern Brazil; and northern Mexico.

Soils within this group are usually very poor in total and available phosphorus; have very low levels of organic matter; are highly weathered; hence, have low CEC and low buffering capacity. External sources of nutrients are required to support commercial agriculture.

Humid and perhumid areas

This agroecosystem is dominated by areas that either once were or still are tropical moist forests and subhumid savannahs. The virgin forests have exceptionally high biological diversity, and generally low soil fertility, but are sustained at a reasonable level of productivity through a very efficient system of nutrient cycling. The major threats to the system are: (i) breaking the nutrient cycling system by cutting the forest, thereby increasing the loss of nutrients as a result of logging; or (ii) increased frequency (shortened fallow) of slash and burn. Most soils in this ecosystem are very fragile. Once the tree cover is removed, the nutrient recycling mechanisms that sustained the ecosystem before are disrupted and soil fertility drops to levels unable to sustain even marginal levels of productivity.

Alfisols, Ultisols, and Oxisols collectively dominate the major portion of the ecosystem. Oxisols occupy 35% of the land area, Ultisols 28%, and Alfisols about 4%.

Areas in which this ecosystem requires special attention are: the Amazon Basin of Brazil/Peru/Ecuador and Bolivia; part of lower Mexico and Central America; southeast Asia, in particular Indonesia, Laos, eastern Malaysia, Myanmar, Papua New Guinea, northern Thailand, and Vietnam; and Central and West Africa (Cameroon, Togo, Ghana, Zaire, and Zambia).

Mountainous and hilly areas

Using available arable land per capita as a means of comparison, this agroecoregion has perhaps the highest density of people per unit of available arable land compared with any agroecosystem in the world. This population pressure, coupled with severe topography, makes the mountainous and hilly areas especially prone to excessive water runoff, soil erosion, landslides, and other such phenomena, and potentially endangers downstream areas if these slopes are deforested and/or cultivation is carelessly practised. For this reason alone, this agroecosystem, although ignored in years past by many researchers, warrants special attention far above what would be the case using population numbers alone as the basis for priority setting. Land use management in this agroecosystem affects people in the neighbouring lowlands at a level far higher than would normally be expected. Sedimentation in waterways and reservoirs, floods, and river channel instability are some of the downstream effects of upland degradation. Furthermore, people living in the mountainous areas are living in an almost unretractable position. Delivery of inputs from outside the area as well as marketing of outputs is generally very difficult or impossible. Similarly, getting food supplies to the region is even more of a challenge. Nevertheless, the growing conditions in the hills and mountains are such that certain products grown in the agroecoregion are in demand by people in the nearby lowlands. This implies that marketing opportunities do exist; hence, the area does have the demand for additional inputs including additional plant nutrients.

The specific geographic areas identified by UNEP (1991) as warranting special consideration include the Himalayan range (Nepal, Bhutan), Indo-Gangetic River Basins, Loess Plateau in the Upper Yellow River Basin (China), the Atlas and Rif Mountains in Morocco, the Fouta Ajallon Mountains in West Africa (Guinea), the East African Highlands (Burundi, Ethiopia, Kenya, Rwanda, and Tanzania), the Highlands of Southern Africa (Lesotho and Swaziland), and the Andes of South America (Argentina, Bolivia, Chile, Colombia, Ecuador, Peru, and Venezuela). Because of severe shortages of energy and fodder, the landscape in some of these areas has been totally denuded of trees. Major examples of landscape stripping exist in Nepal, Burundi, Rwanda, Lesotho, and Morocco. In other regions the pressures, while mounting, have not yet reached such drastic levels.

Megacities on or near coastal zones, small islands, and navigable rivers

Megacities serve as a magnet to attract people as well as foster commercial and industrial development. It is here that population pressures and resultant human activities are brought to bear on land use. Furthermore, there is strong interaction between land and water forces that involves not only the challenges of disposing of nutrients from human waste but also the need to contend with the disposal of nutrients resulting from industrial activities such as animal slaughter and processing, fisheries, and processing of lumber. Disposal of these nutrients is frequently done on a random, uncontrolled basis. This has an increasingly severe impact on the quality of the environment in the immediate and surrounding land and marine areas. The areas involved in these ecosystems provide one of the most challenging fields for planners and developers who aim at the establishment of sustainable nutrient and land-use management systems. The areas include low-lying crop lands (wetlands) and forests, deltas, coastal strips, lagoons, beaches and intertidal areas, mangrove swamps, coral reefs, estuaries, and small coastal watersheds.

The geographic regions of particular concern are: Gulf of Bengal and surrounding land (India, Bangladesh, The Maldives), Mekong Delta (Cambodia, Thailand, and Vietnam); the Gulf of Guinea (West Africa); East African-Indian Ocean (Madagascar and Seychelles); the Caribbean (Greater and Lesser Antilles); the South Pacific Islands, Philippines, and Indonesia; and the Pacific Coast of northern South America and Central America.

Effective control of water, industrial waste, and nutrient discharge is crucial to the sustainability of a number of industrial practices in this agro-ecological system. These industries range from rice farming, shrimp production, fishing, vegetable and fruit production, to tourism. The systems affected need to be monitored in terms of nutrient and water flow fluctuations, water quality and sedimentation, including discharge from watersheds, tidal fluctuations, and sedimentations in estuaries, deltas, and mangroves, brackish water deposits, and reservoir and groundwater regimes.

Nowhere is there greater pressure on effective nutrient supply and sustainable land-use management processes than in and around the mega-cities located on or near the coastal zones, small islands, and navigable rivers of this agroecosystem. Faced foremost with the huge demands for food and fibre from the population of megacities that are literally exploding in size, coupled with industrial and urban wastes of large proportions and prospects that these pressures will double or triple in the next 30 years, this ecosystem is unique as a candidate for development of effective systems of nutrient cycling.

Irrigated farmland

Water and plant nutrients are the principal factors contributing to increased agricultural production in much of the developing world. It is no accident, therefore, that the main pillars of the Green Revolution strategies included the high-yielding varieties and land where plant nutrients and water were near optimal-irrigated farmland. Currently about 17% of the world's total arable land is irrigated (FAO, 1992). However, this accounts for nearly 45% of the world crop production. Irrigation both increases and ensures crop yield and may permit harvesting of two to three crops per year instead of one and at best two under rain-fed conditions. Asia accounts for about 65% of the world total of irrigated land, followed by Latin America, with sub-Saharan Africa accounting for less than 5%. The greatest opportunities for increases in irrigated area lie in Asia (particularly India). It is in the irrigated area of the developing countries where fertilizer use is greatest.

The issues of irrigated land are economic, social, and technical in nature and relate basically to the following untenable features of irrigation: water-logging and salinization, excessive lowering of water tables; build-up of pollutant concentration in groundwater; and low water use efficiency. All of these impact nutrient use efficiency in one or more ways.

The areas that need concentrated efforts regarding maintenance of nutrient status of irrigated soils are arid and semiarid lands and drought-prone areas; in general, areas where fresh water is limited or where sources are unreliable; and areas affected by waterlogging and salinity. Examples of the irrigated zones in Asia are Sind and Baluchistan in Pakistan, Rajasthan in India, and the Middle East (the lower Nile of Egypt, the east bank of Jordan), and Latin America (northeast Brazil).

Strategies to Maintain Nutrient Status of Soils: an Example from Africa

The continent of Africa contains parts of all of the agroclimatic zones described earlier. In addition, all the other external factors including population pressure and concern for the environment are also experienced to varying degrees by the farmers of the continent. This discussion focuses on sub-Saharan Africa and more specifically the zone of the West African subregion south of the Sahel proper but north of the humid Guinea Savanna.

General situation

The entire semiarid (dry) zone to the south of the Sahara is characterized by a long dry season with one short season of rain. Historically, the balance between soil, climate, vegetation cover, and man has generally been in equilibrium. The area contains:

- A dry pastoral zone with a natural cover of dry steppe and annual rainfall of 200 mm or below.
- An area of marginal farming with natural vegetation typical of dry savannah. Annual rainfall ranges from 400 to 600 mm and has been highly variable over the past 10 years. Collectively this zone and the preceding one make up the Sahel proper.

- A zone roughly comprising the Sudano-Sahelian, Sudan, and Sudano-Guinean (northern Guinea Savanna) areas carrying a cover of shrubs and trees. Rainfall ranges from 500 to 1200 mm, all falling within 60–120 days (Pieri, 1992).

The region contains about 25 million people and based on current growth rates, the population is expected to double by the end of the first decade of the 21st century.

The soils of the zones are old (greater than 1 million years), highly weathered, and low in fertility. Organic matter is limiting (usually 0.4–0.6% or less); CEC is low (< 5 meq 100 g^{-1} soil); and the colloidal fraction of the soil is very low. Total phosphorus is low (< 200 ppm) as is available P (8–3 ppm Olsen method). In general the soils of these zones are very fragile and do not withstand intense pressures from population and/or cultivation (Bationo and Mokwunye, 1991; Pieri, 1992).

Besides nomadic cattle breeding on the driest range lands and limited horticulture on the banks of the main rivers, rainfed agriculture was based on shifting cultivation: 3–4 years of cropping after burning, followed by several years of natural fallow during which, hopefully, the soil recovered its physical and chemical properties, thus making possible a further period of cropping. This method served well to maintain agriculture for centuries though at a low level of productivity.

Population pressures (both man and livestock) and scarcity of good agricultural land, coupled with general shortages of energy in the region, have forced the shortening of the fallow period and increasingly its total abandonment.

Although data on nutrient availability to crops grown within Africa, and particularly in the dryland ecosystem, are incomplete, there is ample evidence that energy, fodder, and building materials compete for crop residues produced on the farm (Fig. 12.2). As a result, the nutrients available to the crop are frequently diverted to non-farm uses with the net effect that basic soil fertility and overall crop production declines. The reasons for this have been well documented by a number of researchers – most recently in identifying agricultural production constraints in the Sudano-Sahelian zone of sub-Saharan countries of Africa (Pichot *et al.*, 1981; Pieri, 1986, 1989, 1992; Bationo and Mokwunye, 1991).

Millet straw or maize stover grown on a phosphorus-deficient field cannot serve as a soil fertility building source of that nutrient when the straw/fodder is returned to the same soil to serve as a nutrient source to the next crop. Moreover, the notion that manure is a free resource is not true. Most farmers do not produce crop residues for the purpose of restoring soil fertility. There are often competing uses, such as for fuel, fodder, and fencing. Furthermore, to contain and process these materials requires time, resources, and energy. Therefore, producing enough residue for direct appli-

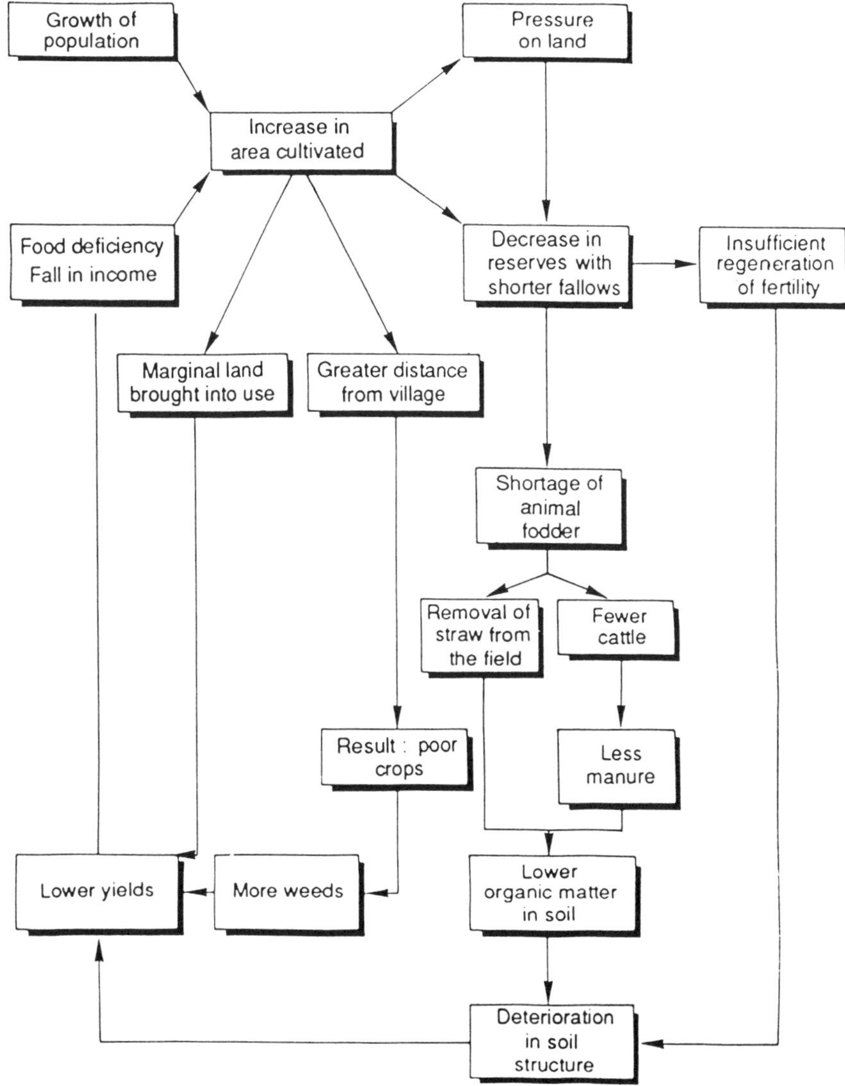

Fig. 12.2. Degradation in a traditional farming system when pressure on land approaches saturation.

cation to the soil or for composting is in itself a challenge which requires the farmer's commitment of resources (Fig. 12.2).

The effects of this drain of feed and energy, and hence movement of nutrients to off-farm sources, are particularly striking. Stoorvogel and Smaling (1990) have estimated that there was a net loss of 49 kg ha^{-1} or 9.3

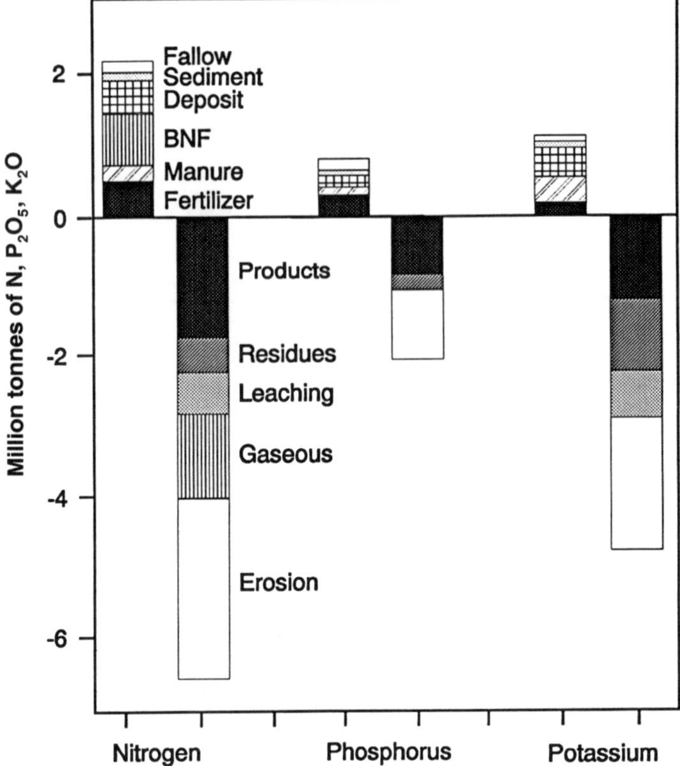

Fig. 12.3. NPK balance for sub-Saharan Africa.

million tonnes of plant nutrients removed from the system (average) in sub-Saharan Africa (Fig. 12.3; Table 12.2) in 1983. Furthermore, these quantities of annual nutrient removal are projected to increase to a rate of 60 kg ha^{-1} and 13.2 million tonnes of plant nutrients by the year 2000 unless some major shifts in nutrient management occur soon and/or off-farm demand is curtailed.

A number of researchers (Pieri, 1986, 1989, 1992; Bationo and Mokwunye, 1991) have concluded that phosphorus, nitrogen, and sulphur (in that order) are the nutrients most limiting crop growth in the Sudano-Sahelian zone. Furthermore, these workers have recognized that commercial fertilizer is an effective and essential means to increase yields in arable farming systems without fallows. However, Pieri (1989, 1992) cautions that long-term problems may arise, especially in the more arid areas if only chemical fertilizers containing N, NP, or NPK are used. Most of the soils are low in exchangeable Mg and Ca and generally acid. To substantiate this observation, he cites Pichot *et al.* (1981):

Table 12.2. Estimated net loss of nitrogen, phosphate, and potash from soils of sub-Saharan Africa, 1983 and 2000.

	Losses (kg ha^{-1})			Losses for the region (10^6 t)		
Year	N	P_2O_5	K_2O	N	P_2O_5	K_2O
1983	23	7	19	4.4	1.2	3.7
2000	27	8	25	6.1	1.7	5.5

Source: Stoorvogel and Smaling (1990).

> Application of NP and NPK fertilizer results in increased yields for some years, but in the long run it leads to decreasing base saturation, acidification of the soil, and a significant drop in soil pH. These phenomena, associated with the use of only N fertilizers, are characterized by increasing K deficiency, decreasing pH and occurrence of Al-toxicity. Application of organic matter, such as green manures, crop residue, compost or animal manure, counteracts the negative effects of chemical fertilizer.

Pieri (1989) accepted the basic premise of Pichot *et al.* (1981) that soil fertility in intensive arable farming systems in the Sudano-Sahelian zone can only be sustained through efficient recycling of organic material, in combination with effective use of N-fixing leguminous species and chemical fertilizers; however, he took issue regarding the benefits of green manure and crop residues as to their real contribution in building stable levels of organic carbon in soil. These materials may overstimulate microbial activity which results in excessive mineralization of native soil organic matter and eventually the immobilization of nitrogen during the decomposition process (de Redder and van Keulen, 1990). Crop residues and forages should be composted or preferably recycled through the animal. Bationo and Mokwunye (1991), Stoorvogel and Smaling (1990) and Pieri (1992) further pointed out the difficulty of recycling the crop residue or collecting adequate supplies of animal manure. They felt that until a more stable livestock system was introduced into the zone and alternate sources were found for competing off-farm uses (for energy and building material) for crop residues, it would be impossible to sustain, much less build, the base fertility of the soils within the region without heavy reliance on the addition of external inputs, such as chemical fertilizers, agrominerals, and various soil amendments. They have shown that much of the Sudano-Sahelian zone possesses an abundance of naturally occurring phosphate rock which when finely ground can serve as an effective source of phosphate for the crop. Depending upon the source, this form of phosphate is 50–100% as effective as that obtained by the plant from the processed high-analysis phosphate fertilizers.

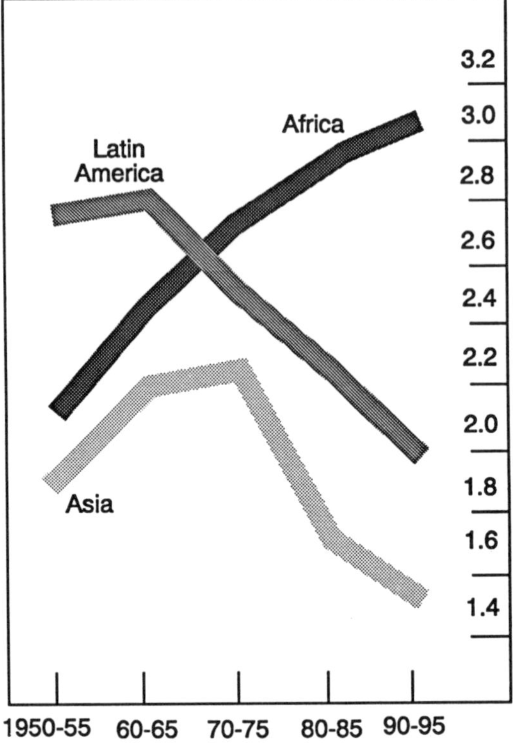

Fig. 12.4. Population growth annual percentage increase.

Researchable areas

The soils in this agroecosystem are very fragile. They are characterized by low cation exchange capacity, extremely low levels of organic matter, and low soil fertility, particularly due to low available phosphate levels. Increased population pressure on these soils to produce more food results in depletion of native amounts of nutrients. Even where fertilizer is applied to replace part of the nutrients removed by crops, severe nutrient imbalances can be a major cause of declining per capita food production, particularly in sub-Saharan Africa (Figs. 12.3–12.5). Considerable research is needed to avoid or correct nutrient imbalances as well as nutrient-related soil degradation.

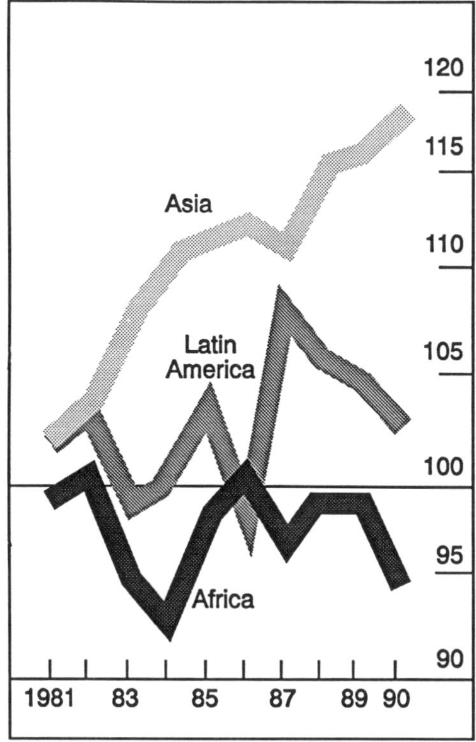

Fig. 12.5. Food production per person, 1979–1981 = 100.

Nutrient balance studies

There is increasing evidence that nutrient balance studies of the type conducted by Stoorvogel and Smaling (1990) and IFDC-Afrique (1990) need to be done at the field, watershed, and national levels (Fig. 12.3). This approach is extremely important for soils with low nutrient-buffering capacity in which nutrient deficiency abounds. In West Africa and the sheep hillside/hilly regions (Burundi, Rwanda, Malawi, as examples) there are also strong forces (energy, fodder, building materials) competing for limited nutrients and biomass.

There are a number of existing experiments established by the French organization CIRAD in the Sudano-Sahelian zone of West Africa, the results of which would contribute to the scientific knowledge of how to manage these fragile soils. In addition IFDC, through its Africa Division, has built on this experience and established several long-term experiments in the key agroecological zones of West Africa.

Finally, it should be stressed that as a result of nutrient balance studies conducted on some research stations and in a limited number of farmer fields, more attention has to be directed toward a better and more precise assessment of the impact of recommended fertilizer systems (organic and inorganic) on soil acidity across the whole Sudano-Sahelian region.

Management of soil organic matter

Maintaining soil organic matter in equilibrium is extremely important to sustainable land-use management in the dry zones. Organic matter acts as a source and sink for plant nutrients. Efforts should be made to find ways to build soil organic matter and identify various management systems needed to sustain levels at 50%, 100%, or 150% above those experienced under conventional management.

The potential of soil organic matter to counter the effects of soil acidity resulting from use of fertilizers also needs to be investigated. Preliminary results from a long-term crop residue management experiment set up by IFDC in the Sahel show that the decomposition of millet residue leads to a reduction in the amount of exchangeable aluminium, which improves P nutrition of crops (Bationo and Mokwunye, 1991). An examination of the way farmers manage their crop residue and manures would be warranted.

The contribution of root mass distribution in soil is an often overlooked means to increase the organic matter content of many soils. The work by Lamotte and Bourliere (1978) and subsequently by Chopart (1980) represents classic research experiments showing that the naturally occurring species of the shrub and grassy savannah contribute 10–30 times more root mass than do cultivated crops such as sorghum, pearl millet, and groundnut-maize (Fig. 12.6). More of such data were collected by IFDC while carrying out a soil fertility restoration project in both the Savanna and Sahel regions of West Africa.

Nutrient supply systems based primarily on indigenous resources

Whereas the agronomic potential is considerable, the actual demand for commercial quantities as fertilizer is comparatively small (only 5000–50,000 tonnes of product per country per year), far below a level to warrant conventional-scale domestic production. Due to infrastructural weaknesses, demand levels are only a fraction of those existing in similar areas of Asia or Latin America. As a result, commercial fertilizers, where available, are imported usually at high cost and, hence, are not affordable for use by other than those growing priority crops like cotton, sugarcane, etc. Nevertheless, a nutrient supply system based on the use of indigenous resources must be found that has the potential not only to arrest the mining of soil fertility

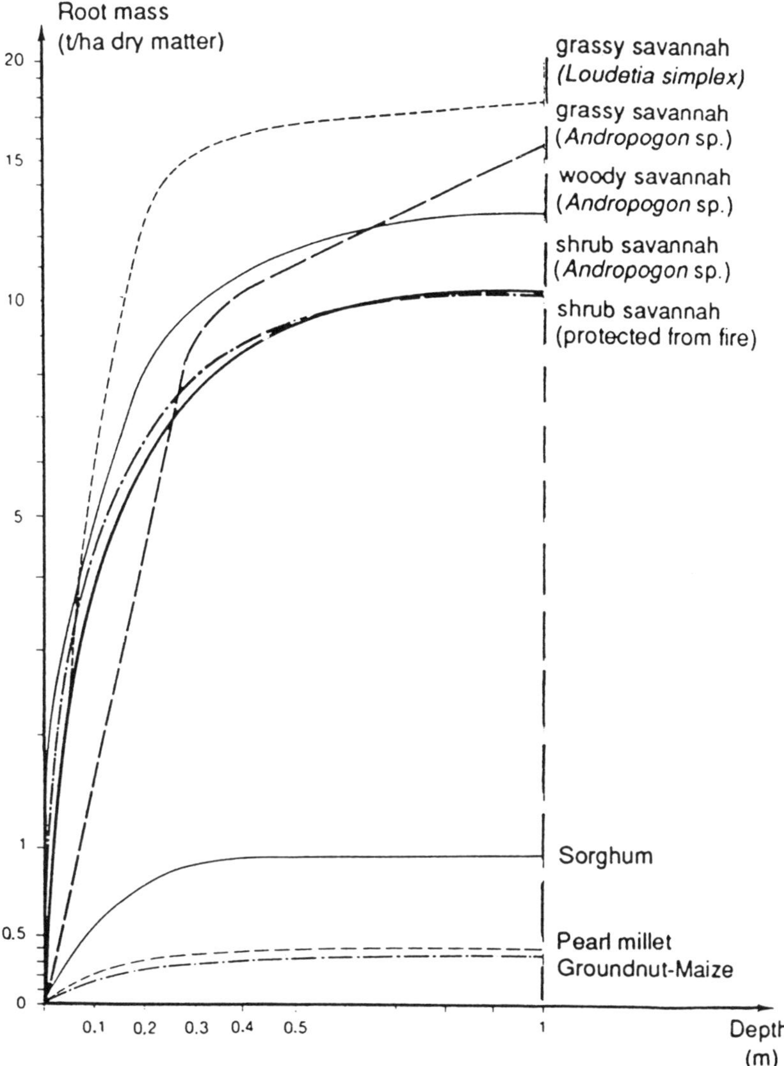

Fig. 12.6. Root mass distribution under natural cover compared to crops (Lamotte and Bourliere, 1978; Chopart 1980).

taking place in much of sub-Saharan Africa but also build soil fertility if a sustainable land-use system is to be realized.

The suggestions of Pieri (1989) and Bationo and Mokwunye (1991) provide much of the solution. Separately, they proposed a system based on biologically fixed nitrogen (BNF) from legumes (trees or annual crops), combined with the judicious use of phosphate (mainly from local sources of

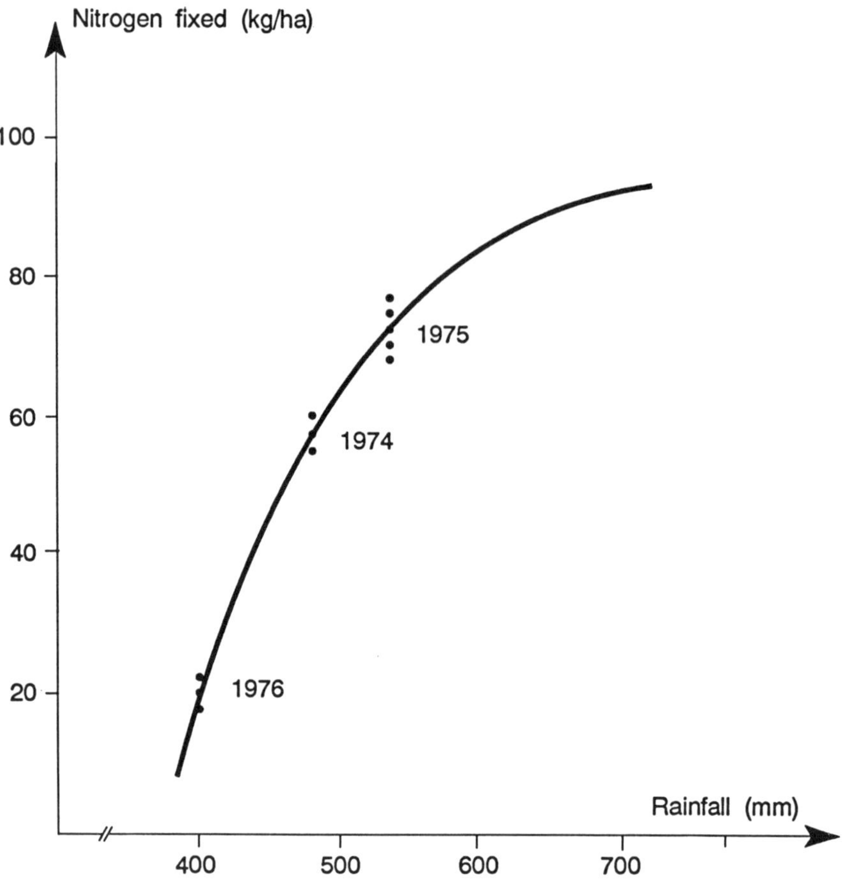

Fig. 12.7. Effect of rainfall on nitrogen fixation by groundnuts in Senegal (Wetselaar and Ganry, 1982).

phosphate rock and return to the soil of significant quantities (2–5 t ha^{-1} of manure or crop residue) supplemented by some imported fertilizer (compounds). Such practices result in a nutrient management system that makes considerable use of cycling, and is balanced, and sustainable. The use of chemical fertilizer depends on how well the farmer is able to secure animal manure and ensure the return of crop residues.

The effectiveness of various nutrient supply systems of the type mentioned above needs to be researched, including: (i) evaluation of various sources of BNF (trees or annual legumes) to supply the entire N needs of the crop; (ii) the value of local phosphate rocks in meeting the P needs of the primary crop and of the legumes supplying the N to the system; (iii) the ability of farmers to retain the crop residue as well as

Table 12.3. Nutrient losses by leaching under crops in Senegal.

Place	Annual rainfall (mm)	Crop (variety)	Drainage (mm)	Losses (kg ha^{-1} $year^{-1}$)				
				N	CaO	MgO	K_2O	P_2O_5
Bambey (Senegal)	507 (1981)	Millet (Souna III)	9.5	0.3	0.8	0.4	0.3	Tr
		Groundnut (55-437)	100.6	25.1	54.1	13.6	5.2	Tr
Maroua (Cameroon)	705 (1975)	Sorghum (IRAT 55)	2	Tr	0.1	0.1	Tr	Tr
	683 (1977)	Cotton (BJA)	83	2.1	43.7	12.3	1.7	Tr
Bouaké (Ivory Coast)	633 (1971)	Maize (CJB)	210	6.1	36.4	26.2	2.4	Tr
	532 (1972)	Cotton (BJA)	260	7.1	18.0	6.6	2.0	Tr

Source: Senegal: Pieri (1983); Cameroon: Gigou (1982); Ivory Coast, Chabalier (1976).

recycle manure on or to the field; and (iv) the need for and the availability of external inputs (chemical fertilizer) to make up the remaining nutrient deficits. These systems need to be studied in their entirety over a 5- to 10-year period to measure their impact on not only maintaining but also building soil fertility as the need arises.

The enhancement of BNF in the system needs careful attention taking into account the work done by Wetselaar and Ganry (1982) in Senegal. The true benefit of nitrogen fixation by legume crops should not be overestimated when establishing the real N balance in environments having erratic rainfall (Fig. 12.7).

Reduce nutrient losses due to runoff, leaching, and erosion

The loss of nutrients depends on rainfall (amount, intensity, and distribution), soil infiltration rate, cropping mix, and anti-erosion measures (crop cover, water-harvesting techniques, and overall management) (Lal, 1990). Although considerable research has been carried out in each of these areas, there is still a great need to investigate the impact that specific crops have on reducing nutrient losses. Examples of the type of work needed have been reported by Chabalier (1976), Gigou (1982), and Pieri (1983) and are summarized in Table 12.3. It is clear that crops vary markedly in their efficiency to capture water and curb nutrient losses particularly through reduced losses due to leaching. Millet and sorghum appear to be much more effective at this than groundnut, cotton and maize. Quick establishment of a dense and deep root system is absolutely essential if nutrient losses due to erosion, runoff, and leaching are to be curbed. More research in this broad area is needed.

Restoration of degraded soils

As stated earlier, cropping of many of the soils in the region has led to a decline in soil fertility, a drop in pH, and a rise in exchangeable aluminium thus rendering the soils unusable for further crop production. According to Oldeman *et al.* (1991), the total degraded area in the world is about 1.964 billion ha or about 17% of the total land area supporting some form of vegetative growth. They estimate that about 494 million ha of soil in Africa is degraded. This is roughly 22% of the land supporting vegetative growth. These are dramatic figures and indicate the seriousness of the problem. In most cases lack of adequate supplies of soil phosphorus is the overriding cause. Special programmes that have nutrient management as their core need to be developed and designed not only to restore the original fertility of these degraded soils but also to increase it in order to meet the social as well as economic requirements necessary for a sustainable land use system.

Nutritional needs of small-scale irrigated systems

There are extensive areas (bas-fonds, inland valleys) traditionally not cultivated for health reasons (animal and human) which are now open for development. In addition, there is increasing evidence that groundwater is near the soil surface (5–15 m) in numerous areas within the Sudano-Sahelian agroecosystem. Effective use of these agroecosystems will allow multiple cropping even though the areas within the region are too remote to arrange large-scale irrigation systems. These areas present opportunities for small-scale, simple irrigation systems (tube wells and drip systems) that are cost effective and could offer a limited number of farmers opportunities to grow specialty high-value crops. The nutritional needs of such crops grown in the more remote regions and attendant nutrient supply systems to sustain such systems need to be studied.

On a more general basis for the whole Sudano-Sahelian region, more research should be directed toward the improvement of the water-nutrient efficiencies.

Summary and Conclusions

The soil supplies 13 of 16 elements essential to plant growth. The actual quantities of nutrients needed to sustain growth depend on the production goals and the amounts of nutrients available from a range of sources including those from soil reserves, crop residues, animal and human wastes, biologically fixed nitrogen, commercial fertilizers, and a number of site-specific sources such as urban and industrial waste and deposition from rainfall as influenced by air emissions near industrial areas.

Strategies used by farmers as well as community and national planners to maintain and/or build the nutrient status of soils are influenced by a range of external pressures, the impact of which are strongly influenced by the resource endowments of a specific agroecological zone. The external factors are population pressure, migration of people, income levels, degree and level of government intervention, importance society places on conservation of the soil resource base, and public concern over the impact of agriculture on the environment. The important agroecological zones are drylands and other areas of uncertain rainfall, irrigated farm lands, humid and perhumid areas, mountain and hilly regions, and megacities on or near coastal zones, small islands, and navigable rivers.

Three strategies to maintain and/or build the nutrient status of soils have been identified. The Sudano-Sahelian region of sub-Saharan Africa (where population pressures are high compared to availability of good land, and fertilizer use is low), typifies the problem facing many developing countries having difficulty in maintaining, much less increasing, the nutrient status of soils. Under the present agricultural system soils are suffering a net loss of essential plant nutrients averaging about 50 kg ha^{-1} of nutrients per year. Research and technology adoption programmes are badly needed to arrest and reverse this trend. Researchable areas include: (i) long-term nutrient balance studies designed to assess the impact of plant nutrient additions on the soil resource base and the environment; (ii) strategies to improve the management of soil organic matter; (iii) effective nutrient supply systems based primarily but not exclusively on the use of indigenous resources; (iv) acceptable strategies to reduce nutrient losses; and (v) strategies to restore degraded soils.

In eastern Asia and northwestern Europe some of the most intensive agriculture is practised, and conflicting economic, social, and environmental pressures are emerging. Both regions are characterized by high population densities, scarcity of good arable land, high levels of industrial and agricultural activity, and very high usage rates of all agricultural inputs including organic and inorganic sources of plant nutrients. The two regions differ in the type of infrastructure already in place to control pollution, population pressure, and economic, social, and ecological goals. Research that integrates these issues is badly needed and must be incorporated into both strategies of supplying nutrients for intensive agriculture as well as strategies of reaching urban wastes in megacities. There is an urgent need to develop methodologies that quantify the economic costs associated with the goals of personal profit versus social equity and environmental values.

References

Arndt, H.W. (1983) Transmigration: achievements, problems, prospects. *Bulletin of Indonesian Economic Studies* 19(3), 40–73.

Bationo, A. and Mokwunye, A.U. (1991) Role of manures and crop residue in alleviating soil fertility constraints to crop production with special reference to the Sahelian and Sudanian zones of West Africa. *Fertilizer Research* 29, 117–125.

CABO-DLO (1992) *Integrated Farming Systems – Design and Optimization.* Agricultural Research Department, Centre for Agrobiological Research, Wageningen, The Netherlands.

CEQ (1980) *The Global 2000 Report to the President,* Vol. 2, US Department of State, Council on Environmental Quality, Washington, DC, USA, pp. 73–104.

Chabalier, P.F. (1976) Comparaison de deux méthodes des mesures de la lixiviation en sol ferrallitique. *Agronomie Tropicale* 39(1), 22–31.

Chopart, J.L. (1980) Etude au Champ des Systèmes Racinaires des Principales Cultures Pluviales au Sénégal (Arachide-Milsorgho-Riz Pluvial). Thése Doct. Prod. Veg. Qual. Prod., Inst. Nat. Polytech Toulouse, 160 pp.

Desai, G. and Stone, B. (1987) Factors affecting fertilizer use in China and India, Paper given at Joint Meeting of International Fertilizer Association, International Food Policy Research Institute, and World Bank. September 1987, Washington, DC, USA.

de Redder, N. and van Keulen, H. (1990) Some aspects of the role of organic matter in sustainable intensified arable farming systems in the West African Semi-Arid Tropics (SAT). *Fertilizer Research* 26, 299–310.

FAO (1981) *Agriculture: Toward 2000.* Food and Agriculture Organization of the United Nations, Rome, Italy.

FAO (1992) *FAO Production Yearbook,* Vol. 45. Food and Agriculture Organization of the United Nations, Rome, Italy.

Freudenberger, C.D. (1988) The agricultural agenda for the twenty-first century. *KIDMA, Israel Journal of Development* 10(2), 32–36.

Gigou, J. (1982) Dynamique de l'azote minéral en sol nu ou cultivé de region tropicale séche du nord-Cameroun. These Doct. Ing. Univ. Sci. Tech. Languedoc, Montpellier, 130 pp.

IFDC (1992) *Fertilizer Statistics.* International Fertilizer Development Center, Muscle Shoals, Alabama, USA.

IFDC-Afrique (1990) *Approvisionnement, Commercialisation et Demande des Engrais en Republique du Togo,* Centre International pour le Développment des Engrais-Afrique, Lomé, Togo.

IMF (1991) *Annual Report,* International Monetary Fund, Washington, DC, USA.

Lal, R. (1990) Soil erosion and land degradation, *Advances in Soil Science* 11, 129–172.

Lamotte, M. and Bourliere, F. (1978) *Problémes d'Écologie; Structure et Fonctionnement des Écosystémes Terrestres.* Masson, Paris, 345 pp.

Oldeman, L.R., Hakkeling, R.T.A. and Sombroek, W.G. (1991) *World Map on Status of Human-Induced Soil Degradation – An Explanatory Note.* International Soil Reference and Information Centre, Wageningen, The Netherlands.

Pichot, J., Sedogo, M.P., Poulain, J.F. and Arrivets, J. (1981) Evolution de la fertilité d'un sol ferrugineux tropical sou l'influence des fumures minerales et organiques. *Agronomie Tropicale* 36, 122–133.

Pieri, C. (1983) Nutrient balances in rainfed farming systems in arid and semi-arid regions. In: *17th Coll. Int. Potash Inst. Rabat, Marrakech,* Maroc, May 2–6. IPI, Berne, pp. 181–209.

Pieri, C. (1986) Fertilisation des cultures vivrieres et fertilité des sols en agriculture paysanne subsaharienne. *Agronomie Tropicale* 41, 1–20.

Pieri, C. (1989) Fertilité des terres de Savannes. *Bilan de Trente ans de Recherche et de Developpement au Sud du Sahara.* Agridoc-International, Paris, France.

Pieri, C. (1992) *Fertility of Soils – A Future for Farming in the West African Savannah.* Springer-Verlag, Paris, France, 346 pp.

PRB (1992) *World Population Data Sheet.* Population Reference Bureau, Inc., Washington, DC, USA.

Reardon, T. and Islam, I. (1990) *Issues of Sustainability in Agricultural Research in Africa,* Reprint No. 180, International Food Policy Research Institute, Washington, DC, USA.

Stapel, C. (1982) Ecologic agriculture in global and national context (in Danish). *Ugeskrift for Landrug* 127(26), 495–500.

Stone, B. and Desai, G.M. (1989) China and India: A comparative perspective on fertilizer policy requirements for long-term growth and transitional needs. In: Longworth, J.W. (ed.), *China's Rural Development Miracle, With International Comparisons.* University of Queensland Press, Australia, pp. 274–293.

Stoorvogel, J.J. and Smaling, E.M.A. (1990) *Assessment of Soil Nutrient Depletion in Sub-Saharan Africa: 1983–2000* (Three Volumes). Wageningen, The Netherlands.

UNCED (1992) *Preliminary Report Rio de Janeiro Summit.* June 1992 United Nations Conference on Environment and Development.

UNEP (1991) *Protection and Management of Land Resources: Alternative Sustainable Systems of Production Conference,* March 18–April 5, Geneva, Switzerland.

Wetselaar, R. and Ganry, F. (1982) Nitrogen balance in tropical agrosystems. In: Dommergues, Y.R., and Diem, H.G. (eds) *Developments in Plant and Soil Sciences,* Vol. 5: *Microbiology of Tropical Soils and Plant Productivity.* Nijhoff, Junk, The Hague, pp. 1–36.

Chapter 13
Maintaining Soil Micronutrient Status

A. Kabata-Pendias

Institute of Soils Science and Cultivation of Plants, 24–100 Pulawy, Poland

Introduction

'Micronutrients' are 'trace elements' essential, at certain concentrations, to vital processes, and their behavioural properties are related to general rules governing the biogeochemistry of these elements. The behaviour of micronutrients in soils depends upon complex reactions among cations and anions and different components of soil phases: solid, aqueous and gaseous. The main features of the soil biogeochemical system governing the behaviour of micronutrients are: (i) heterogeneous distribution of compounds and components; (ii) seasonal and spatial alteration of major soil variables; (iii) transformation of element species; (iv) transfer between phases; and (v) bioaccumulation.

Soil Factors Influencing Trace Element Behaviour

Present-day soils contain trace elements of various origins. Lithogenic elements are those which are directly inherited from the lithosphere (parent material). Anthropogenic elements in the soils are those deposited as direct or indirect results of human activities. Pedogenic elements are of lithogenic and anthropogenic origins but their distribution in soil horizons and soil particles are changed due to pedogenic processes. Thus the behaviour of trace elements is highly related to their speciation and is likely to be governed by their origin as well as by soil conditions.

The chemical equilibria of a soil can be characterized by dissolution, diffusion, sorption and precipitation reactions. Depending on the variability in physicochemical characteristics of metals, their affinity to soil com-

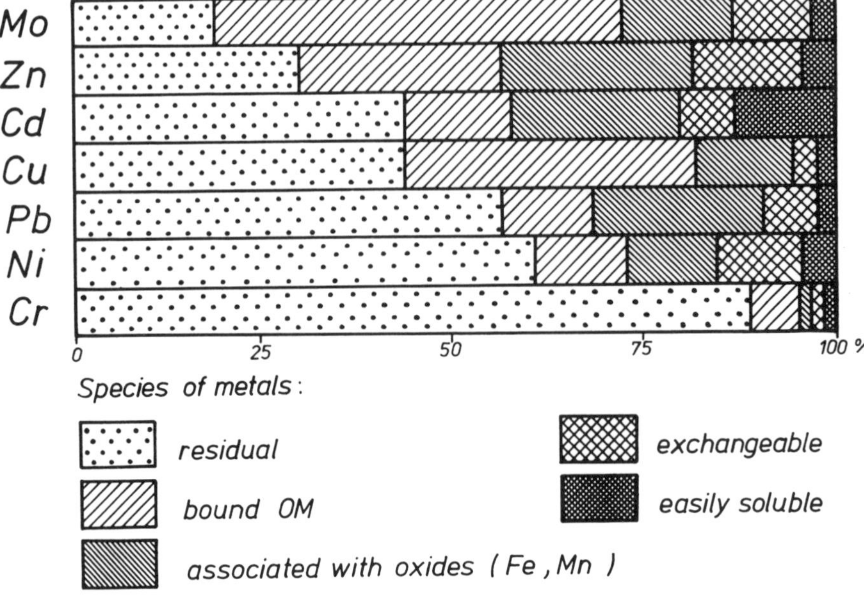

Fig. 13.1. Speciation of trace elements in soils of Poland (% of total content) (Kabata-Pendias, 1992).

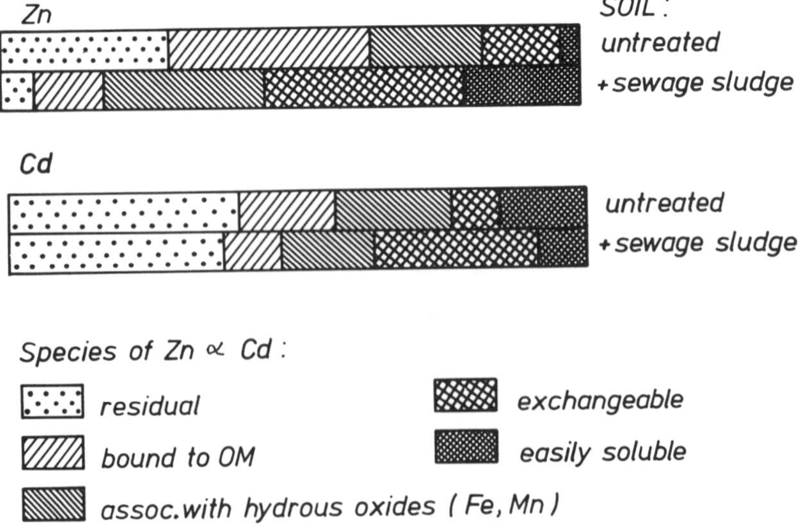

Fig. 13.2. Impact of sewage sludge on Cd and Zn speciation in soil (Dudka and Chlopecka, 1990; Chlopecka, 1992).

ponents governs their speciation (Fig. 13.1). Soluble, exchangeable, and chelated species of trace elements are the most mobile in soils. The behaviour of trace elements reflected in their speciation depends greatly on the form or compounds in which they occur, as well as on soil conditions, as illustrated by the diversity of Zn and Cd species after sewage sludge application (Fig. 13.2).

Pedogenic processes, soil management, and various other anthropogenic factors are likely to control the total pool and phytoavailability of micronutrients in a soil. There are great differences in these processes between soils at the regional scale and sometimes even on individual farms. The main soil variables involved in micronutrient status are: pH, Eh, organic matter, hydrous oxides, clays, carbonates and salt contents. These factors will determine soil resilience to trace element stress. The role of parent material and water regime are also of great importance (Kabata-Pendias and Pendias, 1992).

Agricultural Exploitation and Micronutrient Status of Soils

Unlike some macronutrient reserves, micronutrients are not exploited in most cultivated soils. However, their accessibility can be limited due to soil deterioration. Several soil factors inherited from the parent material and created by pedogenic processes as well as induced by anthropogenic impact contribute significantly to micronutrient deficiency (Table 13.1). The use of a soil by cropping also affects (although should not) some changes in the micronutrient supply to plants.

The Eh–pH system is most variable in soils, and is also most important in controlling the solubility and availability of micronutrients. Deficiencies are likely to occur in most soils with extreme properties; too acid (e.g. light sandy podzols) or too alkaline (e.g. calcareous soils), with improper water regime, and with an excess of phosphates, carbonates, as well as Fe and Mn hydrous oxides. Also an excess of organic raw material is likely to induce deficiency of several micronutrients. Especially rapid changes in redox potential affect significantly the availability of some trace cations to plants (Fig. 13.3).

Micronutrient fertilizers (or multiple-nutrient fertilizers) are commonly used to provide plant needs for optimum growth (Finck, 1982; Wild and Jones, 1988). Different soil tests and plant diagnostic methods have been used to assess micronutrient availability status. Soil chemical tests are based on the determination of water-soluble, acid-extractable, neutral salt-extractable, exchangeable, and complex or chelated trace elements. The interpretation of these results is by no means an easy problem, and usually gives reliable information for only a particular soil–plant system. The evaluation of micronutrient status based on soil and plant testing is presented by

Table 13.1. Soil factors contributing to micronutrient deficiency.

		Factors			
Element	Soil units	pH	OM	Water regime	Others
B	Podzols, rendzinas, gleysols, ferralsols	Acid, neutral	Very high or low	Flooded soils	$CaCO_3$, light texture
Co	Podzols, histosols, rendzinas, solonetz	Alkaline, acid	High	High moisture	High Mn, Fe, $CaCO_3$
Cu	Histosols, podzols, rendzinas, solonetz	–	Low or high	High moisture	High N, P, Zn, light texture
Fe	Rendzinas, ferralsols, solonetz	Alkaline	High or low in acid soil	Poor drainage	High P, Mn, $CaCo_3$
Mn	Podzols, rendzinas, histosols	Strongly acid or alkaline	Very high	Moisture extremes	High Fe, $CaCo_3$
Mo	Podzols, ferralsols	Strongly acid or acid	–	Good drainage	High Fe, Al, S
Se	Podzols, histosols, ferralsols	Acid	High	Waterlogging	High Fe, S
Zn	Podzols, rendzinas, solonetz	Strongly acid or alkaline	Low	–	High P, N, $CaCO_3$

Sillanpää (1982) for different countries. A generalization of this report for soils of different continents indicates that in Europe some deficiencies of B, Cu, Mn and Fe may occur, but an excess of Cu, Fe, Zn, Mn and Mo is also quite frequent. Soils of Africa often contain excessively available Mn, while a deficiency of B, Cu, Mo and Mn is likely to occur. Soils of the Far and Near East may have too high levels of available B, Mo and Mn. All possible deficiencies (mainly of B, Cu, and Zn) are related in most cases to the low availability of these elements.

The correction of micronutrient deficiency through the application of microfertilizers is common. These treatments are effective if the availability of the micronutrient is low, regardless of its total content in a soil. However, it may build up the proportion of hardly soluble species of a given element. Usually microfertilizer rates are much higher than the requirements of the plants for yield and quality, e.g. (in kg ha^{-1} $year^{-1}$) Mo 10–30, Zn 10–20, Cu 5–10, B 2–3, Mo 0.5–1. At such a level of micronutrient supply plants can take up even luxurious amounts of some elements (Fig. 13.4). The yearly trace element uptake by crops represents a very small proportion of the total contents of the soils. Under proper chemical and physical soil conditions the natural (and anthropogenic) sources of micronutrients will meet plant requirements. The removal of micronutrients calculated for a high yield of

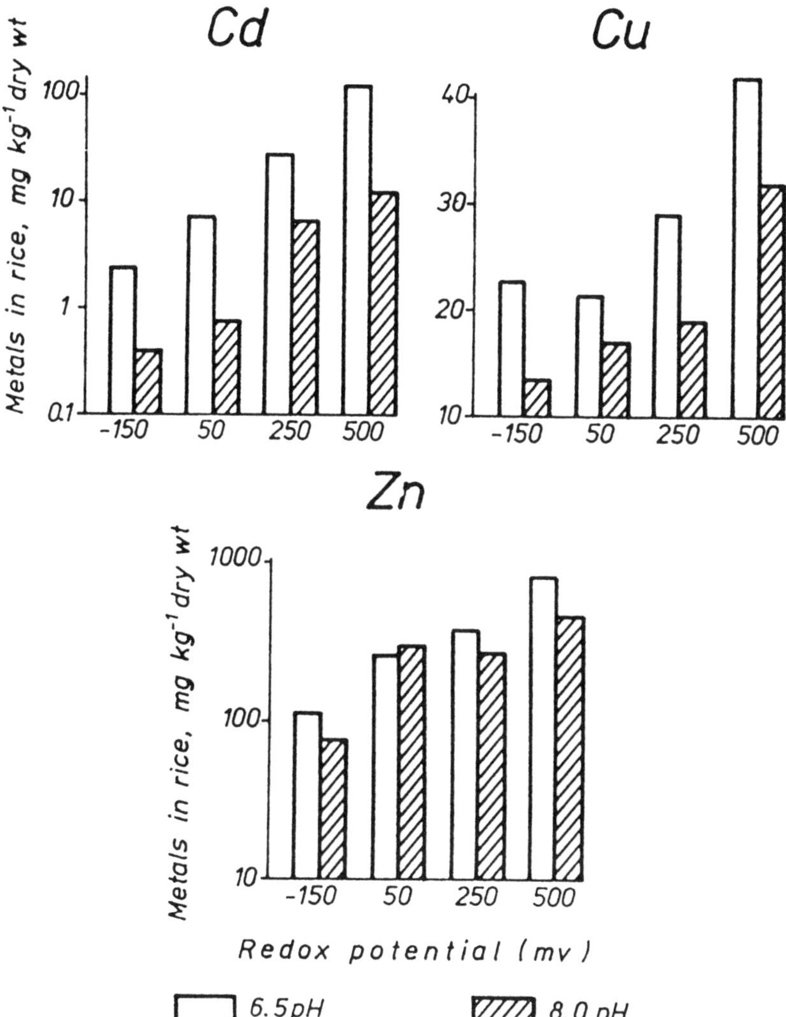

Fig. 13.3. Trace metal absorption by rice as influenced by pH and redox potential of soils (Gambrell and Patrick, 1989).

various crops range from 30 to 100 g ha^{-1} $year^{-1}$, being the lowest for Mo and the highest for Mn (Table 13.2).

The concentration–dilution phenomenon is related to an absorption of micronutrients by the plants which is higher at low yield levels than at high yield levels. In most cases, however, the restriction of plant growth is due to some factor other than the micronutrient supply, and their concentration is a secondary effect (Sillanpää, 1982).

The budget for some micronutrients in a light loamy soil in Poland

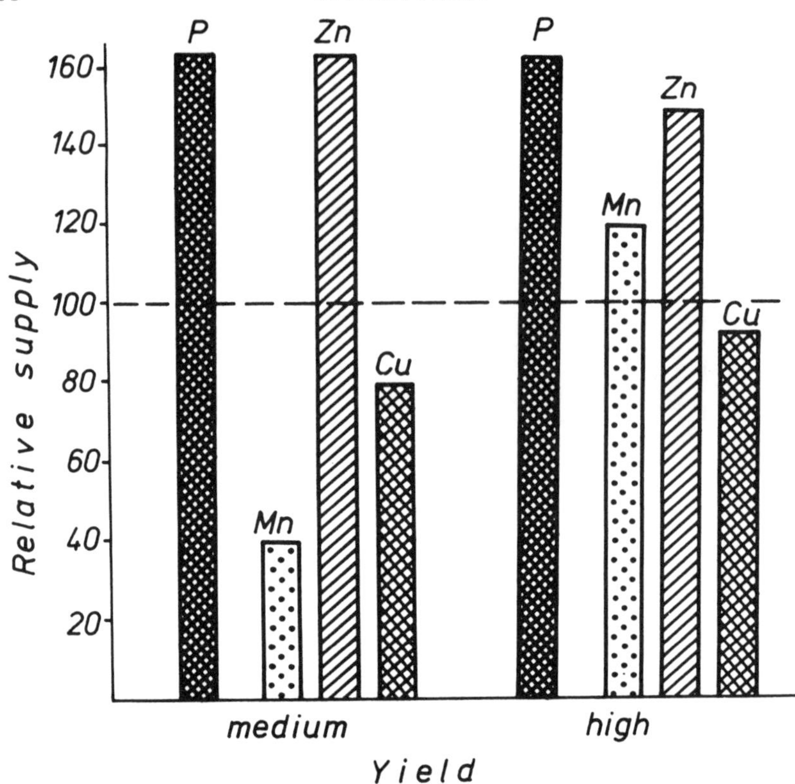

Fig. 13.4. Micronutrient supply to young wheat plants as compared to P-supply at two yield levels (Finck, 1982).

(rural region), based on the total atmospheric input and leaching with seepage water is (in g ha^{-1} $year^{-1}$): Zn 360, Mn 90, B 30, and Cu 15. By including the input from other sources, e.g. fertilizers (mineral and organic) and harvest residues, plant needs for micronutrients may be covered by man-induced cycling in some agroecosystems.

Excess of Trace Elements in Soils

Agricultural soils, nowadays, are exposed to both input and losses of micronutrients at a much higher rate, and on a larger scale than in previous periods. Soils of Europe have been exposed to pollution for a long time, and the emission of metals has been greater in Europe than in other continents. However, in industrial regions of all continents and countries, soils are polluted, especially with trace metals (Table 13.2). Soils which have deteri-

Table 13.2. Excessive levels (maximum values) of trace elements in surface soils reported for various countries (mg kg^{-1}).

		Source	
Element	Country	Agricultural	Industrial
As	Japan	400	2470
	Poland	120	2000
Cd	Great Britain	167	336
	Poland	107	270
Cu	Japan	300	2020
	Netherlands	265	1090[a]
Hg	Canada	1.14	1.9
	USA	0.6	2.4
Pb	W. Germany	>800	3075
	Great Britain	390	4560
Zn	Netherlands	760	3625
	USA	765	12400

[a]Belgium.
Source: Kabata-Pendias and Pendias (1992).

orated because of industrial emissions are broadly investigated, and remedial and land use programmes have been elaborated. The evaluation of soil degradation is of vital importance for establishing guidelines for the control of agricultural environment pollution and for proper land use (Fig. 13.5). Non-point source agricultural pollution can also cause a high accumulation of several trace metals in plants (Table 13.3). This kind of pollution, although at much lower levels than industrial pollution, can create a real problem in the sustainable management of soils. Different features of soil pollution from industrial (aerial) and agricultural sources are summarized in Table 13.4).

The accumulative impact of soil pollution is well illustrated by Cd which is an easily phytoavailable metal. The increased level of this metal, due to its common presence in some phosphate fertilizers, is reflected in the Cd content of wheat grain (Fig. 13.6). However, even in unfertilized soils, the concentration of the metal has built up slowly due to long-distance aerial pollution (Fig. 13.7).

The continuous increase of metal pollutants in the soils of Europe is a real concern (De Bruijn and De Walle, 1989). The agricultural and environmental significance of this pollution is rather unpredictable, especially as the residence time in soils depends on soil conditions. The behavioural properties described for trace metals in soils in humid temperate climates

Table 13.3. Trace element removal with plants from soils.

Element	Content of soil (kg ha^{-1})	Output with plants			
		Reference		Accumulator	
		kg ha^{-1}	%[a]	kg ha^{-1}	%[a]
Mn	810	1000	0.1	5000	0.6
Cr	150	50	0.03	500	0.3
Zn	135	400	0.3	1500	1.0
B	90	100	0.1	2500	2.8
Cu	45	100	0.2	500	1.0
Ni	39	50	0.1	100	0.3
Mo	6	30	0.5	250	4.0
Cd	1.5	1	0.06	100	10.0

[a]Percentage of the total content of soils.

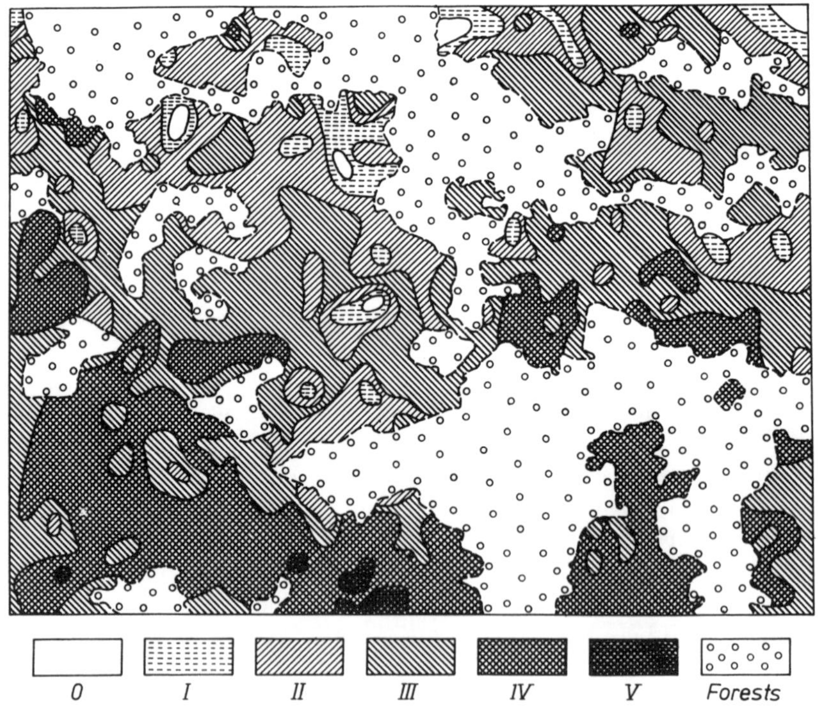

Fig. 13.5. Soil pollution classes in an industrial area (ca 3000 ha) of Lower Silesia in Poland. Numbers denote classes of soil pollution: 0, unpolluted; I, very slightly polluted; II, slightly polluted, III, moderately polluted; IV, heavily polluted, V, very heavily polluted (Kabata-Pendias *et al.*, 1992).

Table 13.4. Some features of soil contamination with metals from aerial and agricultural sources.

Aerial pollution	Agricultural pollution
Widespread	Localized
Very recent	Beginning with fertilizer and waste application
Slow cumulative rise in concentration	Variable loads
Accumulation in thin top layer, unless mobilized	Distribution in organic and/or ploughed horizons
Increased pollution of plant tops then roots	Increased concentration in roots then in plant tops

(Table 13.5) are different from those for the soils in the humid and arid tropics. In humid regions of the tropics micronutrient cycling is very rapid, and the residence time should be calculated for much shorter periods. In arid and semiarid tropical soils elements of easy mobility (e.g. B, I, F) as well as trace metals are likely to concentrate at the surface horizon together with salts. Cycling in arid soil conditions is controlled largely by wind erosion or by irrigation water.

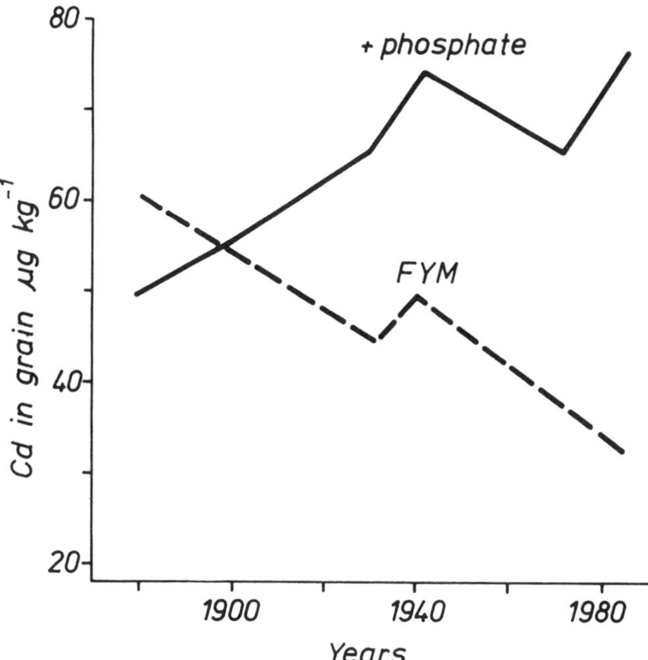

Fig. 13.6. Changes in wheat grain Cd under various fertilization regimes (Jones *et al.*, 1987).

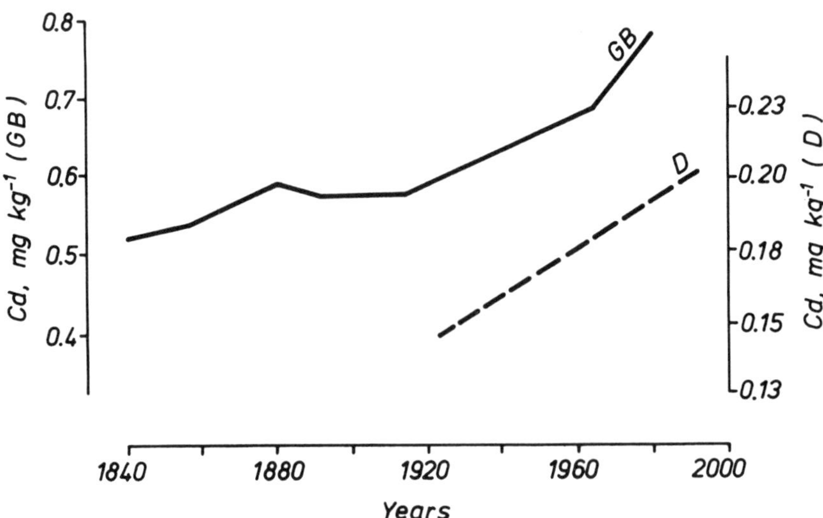

Fig. 13.7. Changes in soil Cd levels in the unfertilized plots of long-term experiments in Great Britain and Denmark (Jones *et al.*, 1987; Christensen and Tjell, 1991).

Contamination of agricultural soils, especially in Europe, has already become relatively common and is likely to continue. Noticeable is the fact that most often the soils become contaminated by several inorganic pollutants that are accompanied quite frequently by acid rain (mainly SO_2, NO_2 and HF). The long-term effects of such pollution on the deterioration processes in soils are not yet fully predictable.

Phytoavailability of Trace Elements and Plant Tolerance

The soluble plus exchangeable fractions characterize the mobile, and therefore phytoavailable species of trace elements in soils. A highly significant correlation is always observed between the metal contents and the concentration of mobile species in soils. However, root uptake is a complex metabolic and/or non-metabolic process and is controlled by several plant and soil factors. There is a great diversity in plant ability either to accumulate or to exclude several trace elements from the root media. The stress of both deficiency and excess of these elements alters plant reactions, and is known to be able to produce mutagenic changes in plants (Table 13.6). Among several soil factors the form in which trace elements are added to the soil also has significant impact on their availability. In most cases, trace metals added in the form of soluble compounds are easily taken up by cereal plants and transported to grain (Fig. 13.8).

Table 13.5. Mobility of trace elements in soils.

Element	Conditions	Plant/soil coefficient	Residence in soils (years)[a]
Cd	pH (−), Eh (−)	10	70–1000
Zn	pH (−), OM (+)	1	70–3000
Hg	alkylation	1 (org.) 0.01 (min.)	500–1000
As	Eh (−), alkylation	0.1	1000–3000
Pb	pH (−), Eh (−), OM (−)	0.01	700–6000
Be	Eh (−)	0.1	500–1000

[a]After Bowen (1979).
The identified effects are in brackets.

The adaptability of plants to deficient or toxic levels of trace elements in the growth medium may allow new varieties to be selected or bred. However, tolerant plants due to their ability to grow in contaminated soils, and due to the accumulation of sometimes extremely high amounts of trace metals, may create a great health risk by forming a polluted link in the food chain.

The possibility of removing an excess of some trace elements by plants (so-called 'biological decontamination' of soils) can be evaluated from the figures given in Table 13.3. The relative proportion of trace elements accumulated in a high yield of a species with a very high trace element content hardly exceeds 1% of the soil content. The estimation is related to the background concentrations in soils. When a soil becomes polluted, the proportion of trace elements that may be removed by accumulation in vegetation is even less.

Table 13.6. Soil–plant interface processes.

	Soil micronutrient status	
Plant reaction	Deficiency	Excess
Tolerance	Low	Moderate or high
Root	Increased exudates	Selective uptake or immobilization in tissues
Compound produced	Cations and acids (e.g. mugineic)	Proteins: metalothioneins, phytochelatins
Transmutation	Mostly induced	Natural and induced

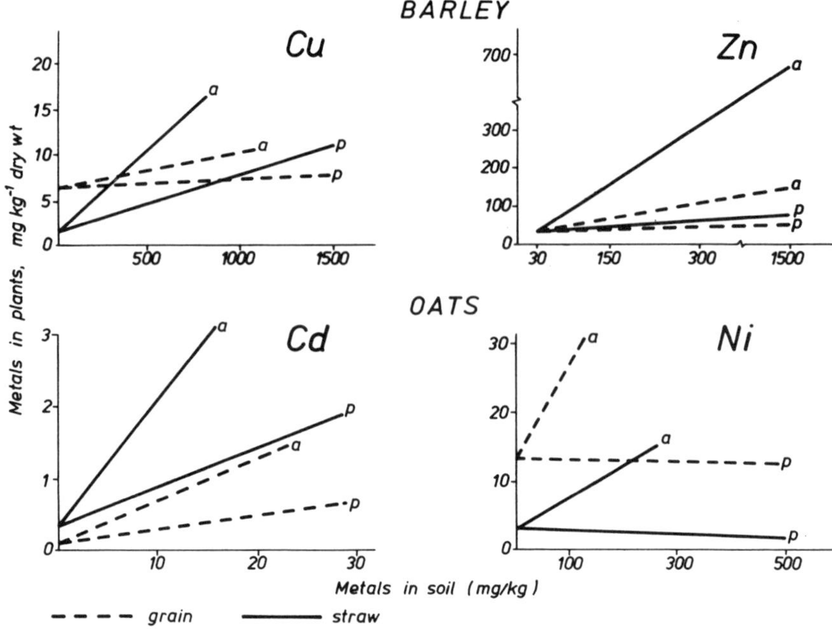

Fig. 13.8. Metal uptake by barley and oats as influenced by their origin (Grupe and Kuntze, 1988; Chlopecka, 1992).

Sustainable Management of Micronutrients in Soils

The maintenance and improvement of soil fertility in relation to micronutrient status should be based on understanding the delicate balance of processes at the soil-solution–root interface. The total content and phytoavailability of micronutrients is subject to both natural and anthropogenic factors. The role of these factors differs (Fig. 13.9) but any soil system may be degraded to a non-sustainable stage.

The sustainable management of micronutrients for the maintenance of soil fertility should meet several environmental requirements. The principal data necessary for the determination of acceptable application rates are:

1. input–output balance based on the initial micronutrient contents of a soil and the total amount added;
2. threshold values of micronutrient concentrations;
3. relative ratios between interacting elements (macro, and micro);
4. soil characteristics (e.g. pH, Eh, water regime etc.);
5. equivalent trace element toxicity to plants;
6. phytoavailability and phytoaccumulation coefficients.

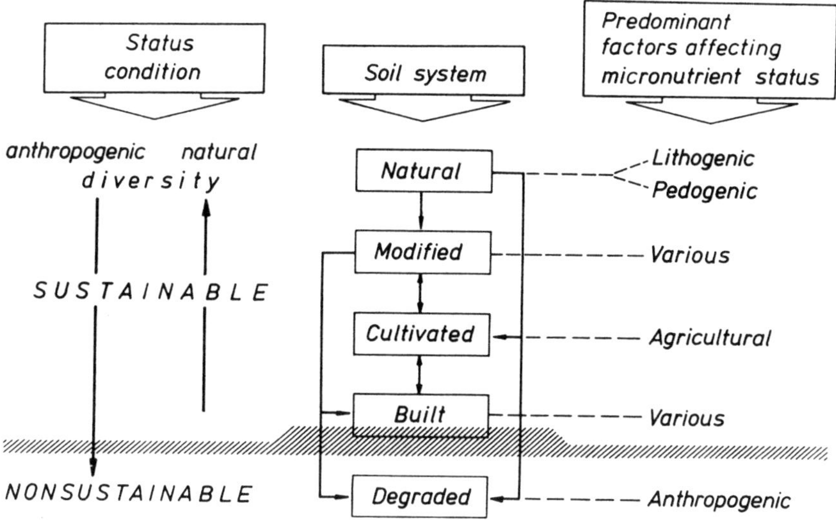

Fig. 13.9. Factors affecting the micronutrient status of various soil systems.

The limitation of trace element loads in soils is the most effective way of protecting soils against contamination. There are several guidelines for the threshold levels of trace elements in soils. These differ from place to place to meet the variable soil ecological conditions in each region or country. A number of ecological standards are used for this purpose:

- NOEC – No Observed Ecological Consequences
- TTL – Threshold Trigger Level
- LKE – Lowest Known Effect
- PAA – Permissible Annual Application
- MAL – Maximum Allowable Loading
- MPL – Maximum Permissible Loading
- MCA – Maximum Cumulative Amounts
- RMCL – Recommended Maximum Contaminant Levels
- ATL – Action Trigger Level
- ELD – Ecosystem Lethal Dose

In addition to general ecological effects, when setting up criteria the possible damaging effects on crops, injury to animals (particularly the hazards associated with the ingestion of soil during grazing) and effects that may occur in the food chain must be taken into account. To ensure the safe quality of food-chain products it is useful to standardize the maximum allowable loading (MAL) of pollutants in agricultural land. The proposed MAL values in European countries do not differ significantly (Table 13.7).

Table 13.7. Maximum allowable loading (MAL) of trace elements into arable soils.

	MAL (kg ha^{-1} $year^{-1}$)		
Element	A	B	C
Zn	15	30	10
Cr	–	–	15
Pb	15	10	10
Cu	7.5	5	5
Ni	3	3	3
Cd	0.15	0.15	0.2
Hg	0.1	0.1	0.2

A – Code of Practice GB (1989)
B – Council Directive EC (1986)
C – Kabata-Pendias and Piotrowska (1987).

However, any value that is established at present should be regarded as tentative and may be changed with improvements in the database and understanding of the significance of particular trace elements in different soils, farming practices, and/or ecosystems. Agreements on criteria based on the available amounts of trace elements present more problems because of the difficulty in measuring 'availability' either by biological or chemical methods.

Reclamation of soils contaminated by some trace metals is usually based on the application of lime and phosphates and the addition of organic matter. Although the addition of lime does not always bring the expected results in the immobilization of some trace metals, in most cases these treatments are quite effective in lowering trace metal concentrations in plants. There are also other techniques of soil purification/rehabilitation based on the principles of soil excavation, soil enclosure (quarantine), and specific curative treatments (based on biological, chemical and electrochemical extraction of metals). The sustainable and acceptable land use at different degrees of pollution always needs to be designed for a specific plant–soil system.

Each standard value for trace element levels in soils, as well as any reclamation and land use systems should consider the great risk of water pollution. The excess of several trace elements (including micronutrients) in groundwater, wells, lakes, and rivers due to either industrial or agricultural (including both crop and animal husbandry systems) pollution has already created great environmental and health problems.

Conclusions

Sustainability and environmental quality are inseparable preconditions for the successful management of long-term micronutrient levels in agrosystems.

Contemporary industrial technology and farming have already contributed to non-sustainable soil conditions in certain regions. Some changes in certain soils have reached an irreversible stage, or a point where the reclamation is very difficult or uneconomic.

In most of the developing countries the soils of some regions are seriously threatened by unsustainable farming practices and environmental pressure. Soils in the tropics are especially subjected to deterioration. In developed countries agriculture has slowly become more sustainable. However, manufacturing practices and energy production still threaten the balance of trace elements in most ecosystems.

The global problem with respect to micronutrient status is how to control by soil management and by correct fertilization the increase in their mobility. When deficiencies occur the technique of precise leaf application of a particular micronutrient should be more commonly adopted.

With regard to the environmental balance of trace elements and especially to protection against water pollution and ensuring that excesses do not enter the food chain, guiding principles and regulations for acceptable levels of inorganic trace pollutants must be adopted at national and international levels.

References

Bowen, H.J.M. (1979) *Environmental Chemistry of the Elements.* Academic Press, London, 333 pp.

Chlopecka, A. (1992) Forms of trace metals from inorganic sources in soils and amounts found in spring barley. *Water, Air, and Soil Pollution,* 66.

Christensen, T.H. and Tjell, J.Chr. (1991) Sustainable management of heavy metals in agriculture, example: cadmium. *Heavy Metals in the Environment* 1, 40–48.

Code of Practice for Agricultural Use of Sewage Sludge (1989) Department of the Environment, HMSO, London.

Council Directive on the Protection of the Environment and in Particular of the Soil, when Sewage Sludge is used in Agriculture. (1986). Official Journal of the European Communities, no. L 181/6.

De Bruijn, A.J. and De Walle, V.P. (1989) Standards for soil protection and remedial action in the Netherlands. In: Wolf, K., van den Brink, W.J. and Colon, F.J. (eds) *Contaminated Soil '88.* Kluwer Academic Publishers, Dordrecht, pp. 339–349.

Dudka, S. and Chłopecka, A. (1990) Effect of solid-phase speciation on metal mobility and phytoavailability in sludge-amended soils. *Water, Air, and Soil Pollution* 51, 153–160.

Finck, A. (1982) *Fertilizers and Fertilization.* Verlag Chemie, Weinheim, 438 pp.

Gambrell, R.P. and Patrick, W.H. (1989) Cu, Zn, and Cd availability in sludge-amended soil under controlled pH and redox potential conditions. In: Bar-Yosef, B., Barrow, N.J. and Goldshmid, J. (eds) *Contaminants in the Vadose Zone.* Springer-Verlag, Berlin, pp. 89–106.

Grupe, M. and Kuntze, H. (1988) Zur Ermittlung der Schwermetallverfügbarkeit lithogen und Anthropogen Belasteter Standorte, I. Cd und Cu. Z. *Pflanzenernährerung und Bodenkunde* 151, 319–324.

Jones, K.C., Symon, C.J. and Johnston, A.E. (1987) Long-term changes in soil and cereal grain cadmium: Studies at Rothamsted Experimental Station. *Trace Substances in Environmental Health* 21, 450–460.

Kabata-Pendias, A. (1992) Trace elements in soil of Poland – occurrence and behavior. *Trace Substances in Environmental Health* 25, 53–70.

Kabata-Pendias, A. and Pendias, H. (1992) *Trace Elements in Soils and Plants,* 2nd edn. CRC Press, Boca Raton, FL, 365 pp.

Kabata-Pendias, A. and Piotrowska, M. (1987) *Trace Elements as Criteria for Waste Use in Agriculture.* Seria P33, IUNG, Puławy, 46 s. (published in Polish).

Kabata-Pendias, A., Dudka, S., Chłopecka, A. and Gawinowska, T. (1992) Background levels and environmental influence on trace metals in soils of the temperate humid zone of Europe. In: Adriano, D.C. (ed.) *Biogeochemistry of Trace Metals.* Lewis Publisher, Boca Raton, FL, pp. 61–84.

Sillanpää, M. (1982) *Micronutrients and the Nutrient Status of Soils: A Global Study.* Food and Agriculture Organization of the United Nations, 444 pp.

Wild, A. and Jones, L.H.P. (1988) Mineral nutrition of crop plants. In: Wild A. (ed.) *Russell's Soil Conditions and Plant Growth.* Longman Scientific and Technical Publishers, Harlow, Essex, pp. 69–112.

Chapter 14
Maintaining Soil Physical Conditions

R.I. Papendick

Soil Scientist, US Department of Agriculture, Agricultural Research Service, 215 Johnson Hall, Washington State University, Pullman, WA 99164-6421, USA

Introduction

The soil physical condition controls several key functions in a given ecosystem that are important to biological processes and environmental stability. These include the capacity of the soil to accept, retain and transmit water, provide for energy and gas exchange, recycle nutrients, resist erosion and nutrient loss, and overall, provide a hospitable medium for the growth of plant roots, and beneficial soil microorganisms, insects, and animals. Physical factors that control these functions are often highly interactive and effects on biological processes are not always easily defined. For example, poor root growth in compact soil may be the single or combined result of physical impedance, lack or excess of water, or oxygen deficiency that restricts root proliferation and nutrient absorption. Poor physical conditions can increase root diseases and soil insect damage and decrease crop competition against weeds. As a result plants may die or grow poorly because of these secondary effects.

Soil properties that have most influence on the soil physical condition include texture, water status, bulk density, aggregation, porosity, organic matter content, strength and consistency. Although not a strict soil physical property, soil organic matter, directly or through stimulating biological activity, can affect, usually positively, several of these properties to varying degrees (Table 14.1). In many cases the properties are not independent; a change in one will cause a change in others. For example, a change in aggregation will cause a change in porosity and possibly in bulk density, strength and consistency. Apart from the texture the soil properties are influenced mainly by structure or aggregation of the primary particles. The characteristics of structure that are most important for plant growth and organism

From *Soil Resilience and Sustainable Land Use*
(eds D.J. Greenland and I. Szabolcs). CAB INTERNATIONAL Wallingford

Table 14.1. Effect of organic matter on soil physical properties[a].

Property	Effect	Remarks
Aggregation	+	Most effect on sandy soils and clays
Aggregation stability	+	Plays a primary role but not well quantified
Bulk density	+	Nearly always decreases bulk density
Available water	N	Some improvements for sandy soils
Water movement	+	Improves infiltration

[a]Interpretations from review paper by MacRae and Mehuys (1985).
+ = improvement; N = neutral.

habitat are the pore size distribution (relative numbers of pores in different size classes), and aggregate stability (the ability of individual aggregates to hold their shape upon wetting or reform upon drying). Many physical problems of soils (poor aeration and drainage, compaction, high strength) relate to poor structure, either inherent or caused by management practices.

Soil Degradative Processes

Soil degradation is a widespread and increasingly serious global problem (Oldeman *et al.*, 1990; Sanders, 1992). Huge areas of the world's land resources in both technically advanced countries and developing countries are being undermined through improper land use and poor management. The most apparent effects of degradative processes on the soil physical condition are accelerated erosion, lower infiltration rates, crusting, lower water-holding capacity, and poor structure.

Soil erosion

On a global scale soil erosion is by far the most widespread cause of soil degradation. Wind and water are the primary erosive forces but on steep slopes downhill movement of soil by tillage can be significant. The most common effect of erosion is loss of topsoil which exposes a less-productive subsoil with reduced ability to hold and recycle nutrients, and absorb, retain and freely circulate air and water. It has been estimated that erosion is causing a loss of the world's topsoil at a rate of 7% each decade (Brown and Wolf, 1984). This layer which contains most of the organic matter and nutrients averages only 15–20 cm deep for most soils. Experiments show that yields of crops where the topsoil has been removed or severely eroded are 20–65% less than where the soils are only slightly eroded (Langdale *et al.*, 1979; Massee, 1990). Erosion is a selective process, normally removing organic matter, clay, and fine silt first and carrying these furthest away. As a result eroded soils are generally less favourable for seedling establishment and root proliferation which are important for early plant development. There are many adverse secondary effects of erosion, such as gully formation and sedimentation in dams and waterways.

Organic matter decline

Organic matter binds soil into aggregates that are responsible for good soil structure and porosity distribution, allowing for aeration, root penetration, and movement and retention of water. Natural levels of organic matter inevitably drop following cultivation but the rate of decline is dependent on management practices and inputs such as crop rotation, residue management and application of organic amendments. In the USA many prairie soils have lost over one-third of their initial organic matter contents after 100 years or more of cultivation; the levels of organic matter now appear to be stabilizing (Sampson, 1981). However, it is not known whether the levels reached are sustainable and will hold productivity at satisfactory levels in the future. Continuing declines in organic matter lead to subtle deterioration of soil structure which in turn creates difficulties with seedbed preparation, seedling emergence and root growth.

Farming itself can increase organic matter levels and improve the physical condition of some soils over native conditions. Under natural conditions desert soils in western USA have very low organic matter contents and poor structure. With good management under irrigation organic matter levels may increase by several percentage points, accompanied by improved water infiltration rates and retention capacity. The key to correcting the deficiencies with the original soil is organic matter management.

Soil compaction

Compaction caused by farm equipment or animal traffic increases bulk density and reduces pore space that provides for passage of air, water, roots, and functioning of fauna and microorganisms. The trend toward larger equipment with modern agriculture not only increases axle weight but also the total percentage of the soil surface affected (Voorhees, 1979). Heavier equipment increases traction which enables earlier tillage in the spring when the soil is too wet and is most subject to compaction by wheel traffic. Besides adverse effects on root penetration and water and air movement, the hard and compacted layers require more energy for tillage. For example, tractor wheel slippage increased from 14 to 25% when the percentage of land previously wheel-tracked increased from 20 to 40% (Voorhees, 1979).

Compaction depth by equipment may exceed 30 cm and varies with soil type, moisture content, organic matter content and texture. Soils that are low in organic matter with a high content of silt or fine sand are most susceptible to compaction. Increases in compaction occur slowly with changes in farming practices and direct affects on crops are not easily measured because of complex relationships with other factors. However, yield reductions in some cases can be dramatic. For example in Britain yield of peas (*Pisum sativum*) were reduced to 50% under compacted soil conditions (Hebblethwaite and McGowan, 1980). Cotton (*Gossypium* sp.) yields in trafficked plots were 6.4 bales ha^{-1} compared with 7.7 bales ha^{-1} in untrafficked areas (Gill and Trouse, 1972).

Alkalization

Alkalization is a process by which sodium becomes the dominant cation in the soil solution owing to the precipitation of calcium and magnesium compounds (US Salinity Laboratory Staff, 1954). This occurs when excess soluble salts accumulate in a soil, usually from irrigation water and the solubility limits of calcium and magnesium salts are exceeded leading to a corresponding increase in the relative proportion of exchangeable sodium. If the excess neutral salts are leached the exchangeable sodium is free to hydrolyse and the pH may rise to between 8.5 and 10. The highly alkaline condition results in dissolution and dispersion of organic matter, dispersion of clay particles, and loss of structure that makes the soil unfavourable for entry and movement of water, tillage and plant growth. Alkalization is generally a problem with irrigation systems or in some cases where the hydrology of a watershed is adversely affected by management practices.

Methods for controlling alkalization have been developed largely through irrigation research and are reported in a number of handbooks and textbooks. The key lies in removing excess salts, replacing exchangeable sodium with calcium, and improving the soil physical condition. If the soil contains gypsum, or gypsum is added, leaching will replace sodium by calcium concurrently with removal of excess salts (US Salinity Laboratory Staff, 1954). Where a high water table and salts are present the upward movement of salts can be checked by use of surface mulches, and in some circumstances by growing a deep-rooted crop such as alfalfa (*Medicago sativa*) (Thorne and Peterson, 1954). Application of organic materials such as farm manure and the root action of salt-tolerant crops are helpful for improving the physical condition of alkali soils.

Characterizing the Soil Physical Condition

It is impossible to evaluate the status of the soil physical condition in relation to productivity and environmental issues, and more importantly changes due to land use and management unless its status at different points in time can be quantified. Methods or procedures are needed to assess the effects of management systems on the physical condition of the soil resource similar to the way the Universal Soil Loss Equation (USLE) and the Wind Erosion Equation (WEQ) are used to assess the effects of conservation management systems on wind and water erosion. Obviously a number of factors need to be considered to create an index which can be used to track the condition of the soil in response to land use and management inputs. Several attempts to do this are being made.

Soil tilth

Tilth is a term that is used to describe qualitatively the soil physical condition as to its fitness as an environment for soil microbial and faunal populations and processes, and plant growth (Karlen *et al.*, 1990). Tilth is regarded as a dynamic characteristic that is subject to change by both natural forces and modifications caused by management and land use. Karlen *et al.* (1990) also defined a new term 'tilth forming processes' as 'the combined action of physical, chemical, and biological processes that bind primary soil particles into simple and complex aggregates and aggregate associations that create specific structural or tilth conditions'.

Tilth index

Though the soil tilth concept is not new there has been no method available whereby it could be measured and quantified. Singh *et al.* (1992) proposed a method to calculate a tilth index with values ranging from zero for plant unusable conditions to 1 for non-limiting soil tilth. They chose bulk density, cone penetrometer index, uniformity coefficient, organic matter content and plasticity index to represent soil properties affecting tilth. The procedure was to compute a tilth coefficient for each soil property using a polynomial relationship for values between 1 (a minimum value below which soil is considered non-limiting to roots) and 0 (a maximum value above which soil is considered to be unusable by plants). For example, they concluded from the literature that any soil with a bulk density less than or equal to 1.3 Mg m^{-3} was considered non-limiting to maize (*Zea mays*) roots. A maximum value of 2.1 Mg m^{-3} was chosen as being unusable by plants. The tilth index TI was computed by the following equation.

$$TI = CF_1 \times CF_2 \times \ldots \times CF_n \quad (14.1)$$

where: $CF_1 \ldots CF_n$ are tilth coefficients for each of the n soil properties.

Field verification of the tilth index for two soils in Iowa, USA, each with two different cropping systems (continuous maize, and maize–soybean (*Glycine max*) rotation) and five tillage systems showed strong positive correlations between tilth index and crop yield (Singh *et al.*, 1992). The effects of sampling time showed that the tilth index was increased by tillage and planting operations and then decreased with time until harvest. Highest correlations between tilth index and yield were obtained from tilth index measured after tillage in the maize–soybean rotation and after planting in continuous corn. This is the first known attempt to quantify soil tilth. The authors recommend that the procedure needs to be tested over a wide range of soil, climatic, and management conditions.

Soil quality

Soil quality is a concept that is gaining acceptance as a way of describing the capacity of the soil for sustainable production of healthy and nutritious crops, and its structural and biological integrity which helps resist erosion and pollution, and environmental and biological stresses on plants (Parr *et al.*, 1992). Larson and Pierce (1991) define soil quality 'as the capacity of the soil to function within ecosystem boundaries and interact positively with the environment external to that ecosystem'. They propose that indicators of soil quality should reflect the ability of the soil to perform three critical functions, i.e., provide a medium for plant growth; regulate water flow in hydrological units; and serve as an environmental filter. In this context the soil physical condition would be a major determinant of soil quality.

A conference was held in July 1991, at the Rodale Research Center, Kutztown, Pennsylvania, USA on the 'Assessment and Monitoring of Soil Quality' (Rodale Institute, 1991). One of the recommendations of the conference was that the attributes of soil quality should not be limited to soil productivity *per se* but should include environmental quality, human and animal health, the food safety and quality (Parr *et al.*, 1992). Thus, soil tilth could be viewed as a subset of soil quality. However, the soil physical condition would be an important component in the formulation of a soil quality index.

Characterization of soil quality

The Rodale Conference proposed a hierarchical or multilevel operational framework for defining soil properties to monitor changes in soil quality on a regional, watershed or field-scale basis (Rodale Institute, 1991). This consisted of the following components: (i) agroecosystem (major land or crop area); (ii) the three soil quality issues (productivity, environment, and health); (iii) a mesolevel soil quality index (tilth, fertility, toxicants); (iv) first order soil properties; and (v) second-order soil properties.

The conference made a good start on defining components of soil quality and a preliminary list of base-line soil properties which should be measured. The next step is to develop a method or procedure for integrating the interactions of the physical, chemical, and biological components of soil quality indices and testing the usefulness of indicators as related to the productivity, environmental, and health components.

Key indicators–minimum data set

Larsen and Pierce (1991) proposed that soil quality can be measured by selection of key indicators which they referred to as a minimum data set. Changes in soil quality can be evaluated as the sum of changes in individual soil quality attributes from some reference point in time. Soil attributes may include such properties as texture, organic matter, pH, nutrient status, bulk density, electrolytic conductivity, and rooting depth. They suggest that other properties can be obtained from the minimum data set through functional relationships called pedotransfer functions that relate different soil characteristics and properties with one another for use in evaluating soil quality. The minimum data set and pedotransfer functions can be used to determine whether the soil quality is changing under a given land use or management practice.

Soil condition ratings

The Soil Conservation Service of the US Department of Agriculture has attempted to develop a method to provide soil quality ratings for crop management systems (USDA, 1991). A Midwest Soil Condition Rating Committee was formed in 1987 to adapt the concept of rating indices to quantify the effect of cropland management systems on the soil physical condition and productivity. The approach was to develop a soil quality rating model that estimates the combined effect of three variables on trends in soil quality. These include: (i) the amount of organic matter that must be returned to the soil in order to maintain or increase the organic matter content; (ii) the effect of predicted erosion associated with the management system; and (iii) the physical effects of various tillage systems. The form of the soil condition rating model is

$$SQR = OM + TP + ER \tag{14.2}$$

where: SQR is the soil quality rating.

OM is the organic matter subfactor which includes all undecomposed organic material from plant sources either grown or imported to the site and returned to the soil.

TP is the subfactor which accounts for all field operations which break down residues and aerate the soil including tillage, planting and fertilizer injection.

ER is the erosion subfactor which accounts for the effect of soil removal and sorting of surface soil material by erosion as predicted by the USLE or WEQ.

The SQR is intended for use as a tool to evaluate farming practices which maintain or increase soil organic matter and enhance biological processes that improve the soil physical condition. The rating expresses whether the cropping sequence, soil tillage, and other management inputs cause positive or negative trends in soil quality, that is, the direction of change. Its use is much like that of the USLE and WEQ for erosion prediction, that is to predict the long-term effect of management systems on the physical condition of the soil resource.

Multiple variable indicator kriging

Smith *et al.* (1993) have developed a computational procedure called multiple variable indicator kriging (MVIK) that integrates a set of continuous variables into a single indicator value that can be used to map soil quality on a landscape basis. Threshold values of soil quality parameters are determined independently by a judgement process. For example, for good soil quality pH thresholds of 4 and 7 may represent minima and maxima, respectively, for crop production. Similarly, values which may be crop specific are established for infiltration rate, bulk density, etc. Each datum is transformed to an indicator variable (zero or 1) depending on its relationship to the chosen threshold value (Journel, 1988; Isaaks and Srivastava, 1989). Such a transformation allows skewed or highly variable data sets to be analysed by non-parametric geostatistics.

The MVIK procedure, which is an extension of indicator kriging provides a powerful way to integrate several indicators (soil properties) into a single new indicator data set which is then used to develop a variogram. The variogram is used to estimate indicator values at other unsampled locations via kriging. Maps are then developed that show the probability of meeting the specified soil quality criteria on a landscape basis.

The power of the MVIK technique is that any number of independent parameters that are being used as measures of soil quality can be evaluated individually or integrated to form a single map. If one or more of the selected parameters fails to meet specified criteria the probability for meeting an acceptable soil quality level will be relatively low on the contour map. In addition, locations on the landscape that have a low probability of being 'good' quality soil can be analysed further to identify which soil parameter is responsible. The MVIK method is flexible enough to allow for adjustments in threshold levels for different soil properties which may be required for different cropping and agroecosystems. A critical aspect of MVIK is the judgement process in selecting threshold values for the different soil properties. A small change in even one variable can cause major changes in a soil quality map (Smith *et al.*, 1993).

The MVIK approach can be applied on a field, farm, watershed or regional basis depending on the number and location of soil samples. Repeated sampling and evaluation should provide a means to determine changes in the soil physical condition, i.e., whether the soil is being degraded, aggraded, or remaining unchanged for a given land use or set of management practices. The technique will also help to identify the cause and location or area of a specific problem. From this information treatments or management practices can be prescribed that will remedy the causes of degradation and loss of soil quality.

Strategies to Maintain and Improve the Soil Physical Condition

Conservation and organic matter management systems are being advocated that are designed to maintain or improve the soil physical condition. These are generally achieved through tillage and cropping sequence, cover and green manure crops, crop residue use, nutrient management, and application of organic amendments. The critical factor that is common to all the proposed management strategies is maintenance of soil organic matter.

Conservation tillage

Conservation tillage is defined as any 'tillage sequence which reduces loss of soil and water relative to conventional tillage', and conventional tillage is considered to be use of a disc or moldboard plough, followed by harrowing. Surface residue management is an important component of most conservation tillage systems because of their effectiveness for controlling erosion, reducing runoff and evaporation, and retarding the rate of organic matter decline. Figure 14.1 shows that a 30% residue cover reduces wind or water erosion by nearly 70% compared with bare soil. About 600 kg ha^{-1} of wheat straw will provide 30% surface cover (Yoder *et al.*, 1993) which illustrates the effectiveness of relatively small amounts of residue for erosion control.

Much progress has been made in the USA and many parts of the world in developing practices for different climate, soil, and cropping conditions but factors remain that limit their adoption by farmers. In the USA nearly one-third of the planted cropland area is currently under some form of conservation tillage system (Conservation Technology Information Center,

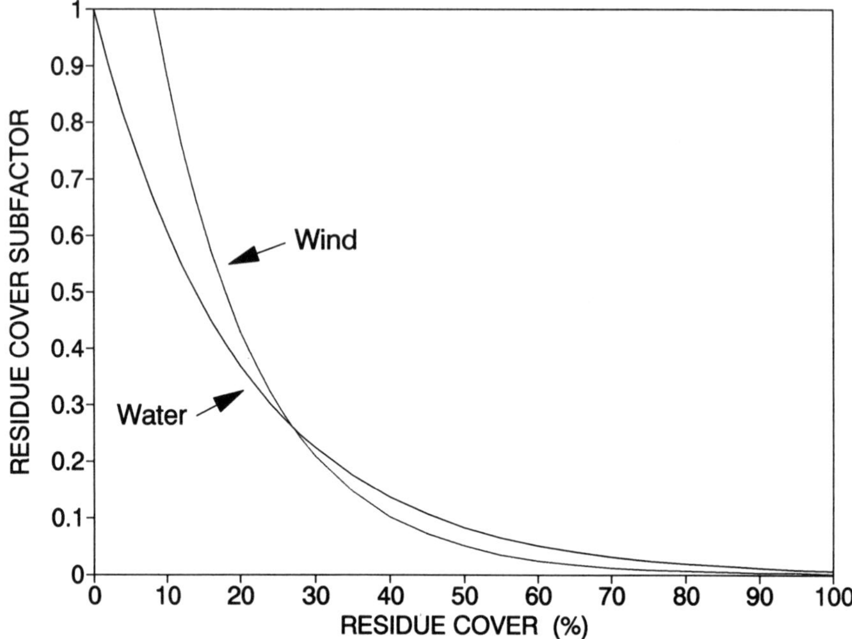

Fig. 14.1. Effect of residue cover on soil erosion by wind (Fryrear, 1985) and water (Yoder, 1992).

1991). It is projected that by the year 2000 at least 75% of the cropland area will be in some form of conservation tillage.

The main features of conservation tillage are increased soil surface roughness and maintenance of crop residues on the soil surface. This is achieved by non-inversion tillage which undercuts or mixes some residues in the shallow layers, and fewer tillage operations, especially secondary tillage. These practices reduce runoff, and shading by residues usually results in lower soil temperatures during warm weather. The extreme case of conservation tillage is no-till where the crop is planted and fertilizer is applied in a one pass operation with minimal disturbance of soil and surface residues. Most studies indicate that no-till planting in surface residues reduces soil erosion to almost zero over a wide range of conditions. In the USA 26% of the cropland areas farmed in a conservation tillage system is planted to no-till and the trend is increasing (W. John Becherer, Conservation Technology Information Center, personal communication).

Soils that are severely degraded or in poor physical condition may not be amenable to reduced tillage or no-till farming (Lal, 1980). Some tillage may be necessary on these soils to break up hard, compact layers, and crusts to improve drainage or to increase infiltration rates. Breaking up high density soils can also increase the rooting depth and markedly increase crop production, especially under low rainfall conditions (Willcocks, 1981). The soil condition may be the result of low organic matter contents and a history of limited residues being returned to the soil. Unger *et al.* (1991) cite a number of studies in semiarid regions where tillage was superior to no-till in increasing crop yields but in most cases these involved problem soils or soil conditions related to past management practices. However, benefits with tillage are usually greatest when done in combination with surface residue management.

Conservation tillage generally promotes an increase in soil organic matter as a result of less tillage, greater return and less disturbance of residues, and less erosion compared with conventional clean tillage systems (Bruce *et al.*, 1990). These new levels of soil carbon should improve the soil physical condition as a medium for plant growth and its resilience to environmental stresses and management inputs.

A limitation to conservation tillage is in the semiarid regions where little residue is produced or residues are removed for other purposes. When cover is not present wind erosion on some cropland soils can be controlled by tillage that forms ridges or leaves non-erodible clods on the surface. On some soils tillage must be repeated after major rainstorms to roughen the surface again (Fryrear, 1990). To control water erosion infiltration must be maintained above rainfall rates through adequate surface macroporosity or microcatchment of rain, or excess water must be conveyed from the land at non-erosive velocities (Unger *et al.*, 1991). However, without return of adequate residues it is difficult to maintain the soil physical condition through tillage alone.

Cropping systems

The cropping system used in combination with conservation tillage has a major influence in maintaining soil organic matter levels and the soil physical condition. High residue-producing crops in a no-till or reduced tillage system and legume or other forage crops in the rotation are especially effective for slowing the decline of organic matter or increasing the equilibrium level in the soil. A higher net return of organic matter and improved soil structure from grass and legume pastures compared with cash and grain crops is presumably due to the activity of the soil fauna and flora, steady root growth and lack of soil disturbance during the time the pasture is in the rotation. Sod-based rotations can reduce erosion to negligible levels during the period of sod cover and the residual effects improve infiltration and leave the soil less erodible for several years after the sod is replaced with other crops (Wischmeier, 1960).

Many studies have shown that the amount of crop residue available for return to the soil is a key determinant influencing the equilibrium soil

Table 14.2. The influence of cropping sequence on wheat yield, and organic carbon in the surface 7.5 cm of soil at three locations in Canada[a].

Location	Rotation	Mean annual yield (kg ha^{-1})	Organic carbon (%)
Swift Current	W–F	949	1.9
1967–1982[b]	W–W–F	1105	1.7
1967–1984	W–W	1354	2.2
Indian Head	W–F	1230	2.2
1958–1984	W–W–F	1464	2.3
1960–1984	W–W	1810	2.4
Melfort			
1957–1984	W–F	1392	3.8
1960–1984	W–W	1808	4.2

[a]Adapted from Campbell *et al.* (1990).
[b]The first period is the duration of the cropping sequence for measurement of organic carbon. The second period is the duration for measurement of yield.
W = spring wheat; F = fallow.

organic matter content. For example, Larson *et al.* (1972) showed that soil organic carbon was linearly related to the amount of maize stover returned to the soil over an 11-year period. The average increase was 10 kg of organic carbon for each Mg of residue returned. Research in Canada with cropping sequences in long-term experiments also shows that increases in organic carbon levels are generally associated with higher grain yields indicating higher residue production and return to the soil (Table 14.2). Similar results were obtained in a 50-year study at Pendleton, Oregon, USA where a continuous wheat or wheat–pea sequence produced 2300 kg ha^{-1} of residue annually and resulted in a soil organic carbon content of 1.6% compared with wheat–fallow which produced 1700 kg ha^{-1} of residue annually and resulted in a soil carbon content of 1.1% (Table 14.3).

A long-term study (1910–1980) at Lethbridge, Canada, established immediately after breaking the soil showed a rapid decline in soil organic carbon after the first 12 years (Freyman *et al.*, 1982). From then on carbon levels gradually increased and by 1980 concentrations approached original values. The resurgence of organic carbon was attributed to increased additions of crop residues resulting from improved yields and management practices including the introduction of combine harvesters, which, unlike early threshing methods, returned all the residues to the soil.

Crop rotations can affect soil aggregation, bulk density, microbial biomass, water infiltration and other soil properties. Francis and Clegg (1990) cite research showing higher bulk densities, reduced water infiltration and lower soil organic matter content and microbial biomass in monoculture

Table 14.3. Annual residue production for different cropping systems and effect after 50 years on organic carbon content in the surface 10 cm of soil. Pendleton, Oregon[a].

Cropping system	Annual residue production ($kg\ ha^{-1}$)	Organic carbon (%)
Permanent pasture	ND	2.2
Wheat–pea	2290	1.6
Wheat–wheat	2250	1.5
Wheat–fallow	1680	1.1

[a]Adapted from Rasmussen *et al.*, (1989).
ND, not determined.

soybeans or other crops than when grown in a rotation. Bolton *et al.* (1982) reported reduced compaction in rotated crops compared with continuous cropping. Erosion rates may be considerably higher following a soybean crop than for other crops such as maize, even when grown continuously (Van Doren *et al.*, 1984).

Narrow strip cropping with three or four crops is being evaluated as a way of enhancing rotation effects by creating unique micrometerorological and soil conditions (Garcia *et al.*, 1990). Another management strategy that shows promise for improving soil physical conditions is alley cropping, that is growing alternate narrow strips of complementary crops including deep-rooted woody species. Strips of perennial vegetation, especially leguminous shrubs can add both fertility and organic carbon to the soil during the rotational period. Pruning shrubs and trees during the growing season provides mulch or green manure for the cropped area (Kang *et al.*, 1984). Alley cropping has also been shown to increase water retention, reduce soil bulk density, and maintain a high microbial biomass presumably from woody species (Yamoah *et al.*, 1986, Hulugalle and Kang, 1990).

Cover and green manure crops

The purpose of growing cover and green manure crops is to protect the soil against erosion and improve the growth of subsequent crops. In many areas of the USA, agronomic recommendations prior to the 1950s usually included a cover crop in the cropping sequence. Early studies indicated that returning crops to the soil increased infiltration, reduced runoff, and decreased erosion (Pieters and McKee, 1938). Cover crops were shown to improve the physical condition of clayey soils (Rogers and Giddens, 1957). The potential benefits of cover and green manure crops for soil improvement have been largely ignored as inexpensive synthetic nitrogen fertilizers became increasingly available.

The main effect of returning crop biomass to the soil on soil physical properties arises from the potential to increase soil organic matter levels. The effectiveness of a crop for maintenance and accumulation of organic matter depends on the composition of the crop, soil and climatic factors as they affect decomposition of the dead material, and management practices (MacRae and Mehuys, 1985).

It is generally reported that low-nitrogen green manures are more effective for increasing soil organic matter levels than high-nitrogen crops such as legumes because the latter decompose more rapidly and completely. However, improved soil physical properties may be derived from the higher yields of the main crop following a high-nitrogen green manure crop. For example, the 25-year average yields of seed cotton in a winter cover crop experiment were 2240, 1882, 1646 and 1008 kg ha^{-1} for hairy vetch (*Vicia villosa* Roth.), common vetch (*Vicia sativa* subsp. *sativa* L.), 45 kg N ha^{-1}, and zero N treatments, respectively (Patrick *et al.*, 1957). Soil organic matter values were highest with hairy vetch grown as a winter cover crop and lowest with the fertilizer only and zero N treatments. The physical condition of the soil in terms of aggregation, bulk density, and non-capillary porosity was best with the vetch plots. Similarly, in another study seed cotton yields were higher with hairy vetch grown as a winter cover crop than with nitrogen fertilizer alone but it took 10 years before the benefits of hairy vetch were realized (Millhollen and Melville, 1991). Plots that were winter fallow and fertilized with 67 kg N ha^{-1} at planting lost nearly 300 kg ha^{-1} of organic matter during 1973–1988 whereas plots not supplemented with nitrogen fertilizer but that included a winter cover crop of hairy vetch gained 900 kg ha^{-1} of organic matter during this period.

It is often difficult to differentiate whether improved crop performance following cover and green manure crops is due to improved soil physical conditions or enhanced fertility. Most experiments are not designed to separate these effects because of the complexity of factors involved. One strong indicator that an improvement in physical conditions is important is that a long-term benefit of cover and green manure crops is to stabilize yields of subsequent crops during dry seasons (MacRae and Mehuys, 1985).

Recycling of organic wastes

Many countries have traditionally utilized various sources of organic materials to improve the physical condition and fertility of their agricultural soils (Parr *et al.*, 1986). Available materials included sewage sludge, pit latrine wastes, municipal solid wastes, animal manures, food-processing wastes, and certain industrial wastes. When relatively inexpensive chemical fertilizers became easily available in the 1960s they largely replaced labour-demanding organic cycling practices, and in many cases organic matter management was neglected, or even forgotten (Parr and Papendick, 1983).

This trend raised concern about the potential long-term adverse impacts on soil productivity and environmental quality because of the well-known beneficial effects of organic materials on soil physical properties. For many soils, without regular additions of adequate amounts of organic materials there is increased leaching and erosion, and gradual deterioration of their physical properties. As a result of these concerns interest in organic cycling has been increasing and some efforts are underway to reintroduce established techniques and develop management practices that are compatible with modern farming technology (Parr *et al.*, 1986, 1989).

Some disadvantages of using organic wastes include the presence of odours and human pathogens, and their bulk, which makes them difficult to transport, handle, and store (Parr and Papendick, 1983). One way to resolve these problems is through composting, a process whereby the organic waste is partially decomposed into a stabilized form of organic matter that is odourless, pathogen free, and much less bulky and more easily handled than the original raw material. Technology for composting a range of organic wastes has been developed (Willson *et al.*, 1980) and widely adopted by both large and small municipalities in the USA and some developing countries. The method is simple, relatively inexpensive, and allows for trade-off between labour and capital.

Some organic wastes have certain properties such as extreme acidity or alkalinity, high or low carbon/nitrogen ratio or other characteristics which make them less suitable for composting. In such cases co-composting or blending with other materials may alleviate these deficiencies and provide a more usable and higher quality product (Parr and Papendick, 1983).

Composts have been successfully used for soil improvement and plant growth including the production of vegetables, field forage and nursery crops, and for the reclamation and vegetation of disturbed lands (Hornick *et al.*, 1979, 1984). Because of their properties the addition of sludge compost to sandy soils will increase their ability to retain water and reduce susceptibility to drought. Addition of sludge compost to clay soils has been reported to reduce soil compaction, lower the bulk density, and increase the rooting depth (Parr and Papendick, 1983).

The main limitation to use of organic waste for agriculture either as composts or raw materials appears to be a reliable delivery system for getting the materials to the land and applied to the soil. In many countries, including the USA, a well-defined infrastructure has not been developed between the routine sources of organic wastes and getting them processed and applied where needed on agricultural lands. In some cases application technology still needs to be developed especially if quantities of compost are limited. There is also some concern about environmental pollution associated with the application of certain wastes such as those containing toxic or heavy metals (Parr and Papendick, 1983). Thus, proper processing and recycling of organic wastes are critical for their safe use in agriculture.

Research Challenges

There is ample documentation that maintaining and improving the soil physical condition is paramount for sustaining crop production capability to meet future food needs and economic goals. Much information is available on how the soil physical condition regulates certain processes such as energy and water flow, and retention in the soil. However, less is known about quantitative relationships between the soil physical condition and crop productivity. Methods or procedures are needed that can quantitatively characterize land use and management-induced changes in the soil physical condition in relation to crop growth. Even less is known about how the soil physical condition impacts the environment such as in filtering, retaining and recycling nutrients, and recycling and detoxifying organic wastes added to the soil. Such information is fundamental for developing management approaches for maintaining and improving the soil physical condition for crop production and environmental protection. Future research should focus on the following goals.

1. The development of simplified methods and procedures for quantifying the physical condition of the soil with regard to effects on plant growth and environmental impact.
2. The determination of the role of biological processes on the soil physical condition and how these can be influenced by management to improve soil properties such as hydraulic conductivity, bulk density, aggregation, and non-capillary porosity.
3. The determination of the levels of organic matter required to maintain the productivity of a soil at satisfactory or sustainable levels.
4. The development of holistic cropping and soil management methods that will maintain organic matter at the levels required for sustained crop productivity.
5. The determination of the effects of conservation tillage, crop residue use, and cover crops on soil physical properties and organic matter content.
6. The determination of the impacts of soil compaction on infiltration, percolation, and crop yields as affected by soil management, and methods for alleviating compaction on problem soils.
7. The evaluation of the effect of organic waste or composts on soil physical properties with regard to rates of application, mode of application, timing, and composition of the organic materials.
8. The evaluation of relationships between green manuring, soil organic matter levels, and soil physical properties.

References

Bolton, E.F., Dirks, V.A. and McDonnell, M.M. (1982) The effect of drainage, rotation and fertilizer on corn yield, plant height, leaf nutrient composition, and physical properties of Brookston clay soil in southwestern Ontario. *Canada Journal of Soil Science*, 62, 297–309.

Brown, L.R. and Wolf, E.C. (1984) *Soil Erosion: Quiet Crises in the World Economy.* Worldwatch paper 60. Worldwatch Institute, 1776 Massachusetts Avenue, N.W., Washington, DC 20036, 49 pp.

Bruce, R.R., Langdale, G.W. and West, L.T. (1990) Modification of soil characteristics of degraded soil surfaces by biomass input and tillage affecting soil water regime. In: *Transactions of the 14th International Congress of Soil Science,* Kyoto, Japan, VI, pp. 4–9.

Campbell, C.A., Zentner, R.P., Janzen, H.H. and Bowren, K.E. (1990) *Crop Rotation Studies on the Canadian Prairies.* Publication 1841/E. Research Branch, Agriculture Canada, 133 pp.

Conservation Technology Information Center (1991) *Conservation Impact* 9, No. 9. National Association of Conservation Districts, 1220 Potter Drive, Purdue Research Park, West Lafayette, Indiana 47906–1334.

Francis, C.A. and Clegg, M.D. (1990) Crop rotations in sustainable production systems. In: Edwards, C.A., Lal, R., Madden, P., Miller, R.H. and House, G. (eds) *Sustainable Agricultural Systems.* Soil and Water Conservation Society, Ankey, Iowa 50021, USA, pp. 107–122.

Freyman, S., Palmer, C.J., Hobbs, E.H., Dormaar, J.F., Schaalje, G.B., and Moyer, J.R. (1982) Yield trends in long-term dryland rotations at Lethbridge. *Canada Journal of Plant Science* 62, 609–619.

Fryrear, D.W. (1985) *Wind Erosion on Arid Croplands.* Science Reviews, Arid Zone Research. Scientific Publishers, Jodhpur, India, pp. 31–48.

Fryrear, D.W. (1990) Wind erosion: mechanics, prediction, and control. *Advances in Soil Science* 13, 187–199.

Garcia, R., Cruse, R.M., Owen, M. and Erbach, D. (1990) Competition for soil water between components of a strip intercropping rotation as affected by tillage and position in the strip. In: *Farming Systems for Iowa: Seeking Alternatives.* Leopold Center for Sustainable Agriculture, 1990 Conference Proceedings. Leopold Center, Iowa State University, Ames, Iowa, p. 103.

Gill, W.R. and Trouse, A.C. Jr (1972) Results from controlled traffic studies and their implications in tillage systems. *Proceedings, No-tillage Systems Symposium,* Ohio State University, Columbus, OH, pp. 126–131.

Hebblethwaite, P.D. and McGowan, M. (1980) The effects of soil compaction on emergence, growth, and yield of sugar beet and peas. *Journal of the Science of Food and Agriculture* 31, 1131–1135.

Hornick, S.B., Murray, J.J., Chaney, R.L., Sikora, L.J., Parr, J.F., Burge, W.D., Willson, G.B. and Tester, C.F. (1979) *Use of Sewage Sludge Compost for Soil Improvement and Plant Growth.* Science and Education Administration, Agricultural Reviews and Manuals, Northeastern Region Series 6. US Department of Agriculture; Beltsville, MD, 10 pp.

Hornick, S.B., Sikora, S.B., Sterrett, S.B., Murray, J.J., Miller, P.D., Burge, W.D.,

Colacicco, D., Parr, J.F., Chaney, R.L. and Willson, G.B. (1984) *Utilization of Sewage Sludge Compost as a Soil Conditioner and Fertilizer for Plant Growth.* Agriculture Information Bulletin No. 464. US Department of Agriculture; Beltsville, MD, 32 pp.

Hulugalle, N.R. and Kang, B.T. (1990) Effects of hedgerow species in alley cropping systems on surface soil properties of an oxic paleustalf in Southwest Nigeria. *Journal of Agricultural Science* 114, 301–307.

Isaaks, E.H. and Srivastava, R.M. (1989) *An Introduction to Applied Geostatistics.* Oxford University Press, New York, 561 pp.

Journel, A.G. (1988) Non-parametric geostatistics for risk and additional sampling assessment. In: Keith, L. (ed.) *Principles of Environmental Sampling.* American Chemical Society, Washington, DC, pp. 45–72.

Kang, B.T., Wilson, G.F. and Lawson, T.L. (1984) *Alley Cropping – A Stable Alternative to Shifting Cultivation.* International Institute of Tropical Agriculture, Oyo Road, PMB 5320, Ibadan, Nigeria. Balding and Mansell Limited, 22 pp.

Karlen, D.L., Erbach, D.C., Kasper, T.C., Colvin, T.S., Berry, E.C. and Timmons, D.R. (1990) Soil tilth: a review of past perceptions and future needs. *Soil Science Society of America Journal* 54, 153–161.

Laflen, J.M. and Moldenhauer, W.C. (1979) Soil and water loss from corn–soybean rotations. *Soil Science Society of America Journal* 43, 1213–1215.

Lal, R. (1980) Crop residue management in relation to tillage techniques for soil and water conservation. *Food and Agriculture Organization Soils Bulletin* No. 43, pp. 72–78.

Langdale, G.W., Box, Jr J.E., Leonard, R.A., Barnett, A.P. and Fleming, W.G. (1979) Corn yield reduction on eroded southern Piedmont soils. *Journal of Soil and Water Conservation* 34, 226–228.

Larson, W.E. and Pierce, F.J. (1991) Conservation and enhancement of soil quality. In: *Evaluation for Sustainable Land Management in the Developing World,* Volume II: Technical Paper. Bangkok, Thailand: International Board for Soil Research and Management, 1991. IBSRAM Proceedings No. 12(2), pp. 175–203.

Larson, W.E., Clapp, C.E., Pierre, W.H. and Morachan, Y.B. (1972) Effects of increasing amounts of organic residues on continuous corn: II. organic C, nitrogen, phosphorus and sulfur. *Agronomy Journal* 64, 204–208.

MacRae, R.J. and Mehuys, G.R. (1985) The effect of green manuring on the physical properties of temperate-area soils. *Advances in Soil Science* 3, 71–94.

Massee, T.W. (1990) Simulated erosion and fertilizer effects on winter wheat cropping in intermountain dryland area. *Soil Science Society of America Journal* 54, 1720–1725.

Millhollen, E.P. and Melville, D.R. (1991) The long-term effects of winter cover crops on cotton production in Northwest Louisiana. *Bulletin No. 830* Louisiana Agricultural Experiment Station, Louisiana State University, Baton Rouge, 34 pp.

Oldeman, L.R., Hakkeling, R.T.A. and Sombroek, W.G. (1990) *World Map on the Status of Human-Induced Soil Degradation, at an Average Scale of 1:10 Million.* United Nations Environment Programme and the International Soil Reference and Information Centre. UNEP/ISRIC. Nairobi, Kenya and Wageningen, The Netherlands.

Parr, J.F. and Papendick, R.I. (1983) *Strategies for improving soil productivity in developing countries with organic wastes.* In: Lockeretz, W. (ed.) *Environmentally*

Sound Agriculture. Praeger Scientific, New York, pp. 131–143.
Parr, J.F., Papendick, R.I. and Colacicco, D. (1986) Recycling of organic wastes for a sustainable agriculture. *Biological Agriculture and Horticulture* 3, 115–130.
Parr, J.F., Papendick, R.I., Hornick, S.B. and Colacicco, D. (1989) Use of organic amendments for increasing the productivity of arid lands. *Arid Soil Research and Rehabilitation* 3, 149–170.
Parr, J.F., Papendick, R.I., Hornick, S.B. and Meyer, R.E. (1992) Soil quality: attributes and relationship to alternative and sustainable agriculture. *American Journal of Alternative Agriculture* 7, 5–11.
Patrick, W.H., Haddon, C.B. and Hendrix, J.A. (1957) The effects of longtime use of winter cover crops on certain physical properties of Commerce loam. *Soil Science Society of America Proceedings* 21, 366–368.
Pieters, A.J. and McKee, R. (1938) The use of cover and green-manure crops. In: *Soils and Men, the Yearbook of Agriculture*. US Department of Agriculture. US Government Printing Office, Washington, DC, pp. 431–444.
Rasmussen, P.E., Collins, H.P. and Smiley, R.W. (1989) *Long-Term Management Effects on Soil Productivity and Crop Yield in Semi-Arid Regions of Eastern Oregon*. Station Bulletin 675. US Department of Agriculture, Agricultural Research Service and Columbia Basin Agricultural Research Center, Pendleton, Oregon, 53 pp.
Rodale Institute (1991) *Conference Report of the International Conference on Assessment and Monitoring of Soil Quality*. Rodale Institute, Emmaus, Pennsylvania, 38 pp.
Rogers, T.H. and Giddens, J.E. (1957) Green manure and cover crops. In: *Soil, the Yearbook of Agriculture*. US Government Printing Office, Washington, DC, pp. 252–257.
Sampson, R.N. (1981) *Farmland or Wasteland*. Rodale Press, Emmaus, Pennsylvania, 422 pp.
Sanders, D.W. (1992) International activities in assessing and monitoring soil degradation. *American Journal of Alternative Agriculture* 7, 17–24.
Singh, K.K., Colvin, T.S., Erbach, D.C. and Mughal, A.Q. (1993) Tilth index: an approach to quantifying soil tilth. *Transactions of the American Society of Agricultural Engineers* (in press).
Smith, J.L., Halvorson, J.J. and Papendick, R.I. (1993) Evaluating soil quality using multivariable indicator kriging. *Soil Science Society of America Journal* (in press).
Thorne, D.W. and Peterson, H.B. (1954) *Irrigated Soils*. The Blakiston Company, Inc., New York, 392 pp.
Unger, P.W., Stewart, B.A., Parr, J.F. and Singh, R.P. (1991) Crop residue management and tillage methods for conserving soil and water in semi-arid regions. *Soil and Tillage Research* 20, 219–240.
USDA (US Department of Agriculture) (1991) *Soil Quality Ratings for Cropland Management Systems in the Midwest States*. Soil Conservation Service, Washington, DC.
US Salinity Laboratory Staff (1954) *Diagnosis and Improvement of Saline and Alkali Soils*. Agriculture Handbook No. 60. United States Department of Agriculture, Government Printing Office, Washington, DC.
Van Doren, Jr D.M., Moldenhauer, W.C. and Triplet, Jr G.B. (1984) Influence of long-term tillage and crop rotation on water erosion. *Soil Science Society of America Journal* 48, 636–640.
Voorhees, W.B. (1979) Soil tilth deterioration under row cropping in the northern

cornbelt: influence of tillage and wheel traffic. *Journal of Soil and Water Conservation* 34, 184–186.

Willcocks, T.J. (1981) Tillage of clod-forming sandy loam soils in the semi-arid climate of Botswana. *Soil and Tillage Research* 1, 323–350.

Willson, G.B., Parr, J.F., Epstein, E., Marsh, P.B., Chaney, R.L., Colacicco, D., Burge, W.D., Sikora, L.J., Tester, C.F. and Hornick, S.B. (1980) *Manual for Composting Sewage Sludge by the Beltsville Aerated Pile Method.* Joint publication by US Department of Agriculture/US Environmental Protection Agency. EPA-600/8-80-022. US Government Printing Office; Washington, DC, 64 pp.

Wischmeier, W.H. (1960) Cropping-management factor evaluations for a universal soil-loss equation. *Soil Science Society of America Proceedings* 24, 322–326.

Yamoah, C.F., Agboola, A.A., Wilson, G.F. and Mulongoy, K. (1986) Soil properties as affected by the use of leguminous shrubs for alley cropping with maize. *Agricultural Ecosystems and Environment* 18, 167–177.

Yoder, D.C., Porter, J.M., Jimanton, J.R., Renard, K.G. and McCool, D.K. (1993) Cover-management factor (C) In: *Predicting Soil Erosion by Water – A Guide to Conservation Planning with the Revised Universal Soil Loss Equation.* Agricultural Research Service, US Department of Agriculture. Washington, DC (in press).

Chapter 15
Maintaining the Biological Status of Soil: a Key to Sustainable Land Management?

M.J. Swift

Humid Forest Program, International Institute of Tropical Agriculture, PO Box 5320, Oyo Road, Ibadan, Nigeria.
Present address: TSBF Programme, UNESCO-ROSTA, UN Complex, Gigiri, PO Box 30592, Nairobi, Kenya

Introduction

The concept of sustainable agriculture is a complex one embodying issues of economic viability, the quality of life and human welfare with those of ecological stability and resilience over time. Such a holistic concept does not represent a realistic target for agricultural research because of the sheer complexity, if not impossibility, of measurement that it entails.

A possible approach to incorporating sustainability as a realistic component of agricultural research is to adopt a parsimonious view by extracting the minimum requirements necessary to any given analysis, i.e. that of the well-tried principle of Occam (Swift and Woomer, 1992; Dvorak and Swift, in prep.). By using this approach a minimum working definition of sustainable agricultural production at the cropping system level can be stated (Spencer and Swift 1992; Dvorak and Swift in prep.) as follows:

> a cropping system is sustainable if it has an acceptable level of production of harvestable yield which shows a non-declining trend from cropping cycle to cropping cycle and which is resistant in terms of yield stability to normal fluctuations of stress and disturbance over the long term.

The definition requires elaboration of baseline parameters (acceptable levels of production, timeperiods for cropping cycle and length of trend period) to suit particular local conditions. Its utility is that it provides a measurable concept of sustainability based on yield equilibrium and stability, which can act as a baseline for considering other aspects. For instance the utilization and conservation of the resources used in production (e.g. soil nutrients) can be incorporated by the corollary proposition that sustainable production is dependent on the capacity to replenish resources such that stocks are maintained at constant or increasing levels from crop cycle to crop cycle. This approach to sustainability assessment necessitates study of the processes, both biophysical and economic, which determine resource use efficiency in cropping systems. Soil biological processes are among these. An important challenge is thus to demonstrate the relationship between levels of soil biological activity and sustainable ecosystem function.

When considered in the context of soil resilience this raises questions concerning the role of the soil biological community in maintaining the productive capacity of soil in the face of stress and disturbance. Such perturbation may be imposed by external events such as climatic changes (including extreme events of drought or precipitation) or deposition of pollutants. More pervasively the soil and its biotic community is influenced by agricultural practice itself. This chapter is concerned with the impact of agricultural practices (with particular emphasis on those of tropical regions) on the biological community of soil and on the capacity of the soil community to maintain or restore the productive function of soil following such impact. In attempting to analyse these relationships four questions will be addressed:

1. How can the biological status of soil be assessed?
2. How does agricultural land use influence the biological status of soil and what are the major factors and mechanisms determining variations in biological status?
3. What is the relationship between biological status and soil resilience?
4. What is the relationship between soil biological status and sustainable agriculture?

These questions are addressed in general terms here. More detailed case studies are given in Chapters 17–19.

Assessment of the Biological Status of Soil

The soil is the habitat for a diverse community of microorganisms and invertebrate animals, all of which play some role in determining the primary productivity of the ecosystems they inhabit. These diverse and varied roles may be summarized into five categories:

1. facilitating nutrient acquisition by the vegetation through the agency of mycorrhiza and nitrogen fixing organisms;
2. regulating the retention and flow of nutrients in the system through the processes of decomposition, mineralization and immobilization;
3. mediating the synthesis and breakdown of soil organic matter;
4. influencing the availability of water to the plant by modification of soil

physical structure and water regimes; and
5. modifying the health of the plant by parasitism and pathogenesis.

This chapter concentrates on the second and third of these roles which are those substantially mediated by the decomposer microflora. The maintenance of soil physical quality is mainly brought about by the activities of the macrofauna and is dealt with in greater detail elsewhere in this volume.

As discussed by Pankhurst (Chapter 20), there are numerous potential indices of soil biological status but they can broadly be considered as three categories: specific population indicators (key functional groups); indices of biodiversity; and integrative indices. It is this third category that will be considered here.

The most widely used integrative indices are those associated with soil organic matter (SOM). Some of the specific SOM-based indices that have been proposed are shown in Table 15.1. There is increasing consensus that the most practicable general measure of the biological status of soil is the microbial biomass. Utilization of the techniques of fumigation – incubation, fumigation – extraction or substrate-induced respiration (Jenkinson and Powlson, 1978; Vance *et al.*, 1987; Anderson and Domsch, 1978) gives an estimate of the biomass-C (or N or P) of the mass of organisms below about 5000 μm^3 in volume, i.e. the bacteria, fungi, protozoa and microfauna. It excludes most of the mesofauna, the macrofauna and, most importantly, all organisms that cannot be separated by simple mechanical means from fragments of living or dead plant material. The microbial biomass thus represents the decomposer subcommunity associated with the particulate and colloidal fractions of soil (Table 15.2). It can be hypothesized that the microbial biomass substantially includes much of the free-living microbial community associated with the functions of SOM dynamics and nutrient acquisition but excludes a component of the faunal community associated with soil mixing and pore formation, as well as the organisms involved in the primary decomposition of litter inputs (Table 15.2).

There is increasing evidence that the microbial biomass can be used as a sensitive index not only of the biological status of soil but of a variety of factors influenced by microbial activity such as N-mineralization capacity

Table 15.1. SOM-based indices of biological status (after Swift and Woomer, 1992).

1.	Total SOM (TSOM)	(Larson and Pierce, 1991)
2.	Microbial biomass (MB)	(Sparling, 1991)
3.	Ratio MB/TSOM	
4.	Labile SOM (LSOM)	
5.	Ratio LSOM/TSOM	(Sparling, 1991)
6.	N-Mineralization capacity	
7.	Charge (CEC) contribution to soil	

Table 15.2. Relationship of microbial biomass to soil functional groups.

Functional group	Included	Excluded
1. Mycorrhiza	External mycelium, spores	Intra- and intercellular mycelium
2. N-fixation	Free-living, rhizobial populations	Symbionts in root
3. Decomposition, mineralization immobilization	Populations associated with mineral and organic particles	Populations on leaf, wood, root litter. Macrofauna
4. SOM synthesis and decomposition	Populations associated with mineral and organic particles	Populations on leaf, wood, root litter. Macrofauna
5. Modification of soil physical structure	Microbes associated with soil aggregation	Meso and macrofauna, burrowing, soil particle transport

and C-storage (Sparling, 1991). The labile-SOM approximated by the particulate organic matter or light fraction (Ford *et al.*, 1969; Anderson and Ingram, 1993) is also a useful indicator of biological activity as it represents the equilibrium state between litter decomposition and humification. This fraction is hypothesized as indicative of the short term 'fertilizer' property of soil organic matter (Sparling, 1991).

Influence of Agricultural Management on Microbial Biomass

Sustainable agricultural production can be achieved in many ways, but is ultimately dependent on the availability and efficient use of essential resources. Traditional tropical agricultural practices, such as shifting cultivation, fallow rotation, mixed cropping and integrated livestock–arable systems, utilize the greatest amounts and variety of natural, including biological, resources. In contrast intensive mechanized monocultural systems characteristic of the temperate zone are substantially maintained by the flow of resources from external sources and utilize a relatively small variety of natural resources. Current perceptions suggest that neither of these extremes is a likely option for much of the lowland tropics of Africa in the foreseeable future. Traditional shifting agriculture or fallow rotation practices with their dependence on extensive land resources are no longer viable under changing demographic, social and economic circumstances. On the other hand the economic base for intensive high external input agricultural management seems only a distant prospect for all but a few areas of Africa. Serious questions have also been raised concerning the sustainability of such systems in many tropical environments.

A good deal of current research is therefore targeted on so-called intermediate systems; agroforestry practices, systems incorporating mulch management (including minimum tillage), low-number intercrops or mixed farming. This approach furthermore builds on the fact that the African farmer has often developed his or her own intermediate form of modified fallow management with a low level system of external inputs. Underlying this focus is the supposition that these systems are the best options for combining improved productivity with sustainability in tropical environments. One feature which is advanced in favour of such intermediate systems is that, in contrast to high external input agriculture, system function is regulated to a much higher degree by biological processes. It is hypothesized, explicitly or implicitly, that this promotes greater sustainability.

We can explore this hypothesis by considering the ways in which agricultural practice influences the microbial biomass, drawing on the extensive literature from temperate regions to supplement the limited evidence presently available for tropical cropping systems.

Table 15.3. Examples of microbial biomass in soils under different land uses.

Reference[a]	Land use/practice	Time (years)	Soil type	Location	MB-C ($\mu g\ g^{-1}$)
1.	Bare	27	.	Sweden	230
	Cereal	27	.	Sweden	313
	Cereal + 80 kg N	27	.	Sweden	397
	Cereal + S + 80 kg N	27	.	Sweden	604
2.	Forest	Climax	Ultisol	India	609
	Savannah	Derived	Ultisol	India	397
	Cropfields	15–40	Ultisol	India	250
	Mine spoil	5–20	(Ultisol)	India	332
3.	Miombo Woodland	Climax	Alfisol	Zimbabwe	850
	Maize field	5	Alfisol	Zimbabwe	410
4.	Maize, topsoil	3	Alfisol	Nigeria	142
	Psopho–Maize, topsoil	3	Alfisol	Nigeria	192
	Psopho–Maize, wormcast	3	Alfisol	Nigeria	447
5.	Uncontaminated Sludge	3	–	UK	269
	Sludge + NPK	.	–	UK	249
	Sludge with Cu	.	–	UK	156
	Sludge with Ni	.	–	UK	127
	Sludge with Zn	.	–	UK	265
	Sludge with Cr	.	–	UK	287

[a]1. Schnurer *et al.* (1985).
2. Srivastava and Singh (1991).
3. Eklund (1991).
4. Mulongoy (1986).
5. Brookes and McGrath (1984).

The microbial biomass is influenced to a variable extent by macroclimate (Insam *et al.*, 1989) and soil type (Sparling, 1991) but in particular is responsive to land use and agricultural practice. Significant differences in microbial biomass can, however, be shown between land use systems, cropping system practices and soils subject to different input treatments including those of tropical areas (Table 15.3). Microbial biomass is thus shown to be depressed by decreases in organic input, soil disturbance, and toxic elements.

Resilience of the Soil Community

Resilience, as classically used in population ecology and as generally stated in this volume is defined as 'the capacity of a population (or system) to return to an equilibrium following displacement in response to a perturbation'. This is thus a property beyond stability which can be defined as 'the capacity to resist displacement from an equilibrium condition' but is closely related to it.

There is increasing evidence, however, that biological systems have the capacity to occupy more than one equilibrium state. Holling (1973, 1986) has defined resilient systems as those where displacement from one equilibrium domain does not result in loss of structural and functional integrity. Such systems may show the capacity to occupy a different equilibrium condition such that although quantitative relationships are altered the system remains qualitatively intact. Despite the confusion in terminology this further dimension of system behaviour should be an important consideration in any analysis of system response to perturbation.

A number of examples can be cited to show that the microbial biomass

Table 15.4. Changes in SOM and MB in relation to OM input over 18-year period (from Powlson *et al.*, 1987).

Site[a]	Straw treatment	SOM-C ($t\ ha^{-1}$)	MB-C ($kg\ ha^{-1}$)	SOM-N ($kg\ ha^{-1}$)	MB-N ($kg\ ha^{-1}$)
1.	Burned	59.3	132	2772	25
	Incorporated	61.9	192	3061	40
	% Increase	5	45	10	50
2.	Burned	28.7	273	2945	46
	Incorporated	30.2	374	3193	68
	% Increase	5	37	8	46

[a]Site 1: Studsgaard, Denmark – loamy sand.
Site 2: Rønhave, Denmark – sandy loam.
Both sampled at 23–25 cm depth.

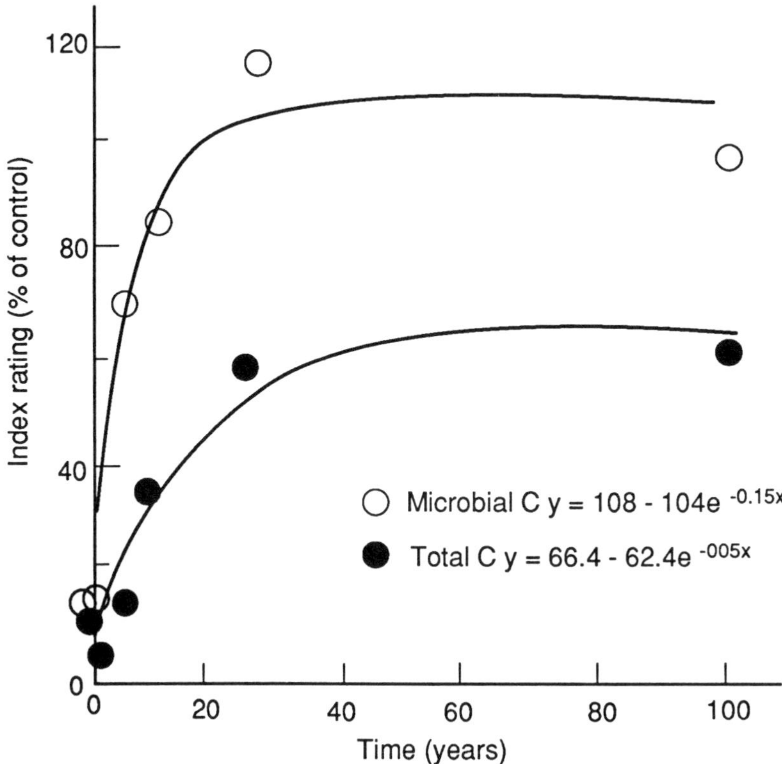

Fig. 15.1. Change in total-C and microbial biomass-C during recovery of a reclaimed eroded hillside in New Zealand. Note the apparent return to the same equilibrium values as the undisturbed control (Sparling, 1991, after Hart *et al.*, 1990).

does have the capacity to respond to changing environmental circumstances. For instance, Bottner (1985) has demonstrated complete recovery of the biomass from cell death caused by drying over a series of wetting and drying cycles. Evidence of this capacity to respond to varying conditions also comes from modelling of the turnover of N through the microbial biomass in the long-term experiments at Rothamsted (Jenkinson and Parry, 1989). Longer-term responses to more extreme change have been shown by Insam and Domsch (1988) and Hart *et al.* (1990). Matson *et al.* (1987) have shown the capacity of the microbial biomass to return to two-thirds of its original level after four years of forest succession following cutting and burning tropical forest, whereas in plots kept bare of vegetation there was a decrease to only 20% of the original value. A similar capacity of response over a short period has been shown for the labile SOM component under

Table 15.5. Microbial biomass as proportion of total SOM in Central European cropping systems (from Anderson and Domsch, 1989).

Cropping system	Input	MB-C/SOM-C $\times 10^2$	Significance[a]
Continuous monocrop	NPK	2.36	a
	S or FYM	2.57	b
	G or G+S	4.04	c
Continuous crop rotation	NPK	2.90	d
	S or FYM	2.51	b
	G or G+S	3.71	e

[a]Values with different letters are significantly different at $P = 0.05$.
S = Straw; FYM = Farm Yard Manure; G = Green Manure.

different fallow regimes in tropical environments (Mulongoy *et al.*, personal communication).

One of the values of the microbial biomass measurement is that it is more rapidly responsive to the changing condition of soil than the total organic matter. This is demonstrated by the data of Powlson *et al.* (1987) in Table 15.4.

Bearing in mind the considerations cited above we should ask whether these changes demonstrate the capacity to return to previous equilibrium values. This is illustrated in Fig. 15.1 (Sparling, 1991) in which both microbial biomass and total-SOM of a degraded soil are shown to return, over a period of about 20 years, to the same level as that in the non-degraded control. Anderson and Domsch (1989) have produced evidence, from the analysis of 134 long-term cropping systems in Europe, that over time the ratio between microbial biomass carbon and total-SOM carbon will reach a characteristic (equilibrium) value for any given type of cropping system. The most important factors determining the level of this equilibrium are the amounts of carbon and nitrogen returned to the soil by the system (Table 15.5).

The clear differences in the MB/SOM ratio between different cropping systems shown in Table 15.5 indicate the capacity of the soil biomass to occupy different equilibria. Although soils in these different conditions show significant differences in the level of production supported they still retain the capacity for production, i.e. functional integrity is maintained. The persistence of such functional equilibria over long time periods has also been shown in the long-term experiments at Rothamsted (Jenkinson and Rayner, 1977; see Fig. 22.5 in Chapter 22).

All this evidence suggests a high degree of resilience of the microbial biomass both in terms of its capacity to recover from disturbance or stress once the retarding factor is removed and in terms of its capacity to maintain functional integrity at a new equilibrium in the face of stress or disturbance. This resilience is no doubt derived from the property of a wide range of microbes to produce survival propagules which can remain dormant for long periods of time under extreme conditions. Indeed it has been cogently argued that the major part of the microbial community assumes this dormant state for most of the time, responding rapidly to the intermittent input of readily utilized organic substrate and then dying down to a survival population (Chapman and Gray, 1986; Nedwell and Gray, 1988). For instance, Schnurer *et al.* (1985) showed that in a bare fallow field that had received no fresh carbon input for 27 years a microbial biomass of 260 $\mu g\ g^{-1}$ was still detectable as compared with 604 $\mu g\ g^{-1}$ for a field that had regularly received straw and N-fertilizer additions (see Table 15.3).

It can therefore be hypothesized that the microbial biomass will retain its functional capacity under all but the most extreme conditions and that the key factor for stimulating recovery to the equilibrium level characteristic of a given climate and soil type is provision of an adequate quantity and quality of organic input. It should be emphasized, however, that there is little current information to enable predictions of the rates of recovery in relation to prevailing conditions.

Although the microbial biomass may be utilized as a good general indicator of the resilience of the soil community it does not tell the whole story. The observed changes in biomass presumably represent an integration of fluctuations in a very large array of species populations. Clear evidence for resilience in response to stress and disturbance can only be taken where it can be shown that all the functions of the microbial community can be restored to former levels. This is a test of system or community resilience in Holling's sense, i.e. of the capacity to maintain system, structure, diversity and function. Evidence can readily be found that at this level some components of the microbial community are highly susceptible to perturbations introduced by cropping system practices. Population sizes of vesicular–arbuscular mycorrhiza for instance are known to be highly susceptible to changes in vegetation, to agricultural practices such as fertilization or to major disturbance or pollution of the soil (Harley and Smith, 1983, Stahl *et al.*, 1988). Natural populations of rhizobium and other components of the N-cycle, are also known to be susceptible to changing soil environments (Woomer *et al.*, 1988). Other chapters in this volume give evidence of the sensitivity of faunal populations to change in soil environments.

Although increasing aboveground inputs of organic matter may indeed be the best way of stimulating the microbial biomass, this restoration process will be disrupted if the distinct macrodecomposer subcommunity of litter and roots is not functioning efficiently. Because of the lack of a good method of assessment equivalent to soil microbial biomass there is much less evidence for the responsiveness of this subcommunity to disturbance and stress. Thus, a case can be made for a high degree of resilience in the microbial community in relation to rather general functions (carried out by a relatively diverse microflora) but the maintenance of a number of specific, and important, functions may be much more susceptible and require specific actions for restoration.

Conclusions: Soil Biological Resilience and Sustainable Agriculture

The examples quoted here suggest that microbial biomass (C, N or P) can be used as a good integrative index not only of an important component of the soil community but also of some biologically mediated functions in soil such as mineralization and immobilization of nutrients and soil organic matter status. However, this should preferably be accompanied by other measures of keystone functions or populations such as litter decomposition, N-fixation, mycorrhiza and macrofauna. Measures of soil biodiversity may also be required for critical assessment of biological status.

Research on microbial biomass and soil fractionation has been dominated by methodological concerns. Emphasis should now be given to determining patterns and sources of variation in space and time to improve the capacity for interpretation of field data. Agricultural practices strongly influence microbial biomass and other indicators of biological status but there is a need for critical analysis of current data to analyse the response of microbial biomass with time to agricultural practice, climate and soil type, and to assess the concept of 'equilibrium biomass'. The study by Anderson and Domsch (1989) gives a good model for this. Thus although the soil microbial community has an apparent capacity for general resilience, specific populations may be vulnerable to some impacts. There is very little specific information on the time scales and patterns of the soil biological response to disturbance. In particular, we do not know if there is a significant hysteresis effect distinguishing decline and recovery. Another critical requirement in research is the investigation of recovery times of the soil community in relation to specific ameliorative practices, such as tree planting or incorporation of cover crops, which have been identified as key components of 'sustainable' systems.

Nonetheless although maintenance of the biological status of soil is generally recognized as a key feature of sustainable production, rigorous means of quantitatively linking these two factors is still lacking. Experimentation is required to investigate the possibilities of critical thresholds of key biological functions. Long-term studies of change in biological status in relation to crop production and other components of agricultural practice are still a priority in soil research. The model of sustainability at the cropping system level, advanced in the introduction to this chapter, may be utilized to assess the relationship between production sustainability and biological status of soil.

Organic matter input and soil disturbance (tillage) are the factors which most affect general biological status whereas pesticides and other toxic materials may also have specific effects on key groups of functions. Careful experimentation is required to investigate the quantitative and qualitative relationship between organic inputs, microbial biomass and microbial activity. In particular the maintenance requirement enigma deserves detailed investigation.

References

Anderson, J.M. and Ingram, J.S.I. (eds) (1993) *Tropical Soil Biology and Fertility: a handbook of methods*, 2nd edn. CAB International, Wallingford.

Anderson, J.P.E. and Domsch, K.H. (1978) A physiological method for the quantitative measurement of microbial biomass in soils. *Soil Biology and Biochemistry* 10, 215–221.

Anderson, T.-H. and Domsch, K.H. (1989) Ratios of microbial biomass carbon to total organic carbon in arable soils. *Soil Biology and Biochemistry* 21, 471–479.

Bottner, P. (1985) Response of microbial biomass to alternate moist and dry conditions in a soil incubated with ^{14}C- and ^{15}N-labelled plant material. *Soil Biology and Biochemistry* 17, 329–337.

Brookes, P.C. and McGrath, S.P. (1984) Effects of metal toxicity on the size of the soil microbial biomass. *Journal of Soil Science* 35, 341–346.

Chapman, S.J. and Gray, T.R.G. (1986) Importance of cryptic growth, yield factors and maintenance energy in models of microbial growth in soil. *Soil Biology and Biochemistry* 18, 1–4.

Eklund, M. (1991) *Soil microbial biomass and soil organic-C distribution in a savanna–woodland and on arable field in Zimbabwe.* Working Paper 171, International Rural Development Centre, Swedish University of Agricultural Sciences, Uppsala.

Ford, G.W., Greenland, D.J. and Oades, J.M. (1969) Separation of the light fraction from soils by ultrasonic dispersion in halogenated hydrocarbons containing a surfactant. *Journal of Soil Science* 20, 291–296.

Harley, J.L. and Smith, S.E. (1983) *Mycorrhizal Symbiosis.* Academic Press, London.

Hart, P.B.S., West, A.W., Ross, D.J., Sparling, G.P. and Speir, T.W. (1990) The importance of organic matter for restoration of soil fertility in drastically disturbed New Zealand landscapes. In: *Issues in the Restoration of Disturbed Land.* Massey University, New Zealand.

Holling, C.S. (1973) Resilience and stability of natural ecosystems. *Annual Review of Ecology and Systematics* 4, 1–23.

Holling, C.S. (1986) The resilience of terrestrial ecosystems: local surprise and global change. In: Clarke, W.C. and Munns, R.E. (eds) *Sustainable Development of the Biosphere,* International Institute for Applied Systems Analysis, Vienna and Cambridge University Press, Cambridge, pp. 292–320.

Insam, H. and Domsch, K.H. (1988) Relationship between soil organic carbon and microbial biomass on chronosequences of reclamation sites. *Microbial Ecology* 15, 177–188.

Insam, H., Parkinson, D. and Domsch, K.H. (1989) Influence of macroclimate and soil microbial biomass. *Soil Biology and Biochemistry* 21, 211–221.

Jenkinson, D.S. and Parry, L.C. (1989) The nitrogen cycle in the Broadbalk wheat experiments: a model for the turnover of nitrogen through the soil microbial biomass. *Soil Biology and Biochemistry* 21, 535–541.

Jenkinson, D.S. and Powlson, D.S. (1976) The effects of biocidal treatments on metabolism in soil. V. A method for measuring soil biomass. *Soil Biology and Biochemistry* 8, 209–213.

Jenkinson, D.S. and Rayner, J.H. (1977) The turnover of soil organic matter in some of the Rothamsted classical experiments. *Soil Science* 123, 298–305.

Larson, W.E. and Pierce, F.J. (1991) Conservation and enhancement of soil quality. In: Dumanski, J., Pushparajah, E., Latham, M. and Myers, R.J.K. (eds) *Evaluation for Sustainable Land Management in the Developing World* Volume 2, Technical Papers, Bangkok, Thailand, International Board for Soil Research and Management, 1991. IBSRAM Proceedings no 12(2), pp. 175–204.

Matson, P.A., Vitousek, P.M., Ewel, J.J., Mazzarino, M.I. and Robertson, G.P. (1987) Nitrogen transformation following forest felling and burning in a volcanic soil. *Ecology* 68, 491–502.

Mulongoy, K. (1986) Microbial biomass and maize nitrogen uptake under a *Psophocarpus palustris* live-mulch grown on a tropical Alfisol. *Soil Biology and Biochemistry* 18, 395–398.

Nedwell, D.B. and Gray, T.R.G. (1987) Soils and sediments as matrices for microbial growth. In: Fletcher, M., Gray, T.R.G. and Zones, J.G. (eds) *Ecology of Microbial Communities,* Cambridge University Press, Cambridge, pp. 21–54.

Powlson, D.S., Brookes, P.C. and Christenson, B.T. (1987) Measurement of soil microbial biomass provides an early indication of changes in total soil organic matter due to straw incorporation. *Soil Biology and Biochemistry* 19, 159–164.

Schnurer, J., Clarholm, M. and Rosswall, T. (1985) Microbial biomass and activity in an agricultural soil with different organic matter contents. *Soil Biology and Biochemistry* 17, 611–618.

Sparling, G.D. (1991) Organic matter carbon and soil microbial biomass carbon as indicators of sustainable land use. In: Dumanski, E., Pushparajah, M. Latham and R.J.K. Myers (eds) *Evaluation of Sustainable Land Management in the Developing World* Volume 2: Technical Papers, Bangkok, Thailand; International Board for Soil Research and Management, 1991. IBSRAM Proceedings No. 12(2), pp. 563–580.

Spencer, D.S.C. and Swift, M.J. (1992) Sustainable agriculture: definition and measurement. In: Mulongoy, K., Gueye, M. and Spencer, D.S.C. (eds) *Biological Nitrogen Fixation and Sustainability of Tropical Agriculture.* John Wiley, Chichester, pp. 15–24.

Srivastava, S.C. and Singh, J.S. (1991) Microbial C, N and P in dry tropical forest soils: effects of alternate land-uses and nutrient flux. *Soil Biology and Biochemistry* 23, 117–124.

Stahl, P.D., Williams, S.E. and Christensen, M. (1988) Efficacy of native vesicular–arbuscular mycorrhizal fungi after severe soil disturbance *New Phytologist* 110, 347–354.

Swift, M.J. and Woomer, P.L. (1992) Organic matter and the sustainability of agricultural systems: definitions and measurements In: Merkx, R. and Mulongoy, K. (eds) *Dynamics of Organic Matter in Relation to the Sustainability of Agricultural Systems.* John Wiley, Chichester.

Vance, E.D., Brookes, P.C. and Jenkinson, D.S. (1987) An extraction method for measuring soil microbial biomass. *Soil Biology and Biochemistry* 19, 703–707.

Woomer, P.L., Singleton, P.W. and Bohlool, B.B. (1988) Ecological indicators of native rhizobia in tropical soils. *Applied and Environmental Microbiology* 54, 1112–1116.

Chapter 16

Sustainable Land Use in the Light of Resilience/Elasticity to Soil Organic Matter Fluctuations

H.W. Scharpenseel and P. Becker-Heidmann

Institute of Soil Science, Hamburg University, Allendeplatz 2, D-2000 Hamburg 13, Germany

Introduction

The source–sink behaviour of the carbon cycle in the soil–plant system is characterized by fluctuations of considerable magnitude. The soil cover, which has probably existed since the end of the Permian glaciation (Dudal, 1990) is characterized mainly by its organic matter content. The level of organic matter in the soil is sustained by a harmonious supply of growth factors for biomass production and suitable soil temperature, moisture and pH (in wetlands also Eh) conditions as well as the nature of inorganic complexing matrices (mainly high activity (HAC) or low activity (LAC) clays or oxides) which are the main edaphic factors and major modifiers of soil organic matter (SOM) decomposition or humus stabilization. Finally erodability of the site has an important bearing in determining whether the soil is part of a more sustainable or more easily degradable ecosystem.

SOM fluctuations in the Phanerozoic and shorter periodic amplitudes are determined (Fig. 16.1) by the erratic trends of atmospheric CO_2 concentration and temperature, the eustatic sea level alterations, and corresponding spells of rising aridity or humidity. System complexity is producing sustainability, which however May (in Lovelock, 1988) rejects. Stable ecosystems require considerable elasticity and resilience to resist the inherent instability of the system parameters, for example SOM, that distinguish sediment from soil.

Obviously, prerequisites for sustainable life systems are numerous. Not all of them existed at the beginning of life on earth. During the anoxic phase the prokaryotes and eukaryotes, early photosynthesizers, had to perform under protection of water until, at the verge of the Phanerozoic with about 0.1% oxygen from photodissociation, early photosynthesis and early

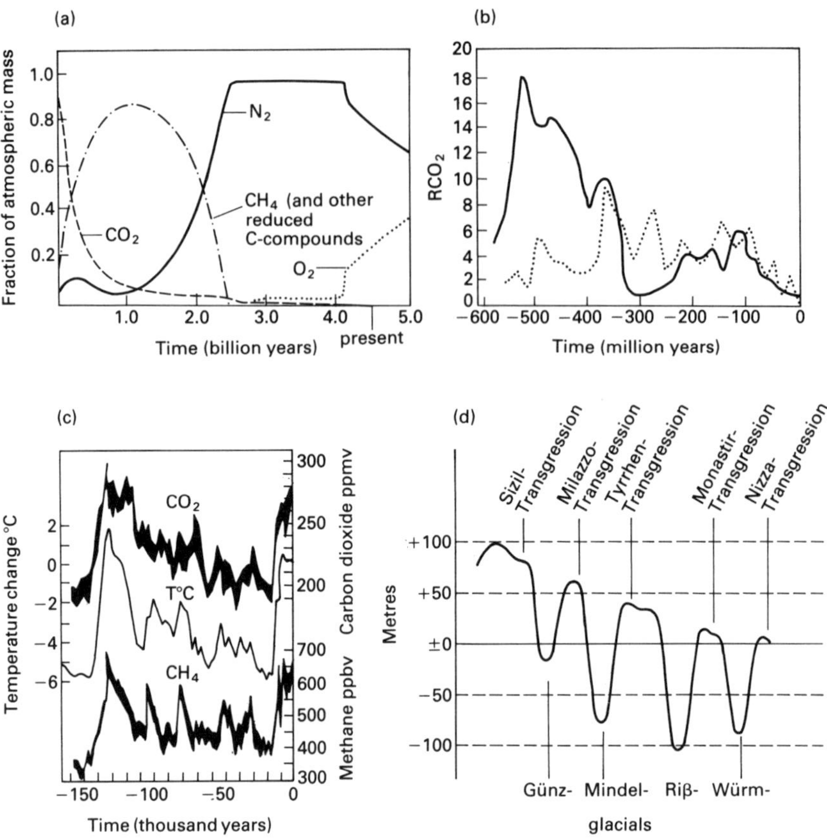

Fig. 16.1. Long-term oscillations in atmospheric constituents, temperature and sea level versus time. (a) Evolution of Earth's atmosphere (Hart, 1978) (CO_2, CH_4, N_2, O_2). (b) RCO_2, ratio of mass of CO_2 in atmosphere at time *t* to that in the present atmosphere between Cambrian and Pleistocene. —— According to Berner and Lasaga (1990); ········· According to Budyko *et al.* (1987). (c) CO_2, CH_4 and temperature versus time. According to 'Global Change' IGBP(1992). (d) Glacials, interglacials-transgressions and eustatic sea level changes (Schwarzbach, 1974).

development of an ozone shield, UVb radiation became increasingly absorbed.

Lovelock (1988) in his approach to 'Geophysiology' presents a comprehensive account of the capabilities of the living biomass on earth to exert some kind of self-regulation to potentially damaging system parameters, such as temperature, composition of the atmosphere including the oxygen concentration, marine salt concentration, and others.

Lovelock reflects on the geophysiological regulation mechanisms producing a 'tendency to constancy', or 'homeostasis', the 'wisdom of the body'. The author also draws attention to an opposing approach to his system, by Lewis and Prinn (1984). Sustainability of land use required by system resilience actively sustained low entropy, i.e. a high level of order and energy conservation, that transcends homeostasis.

From a geophysiological viewpoint, replacement of natural woodland by crop monoculture – 'agricultural mining' according to Dover and Talbot (1987) – is deplorable as it introduces a constraint to the life support system of CO_2 and water pumping and reduces the heat absorption capacity. As a result homeostasis by system resilience is reduced.

Background, SOM as C-sink: Resilience to Changes

Plant nutritionists sometimes argue that a soil does not require humus for sustainable crop production, provided that soil moisture and aeration, porosity, soil pH and plant nutrients are present in harmonious order. Soil organic matter specialists may admit to some resilience and elasticity of agricultural crops towards SOM concentration, at least with regard to some of the functional SOM compartments (Swaby and Ladd, 1963; Jenkinson and Rayner, 1977; Parton *et al.*, 1987). Most heavily cultivated loamy sands, sandy loams and sandy–silty loams have lost close to half the humus from their Ap-horizons compared to their former level as virgin soils. In spite of this, many agroecosystems are sustainable due to soil resilience and elasticity provided by SOM. The SOM level falls from that under autochthonous steppe or forest as it does in allochthonous alluvial soils, losing especially from the Ap-horizon the SOM that was not complexed with inorganic material. As a cropland soil the remaining humus pool, especially the 'passive' SOM functional compartment (Parton *et al.*, 1987), has to be the source of soil resilience.

Theng *et al.* (1992) have isolated a biologically inert SOM fraction from a recent Spodosol in New Zealand. They consider the ^{14}C-age of this fraction, 6700 years BP, as the true age of the soil.

Crop soils with a deep epipedon have conserved most of the humus below the plough depth. Assuming a total soil humus-C pool in terrestrial soils of only ca. 1200 Gt C (about 9.3 kg C m^{-2}), ca. 150 Gt C (about 10 kg C m^{-2}) are in the 11% ice-free continental surface that forms the 1.5×10^9 ha of cropland. Most steppe soils have maintained their humus content below 15 – 30 cm. In the Alfisols and Ultisols part of the humus has migrated in a complex with clay minerals to the argillic horizon. Similarly in Spodosols it has migrated with oxides to the spodic horizon. In both cases a carbon sink is formed in B horizon. Between 30 and 60 Gt of C may have been volatilized during about the last century from the 1.5×10^9 ha cropland (ca. 2 – 4 kg C m^{-2}) due to ploughing and other soil cultivation methods in the course of intensive agriculture (Table 16.1). Nevertheless the increase in

Table 16.1. Carbon compartments in the pedosphere.

Carbon in soil epipedon (kg C m^{-2})			
Terrestrial soils, 1.28 × 10^{14} m^2; 1.5 × 10^{15} kg C			= 11.7 kg C m^{-2} (1500 Gt C)
Wetlands, 6.3 × 10^{12} m^2; 5 × 10^{14} kg C			= 79.4 kg C m^{-2} (500 Gt C)
Cropland, 1.5 × 10^{13} m^2; 3 × 10^{14} kg C			= 20.0 kg C m^{-2} (300 Gt C)
Carbon outside soil epipedon (Gt C in total)			
Argillic horizons, 0.5–2.5 kg C m^{-2}; or ca.1 kg C m^{-2}			
According to Dudal (1990)	Nitisols	250 mha	
	Acrisols	800 mha	
	Alisols	100 mha	
	Lixisols	200 mha	
	Luvisols	600 mha	
	Podzoluvisols	260 mha	
	Total	2210 mha	
			ca. 2.2 × 10^{13} m^2 = 22 Gt C
Spodic horizons, ca. 1.0 kg C m^{-2}			
According to Dudal (1990)	Podzols	480 mha	
			4.8 × 10^{12} m^2 = ca. 5 Gt C
Paleosols, ca. 1.0 kg C m^{-2}			
According to Dudal (1990)	Fluvisols	320 mha	
	Gleysols	620 mha	
	Cherno-, Kastano-, Phaeo-, and Greyzems	830 mha	
	Total	1770 mha	
	Assumed, ca. 1/2 have paleosols of about 1 kg C m^{-2}		
	Ca. 900 mha = 9 × 10^{12} m^2 = ca. 9 Gt C		
	(mainly alluvial, interstadial, interglacial soil formations)		
Carbon in calcretes, ca. 100 kg C m^{-2}			
According to Dudal (1990)	Calcisols	1000 mha	= 10^{13} m^2
			= 10^{12} t C
			= 1000 Gt C
		Total	3036 Gt C
	(In calcretes only ca. 50% of carbon was originally organic carbon)		

the storage of carbon in SOM since the last glacial period (18,000 years BP) is very high (Adams *et al.*, 1990). At the present time the humus content of croplands may be slowly increasing in Entisols, Andisols (as allophane–humus complexes) and flooded rice soils.

Soils with caliche or calcretes are also carbon sinks. Respiratory CO_2 contributes about 50%, and lithogenic C or primary $CaCO_3$ the other half ($\delta^{13}C$ is −7 to −10%). It is estimated (Table 16.1) that the C of calcretes is similar in amount to the C in soil epipedons (Yaalon, 1990). Freytag (1985) identified the mechanisms of formation of calcretes in Tunisia by $\delta^{13}C$ and $\delta^{18}O$ profile scans. A net C-sequestration in calcretes can proceed only to the extent that the Ca-content of the previously existing primary carbonates is increased.

Humus-C sinks in cropland soils are also produced by pedogenetic processes, such as che- and chilluviation, arge- and argilluviation, and by bio-, pelo- and cryoturbation. In addition, considerable SOM-C sinks exist in buried paleosols. An excellent example is the buried Xeroll below the allochthonous sandy top layer in North Africa, especially Tunisia (Scharpenseel and Schiffmann, 1985). Table 16.1 provides estimates of the different soil carbon compartments.

Humus, besides being a source of nutrients, contributes to friability, structure in general, water-holding capacity, and sorption capacity. Particularly in LAC-soils the humus compartment, with its point of zero charge at pH 3.0–3.5, forms the major negatively charged matrix for cation exchange capacity (CEC) in acidic soils (Scharpenseel and Miehlich, 1989). Modifiers of SOM decomposition are temperature, moisture, pH, in wetlands Eh, and the nature of the inorganic complexing matrix. These factors mean that humus accumulates mainly in steppe soils with rigidly alternating dry and wet seasons. During the rainy season the soil moisture is above, and during the dry season it is below, the optimum for biotransformation, which is near the field capacity level. In permanent grassland the residence time of carbon is especially high, often superior to the carbon content of soils under adjacent woodlands (Scharpenseel and Becker-Heidmann, 1993).

Towards Sustainability and Resilience

Sustainability of land use is supposed to be the remedy for growing food scarcity and erratic trends in production. Factors involved are believed to stretch from modern agronomy, integrated plant protection (agrochemistry), and soil conservation to agricultural 'mining' (Dover and Talbot, 1987), slash and burn (Ripley, 1975), unsteady land use and landscape degradation. Considering the rapid increase of the world population (ca. 2% year^{-1}), the cropland area cultivated at present is limited to the dangerously low level of 1.5×10^9 ha. At the most it may reach 3×10^9 ha (the total continental surface of ca. 13.4×10^9 ha comprises 4×10^9 ha forest, 3×10^9 grassland, ca. 1.5×10^9 ha cropland and 5×10^9 ha of near-melting permafrost, mountainous and (semi)desertic soils). Forests and grassland perform most of the carbon sequestration and should as long as possible not be converted into cropland in order to avoid additional greenhouse forcing by CO_2. With 1.5×10^9 ha the constrained cropland area can be fully serviceable for sustainable land use and a lasting adequate food supply only with yield increases comparable to the rate of population growth.

Considering the major growth factors, plant nutrients, soil moisture, light intensity and CO_2 concentration, this seems at first glance difficult to achieve. The trends in these growth factors are erratic, and non-sustainable against time (Fig. 16.1). Only the complexity of the fusion between lithosphere, hydrosphere, atmosphere and biosphere into a pedosphere provides the resilience, stability and source of sustainability that can make the system perform with perfection. The higher the complexity, the more pronounced is the buffer capacity of any system against breaks in performance and dramatic non-linearities, rising disorder and entropy in the sense of chaos theory.

Swaby and Ladd (1963) found that small polymers are usually more easily degraded by soil enzymes than large polymers and that linear copolymers with regular repeating units are more readily decomposed than those with dislocations. Molecules weakly crosslinked by Van der Waals or hydrogen bonds are more easily attacked than those strongly linked by covalent bonds. As one explanation for the several thousand years' existence of some humic substances the authors drew attention to a possible autolytic process of humic substance formation in the decomposing cell. These humic substances would be locked away from attack by microorganisms in clay domains where stable complexes would be formed.

Sustainability of land use, concomitant with the doubling world population, seems likely to be a chimera against the background of adverse factors such as acidification, humus loss, nutrient export, reduction of cropland per capita from 0.5 ha in 1950 to an estimated 0.25 ha in 2000 (Global 2000, Report to the President, 1980).

Factors which will tend to mitigate the increasing demand on the land include: the increase in atmospheric CO_2 which will produce yield increases in C3 plants, and save moisture for C3, C4 and CAM-crops; improved C sequestration by reforestation and conservation of grassland; reduction in the rate of emission of other greenhouse forcing, IR trapping trace gases, especially CH_4 and N_2O (although under the pressure of expanding irrigation and N-fertilization this may be extremely difficult, even in conjunction with strict organic matter recycling); enrichment of cropping systems with legumes (although diazotrophic systems also contribute to N_2O emission). These factors will assist the soil to form a complex system of high resilience and sustainability.

But, is not the growth system's efficiency always dependent on the growth factor being in minimum supply? This certainly applies to yield increases when another growth factor is limiting the response, but the effects may not appear quite so promptly in terms of damage to a system. Here the resilience to changes has a considerable buffering capacity, entrenched in the sorption capacity of clay and humus. This leads to a slow partition, associated with the movement of most factors and processes along the sorption path. Nutrient constraints in subsistence farming are likely to lead to wider C/N ratios of plant products, and hence a need for higher food ingestion across the food web (Bazzaz and Fajer, 1992).

Supporting measures for soil organic matter stabilization, to the advantage of soil resilience, avoidance of degradation of the epipedon catalysis of sustainability, and continuity of land use include:

- preservation of the deep, well-humified grassland humus compartments, resembling forests in efficiency of carbon sequestration (Whittaker and Likens, 1973);
- protection of humus in the Ap horizon of the epipedon by infrequent ploughing;
- opening the subsoil to improve root penetration and biological activity and where necessary by decompaction of pans or argillic horizons, by subsoiling measures;
- curtailment of humus transport by cheluviation and argeluviation by liming;
- enrichment of the humus pool of the soil by legume intercropping, under special conditions even by soil amendment with HAC (high activity clay) or sesquioxides as a complexing matrix.

Examples

The best example of sustained land use, based on resilience and elasticity to soil organic matter fluctuations, is the carbon cycle in highly productive agriculture. The ancient land use systems used shallow hoeing and/or fallow years. These measures protected the SOM level sufficiently. Ploughing, sometimes with periodical subsoiling, in conjunction with high yielding varieties, mineral fertilizer, and in some instances supplemental irrigation, in land use systems with one or two crops per year, depletes the SOM level. This restabilizes at the level where the remaining SOM is firmly protected by complexation with clay and oxides. This latter fraction is identified by its ^{14}C-age, (Fig. 16.2), which is commonly several thousand years. The SOM of the coarse and middle clay fractions (2.0–0.6 μm or 0.6–0.2 μm) and often the fine silt fraction (2.0–6.0 μm) had in our dating of texture fractions the highest ^{14}C-age (Scharpenseel *et al.*, 1986; Tsutsuki *et al.*, 1988; Becker-Heidmann, 1989). The database requires expansion and improvement. But we can tentatively assess the global carbon turnover. The rate of humus-C loss by respiratory volatilization upon change from wood- and grassland into cropland is about 30–60 Gt C (2–4 kg m^{-2}) from the total of about 150 Gt C (10 kg m^{-2}) although Bertram (1986) calculated, based on stable isotope (^{13}C) measurements, that in course of the last 50 years of intensified agricultural production an additional 50 Gt C have been sequestered in the

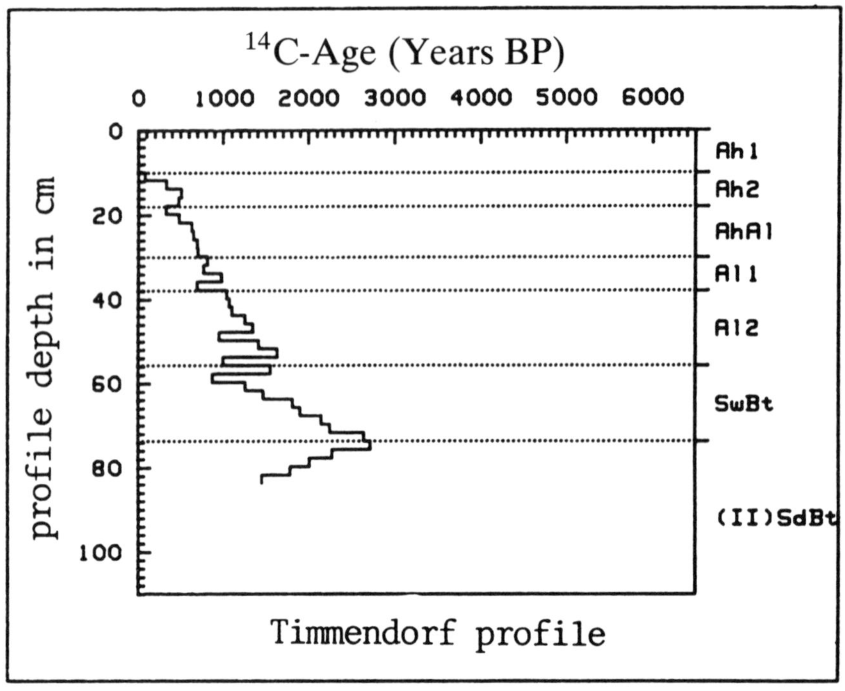

Fig. 16.2. Thin-layer ^{14}C-dating profile scan, Hapludalf 'Timmendorf' (highest residence time of C corresponds with highest clay content).

biomass/SOM pool. This is probably C accumulated in the soil–plant system as a consequence of CO_2 and mineral fertilization in the course of the last 50 years. The contribution of CO_2 fertilization to yield increases in this century is assessed as being of the order of 10–15%. Based on a rise of biomass production of 25 kg kg^{-1} nutrient 100 × 10^6 t of N+P+K added as annual mineral fertilizer input should have enabled the croplands to sequester about 1.25 Gt C $year^{-1}$. These gains are addressed by Esser (1990), whose 'Osnabrück models' show that the annual liberation of ca. 1.5 Gt C by forest clearing and slash and burn is more than compensated by the sequestration of C in the soil–plant carbon-cycle compartments.

Among the oldest identifiable paleosols are the 'root soils', chemically and physically well protected below the Upper Carbonian coal deposits (Roeschmann, 1962). The youngest identifiable humus is probably the 'bomb-carbon-humus', formed since 1956, as a result of thermonuclear tests (Tans, 1981; Scharpenseel *et al.*, 1989.). The gap between terrestrial C sequestration at the last glacial maximum (about 18,000 years BP) and the present level has been estimated for 24 major ecosystems by Adams *et al.* (1990) as amounting to more than 100%.

The exchange of carbon in the pedosphere, accompanied by humus rejuvenation can be estimated as about half of the annual net photosynthetic biomass-C (60 Gt). The substitution involves mostly younger SOM fractions, very little of the stable SOM / clay complexes (Scharpenseel *et al.*, 1989).

Soil dynamic processes and type of organic material are responsible for the SOM forms developed such as mull, mor or raw humus and for carbon sequestration and distribution in the soil profile. Resilience against those dynamic processes which affect SOM is variable in strength. System sustainability in the climax phase of soil and SOM development differs, and depends not least on the system complexity. Different strengths of resilience exist against processes involving SOM because:

- SOM is lost due to turnover (biotic, abiotic, either protolytic or photochemical turnover); biotic turnover is ubiquitous in occurrence, protolytic turnover is strongest in very acid soils, photochemical turnover prevails in the (semi)arid belt with high light intensity.
- SOM dynamics involve processes of leaching and respiration in all but Aridisols, submergence with slightly retarded turnover in paddy soils, and formation of paleosols and fossilization due to burial.
- decline of SOM quality, with widening C/N ratio (eutrophic → mesotrophic → oligotrophic → dystrophic) occurs due to lack of soil protection and increasing acidification.
- when the plant culture changes or woodland or grassland is used as cropland, SOM becomes a carbon source. Only Entisols and riceland, in temperatures up to 28°C, become a mild carbon sink. Andisols, during the phase of SOM–allophane complex formation, may bind large amounts of C, up to 15%.

Typical processes of SOM transfer imply considerable long-term (sustainable) features of carbon source – sink alterations.

Steppe soils, particularly Mollisols and Alfisols, are characterized by bioturbation, sustained in particular by earthworms and termites. Wormcasts or body carbon can reveal by $\delta^{13}C$ scanning past changes between C3 and C4 vegetation, e.g. from gallery forest (C3) to savannah grasses (C4) (Martin *et al.*, 1990). Earthworm body carbon of different depth layers, subjected to ^{14}C-dating, reflects, by comparison with the ^{14}C-age profile of the surrounding soil, the feeding habits of the earthworms (Scharpenseel *et al.*, 1986).

Smectitic Vertisols are characterized by peloturbation, a process of more physical 'self mulching' SOM transport through cracks and fissures, and of colloid–chemical partition along the faces of slickensides.

Cryoturbation in tundra soils due to alternating thawing and freezing gives rise to physical transport of SOM to the subsoil. In acidic, and particularly sandy acidic soils, SOM cheluviation occurs to form spodic or chilluvial

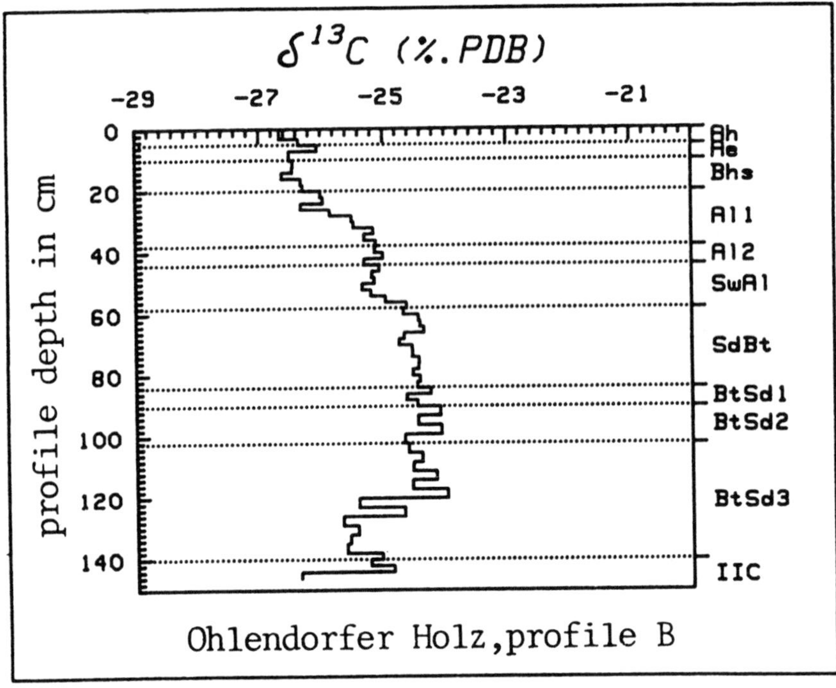

Fig. 16.3. Thin-layer $\delta^{13}C$ profile scan, aquic Hapludalf 'Ohlendorfer Holz' (slight rise of $\delta^{13}C$ in argillic horizon; isotope effect due to humus migration).

horizons. Old Spodosol SOM exists with a 6700 year BP ^{14}C-True Age (Theng *et al.*, 1992).

Middle-textured soils e.g. formed from loess or young basal moraine, when free of carbonates, are subject to the *lessivé* process in which the clay–SOM complex moves down the soil profile to form the argillic horizon. The greatest ^{14}C-ages are always found in the zone of the most advanced clay illuviation (Fig. 16.2). A slight rise of the ^{13}C concentration in the argillic horizon (Fig. 16.3) shows the resistance of the clay–humus complexes once formed.

In savannah soils SOM sequestration in some termite mounds makes them well suited to use as fruit gardens or for vegetable growth on account of their resilience against soil degradation.

Submerged rice soils are about the only cropland which provides a mild carbon sink, with slowly increasing SOM content despite aeration by the puddling process. A host of living organisms (Roger and Kurihara, 1988) sustains intensive mixing of carbon species, reflected in unusually low ^{14}C-dates of most rice soils, down to the depth of land preparation (Scharpenseel *et al.*, 1989) (Fig. 16.4).

Andisols change from C sink to C source in the typical sequence of

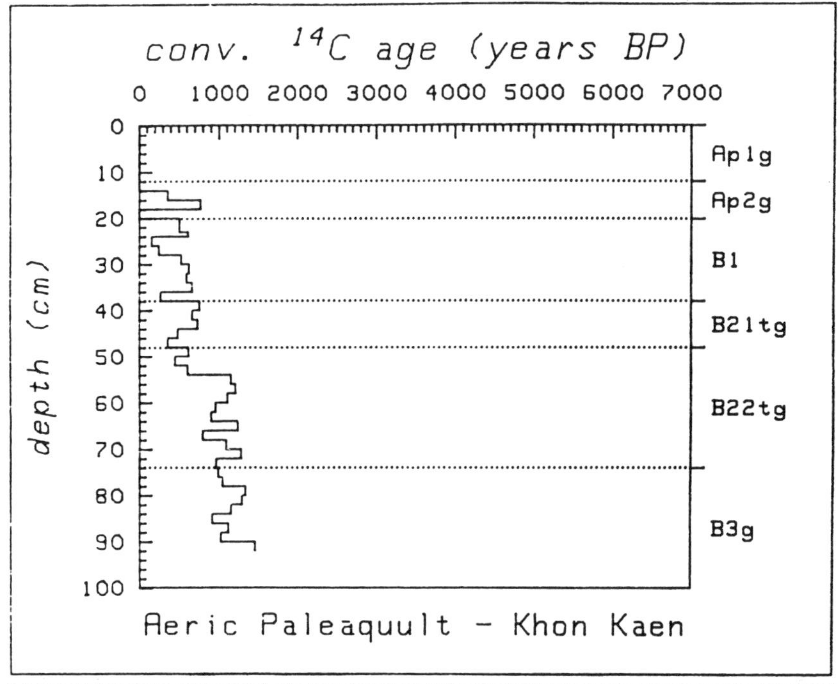

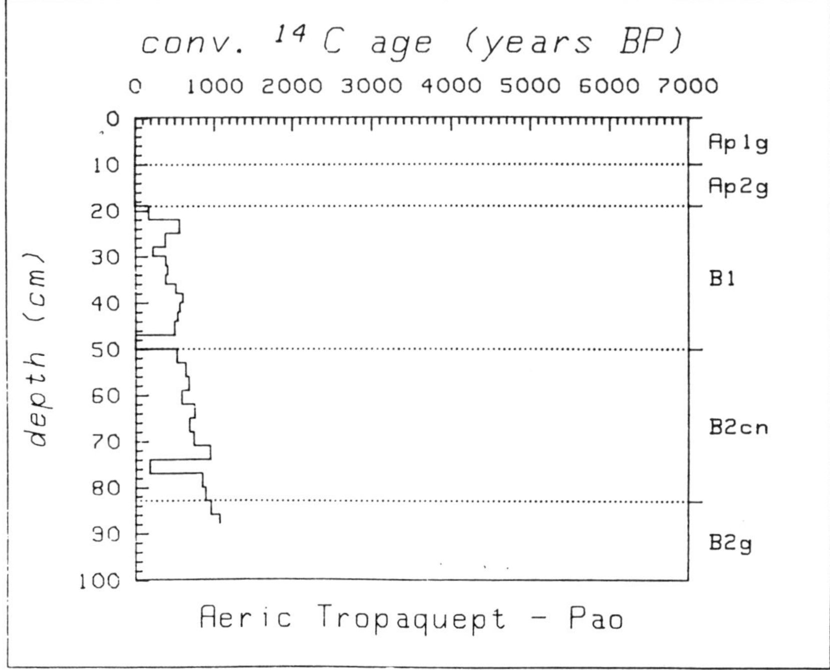

Fig. 16.4. Low ^{14}C-age gradient towards depth in two rice soil profiles of Khon Kaen (Thailand) and Pao (Philippines). (From Terminal Report of GTZ and DFG (contract Scha 48/23), Becker-Heidmann, 1990.)

pedogenetic phases (Miehlich, 1991). Volcanic ashes enter the strongly C sequestering phase of allophane – humus complex formation after about 1000–2000 years, leading to highly humic Andisols (up to 25 or 30% of SOM). After 8000–10,000 years the complexes disintegrate with the alteration of allophane into clay minerals, such as metahalloysite, as they change from Andisol into a zonal soil type. Most of the initially sequestered SOM is thereby decomposed and released as CO_2.

In soil formed under 300–500 mm rainfall on calcareous substrates such as calcrete and caliche crusts about 50% of their total carbon may be sequestered from biomass turnover. The carbon pool of calcretes is one of the largest in the soil carbon cycle, estimated (Yaalon, 1990) to contain 1200–1500 Gt of C (Table 16.1). Catenary mechanisms of formation are revealed by $\delta^{13}C$ and $\delta^{18}O$ scans of the crust (Schleser *et al.*, 1983; Freytag, 1985). Sustainable land use on cropland underlain by a substantial calcareous crust is exemplified in the xeric and aridic belt by olive tree culture (*Olea europaea* L.), characterized by rather shallow and very widespread racination of up to 12 m radius (Scharpenseel, 1965).

Conclusions

Synthesis and breakdown of SOM are integrated by soil profile dynamics, which leads to better-adapted transient or sustainable soil types and land use systems. This may be compared to the self-regulating mechanism of Lovelock's 'geophysiological Gaia hypothesis' of temperature-, oxygen concentration-, ocean salinity-regulation, with positive and negative feedbacks (Lovelock, 1987, 1991a,b). Thus, a state of homeostasis develops with high resilience for conservation of the stable (mostly clay–organic complexed) SOM of the 'passive' SOM compartment (Parton *et al.*, 1987). Carbon residence time of this material considerably exceeds 1,000 years. Cropland under a complex land use system exhibits a dependable defence mechanism in favour of cation exchange, moisture retention and a host of soil physical properties, arising from the stable SOM pool.

Pertinent urgent research needs include:

1. Can we afford in our fight for food security for a world population which will double in 50 years, to ignore the considerable yield increases which will arise from the extra CO_2 concentration in the atmosphere (Idso, 1990)? Field results from different climate belts and agroecosystems are badly needed.
2. Carbon dioxide is both raw material for biomass production and a major factor for IR trapping by trace gases of anthropogenic origin. In consequence, optimization of biomass and SOM recycling to reduce NO_3 pollution of groundwater from N-fertilizers is essential. It is also necessary to develop methods to minimize N_2O and CH_4 emission from soils.

3. Carbon source and sink studies in different agroecosystems and climate belts are badly required to enable rational decisions to be made at the highest policy level regarding both world climate and food production. For taller plants the $\delta^{13}C$ gradient between CO_2 from the soil (ca. −25%), the CO_2 of intermediate $\delta^{13}C$ at the level of the plants and atmospheric CO_2 beyond the tip of plants must be determined to develop estimates of the atmospheric and pedospheric C contribution (Schleser and Jayasekera, 1985).

4. Sustainability and resilience need intensive long-term testing under different levels of system complexity; this should include studies of the N and P economy, and the effects of legume–rhizobium and mycorrhizae on the system.

5. In order to learn from the past, studies of landscape history are important. $\delta^{13}C$ scans of SOM profiles can reveal C3–C4–CAM vegetation changes in the past, and provide information regarding forest–savannah transitions, and occurrences of methanogenesis.

6. Comparisons of the residence time of SOM-C in woodland, grassland, and cropland on the same soil type to assess their C-source-sink contribution are needed.

7. Reliable information is needed concerning the significance of marshes, swamps and thawing permafrost soils for methanogenesis. Are rising temperatures promoting increased methanogenesis or methane oxidation? Could methane from reductive biotransformation of SOM, emitted from the higher latitude landmasses jointly with methane of the tropical wetlands plus the other, better assessable methane compartments, become a major component for global warming, shifting major life-supporting crop belts to higher latitudes?

8. The role and source of the 'missing carbon fraction' (Keeling *et al.*, 1989; Tans *et al.*, 1990; Broecker and Peng, 1992) representing 1.5–2.0 Gt C from the annual anthropogenic CO_2 input of ca. 7–8 Gt C has to be verified regarding its contribution to biomass gains and corresponding C sinks by CO_2 fertilization (Bertram, 1986; Esser, 1990)

References

Adams, J.M., Faure, H., Faure-Denard, L., McGlade, J.M. and Woodward, F.L. (1990) Increases in terrestrial carbon storage from the last glacial maximum to the present. *Nature* 348, 711–714.

Bazzaz, F.A. and Fajer, E.D. (1992) Plant life in a CO_2-rich world. *Scientific American* January, 68–74.

Becker-Heidmann, P. (1989) Die Tiefenfunktionen der natürlichen Kohlenstoff-Isotopengehalte von vollständig dünnschichtweise beprobten Parabraunerden und ihre Relation zur Dynamik der organischen Substanz. *Hamburger Bodenkundliche Arbeiten*, 13.

Becker-Heidmann, P. (1990) Carbon fluxes in important soil classes, with emphasis on Lessivé soils and on soil of the terrestrial, of the hydromorphic and temporarily submerged environment. Terminal Report to GTZ (contract 72.7866.6-01.400/1420), pp. 1–177.

Becker-Heidmann, P. and Scharpenseel, H.W. (1992) The use of natural ^{14}C and ^{13}C in soils for studies on global climate change. *Radiocarbon* 34, 535–540.

Berner, R.A. and Lasaga, A.J. (1990) Semulation des geochemischen Kreislaufs. *Spektrum der Wissenschaft* 5, 56.

Bertram, H.G. (1986) Zur Rolle des Bodens im globalen Kohlenstoffzyklus. *Veröffentlichungen der Naturforschenden Gesellschaft zu Emden v. 1814* 8, 3–03.

Broecker, W.S. and Peng, T.H. (1992) Interhemispheric transport of carbon dioxide by ocean circulation. *Nature* 365, 587–589.

Budyko, M.J., Ronov, A.B. and Yanshin, A.L. (1987) *History of the Earth Atmosphere*, Springer Verlag, Berlin, p. 80.

Craig, H. (1957) Isotopic standards for carbon and oxygen and correction factors for mass spectrometric analysis of carbon dioxide. *Geochimica Cosmochimica Acta* 12, 133.

Dover, M. and Talbot, L.M. (1987) *To Feed the Earth; Agro-ecology for Sustainable Development*. World Resources Institute, New York.

Dudal, R. (1990) Soil cover of the world. In: Arnold, R.W., Szabolcs, I. and Targulian, V.D. (eds) *Global Soil Change: Report of an IIASA-ISSS-UNEP Task Force on the Role of Soil in Global Change*, Chapter III. International Institute for Applied Systems Analysis, Laxenburg, Austria, p. 33.

Esser, G. (1990) Modeling global terrestrial sources and sinks of CO_2 with special reference to soil organic matter. In: Bouwman, A.F. (ed.) *Soils and the Greenhouse Effect* Wiley, Chichester, pp. 247–261.

Freytag, J. (1985) Das $^{13}/^{12}C$ Isotopenverhältnis als aussagefähiger Bodenparameter, untersucht an tunesischen Kalkkrusten und sudanesischen Vertisolprofilen. *Hamburger Bodenkundliche Arbeiten* 3.

Global 2000, *Der Bericht an den Präsidenten* (1980) Zweitausendeins, Frankfurt, p. 279.

Global Change, IGBP (1992) Reducing Uncertainties, *IGBP Publication*, Royal Swedish Academy of Sciences, Stockholm, p. 27.

Hart, M.H. (1978) The evolution of the atmosphere of the Earth. *Icarus* 33, 23–39.

Idso, S.B. (1990) The carbon dioxide/trace gas greenhouse effect; greatly overestimated? *American Society of Agronomy, Special Publication* No 53, pp. 19–26.

Idso, S.B. (1990) Interactive effects of carbon dioxide and climate variables on plant growth. *American Society of Agronomy, Special Publication* No 53, pp. 61–69.

Jenkinson, D.S. and Rayner, J.H. (1977) The turnover of soil organic matter in some of the Rothamsted classical experiments. *Soil Science* 123, 298–305.

Keeling, C.D., Piper, S.C. and Heimann, M. (1989) In: Peterson, D.H. (ed.) Aspects of climate variability in the Pacific and the Western Americas. *Geophysics Monograph 55* American Geophysical Union, Washington DC, pp. 305–363.

Lewis, J.S. and Prinn, R.G. (1984) *The Planets and Their Atmospheres.* Academic Press, Orlando, FL.

Lovelock, J.E. (1988) *The Ages of Gaia. A Biography of Our Living Earth.* W.W. Norton, New York.

Lovelock, J.E. (1991a) *Gaia: A New Look at Life on Earth.* Oxford University Press, Oxford.

Lovelock, J.E. (1991b) *Healing Gaia.* Harmony Books, New York.

Martin, A., Mariotti, A., Balesdent, J., Lavelle, P. and Vuattoux, R. (1990) Estimate of organic matter turnover rate in savanna soil by ^{13}C natural abundance measurements. *Soil Biology and Biochemistry* 22, 517–523.

May (1988) cited in Lovelock, J.E. *The Ages of Gaia. A Biography of Our Living Earth.* W.W. Norton, New York.

Miehlich, G. (1991) Chronosequences of volcanic ash soils. *Hamburger Bodenkundliche Arbeiten,* 15.

Parton, W.J., Schimel, D.S., Cole, C.V. and Ojima, D.S. (1987) Analysis of factors controlling soil organic matter levels in Great Plains Grassland. *Soil Science Society of America Journal* 51, 1173–1179.

Riebesell, U. and Wolf-Gladrow, D. (1992) Das Defizit in der Kohlenstoffbilanz. *Spektrum der Wissenschaft* July, 28–32.

Ripley, P.O. (1975) Shifting cultivation and burning versus crop rotations in the tropics. *Proceedings ISSS Savannah Soils Conference,* Accra, Ghana.

Roger, P.A. and Kurihara, Y. (1988) Floodwater biology of tropical wetland ricefields. *Proceedings International Symposium on Paddy Soil Fertility,* Chiangmai, Thailand, IRRI Publ.

Roeschmann, G. (1962) Die Entstehung der Wurzelböden und kaolinitischen Kohlentonsteine des Ruhrkarbons. *Fortschritte der Geologie Rheinland und Westfalen* 3, Geological Survey Krefeld.

Scharpenseel, H.W. (1965) Die Nährstoffaufnahme des Olivenbaums aufgrund von Markierungsstudien mit $H_3{}^{32}PO_4$ und ^{32}P-Superphosphat. *Landwirtschaftliche Forschung* 18, 200–207.

Scharpenseel, H.W. and Becker-Heidmann P. (1992) Twenty five years radiocarbon dating of soils; paradigm of erring and learning. *Radiocarbon* 34, 541–49.

Scharpenseel, H.W. and Becker-Heidmann, P. (1993) Carbon sequestration by grassland soils of different climate zones, as revealed by (thin) layer ^{14}C-dating. In: *Proceedings, XVII International Grassland Congress,* New Zealand and Queensland (in press).

Scharpenseel, H.W. and Miehlich, G. (1989) Soil fertility and organic fractions. Proceedings Soil Management and Smallholder Development in the Pacific Islands. Pacific-land. *IBSRAM Proceedings* 8, 111–19.

Scharpenseel, H.W. and Schiffmann, H. (1985) Natürliche Radiokohlenstoffmessungen als Beitrag zur Definition rezent-oder paläoklimatischer Leithorizonte in Tunesien. *Zeitschrift f. Pflanzenernährung und Bodenkunde* 148, 113–130.

Scharpenseel, H.W., Tsutsuki, K., Becker-Heidmann, P. and Freytag, J. (1986) Untersuchungen zur Kohlenstoffdynamik und Bioturbation von Mollisolen *Zeitschrift fur Pflanzenernährung und Bodenkunde* 149, 582–597.

Scharpenseel, H.W., Becker-Heidmann, P., Neue, H.U., and Tsutsuki, K. (1989) Bomb-carbon, ^{14}C-dating and ^{13}C-measurements as tracers of organic matter dynamics as well as of morphogenetic and turbation processes. *The Science of Total Environment* 81/82, Elsevier, Amsterdam, pp. 99–110.

Schleser, G.H. and Jayasekera, R. (1985) $\delta^{13}C$-variations of leaves in forests as an indication of reassimilated CO_2 from the soil. *Oecologia* 65, 536–542.

Schleser, G.H., Bertram, H.G., Scharpenseel, H.W. and Kerpen, W. (1983) Aussagen über Bildungsprozesse tunesischer Kalkkrusten mittels $^{13}C/^{12}C$-Isotopen-analysen. *Mitt.Dtsch. Bodenkundl. Gesellschaft* 38, 573–578.

Schwarzbach, M. (1974) *Das Klima der Vorzeit,* Enke Verlag Stuttgart.

Swaby, R.J. and Ladd, J.N. (1963) Stability and origin of soil humus. *Proceedings FAO/ IAEA Technical Meeting.* Pergamon, Oxford, pp. 153–159.

Tans, P. (1981) A compilation of bomb C-14 data for use in global carbon model calculations. In: Bolin, B. (ed.) *Carbon Cycle Modeling, SCOPE 16,* Wiley, Chichester, pp. 131–157.

Tans, P.P., Fung, I.Y. and Takahashi, T. (1990) Observational constraints on the global atmospheric CO_2 budget. *Science* 247, 1431–1438.

Theng, B.K.G., Tate, K.R. and Becker-Heidmann, P. (1992) Towards establishing the age, location, and identity of the inert soil organic matter of a Spodosol. *Z. Pflanzenernähr. Bodenkunde* 155, 181–184.

Tsutsuki, K., Suzuki, C., Kuwatsuka, S., Becker-Heidmann, P. and Scharpenseel, H.W. (1988) Investigation on the stabilization of the humus of Mollisols *Z. Pflanzenernähr. Bodenkunde* 151, 87–90.

Whittaker, R.H. and Likens, G.E. (1973) The primary production of the biosphere. *Human Ecology* 1, 299–369.

Yaalon, D.H. (1990) Personal communication at Workshop 'Soils on a Warmer Earth', February 1990, UNEP, Nairobi.

Part IV

Soil Organisms and Soil Resilience

Chapter 17

Functional Attributes of Biodiversity in Land Use Systems

J.M. Anderson

Department of Biological Sciences, University of Exeter, Exeter EX4 4PS, UK. Present address: Rothamsted International, Rothamsted Experimental Station, Harpenden, Herts, AL5 2JQ

Introduction: Values Placed on Biodiversity

The increasing recognition of the importance of conserving the diversity of plant and animal species on a global basis has been manifested in the UNCED Convention on Biodiversity but the exact meaning of this concern for policy is unclear. Many countries have legislated for the protection of endangered species recognizing that the rate and scale of extinctions attributable to human activities has accelerated to unacceptable rates. During the past few decades rates of species losses have accelerated to probably 1000 times faster than the norm over evolutionary time (di Castri *et al.*, 1992) as a consequence of habitat destruction, pollution and direct exploitation of plant and animal populations. However, given that extinction is a normal evolutionary process the question is raised as to what constitutes the 'acceptable' rate of extinctions (Norton and Ulanowicz, 1992). If all species are not sacrosanct then what criteria should be used for their protection? Three aspects of biodiversity are the focus of this debate: the moral, utilitarian and functional values of species.

The moral basis for conserving species involves both the intrinsic value of species (Ehrenfeld, 1988) as well as the damage to natural resources caused by short-term economic gains with the long-term ecological impact borne by others. Often those others are the least able to bear these costs – the resource-poor farmers of the tropics (Colwell, 1992).

Plant, animal and microbial species have utility values as human resources. Traditional societies have a close dependence on the natural habitats surrounding them for potable water, fuel, food, fodder and a wide range of plant and animal products for medicines, construction materials, containers, minor household items, etc. In permanent settlements, staple foods such as maize, rice or cassava may be grown near the houses but the total biodiversity of the resource base may be very high. Villages in Indonesia have been found to contain 200 species of plant in home gardens which are complex analogues of natural rainforests (Soemarwoto and Soemarwoto, 1982). With intensification of settlement, the low resource/ biomass ratio of many tree crops is supplanted by the higher yield of annuals (Ewel, 1986). Farmers in less-developed regions frequently plant a mosaic of traditional varieties as insurance against outbreaks of disease and weather extremes, to maximize total production and to provide a range of food resources (Beets, 1982; Ewel, 1986). On more intensively managed farms, the demands for high yields and mechanized farming practices have resulted in the domination of the land by monocultures. Only 25 species of plants and five species of animals account for 90% of human food resources and international commerce in foodstuffs. Three cereals, rice, wheat and maize, account for 49% of human calorie intake (FAO in Solbrig, 1992). Accompanying this intensification of farming practices whole landscapes have been grossly simplified in terms of natural species richness and the diversity of land uses to meet the demands for more food. In addition to non-commodity values, such as the loss of cultural heritage, wild species represent an untapped germplasm for new cultivars, sources of active biochemical compounds in plants, animals and microorganisms for drugs, biocides and industrial biochemicals (Farnsworth, 1988; Bull *et al.* 1992; Colwell, 1992). The utility values of species, and their potentials, therefore present a very strong economic imperative for conservation.

The third property of biodiversity is its functional role in maintaining ecosystem processes. Biodiversity of animals and plants generally declines as an inverse function of the intensity with which crops are cultivated using mechanized methods and agrochemicals. This can have very important implications for pest management in temperate and tropical systems (as has been authoritatively reviewed by Perfect, Waage, Altieri and Claridge in Hawksworth, 1991). This review, however, is concerned with functional properties of soils and direct plant pest and pathogenic effects on plants are not considered further here.

The intensity with which soils are cultivated also depletes soil organism communities as a consequence of toxic effects of agrochemicals, the physical disruption of their habitats and the depletion of the litter and soil organic matter resource base. It has also been observed that with reduced tillage and organic husbandry the biodiversity of organisms (including pest species) tends to increase (Crossley *et al.*, 1992). This recovery, however, is also determined by biogeographic effects since a source of organisms must be present within their dispersal range to colonize the improved habitat. These relationships between biodiversity and management practices have been recognized for decades and proposed as bioindicators of soil health by soil biologists worldwide. Potter and Meyer (1990) provide a useful evaluation of biodiversity as an indicator of soil degradation. However, it is remarkable that despite the voluminous literature on the effects of agricultural practices on soil organisms, the significance of these changes in complex communities for maintaining the functioning of soils has not been addressed (for example, in Hawksworth, 1991). This chapter questions the role of biodiversity of plants and soil organisms as a requisite for the healthy functioning of soils.

Scales at which Biodiversity Operates

The first requirement for linking community structure and function is to link the species complement of the system at the same scale for which process measurements are made. There are two components of this relationship: the number of species per unit area and the area of influence of species.

Typically, in soil ecological studies, population densities of animals are expressed per square metre. At this scale these communities seem extremely diverse, possibly comprising thousands of bacteria, fungi, nematodes, mites and other arthropods, but this is partly a question of scale of perception. With small soil invertebrates ($<$ 1 mm diameter), such as mites or nematodes, the number of new species encountered with increasing sample size may plateau at a scale of 1–10 m^2 but up to 1 ha for larger invertebrates such as earthworms. Insects and plants in herb-rich meadows or rainforest show comparable diversity at 1–10 ha, or up to 100 ha for rare species (Swift and Anderson, 1993).

The functional diversity is also high with sizes of organisms ranging over 6–7 orders of magnitude, from bacteria and protozoa of a few micrometres in diameter, to large earthworms and macroarthropods. Although fungal hyphae are microscopic the mycelium of a single genetic individual can dominate decomposition processes over an area of many square metres (Rayner and Boddy, 1988). The scale at which organisms influence soils therefore ranges from bacterial microsite effects to patches influenced by different tree species or forest stands. The functional properties of these different organism effects can be considered as a hierarchy in which complexity operates as a nested series of interacting systems (Allen *et al.*, 1984). The highest order is represented by the landscape or catchment scale ($>$ 1000 ha) containing a mosaic of different plant communities or ecosystem types. At the lowest order ($<$ 1 mm) is the cellular structure of litter, clay domains or soil aggregates at which microorganisms function. Spanning these extremes are organizational scales defined by areas affected by animals and plants of different sizes (1–50 m^2), the patchiness imposed by aggregations within species populations (100 mm^2–1 ha), and the scale of arable or forestry plots (0.5–25 ha). Moving up this hierarchy successive levels accommodate the same processes, but with slower dynamics, and covering a larger area.

'Bottom-up' and 'top-down' approaches

The distal factors regulating carbon and nutrient fluxes are the physiological activities of microorganisms operating in soil microsites. As the spatial and temporal scale of the investigation is expanded, more and more organisms and complex interactions are introduced and additional attributes of higher levels of organization come into play. These include the patchy effects of larger animal and plant species on soil properties affecting microbial activities, weather variation, soil heterogeneity, hydrological factors and other abiotic/biotic factors contributing to system-level fluxes. The 'bottom-up' approach is therefore confounded by the species/area effect of retaining the same unit of measurement across expanding spatial scales. As a consequence, ecosystem-level studies generally measure the net effects of organism communities in terms of process rates regulated by more distal environmental variables.

The alternative, 'top-down', approach is to scale down progressively from higher to lower order systems to determine the point at which the effects of species assemblages on soil properties and processes emerge as regulators of soil fertility. This is illustrated in Fig. 17.1 where the 'top-down' approach could be considered analogous to draining a lake. The first features which emerge are the main determinants of ecosystem functioning (the plant and soil subsystems), then the components of the subsystems (wood, leaves, roots; soil organic and mineral pools) and finally the effects of different species which are at the core of ecosystem structure. It is also possible to encounter situations where species manifest their specific attributes at a much higher level of organization. In terms of the lake analogy, these are 'water-lily' effects where a species rooted in the benthos (i.e. operating at a proximate level of organization) affects the functioning of the lake ecosystem at the surface (i.e. a distal level of organization). Examples, are primary production by a monospecific crop, *Rhizobium* mediating nitrogen fixation, the effects of outbreaks of single herbivore species on nutrient cycling and, as considered below, single species of termites affecting landscape-level processes.

The point to be made is that in order to relate community structure and functioning, the components of the community must be defined at the same scale as process measurements. If these measurements are made at more distal scales than the level at which particular organisms manifest their effects then the contribution of those communities to fluxes will not be apparent. Conversely, relationships between processes and community structure determined at too fine scales may be irrelevant to ecosystem-level

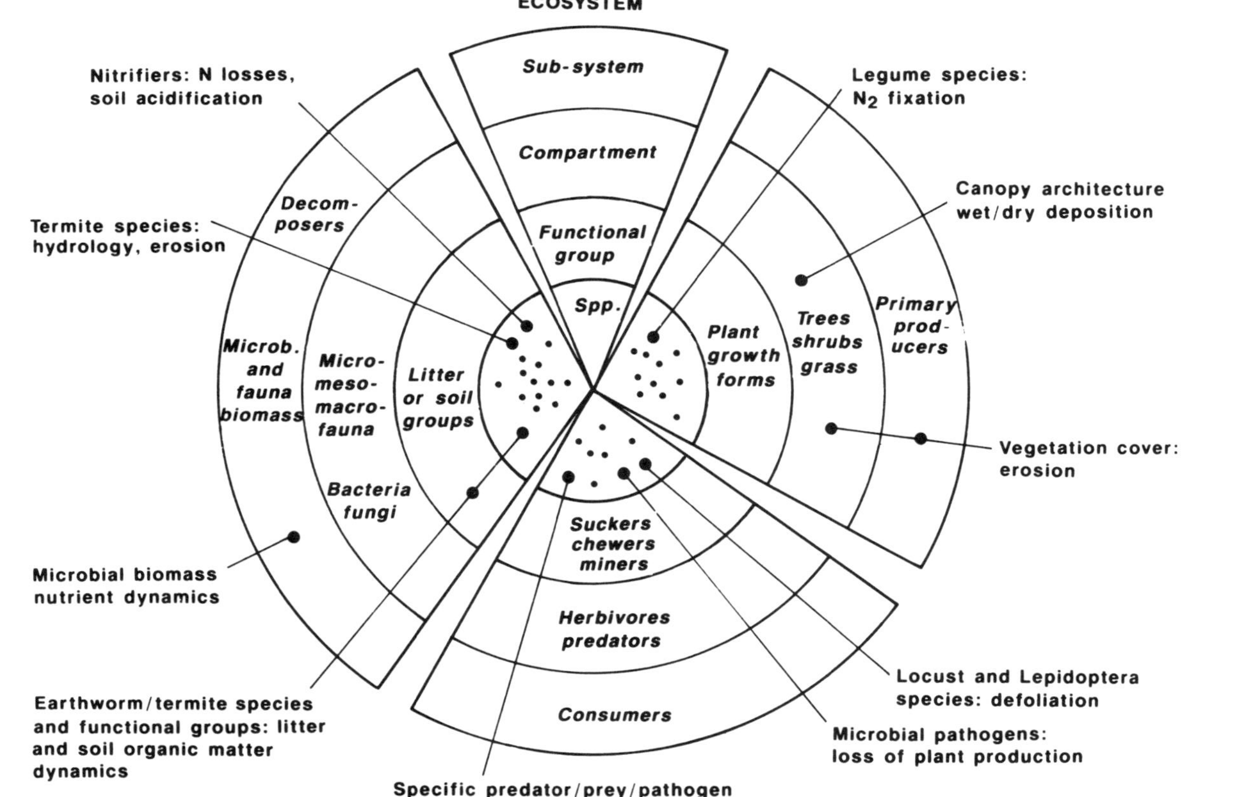

Fig. 17.1. Hierarchical scales of organization in terrestrial ecosystems. The structure and functioning of an ecosystem can be defined at different levels from subsystems down to the activities of specific organisms. At present there is poor understanding of the relationships between community structure and the proximate controls which organisms have on process rates. Hence as processes are measured over larger areas and longer periods of time the processes are related to more distal, physical environmental controls. The figure illustrates the point, however, that some species have effects which determine ecosystem-level functions. Some examples are shown by lines linking species to functions on the outer circle.

fluxes because their total cumulative effects are masked by process controls at a higher level of organization. These scaling relationships will be considered for plant and soil organism communities maintaining soil fertility in natural and agroecosystems.

Biodiversity and Hydrology from Catchments Down to Soil Macropores

Studies of water balances in catchments, which have been conducted in many areas of the world to investigate the effects of vegetation changes on water yield and evapotranspiration (Bosch and Hewlett, 1982), provide a useful insight into the role of plant and soil processes at the ecosystem level. One of the few rigorous and long-term studies, involving the conversion of forest in Kenya into a tea plantation (Edwards and Blackie, 1981), showed that when the forest was felled peak flows increased and water yields were temporarily increased by about 14% over the first three years. The difference diminished as the tea developed and, when summed over the first 15 years, the extra yield was only 9%. When fully mature the tea estate closely matched the water yield of the forested control catchment. Both yielded an average of 800 mm from an average precipitation of 1200 mm. No significant differences in water yield were obtained for the conversion of a montane catchment, containing bamboo and patches of evergreen forest, into a uniform *Pinus patula* plantation (Edwards and Blackie, 1981). Studies on forest catchments in Malaysia, which included some 13% of the area converted into rubber plantations, showed no differences in stream flow from a fully forested catchment, possibly because the plantation area was too small to detect differences (Low and Goh, 1972). These results are consistent, however, with the pattern shown by a large number of short-term studies (Bosch and Hewlett, 1982) that the diverse forest community has similar effects on catchment hydrology to smaller stature, monospecific stands. The effects of structural complexity and plant species composition must be sought at a finer scale of resolution.

Vegetation Structure and Soil Erosion

Infiltration rates are usually high under undisturbed, natural vegetation and litter cover as a consequence of soil surface protection and a good macroporous structure maintained by organic matter and biological activities. Consequently overland flow rarely occurs and soil losses are usually insignificant. When surface cover is removed raindrop impact destroys surface aggregate structure, loosens fine soil fractions (usually containing high concentrations of nutrients), blocks macropores, and increases soil erodibility. Hence reduction in canopy and litter cover often results in increased soil erosion.

Under tree cover this relationship is quite complex. The presence of the canopy reduces the volume of throughfall but the structural characteristics of the canopy determine the size of droplets and the canopy height their fall velocities (Wiersum, 1984). This significantly increases the kinetic energy of droplets. Brandt (1988) showed that splash detachment from throughfall increased by 600% compared with incident precipitation. Studies by Wiersum (1984) in Java showed that experimental removal of trees and undergrowth increased sediment loads from 0.03 kg m^{-2} under undisturbed forest to 0.08 kg m^{-2} with intact litter cover. The removal of all cover increased sediment loads to 1.59 kg m^{-2} but with intact canopy cover and no litter or ground cover erosion increased to 4.32 kg m^{-2}. Ground cover by litter or vegetation is therefore the most critical factor in erosion control. Wiersum (1984) also records similar values for the erosive power of throughfall beneath a bamboo forest and the more complex canopy of a home garden, 180% and 202% of incident rainfall respectively. This suggests that the contribution of the very different species composition in these plant communities is a relatively unimportant variable in these relationships in comparison with the architecture of the vegetation.

Plantations

Soil surface erosion in plantations of different tree species falls to similar levels as undisturbed forest once canopy and ground cover are established and remain undisturbed (Table 17.1). In teak plantations canopy height and the morphology of the large leaves can create throughfall with high kinetic energy which causes severe erosion if the litter layer is harvested or burnt (Bell, 1973).

Plot-scale studies show that conversion of forest into plantations or perennial tree crops generally causes less erosion than conversion into arable agriculture, except during the establishment phase when canopy closure takes longer than for arable crops (Bruijnzeel, 1990). Fast-growing, leguminous cover crops such as *Centrocema, Pueraria* or *Macuna* are commonly seeded during the establishment phase of tree plantations for soil and nutrient conservation (Broughton, 1977) or grass cover may be encouraged (Tinker, 1968). Once ground cover is complete all low-growing cover crops, weeds or natural regrowth show similar levels of surface protection from high intensity rains but the rate of leaf area development varies according to plant morphology and growth as in arable crops. Under drier regimes, little additional reduction in erosion occurs with vegetation cover above 20% across a variety of different rangeland communities (Mbkaya *et al.*, 1988).

Table 17.1. Surface erosion in tropical forests and tree crop systems (t ha^{-1} $year^{-1}$) (Bruijnzeel, 1990 after Wiersum, 1984).

	Minimum	Median	Maximum
1. Natural forest (18/27)[a]	0.03	0.3	6.2
2. Shifting cultivation, fallow period (6/14)	0.05	0.2	7.4
3. Shifting cultivation, cropping (7/22)	0.4	2.8	70
4. Plantations (14/20)	0.02	0.6	6.2
5. Multi-storied tree gardens (4/4)	0.01	0.1	0.15
6. Tree crops with cover crop/mulch (9/17)	0.1	0.8	5.6
7. Agricultural intercropping in young forest plantations (2/6)	0.6	5.2	17.4
8. Tree crops, clean weeded (10/17)	1.2	48	183
9. Forest plantations, litter removed or burnt (7/7)	5.9	53	105

[a]Number of locations/number of treatments.

Arable crops

Arable crops which rapidly establish close canopy cover protect the soil whereas slow-growing crops leave bare areas which are vulnerable to splash erosion. Aina *et al.* (1979) showed that soil erosion on experimental plots in Nigeria decreased exponentially with increases in canopy cover (Table 17.2). Cassava has a very open canopy during the first 3–4 months after planting

Table 17.2. Effect of crop cover and methods of seedbed preparation on soil and sand splash (t ha^{-1}) under different cropping systems (14 June to 12 October). Splash is expressed in absolute units and relative to bare ridges. (From Lal, 1983).

	Soil splash		Sand splash	
Cropping system	Absolute	Relative	Absolute	Relative
Cassava (ridges)	186.8	42.4	61.3	65.8
Maize	11.0	25.2	53.5	57.5
Yam (mounds)	203.6	46.2	81.2	87.2
Sweet potato	237.2	53.9	86.1	92.5
Cassava and sweet potato (ridges)	60.2	13.7	62.3	66.9
Sweet potato and maize	140.0	31.8	76.5	82.2
Yam and maize (mounds)	175.0	39.8	74.3	79.7
Cassava and maize (ridges)	65.4	14.9	68.0	73.0
Ridges (bare)	440.1	100.0	93.1	100.0

and provides less soil cover than maize. Plots with mixed crops of cassava and maize showed less runoff and soil loss than either crop grown alone. Noble and Morgan (1983) showed in carefully controlled laboratory experiments that the morphology of crop plants does affect splash detachment. The effects, however, operate at a much lower order of scale than crop management practices, and these in turn are less effective than soil management practices in controlling erosion. Thus crops which provide poor ground cover in a no-till system with crop residue mulches cause less soil erosion than soil-conserving crops grown with inappropriate soil management practices (Lal, 1987).

Soil invertebrate effects

The most diverse and numerically abundant groups of invertebrates such as the protozoa, nematodes, mites and collembola have small body sizes so that their activities in soil are confined to macropores and voids, or to surface litter habitats. Their effects on the structure of mineral soils are negligible. In contrast, the feeding and burrowing activities of some larger invertebrates, particularly termites and earthworms, can change soil physical properties and process rates through the removal of litter cover, transport of soil, and creation of soil aggregates, macropores and channels (Fig. 17.2). These activities have cumulative effects and can reach thresholds where they control surface water fluxes and the erodibility of soil at the field plot, and even landscape scales. This is best documented for earthworms and termites which forage at the litter/soil interface but, as will be shown, mainly for situations where a few species have dominant effects (Anderson, 1988a).

Some litter-feeding termites (Macrotermitinae) in Africa have been found to have a network of galleries extending for up to 50 m from mounds (Darlington, 1982) and remove most of the dead wood and leaf-litter production within this zone. In overgrazed rangeland in Ethiopia, the scavenging of the little remaining grass and litter in the dry season by *Macrotermes*, *Odontotermes* and *Pseudacanthotermes* leaves the soil bare and susceptible to erosion (Wood, 1991). The bare and consolidated soil around termite mounds can also initiate erosion processes to the extent that high densities of *Trinervitermes* mounds in grassland fallows of Côte d'Ivoire may lead to land degradation (Janeau and Valentin, 1987). Conversely the longer-term elimination of a single termite species, *Gnathitermes tubiformans* in a Chiuhuahuan desert system by Elkins *et al.* (1986) resulted in changes in soil structure, increased runoff and sediment loads in areas with low or no vegetation but insignificant effects under shrub cover (Table 17.3). The reduced water supply to soils in the open areas resulted in a reduction in grasses and increased shrub cover.

Many field studies have reported that earthworm-worked soils in

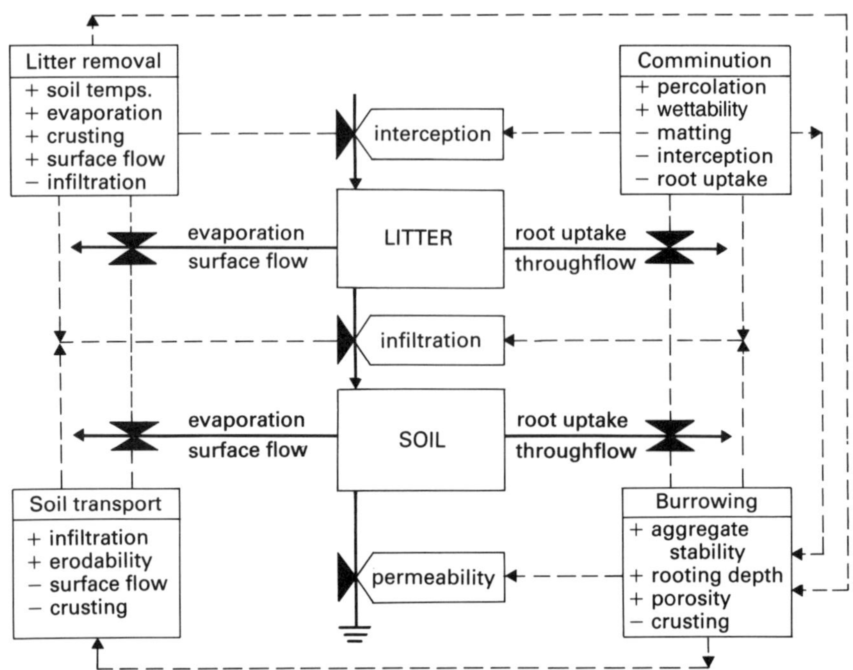

Fig. 17.2. Effects of soil invertebrates on water flux pathways. The consequences of litter removal, litter breakdown (comminution), burrowing and soil transport are shown in terms of positive (+) or negative (−) effects on rainwater transfers through the litter and soil horizons (Anderson, 1988a).

temperate and tropical regions generally have 2–10 times higher porosity, increased field water-holding capacity, and more water-stable aggregates than soils where worms were scarce or absent (Lee, 1985; Lal, 1987; Hulugalle and Ezumah, 1991). The kinetic energy of raindrops required to disperse earthworm casts in a sequence of Nigerian soils was 5–54 times higher than for soil macroaggregates not derived from earthworm activities (De Vleeschauwer and Lal, 1981). Conversely, Sharpley *et al.* (1979) found that in New Zealand pastures the splash dispersal of soil in casts increased the sediment load of surface runoff but burrows acted as a sink for suspended material. As a consequence, net sediment losses (and phosphorus in fines) were lower in the presence of earthworms than when they were eliminated.

The colonization of pastures in New Zealand by European lumbricids is one of the few cases where it has been possible to document the effects of different species of earthworms under field conditions. The improved pastures contained few native earthworm species and built up a substantial turf mat. This accumulated litter became incorporated into soil when *Aporrectodea caliginosa* was introduced resulting in a sustantial increase in grass production and a 100% increase in infiltration rates. Springett (1985)

Table 17.3. Effects of pesticide elimination of a subterranean termite *Gnathitermes tubiformans* on soil properties and hydrology in a Chiuhuahuan desert (Elkins *et al.*, 1986).

	Untreated		Chlordane treated	
	Open area	Canopy cover	Open area	Canopy cover
Bulk density (g cm^{-3})	1.70	1.66	1.99	1.77
Total porosity (%)	35.8	37.4	24.9	25.5
Infiltration (mm h^{-1})	88.4	100.5	51.3	106.4
Runoff (% rainfall)	13.3	4.6	30.5	3.9

documents further effects of introducing *Aporrectodea longa* into pastures where populations of surface-active lumbricids (*A. caliginosa. A. trapezoides* and *Lumbricus rubellus*) were already established. Prior to the introduction of *A. longa* the soil was well structured to a depth of 10 cm but roots only penetrated to the depth of this horizon and were susceptible to rotting when soils were waterlogged in winter. After 18 months the patches where *A. longa* had been introduced showed a doubling in total porosity below 10 cm depth and improved root penetration and survival. The effects of these changes on sward production were small overall but as a consequence of improved drainage and greater grass rooting-depth there was less damage to pastures by stock.

Clements *et al.* (1991) showed that the elimination of earthworms from a fertilized, UK pasture over a period of 20 years caused a dramatic increase in soil bulk density, shear strength and litter accumulation. There was decreased soil carbon content, infiltration rate, pH and soil moisture content. There was no long-term decline in herbage yield indicating that soil structural properties were not a constraint to grass production and that potential checks on nutrient cycling due to the accumulation of a turf mat were compensated by fertilizer applications.

Microorganisms

The importance of water-stable aggregates for the maintenance of soil physical properties is well established. From the soil management perspective, evidence suggests that conventional tillage reduces the size of soil aggregates, which can promote pore clogging and reduce infiltration. Soil structure can be improved and maintained by organic matter additions as crop residues, as organic manures and from root exudates. Evidence shows that the incorporation of organic matter alone does not lead to improvement of soil structural properties; rather it is the subsequent microbial transformations involving the production of polysaccharides together with the physical binding of soil particles by filamentous organisms (and plant roots). These processes have been extensively investigated (Tiessen and Stewart, 1988) but the structure of the microbial community involved has rarely been explicitly considered. Laboratory studies by Aspiras *et al.* (1971 a,b) showed that 17 species of fungi, bacteria and actinomycetes differed widely in their capabilities and mechanisms for binding soil particles and stabilizing aggregates. Cementing of soil particles by aromatic or aliphatic compounds conferred different degrees of stability on the aggregates which were also mechanically reinforced by hyphae. The forms of organic matter added to soil, growth responses by different species in the microbial community, and the determinants of metabolic products may offer opportunities for managing structural properties of soils. Field studies to date, however, have focused on the net effects of microbial processes and the roles of simple and complex communities are undetermined.

The examples considered in this section illustrate two points. First, there is little evidence that controls over hydrological processes are related to biodiversity. Second, management practices which reduce complex biological interactions in soils are not inherently unsustainable. The particular ecological and agronomic context of the system must be taken into account.

Biodiversity and Nutrient Cycling

Natural ecosystems are selected to withstand a wide range of perturbations to nutrient transfers imposed by wetting up after dry seasons, intense rainstorms, tree throw, cataclysmic litter falls and outbreaks of defoliating insects. With the exception of infrequent, extreme events such as hurricanes and wildfires, only small amounts of nutrients are lost through leaching or runoff because many different buffering mechanisms are operating. These can be considered in terms of vegetation characteristics which regulate nutrient uptake and return, controls over nutrient mineralization, and exchange pools which buffer fluxes. Perennials, such as trees, exert powerful stabilizing effects on the variability of nutrient fluxes through internal nutrient re-allocation and the deposition of a wide range of resources (fruits, leaves, twigs, branches, coarse roots, fine roots) with different nutrient contents and decomposition rates. The phenologies of different species spread the timing of inputs as well as nutrient demands. The turnover rates of different soil organic matter fractions, organic and mineral exchange capacities and protection of the soil from erosion all reduce the amount of the labile nutrient 'at risk' to a small proportion of the capital. Small imbalances in excess of plant demand are sequestered by biogeochemical processes and subsequently remobilized over periods of days to decades.

Nutrient conservation in natural systems is therefore regulated by a diversity of mechanisms rather than properties which are a particular attribute function of species diversity. In derived systems the buffering capacity and linkage of some of these mechanisms are reduced with increasing management intensity and can result in nutrient losses when mineralization occurs in the absence of root sinks and the retentive capacity of soil is low.

Relationships between Plant Diversity and Soil Fertility

Many researchers have noted the influence of tree species on the spatial variability of soil properties in temperate and tropical regions. In forested land, patterns in soil variability have been related to the distribution of litter fall, stand density and the dynamics of patches associated with tree falls and regrowth (Charley and West, 1975; Kang and Moorman, 1977; Mollitor *et al.*, 1980). This implies that soil variability should increase with the diversity of vegetation (Grieve, 1977). There is little indication that this is a general phenomenon. In highly diverse rainforest communities, for example, there is more evidence for edaphic determinism of species groupings (Gartlan *et al.*, 1986; Baillie *et al.*, 1987; Ho *et al.*, 1987). This pattern may be partly a problem of scaling in soil sampling because species–soil relationships become clearer as patch dominance increases. In the Cameroons, Newbery *et al.* (1988) showed that three *Microberlinia/Tetraberlinia* species (Cesalpinioidea, Leguminosae) formed groves of variable size up to 600 m across, whereas no such distinct patterns were shown for 182 other species of trees. The groves were characterized by high inputs of litter in the dry season, extensive ecomycorrhizal colonization of surface organic matter and higher soil carbon concentrations than adjacent areas with other species associations. Whether the soil conditions in these groves are the net effect of synergistic interactions between these species has not been determined. Few experiments have been carried out to determine the effects of species mixtures on soils but one temperate study showed quite small differences under various combinations of conifers and broad-leaf species compared with pure stands after 35 years (Moffat and Boswell, 1990).

At least 55 tropical tree species are known to have beneficial effects on crop growth, either when used as a source of green manure or when crops are grown under the canopy (Young, 1989). *Acacia albida* is widely recognized in the semiarid zone of Africa for its soil-improving properties which are traditionally exploited by subsistence farmers. Yields of crops, such as maize or sorghum, may be 50% higher under the canopy, which is leafless during the cropping season, than further away due to a combination of N-fixation by the tree and the effects of litter on soil properties (Young, 1989). Assessment of the mechanistic basis of these effects is generally confounded by the lack of information on below-ground interactions between species in water and nutrient uptake determined by the spatial and temporal patterns of plant roots.

A characteristic of all undisturbed ecosystems is that very low concentrations of nitrogen or phosphorus are leached below the rooting zone irrespective of the diversity of the vegetation cover. Hence it is the presence of an undisturbed root sink rather than the number of plant species which determines nutrient conservation. Where crops are grown together it may be desirable spatially to partition the soil volume so as to reduce competition for water and nutrients. Van Noordwijk (1989) observed that several different tree legume species grown in tea plantations in Indonesia showed broadly similar root distribution patterns which may complement the shallower feeder roots of tea. Similarly in agroforestry systems, such as alley-cropping, where annuals and perennials are growing together it may be desirable to maintain established rooting patterns where the tree provides a deep rooting system to act as a 'nutrient pump' and fine radial roots which act as a 'safety net' for leachates not taken up by the crop. A number of candidate tree species can be identified for alley-cropping systems based on their root geometry (van Noordwijk, 1989) whereas the rooting patterns of annual crops have been extensively studied (Goss, 1991). In principle comparatively simple combinations of mixed tree and crop species could be constructed according to basic rules which could reduce periods of risk when nutrient leaching could occur. Under intensive agricultural practices such strategies may be less effective than better fertilizer management. For example, temperate pastures have swards dominated by the single grass species, *Lolium perenne*, which can maintain very high cattle stocking densities as a consequence of large fertilizer inputs (100–200 kg N ha^{-1} $year^{-1}$). Losses of nitrogen by leaching and denitrification are very high under these regimes because total mineral N, from fertilizer, soil organic matter and excreta, periodically exceeds storage pools and plant demand (Scholefield *et al.*, 1988). Careful N management rather than changes in sward species composition can substantially reduce these losses.

Soil Biota and Nutrient Cycling

The sequences of nutrient transformations carried out by different species of bacteria are well understood. In comparison little is known about the extent to which complex or simplified microbial communities are needed for organic matter transformations. One of the few such studies (Bowen and Harper, 1990) showed that eight species of fungi isolated from decomposing straw varied widely in their ability to degrade lignin. Various combinations of species showed interactions which were greater or smaller than the rates in pure cultures of the most-effective decomposer (Bowen, 1990). The consequences of these interactions in the initial stages of decomposition for processes of soil organic matter formation are unknown. The rather limited information available suggests that microbial communities with different species components function in a similar manner and that compensatory growth can accommodate the effects of massive perturbations to community structure. Intensive fumigation of agricultural soils (to eliminate pathogens), for example, has been shown to eliminate whole classes of bacteria (Martin, 1963; Ridge 1976). Surviving cells, particularly fluorescent pseudomonads protected by soil organic matter, proliferate rapidly and dominate the microbial community over the cropping period of 5 months. Nitrogen mineralization showed transient effects of fumigation and grain yield showed an overall increase in relation to fumigation treatments (Rovira, 1976).

Similar conclusions can be made regarding the relationships between soil fauna community diversity and nutrient cycling. Nutrient budgets calculated for the whole soil organism communities in a variety of different ecosystem types suggest that the direct contribution of soil fauna is about 30% of the flux (Anderson, 1988b; Verhoef and Brussaard, 1990). One of the most detailed studies (Hunt *et al.*, 1987) for a short-grass prairie involved a food-web model containing eight trophic levels and 12 functional groups of fauna. Groups excluded from the model, because of negligible contribution to N turnover, included ciliates, earthworms, termites and insect larvae. In this system the fauna mediated about 38% of the 77 kg ha^{-1} annual flux. Bacterial-feeding nematodes and protozoa accounted for over 84% of the N flux. Running the model with different densities of other groups showed that they had insignificant effects at all reasonable densities. Similar results have been achieved in field experiments with biocides which show N-mineralization rates largely unaffected by major changes in invertebrate community structure (Ingham *et al.*, 1985). These results suggest that soil biological processes, such as N-mineralization, appear insensitive to the diversity of species and invertebrate groups in the soil organism community and short-lived groups, such as fungi, bacteria, protozoa and nematodes, are able to maintain nitrogen fluxes.

This conclusion seems a paradox in view of the wealth of complex interactions which have been shown between invertebrate and microbial species and functional groups (reviewed in Fitter *et al.*, 1985; Edwards *et al.*, 1988). These include the demonstration that grazing by mites and collembola can change the species balance of fungal communities, nematodes and protozoa; that grazing increases the turnover of soil bacteria and increases the availability of N and P to plants, and that mycophagous nematodes can affect the nodulation of legumes by disrupting P uptake by VA mycorrhizas.

As yet, none of these effects has been demonstrated to affect plant production in agroecosystems because, under intensive agriculture, key biological processes under climatic controls are largely overridden by agrochemicals, mechanical tillage and irrigation. Furthermore, agronomic measures of crop performance at a plot scale of hectares, over a cropping season of several months, integrate across small-scale, short-term events such as those influenced by specific activities of the soil biota. For example, earthworms may promote seedling emergence in capped soils (Kladivco *et al.*, 1986) but these initial effects can be obscured by subsequent plant demographic processes such as self-thinning or tillering (Harper, 1977) which are more proximate determinants of yield for the plant population. The question therefore remains as to the systems and circumstances under which the functional attributes of biodiversity in the soil biota may be manifested.

Functional Attributes of Biodiversity in Agroecosystems

Currently concern over the environmental impact of intensive agriculture is shifting the focus away from highly energy-subsidized cropping systems to more ecologically sustainable systems. The management of these systems includes the use of vegetative cover as an effective soil and water-conserving measure (met through the use of no-till practices, mulch farming, use of cover crops, etc.), the regular supply of organic matter (manure or compost), nutrient recycling mechanisms through the use of crop rotations, crop/livestock mixtures, appropriate tree/crop combinations in agroforestry and intercropping based on legumes, etc. (Altieri, 1992). All of these practices will change the pools and fluxes of carbon and nutrient in soils, and increase the biological activity and biodiversity of soil organism communities. The difficulty lies in separating cause and effect. Hendrix *et al.* (1986) showed that different food webs were operating in soils under different cultivation practices in Georgia, USA: the no-till system was dominated by fungi and earthworms, whereas the biota of the conventional till was dominated by bacteria, nematodes and enchytraeids. Nutrient fluxes also differed with higher leaching losses of N and Ca (but not P, Mg or K) under conventional till than no-till plots (Stinner *et al.*, 1984). The driving factors in the conventional till were soil disturbance and residue incorporation, and the relative contributions of biological processes to the differences in nutrient fluxes remain unclear. In the no-till system the biological processes emerge as determinants of soil structure and functioning but the degree to which numbers of functional groups, and species associations, are required in maintaining these properties is undetermined. As considered earlier the complexity of these communities, and their capacity for compensatory activities, has made this a somewhat intractable problem.

In the humid and subhumid/semiarid tropics farmers are faced with the same basic constraints on the maintenance of soil fertility and crop production as in temperate regions but the variables often operate on a greater

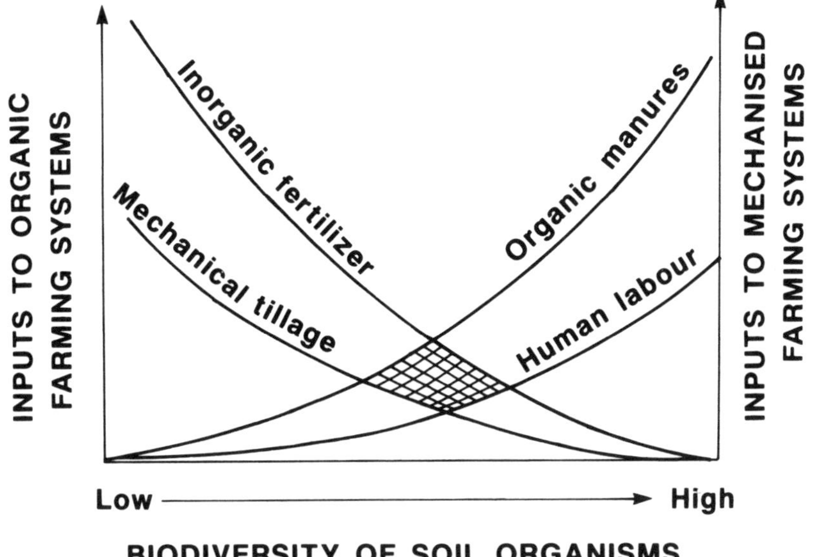

Fig. 17.3. Schematic representation of the emergence of soil fauna effects on soil fertility with the shift from technology-based agriculture management to production based on organic resources. The hatched zone represents the interface at which the biological effects are overridden by intensive management practices.

range of extremes: storm events are more erosive, leaching may be more intensive, dry periods are more often limiting to plant growth, and highly weathered soils, with low inherent fertility, are more prevalent. In addition, rates of organic matter decomposition are high so that management of plant materials, for mulching, maintenance of soil organic matter and as a nutrient source, often requires large inputs to maximize effects. Hence although the principles of sustainable farming practices generally hold for temperate and tropical regions, in practice the opportunities for many tropical farmers to optimize nutrient and organic matter management are limited by environmental, as well as social and economic, factors.

Many farming systems in the seasonal tropics therefore operate within a realm where mechanized tillage and the use of agrochemicals is at a low intensity, but organic inputs are frequently limited. As a result the increased soil biological activity and community diversity which results from reduced tillage is offset by the low organic resource base maintaining the biota and soil structure (Fig. 17.3). Rates of biological processes are also faster, more seasonally related to crop growth conditions, and less buffered by organic matter pools than in temperate soils. Under these conditions biological processes may operate in a more discrete manner in relation to plant phenology as illustrated for a grain crop in Fig. 17.4. During the phases of germination, seedling emergence and root development, soil physico-

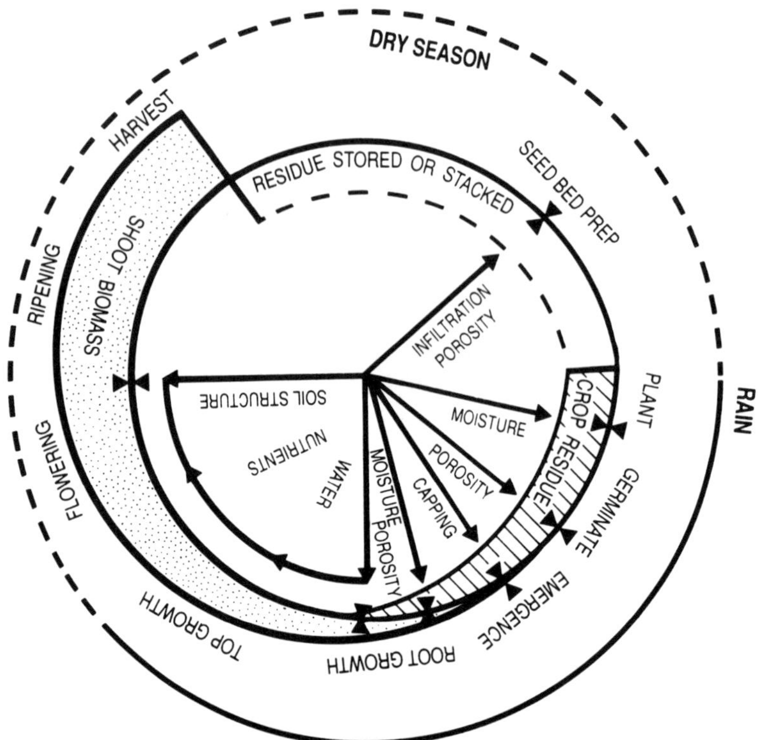

Fig. 17.4. Representation of the growth cycle of a crop such as maize and the biological processes (shown as gates) regulating critical soil conditions for plant development and nutrition. Seed germination and decomposition of crop residues (or organic manures) start with the onset of rain; though termites can have a considerable effect on residue removal, and hence on soil properties, during the dry season. The crop has various demands which must be met during its growth cycle. These must be met, but not exceeded, to optimize resource utilization by smallholder farmers. Hence the management of scarce resources should take into account the 'window of opportunity' to meet soil conditions for crop growth and not mean values for soil properties over the cropping season.

chemical conditions (compaction, crusting, temperature, free aluminium in acid soils) may be critical. Biological factors such as mycorrhizal or rhizobia inocula may then have a narrow window for successful plant establishment. Nutrients and water are the main requisites for the rapid growth phase, and water rather than nutrients may be critical in the postanthesis maturation phase. The plant is then largely independent of the soil conditions as the grain matures.

Each of these critical phases in crop growth has been shown to be influenced by soil fauna or microorganisms but, with the exception of the root symbiotic associations, the degree to which they can be supported by one, two or more invertebrate or microbial species in different soils and management systems has not been investigated. Even less is known about the role of microbial diversity in soil organic matter transformations which are so critical to the maintenance of soil structure and functioning. This is a remarkable gap in current understanding, given the extensive soils literature, and limits the potential for management of soil biological processes. Until the focus on biodiversity shifts from the effects of land use practices on species losses, to placing greater emphasis on the consequences of species losses for maintaining the functional properties of soils, the criteria for policy on biodiversity conservation in relation to soil management will remain undefined.

Conclusions

The criteria for biodiversity conservation were initially defined in terms of the moral, utility and functional roles of species. All these roles can ideally be met within a landscape comprising a mosaic of protected natural habitats, extensively managed systems for hunting, timber and non-food products, and areas of arable cultivation and plantations, for food and cash crops. The relative weighting given to these different land uses and the associated biota depends on the needs of different societies. Ultimately these choices may depend on economic assessments.

The shift from extensive to intensive methods of food production led to widespread reduction in the biodiversity of temperate and tropical systems. This review has shown that both natural and derived ecosystems can function in a sustainable manner provided that a basic complement of functional components of the plant/soil system are present, and careful soil management is maintained in arable cropping systems.

In the more technologically developed countries the aesthetic and utility values of species are given increasing weighting in land use policies. The environmental and economic costs of excess food production are also directing research and management towards less-intensively cultivated, more environmentally sensitive, agricultural systems. Coupled with reduced energy and agrochemical inputs to farmland the biodiversity of the soil organism communities may increase. But where intensive cultivation has been carried out over whole landscapes the sources of species for recolonization may be remote. Should soil conditions be managed to increase their specific activities or species of invertebrates or microorganisms inoculated, as for earthworms in New Zealand and elsewhere?

These activities have obvious economic implications which will have to be assessed in terms of costs and benefits. Such an analysis has been carried out for *Rhizobium* inoculation and for biological control of crop pests where the activities of the organisms have direct relationships to crop performance. Economic models are also available for nitrogen cycling in soils though the roles of organisms are not explicit. No such cost–benefit analysis has been made for soil organism activities which contribute to soil properties and processes such as aggregate stability, erosion, nutrient leaching, denitrification, etc. Even for single key species, such as the earthworm, *Lumbricus terrestris,* there is little quantitative information on the agronomic value of their activities in different soils and cropping systems.

In many, less technologically developed, regions of the world, food surpluses are a primary goal and the demands of burgeoning populations continue to shift from extensive to intensive land-use with concomitant losses of biodiversity and decreased utilization of plant and animal products from natural systems. Most smallholder farmers, however, are not in a position to achieve the genetic potential of high-yield crop varieties because of limited availability of water, fertilizers and pesticides. Under these conditions improvements in traditional farming practices are increasingly recognized as being more environmentally acceptable goals by optimizing the uses of energy, nutrients and organic matter.

As shown earlier the basic principles for designing cropping systems containing a few plant species which conserve soil fertility are known. But the details of how mixtures of more than two or three plant species, cultivars, genotypes, annuals and perennials interact above and belowground under different climate regimes and soil conditions have hardly been studied (Swift and Anderson, 1993).

Farming systems, employing zero or low intensities of tillage, with some dependence on organic matter inputs to maintain soil fertility, are in a realm where the mechanistic understanding of soil biological processes can be used to underpin improved cultivation practices. The need to conserve plant, animal and microbial biodiversity in these systems is even more critical than in temperate regions because the ecology of the biotas as well as the species associations required for developing sustainable agroecosystems have been so little studied. The IUBS/UNESCO Tropical Soil Biology and Fertility Programme (TSBF) (Seward and Woomer, 1992) is one of the few strategic research initiatives concerned with investigating soil biological processes. The main objective of TSBF is to develop and evaluate the options for maintaining or improving soil fertility through the management of the biological processes regulating the synchrony of nutrient release and plant uptake, and soil organic matter turnover. The need for this understanding is not unique to the tropics but the degree of urgency is and constitutes a critical focus for biodiversity research.

References

Aina, P.O., Lal, R. and Taylor, G.S. (1979) Effects of vegetal cover on soil erosion on an Alfisol. In: Lal, R. and Greenland, D.J. (eds) *Soil Physical Properties and Crop Production in the Tropics.* John Wiley, Chichester, pp. 501–507.

Allen, T.F.H., O'Neill, R.V. and Hoekstra, T.W. (1984) Interlevel relations in ecological research and management: some working principles from hierarchy theory. *USDA Forest Service General Technical Report RM-110.* Rocky Mountain Forest and Range Experimental Station, Fort Collins, CO.

Altieri, M.A. (1992) Sustainable agricultural development in Latin America: exploring the possibilities. *Agriculture, Ecosystems and Environment* 39, 1–21.

Anderson, J.M. (1988a) Invertebrate-mediated transport processes in soils. *Agriculture, Ecosystems and Environment* 24, 5–19.

Anderson, J.M. (1988b) Spatiotemporal effects of invertebrates on soil processes. *Biology and Fertility of Soils* 6, 216–227.

Aspiras, R.B., Allen, O.N., Harris, R.F. and Chesters, G. (1971a) Aggregate stabilization by filamentous microorganisms. *Soil Science* 112, 282–284.

Aspiras, R.B., Allen, O.N., Harris, R.F. and Chesters, G. (1971b) The role of microorganisms in the stabilization of soil aggregates. *Soil Biology and Biochemistry* 3, 347–353.

Baillie, I.C., Ashton, P.S., Court, N.M., Anderson, J.A.R., Fitzpatrick, E.A. and Tinsley, J. (1987) Site characteristics and the distribution of tree species in mixed dipterocarp forest on Tertiary sediments in central Sarawak, Malaysia. *Journal of Tropical Ecology* 3, 210–220.

Beets, W.C. (1982) *Multiple Cropping and Tropical Farming Systems.* Gower, Aldershot.

Bell, T.I.W. (1973) Erosion in the Trinidad teak plantation. *Commonwealth Forestry Review* 52, 223–233.

Bosch, J.M. and Hewlett, J.D. (1982) A review of catchment experiments to determine the effect of vegetation changes on water yield and evapotranspiration. *Journal of Hydrology* 55, 3–23.

Bowen, R.M. (1990) Decomposition of wheat straw by mixed cultures of fungi isolated from arable soils. *Soil Biology and Biochemistry* 22, 401–406.

Bowen, R.M. and Harper, S.H.T. (1990) Decomposition of wheat straw and related compounds by fungi isolated from straw in arable soils. *Soil Biology and Biochemistry* 22, 393–399.

Brandt, J. (1988) The transformation of rainfall energy by a tropical rainforest canopy in relation to soil erosion *Journal of Biogeography* 15, 41–48.

Broughton, W.J. (1977) Effects of various covers on soil fertility under *Hevea braziliensis* and on growth of the tree. *Agroecosystems* 3, 147–170.

Bruijnzeel, L.A. (1990) *Hydrology of Moist Forests and Effects of Conversion: A State of the Art Review.* UNESCO, Paris.

Bull, A.T., Goodfellow, M. and Slayter, J.H. (1992). Biodiversity as a source of innovation in biotechnology. *Annual Review of Microbiology* 46, 219–252.

Charley, J.L. and West, N.E. (1975) Plant-induced soil chemical patterns in some shrub-dominated, semi-desert ecosystems of Utah. *Journal of Ecology* 63, 945–963.

Clements, R.O., Murray, P.J. and Sturdy, R.G. (1991) The impact of 20 years' absence of earthworms and three levels of N fertilizer on a grassland soil environment. *Agriculture, Ecosystems and Environment* 36, 75–85.

Colwell, R.K. (1992). Human aspects of biodiversity: an evolutionary perspective. In: Solbrig, O.T., van Embden, H.M. and van Oordt, P.G.W.J. (eds) *Biodiversity and Global Change,* IUBS Monograph 8, Paris, pp. 209–222.

Crossley, D.A., Mueller, B.R. and Perdue, J.C. (1992) Biodiversity of microarthropods in agricultural soils: relations to processes. *Agriculture, Ecosystems and Environment* 40, 37–46.

Darlington, J.P.E.C. (1982) The underground passages and storage pits used in foraging by a nest of the termite *Macrotermes michaelseni* in Kajaido, Kenya. *Journal of Zoology* 198, 237–247.

de Vleeschauwer, D. and Lal, R. (1981) Properties of worm casts under secondary tropical forest regrowth. *Soil Science* 132, 175–181.

di Castri, F., Vernhes, J.R. and Younès, T. (1992) The network approach for understanding global biodiversity. *Biology International* 25, 3–9.

Edwards, C.A., Stinner, B.R., Stinner, D. and Rabbitin, S. (1988) *Biological Interactions in Soil.* Elsevier Press, New York.

Edwards, K.A. and Blackie, J.R. (1981). Results of the East African Catchment Experiment: 1958–1974. In: Lal, R. and Russell, E.W. (eds) *Tropical Agricultural Hydrology.* John Wiley, Chichester, pp. 163–188.

Ehrenfeld, D. (1988). Why put a value on biodiversity? In: Wilson, E.O. (ed.) *Biodiversity,* National Academy Press, Washington DC, pp. 212–216.

Elkins, N.Z., Sabol, G.V., Ward, T.J. and Whitford, W.G. (1986) The influence of subterranean termites on the hydrological characteristics of a Chiuhuahuan desert ecosystem. *Oecologia* 68, 521–528.

Ewel, J.J. (1986) Designing agricultural ecosystems for the humid tropics. *Annual Review of Ecology and Systematics* 17, 245–271.

Farnsworth, N.R. (1988) Screening plants for new medicines. In: Wilson, E.O. (ed.) *Biodiversity.* National Academy Press, Washington DC, pp. 83–97.

Fitter, A.H., Atkinson, D., Read, D.J. and Usher, M.B. (eds) (1985) *Ecological Interactions in Soil.* Blackwell Scientific Publications, Oxford.

Gartlan, J.S., Newberry, D. McC., Thomas, D.W. and Waterman, P.G. (1986) The influence of topography and soil phosphorus on the vegetation of Korup Forest Reserve, Cameroon. *Vegetatio* 65, 131–148.

Goss, M.J. (1991) Consequences of the activity of roots on soil. In: Atkinson, D. (ed.) *Plant Root Growth: An Ecological Perspective.* Blackwell Scientific Publications, Oxford, pp. 171–186.

Grieve, I.C. (1977) Some relationships between vegetation patterns and soil variability in the Forest of Dean, UK. *Journal of Biogeography* 4, 193–200.

Harper, J.L. (1977) *The Population Biology of Plants.* Academic Press, London and New York.

Hawksworth, D.L. (ed.) (1991) *Biodiversity of Microorganisms and Invertebrates and its Role in Sustainable Agriculture.* CAB International, Wallingford.

Hendrix, P.F., Parmelee, R.W., Crossley, D.A., Coleman, D.C. Odum, E.P. and Groffman, P.M. (1986) Detritus food webs in conventional and no-tillage agroecosystems. *Bioscience* 36, 374–380.

Ho, C.C., Newbery, D. McC. and Poore, M.E.D. (1987) Forest composition and inferred dynamics in Jengka Forest Reserve, Malaysia. *Journal of Tropical Ecology* 3, 25–26.

Hulugalle, N.R. and Ezumah, H.C. (1991) Effects of cassava-based cropping system on physicochemical properties of soil and earthworm casts in a tropical Alfisol. *Agriculture, Ecosystems and Environment* 35, 55–63.

Hunt, H.W., Coleman, D.C., Ingham, E.R., Ingham, R.E., Elliott, E.T., Moore, J.C., Rose, S.L., Reid, C.P.P. and Morley, C.R. (1987) The detrital food web in a short-grass prairie. *Biology and Fertility of Soils* 3, 57–68.

Ingham, R.E., Trofymow, J.A., Ingham, E.R. and Coleman, D.C. (1985) Interactions of bacteria, fungi and their nematode grazers: effects on nutrient cycling and plant growth. *Ecological Monographs* 55, 119–140.

Ingham, E.R., Trofymow, J.A., Ames, R.N., Hunt, H.W., Morley, C.R., Moore, J.C. and Coleman, D.C. (1986) Trophic interactions and nitrogen cycling in a semi-arid grassland soil. II. System responses to the removal of different groups of soil microbes or fauna. *Journal of Applied Ecology* 23, 615–630.

Janeau, J.L. and Valentin, C. (1987) Relations entre des termitières *Trinervitermes* spp. et la surface du sol: réorganizations, ruissellment et érosion. *Revue Écologie et de Biologie du Sol* 24, 637–647.

Kang, B.T. and Moorman, F.R. (1977) Effect of some biological factors on soil variability in the tropics. *Plant and Soil* 47, 441–449.

Kladivco, E.J., MacKay, A.D. and Bradford, J.M. (1986) Earthworms as a factor in the reduction of soil crusting. *Soil Science Society of American Journal* 50, 191–196.

Lal, R. (1983) Soil erosion in the humid tropics with particular reference to agricultural land development and soil management. *IAHS Publications* 140, 221–239.

Lal, R. (1987) *Tropical Ecology and Physical Edaphology.* John Wiley, Chichester.

Lee, K.E. (1985) *Earthworms: Their Ecology and Relationships with Land Use* Academic Press, Sydney.

Low, K.S. and Goh, G.C. (1972) The water balance of five catchments in Selangor, West Malaysia. *Journal Tropical Geography* 35, 60–66.

Martin, J.P. (1963) Influence of pesticide residues on soil microbiological and chemical properties. *Residue Revue* 4, 96–129.

Mbakaya, B.S., Blackburn, W.H., Skovlin, J.M. and Child, R.D. (1988) Infiltration and sediment production of a bushed grassland as influenced by livestock grazing systems, Buchuma, Kenya. *Tropical Agriculture (Trinidad)* 65, 99–105.

Moffat, A.J. and Boswell, R.C. (1990) Effects of tree species and species mixtures on soil properties at Gisburn Forest, Yorkshire. *Soil Use and Management* 6, 46–51.

Mollitor, A.V., Leaf, A.L. and Morris, L.A. (1980) Forest soil variability on north-eastern flood plains. *Soil Science Society of America Journal* 44, 617–620.

Newbery, D.N., Alexander, I.J., Thomas, D.W., and Gartlan, J.S. (1988) Ectomycorrhizal rainforest legumes and soil phosphorus in Korup National Park, Cameroon. *New Phytologist* 109, 433–450.

Noble, C.A. and Morgan, R.P.C. (1983) Rainfall interception and splash detachment with a brussels sprout plant: a laboratory simulation. *Earth Surface Processes and Landforms* 8, 569–577.

Norton, B.G. and Ulanowicz, R.E. (1992) Scale and biodiversity policy: a hierarchical approach. *Ambio* 21(3), 244–249.

Potter, C.S. and Meyer, R.E. (1990) The role of biodiversity in sustainable dryland farming systems. *Advances in Soil Science* 13, 241–251.

Rayner, A.D.M. and Boddy, L. (1988) *Fungal Decomposition of Wood; Its Biology and Ecology.* John Wiley, Chichester.

Ridge, E.H. (1976) Studies on soil fumigation. II. Effects on bacteria. *Soil Biology and Biochemistry* 8, 249–253.

Rovira, A.D. (1976) Studies on soil fumigation. I. Effects on ammonium, nitrate and phosphate in soil and on the growth, nutrition and yield of wheat. *Soil Biology and Biochemistry* 8, 241–247.

Scholefield, D., Garwood, E.A. and Titchin, N.M. (1988) The potential of management practices for reducing losses of nitrogen from grazed pasture. In: Jenkinson, D.S. and Smith, K.A. (eds) *Nitrogen Efficiency in Agricultural Soils,* Elsevier Applied Science, Amsterdam.

Seward, P.D. and Woomer, P.L. (1992) *The Biology and Fertility of Tropical Soils.* Report of the Tropical Soil Biology and Fertility Programme, TSBF, UNESCO-ROSTA, Nairobi.

Sharpley, A.N., Syers, J.K. and Springett, J.A. (1979) Effect of surface casting earthworms on the transport of phosphorus and nitrogen in surface runoff from a pasture. *Soil Biology and Biochemistry* 11, 459–462.

Soemarwoto, O and Soemarwoto, I. (1982) Homegarden: its nature, origin and future development. In: Awang, K., See, L.S. See, L.F, Derus, A.R.M and Sheikh Ali Abod (eds) *Ecological Basis for Rational Resource Utilization in the Humid Tropics of South East Asia.* UNESCO, Paris, pp. 130–139.

Solbrig, O.T. (1992) Biodiversity: an introduction. In: Solbrig, O.T., van Embden, H.M. and van Oordt, P.G.W.J. (eds) *Biodiversity and Global Change, IUBS Monograph* 8, IUBS, Paris pp. 13–20.

Springett, J.A. (1985) Effects of introducing *Allolobophora longa* Ude on root distribution and some soil properties in New Zealand pastures. In: Fitter, A.H., Atkinson, D., Read, D.J. and Usher, M.B. (eds) *Ecological Interactions in Soil.* Blackwell Scientific Publications, Oxford, pp. 39–45.

Stinner, B.R., Crossley, D.A., Odum, E.P. and Todd, R.L. (1984) Nutrient budgets and internal cycling of N, P, K, Ca and Mg in conventional tillage, no-tillage and old-field ecosystems on the Georgia Piedmont. *Ecology* 65, 354–369.

Swift, M.J. and Anderson, J.M. (1993) Biodiversity and ecosystem function in agricultural systems. In: Schultz, E.D. and Mooney, H.A. (eds) *Biodiversity and Ecosystem Function.* Springer Verlag, Berlin, pp. 17–38.

Tiessen, H. and Stewart, J.W.B. (1988) Light and electron microscopy of stained micro-aggregates: the role of organic matter and microbes in soil aggregation. *Biogeochemistry* 5, 312–322.

Tinker, P.B.H. (1968) Changes occurring in sedimentary soils of southern Nigeria after oil palm establishment. *Journal of West African Institute of Oil Palm Research* 4, 66–81.

van Noordwijk, M. (1989). Rooting depth in cropping systems in the humid tropics in relation to nutrient use efficiency. In: van der Heide, J. (ed.) *Nutrient Management for Food Crop Production in Tropical Farming Systems,* Institute of Soil Fertility, Haren, pp. 129–144.

Verhoef, H.A. and Brussaard, L. (1990) Decomposition and nitrogen mineralization in natural and agroecosystems: the contribution of soil animals. *Biogeochemistry* 11, 175–211.

Wiersum, K.F. (1984) Surface erosion under various tropical agroforestry systems. In: O'Loughlan, C.L. and Pearce, A.J. (eds) *Proceedings Symposium on Effects of Forest Land Use on Erosion and Slope Stability* IUFRO, Vienna, pp. 231–239.

Wood, T.G. (1991) Termites in Ethiopia: the environmental implications of their damage and resultant control measures. *Ambio* 20, 136–138.

Woomer, P. and Ingram, J.S.I. (1990) *The Biology and Fertility of Tropical Soils. 1990 Report.* Tropical Soil Biology and Fertility, UNESCO-ROSTA, Nairobi.

Young, A. (1989) *Agroforestry for Soil Conservation.* CAB International, Wallingford.

Chapter 18

Soil Fauna and Sustainable Land Use in the Humid Tropics

P. Lavelle[1], C. Gilot[1], C. Fragoso[2] and B. Pashanasi[3]

[1]*Laboratoire d'Ecologie des Sols Tropicaux, Université Paris VI/ORSTOM, 72 route d'Aulnay, 93143-Bondy Cedex, France:* [2]*Instituto de Ecologia, Xalapa, Ver., Mexico:* [3]*Estacion Experimental San Ramon, Yurimaguas, Loreto, Peru*

Introduction

It has been long claimed that soil fauna are key actors in sustainable land use. Long after Aristoteles called earthworms the 'intestine of the earth', Darwin (1881) said that '*It may be doubted whether there are many other animals which have played so important a part in the history of the world, as have these lowly organized creatures*'. Since that time, considerable efforts have been made to substantiate these statements and extend them to the whole soil fauna community. Their importance and role in energy cycling has been widely studied during the IBP Programme (Petersen and Luxton, 1982). These studies reinforced the feeling that they had significant impacts on major soil processes. During the last decade, considerable efforts have been made to describe and quantify these effects and assess the influence of land use on their communities (see e.g. Veeresh *et al.*, 1991; Andren *et al.*, 1988 and syntheses by Lee, 1985; Anderson and Flanagan, 1989; Lavelle *et al.*, 1992b).

Processes whereby soil fauna may affect the dynamics of soil fertility have been described and quantified, mostly at the 'micro' and 'meso' scales at which these organisms operate. These studies gave a clear understanding of the processes involved; none the less, the lack of experiments at the scale of a farmer's plot have not so far allowed the evaluation of the exact role of these processes, and the potential to manipulate them to improve soil fertility.

This chapter gives an overview of the effects of land management practices on soil fauna communities in tropical soils. The effects of soil fauna on processes relevant to sustainability are discussed, and finally, recent results of manipulative experiments are presented.

 Soil Resilience and Sustainable Land Use (eds D.J. Greenland and I. Szabolcs)

Soil Fauna Communities

Soil fauna comprise a large variety of organisms with contrasted sizes and adaptive strategies. The abundance and composition of their communities, and hence their impact on soil processes, vary greatly depending on vegetation and land use practices.

Invertebrates in the soil system

Adaptive strategies

Microfauna comprise hydrobiont invertebrates which live in free soil water and water films that cover soil particles; they are usually either micropredators of microorganisms and other microinvertebrates, or plant parasites. Their average size is less than 0.2 mm. Protozoa and nematodes are the main representatives of this group. The spatial range of their activities is of micro- to millimetres. They usually form foodwebs of micropredators with a significant impact on nutrient cycling, e.g. in the rhizosphere (e.g. Trofymow and Coleman, 1982; Setalä *et al.*, 1991a);

Mesofauna comprise microarthropods and small Oligochaeta Enchytraeidae. Average length ranges from 0.2 to 2 mm. They are typical inhabitants of litter systems where they feed on litter and microorganisms. None the less, they may also colonize the whole profile of soil, although with reduced densities. They may have a significant impact on litter comminution and dispersal of fungal spores (Persson *et al.*, 1980; Swift and Body, 1985).

Macrofauna operate at much larger scales of time and space. They are large-sized invertebrates that may disrupt the soil and modify its structure through their movements and feeding behaviour. They are 'ecosystem engineers' (Stork and Eggleton, 1992) that may transport and mix soil and organic residues in the whole soil profile and create diverse and conspicuous structures, e.g. mounds, galleries and soil aggregates.

These groups adapt differently to the three main constraints that soil organisms face, i.e., feeding on relatively poor quality resources, resisting occasionally unfavourable microclimatic conditions and moving in the limited and discontinuous pore space of soil. Small invertebrates have a much higher resistance to environmental stress than larger ones. They are unable to develop mutualistic associations with microflora to easily digest organic resources. As a consequence, they rely mainly on predation or plant parasitism. On the other hand, large invertebrates have better abilities to develop external ('exhabitational' *sensu* Lewis, 1985) or internal ('inhabitational') mutualistic associations with microflora. These associations are efficient at using low-quality resources, e.g. woody or humified material.

Although these invertebrates may escape from unfavourable microclimatic conditions by building suitable structures for their shelter and movement, they are limited to non-extreme conditions.

Soil food webs

Food webs in the soil are fundamentally based on relationships of invertebrates with microorganisms. These relationships are organized at different levels which depend on the size of the organisms (Fig. 18.1).

Microorganisms directly exploit resources; they comprise the first level. Microfauna mainly act as predators of microflora; this is the only kind of interaction that they can develop since their small size does not allow any significant mutualism.

Mesofauna comprise a mixture of microbe and fungal grazers and invertebrates, which rely on the 'external rumen' feeding strategy. They create suitable conditions for microbial life in their faecal pellets and an intense microbial activity results in the digestion of part of the undigested organic matter. By reingesting their faeces at that stage, invertebrates may make use of assimilable organic matter, and possibly feed, at least partly, on the microbial biomass.

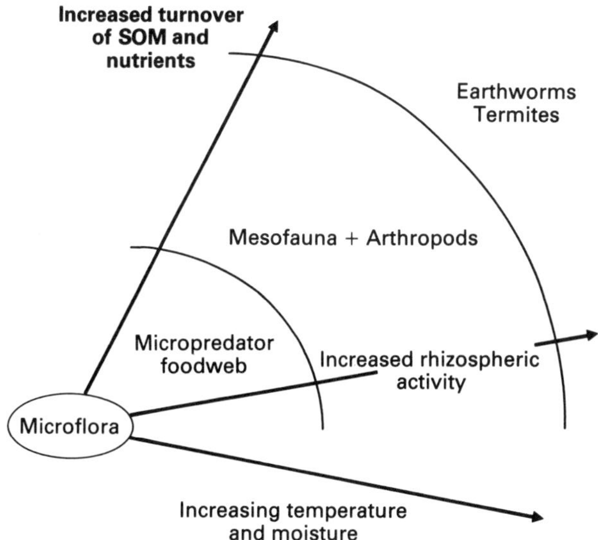

Fig. 18.1. Relationships among soil microflora and soil fauna: a conceptual model. For explanations see text.

Macrofauna rely on mutualistic interactions with microorganisms to extract assimilable compounds from the decomposing materials. They may use the external rumen strategy. Some of them, especially the termites, and earthworms, have developed with microorganisms facultative or obligate internal mutualist systems of digestion. This is the case with earthworms which stimulate the activity of microflora ingested with soil by adding large amounts of water (up to 100%) and mixing significant amounts of readily assimilable organic compounds (Barois and Lavelle, 1986). These additions and the intense mixing in the gut trigger intense microbial activity. In the posterior part of the gut, organic matter has been digested and the worm may absorb part of the assimilable compounds. Termites have even more sophisticated inhabitational (*sensu* Lewis, 1985) mutualist relationships with microflora; in the lower termites obligate associations with protozoa have been described.

In soils where climatic constraints do not allow large invertebrates to live, microbial activity is mainly regulated by micro foodwebs (e.g. Sestedt and James, 1987). When these constraints are released, large invertebrates become dominant regulators of the microorganisms that they influence directly, as explained above, or via predation on micro and mesofauna. The behaviour of earthworms feeding partly on protozoa (Piearce and Phillips, 1980; Rouelle *et al.*, 1985) and nematodes (Dash *et al.*, 1980; Yeates, 1981) is an example of this type of interaction. In these conditions, micro foodwebs are still an essential component of the soil system, but they tend to restrict their activities to specific microsites (e.g. root tips in the rhizosphere). Macrofauna operate at much larger scales of time and space, and behave as ecosystem engineers (Stork and Eggleton, 1992). When present, effects of microorganisms and smaller invertebrates are largely dependent on their physical activities (mixing litter and soil, building structures and galleries and aggregating the soil) as well as metabolic activities (utilization of the available organic resources, development of mutualist or antagonist relationships).

In the humid tropics there are seldom strong climatic or edaphic limitations to colonization of the soil by macroinvertebrates. They become predominant regulators of soil biological activities. Voluntary or involuntary modifications of their communities thus affect other soil organisms within biological systems of regulation. These systems link key macroinvertebrates (e.g. termites or earthworms) to roots, to the whole microflora, and to mesofauna living in the part of the soil that they affect by their activities. Four such systems have been identified, i.e. the rhizosphere, the litter system, the drilosphere (earthworms) and the termitosphere (termites).

Soil macro-invertebrate communities and types of land use

General effects

Types of land use deeply affect the composition and abundance of soil macrofauna communities. Microclimate and food resources together with application of pesticides are major factors that affect the diversity and abundance of communities. The initial disturbance linked to the clearance of the original ecosystem rapidly eliminates a large number of species especially the ones with narrow niches and slow population turnover rates (Fig 18.2).

Annual cropping is the most detrimental practice as regards soil fauna communities. Within a few weeks of initial cultivation, biomass dramatically decreases. Traditional cultivation techniques in general may be more conservative and in some areas, local fauna may better resist the disturbance. None the less, the average biomass recorded in such environments (ca. 10 g fresh weight m^{-2}) is more than two to four times as low as average values measured in original forest and grassland ecosystems.

Pastures often have a much higher overall biomass of macroinvertebrates than the original ecosystems. This is mainly due to the proliferation of populations of one or two indigenous or exotic earthworm species, and

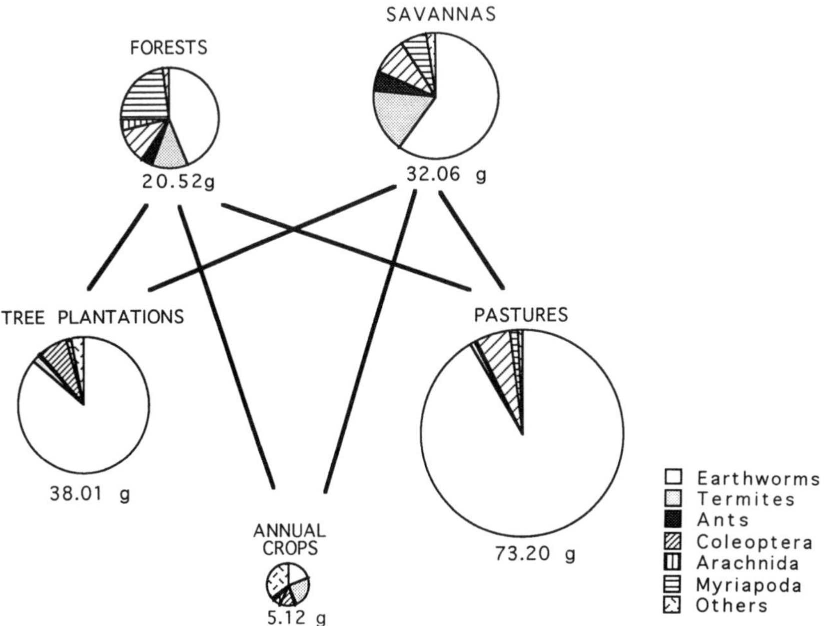

Fig. 18.2. Structure of soil macrofauna communities in major types of land use in the humid tropics.

the local development of Coleoptera larval communities, e.g. in Central American pastures (Villalobos and Lavelle, 1990). In pastures of tropical America, Australia and Asia two species of earthworm, *Pontoscolex corethrurus* and *Polypheretima elongata,* may build a large biomass of 1–4 t ha^{-1} fresh weight. In regions of natural savannah in Africa, local species adapt naturally to pastures and their biomass may be enhanced by grazing (Kouassi, 1987). However, overgrazing may result in the decline and disappearance of earthworm communities (Castilla and Sanchez, unpublished).

Finally, perennial crops and agroforestry systems generally have large macrofauna biomasses which may be higher than in the original ecosystem. Tree plantations with legume covers may be suitable environments for both indigenous forest species and exotic colonizers, as they offer abundant and diverse resources for decomposers. As a result, these ecosystems often have diverse and high biomass of macroinvertebrates and represent types of land use which sustain both acceptable levels of biodiversity and overall activity.

The energetic bottle neck

These systems, however, may not be able to sustain active macroinvertebrate communities for indefinite periods of time. Sampling conducted

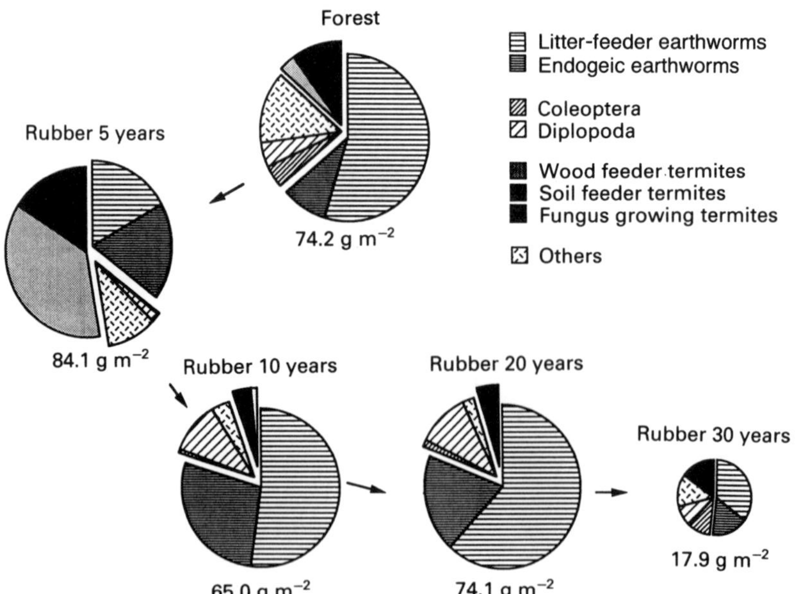

Fig. 18.3. Soil macrofauna communities in Hevea plantations of different ages and in the original forest (Gilot *et al.*, in press).

in Hevea plantations of different ages showed that communities change with time and tend to be severely depleted after 30 years. At Bimbresso (Côte d'Ivoire), soil macrofauna biomass was large in the original forest (74.2 g m^{-2}); earthworms comprised most of this biomass, but termites and myriapods were significant elements in this community (Fig. 18.3). In the 5-year-old plantation, biomass was still high, but composition was different. Termites, especially xylophagous termites, were overdominant. In 10- and 20-year-old plantations, biomass of macroinvertebrates was sustained at high levels (65.0 and 74.1 g m^{-2} respectively) but xylophagous termites had nearly disappeared. They had been replaced by large earthworm communities, especially of large endogeic species feeding on soil organic matter. After 30 years the community was highly depleted as was the production of rubber. These results demonstrate the dynamic changes in the community as the Hevea plantation ages. It is likely that this community had been maintained for 20 years by the large energy input from the wood of trunks, branches and large roots in the early years of the plantation. Xylophagous termites first exploited these resources. Earthworms came later in the succession to feed on organic material transformed by the termites. After 20 years, the initial flux of energy faded and organic resources sustained a lower biomass. Changes in the herbaceous cover may also explain part of the observed variations. During the first few months, a legume cover (*Pueraria phaseoloides*) had been maintained. This cover was rapidly replaced by a poor herbaceous stratum which was reduced to a minimum as the canopy closed.

These results clearly emphasize the need for significant amounts of energy sources to maintain active soil macrofauna communities.

The colonization issue

Energetic deficiencies may not be the only process whereby communities are depleted. In some circumstances species which might adapt to the newly created conditions may not be able to colonize because they are not present and/or do not have the ability to invade the site rapidly. This is especially true for earthworms and termites as not all the existing functional groups are present in a given biogeographical area. Such a situation gives scope for a wide range of manipulative experiments. In temperate soils, inoculation of well-adapted earthworm species has been implemented in a number of situations, e.g. for the improvement of pastures in New Zealand, and Australia (Syers and Springett, 1984), reclamation of degraded sites (Curry and Boyle, 1987) or improvement of polder soils (Hoogerkamp *et al.*, 1983). In soils of the humid tropics critical targets might be to introduce species with expected favourable effects on the soil function in low input agricultural systems (termites).

Manipulative Experiments: Introduction of Endogeic Earthworms in Low Input Agricultural Systems

In Peru, Mexico and Côte d'Ivoire, situations have been identified where the introduction of earthworms was feasible. They comprise traditional systems based on slash-and-burn agriculture. Research has been conducted which demonstrates that carefully chosen adapted species may build up sizeable populations. These populations affect plant production and soil fertility as assessed in terms of soil organic matter (SOM) and nutrient conservation, and conservation of soil physical properties. Most of the results presented here have been obtained at Yurimaguas (Peru) in acid Ultisols under a traditional rotation with no fertilizers, or continuous maize cropping fertilized after the third cropping cycle.

Experimental design and general methodology

Soil monoliths 60 cm in diameter and 50 cm deep were isolated from the soil of a 20-year-old secondary forest and a nylon mesh was used to prevent any movement of earthworms. After clearing and burning the forest, these units were cleaved of native earthworms by an application of carbofuradan to create 'no earthworm' situations. After six weeks, earthworm populations were introduced. The species chosen, *Pontoscolex corethrurus* is very widespread in all disturbed soils of the humid tropics; it has a large tolerance for a wide range of edaphic factors, except for drought. It has also the ability to build numerous populations rapidly when placed in favourable conditions (Lavelle *et al.*, 1987). At Yurimaguas, 120 such experimental units were installed. 108 of them were cropped to a traditional rotation system with no fertilizers; 12 were cropped to maize with fertilizers from the fourth cropping cycle onwards. Six different treatments were applied, i.e. three levels of organic treatments (no organic residues) (C); organic residues produced on the experimental unit (CR), and CRV i.e. CR + legume green manure of *Centrosema macrocarpum* (2.5 t ha^{-1} dry weight), with or without earthworms.

In 'with earthworm' treatments, a relatively low initial (36 g m^{-2}) biomass was introduced. It was believed that biomass would rapidly reach an equilibrium representing the actual carrying capacity of the system for the species. The experiment was designed to have six successive cropping cycles. At each harvest, aboveground production was measured in all units. 'Internal' parameters, i.e. characterization of SOM, earthworm and root biomass, bulk density, infiltration and aggregation were measured on three experimental units which were destroyed for sampling. During the growth period, soil water content and the decomposition rates of organic residues were monitored. Methods are basically the ones proposed in the Tropical

Soil Biology and Fertility programme (TSBF).

In a nearby plot, 20 units were allocated to continuous maize cropping to monitor dynamics of SOM using natural ^{13}C labelling, by shifting from a pure C3 vegetation (forest) to a pure C4 maize culture to sustain significant production in the system. Fertilizers (100 N, 20 P, 20 K) were applied from the fourth cropping cycle onwards to sustain production.

At Lamto, similar experiments were installed with a native species (*Millsonia anomala*) under continuous maize cropping in a forest soil, and continuous yam cultivation in a savannah soil.

Establishment of earthworm populations

The establishment of introduced populations was not equally successful at all sites. At Lamto (Côte d'Ivoire), populations were sustained at very low biomass (3–4 g m^{-2} after four crops). There was clear indication from a specific experiment that the limitation to earthworm settlement was the lack of food (Gilot *et al.*, 1992).

At Yurimaguas, populations actively reproduced in all treatments. They responded significantly to organic inputs, with the lowest biomass in treatments without residues and highest biomass in treatments receiving either crop residues alone or crop residues + legume green manure (Fig. 18.4).

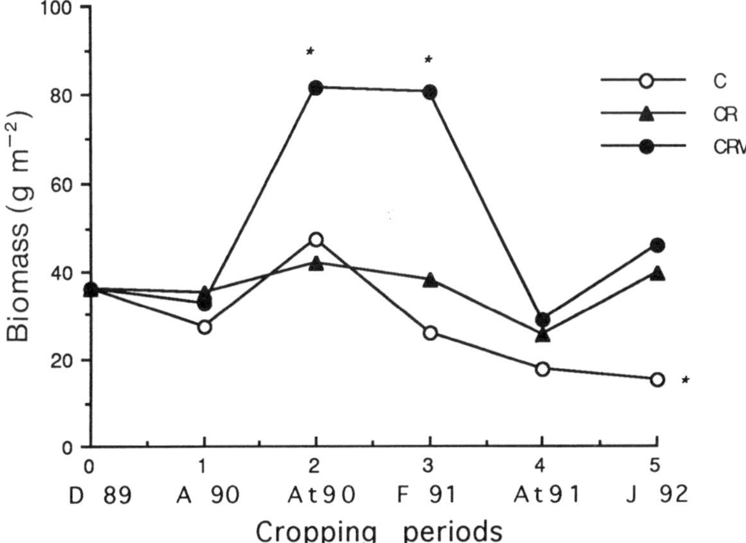

Fig. 18.4. Changes in biomass of *Pontoscolex corethrurus* in a traditional cropping rotation submitted to three different organic treatments at Yurimaguas (* indicate significant differences with other treatments at the same date). (C: no organic inputs; CR : application of crop residues produced at the site ; CRV : crop residues + 2.5 t ha^{-1} legume green manure).

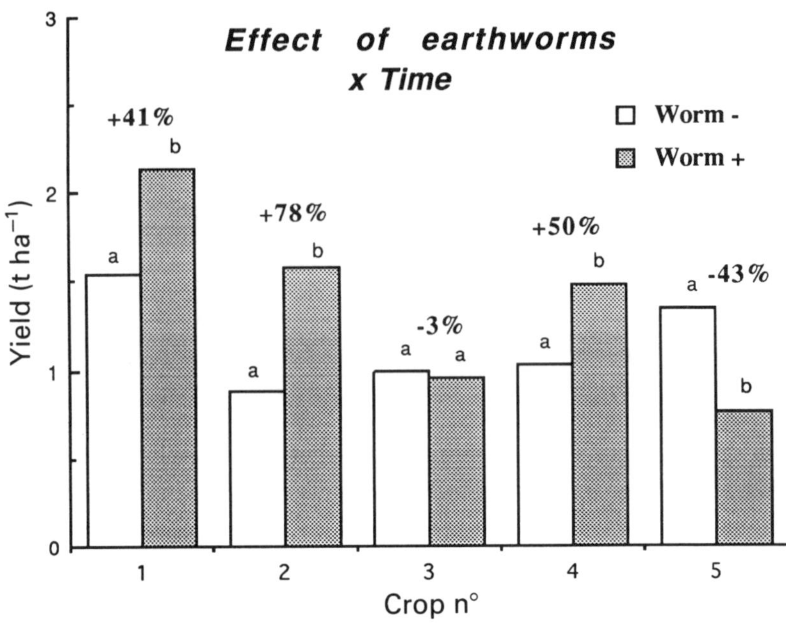

Fig. 18.5. Effect of earthworm activities on grain production at five successive harvests irrespective of organic treatments.

Crop production was sustained at acceptable levels according to local standards with grain production in the range 0.8–2.4 t ha^{-1}. There were significant effects of earthworms on crop production.

Grain production was increased on the average by 27% during the first five cropping cycles (Fig. 18.5). A maximum increase of 78% was obtained at the second harvest, whereas a significant decrease of 43% was observed at the fifth harvest. The next crop will show whether this decrease indicated a significant inversion of the trend, or simply was the consequence of accidental effects of earthworms, e.g. on water availability at a critical stage of plant growth. The response was higher in treatments without residues (+ 36%) or receiving crop residues + legume green manure (+ 36%) than in treatments with crop residues only (+ 8%).

In the plot with continuous maize, the average increase of production due to the introduction of earthworms amounted to 130%. This effect was especially spectacular at the second harvest with production of 0.8 t and 3.2 t ha^{-1} in treatments 'without' or 'with' earthworms. After three successive cropping cycles, production had dramatically decreased in both systems. Therefore, to maintain the crop, fertilizers were applied. At the fifth crop, production was twice as high in the earthworm treatments as in the no-earthworm treatments (Fig. 18.6).

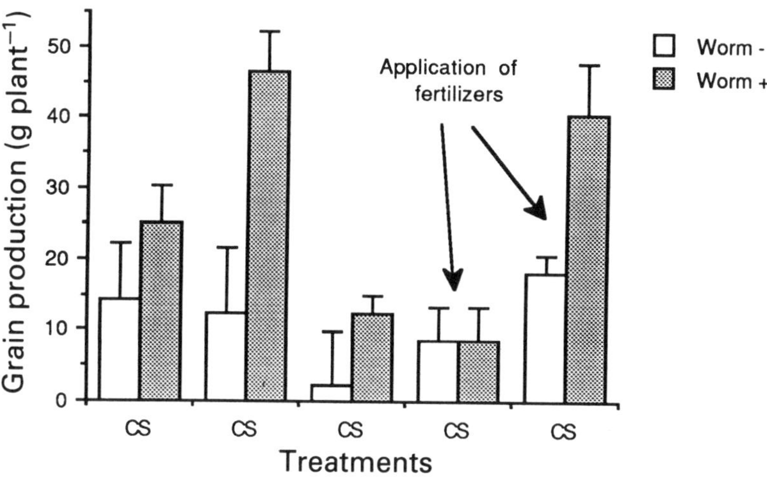

Fig. 18.6. Effect of earthworm inoculation on grain production in a continuous maize crop at Yurimaguas. CS, surface application of stubble.

These results indicate that: (*i*) earthworms can prolong and increase production; and (*ii*) their positive effects may not be explained simply in terms of the improvement of the nutrient supply to plants, since their effect remains significant when fertilizers are used to correct nutrient deficiencies.

Effects of earthworms on SOM and nutrient dynamics

After five cropping cycles, earthworm activities had not prevented losses of soil organic matter (Fig. 18.7). Nevertheless, the quality of organic matter as assessed by particle size fractionation had significantly changed (Feller, 1979). At Lamto, the proportion of SOM in the coarse fraction ($> 50\ \mu m$) decreased less in the earthworm than in the no-earthworm treatment. On the other hand, more organic matter from the fine fractions seemed to have been mineralized (Fig. 18.8). Earthworm activities did not help maintain microbial biomass as this was significantly lower (by 23%) in treatments with earthworms.

Neither did earthworm activities impede the depletion of nutrients. A progressive return towards acid pH and high levels of A1 saturation occurred in all treatments. None the less, ^{15}N from labelled legume green manure applied at the soil surface was better recovered by plants in the presence than in the absence of earthworms. This demonstrated significant qualitative differences in nutrient cycling, oriented towards a more efficient use of available nutrients.

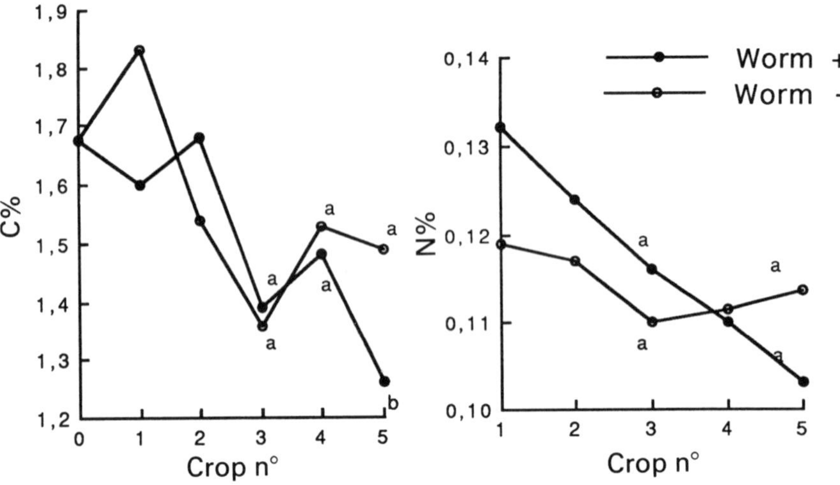

Fig. 18.7. Variation of C and N contents of the 0–10 cm strata with time and treatments.

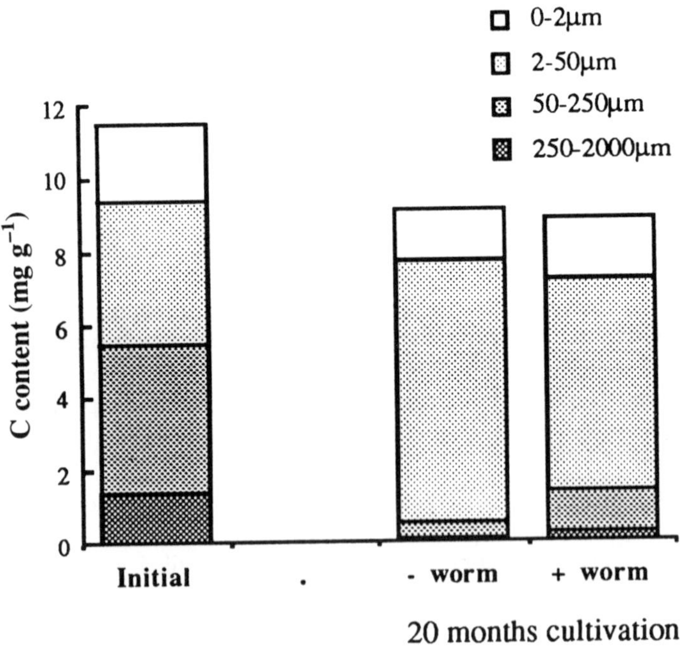

Fig. 18.8. Partition of the soil organic matter between granulometric fractions in the maize plot at the initial time and after the fourth crop.

Changes in soil physical properties

All soil physical properties measured were significantly affected by earthworm activities. Rough calculations based on assessments of earthworm biomass and on the results of past laboratory experiments gave an estimate of 150 t ha^{-1} $year^{-1}$ dry soil ingested, equivalent to 10–15% of the upper 10 cm of soil. *P. corethrurus* ingests small aggregates and organic debris of a size smaller than the size of the mouth (i.e. up to 2 mm). The effects on soil aggregation were highly significant after five crops at Yurimaguas. In 'with earthworm' treatments, the relative proportion of large (> 10 mm) and intermediate (2–5 mm) aggregates was significantly increased, whereas the reverse effect was observed in the absence of earthworms. Earthworms clearly increased soil macroaggregation and they counteracted the trend for soil disaggregation observed in the conventional system.

These changes of macroaggregation had effects on other global physical parameters. In the presence of earthworms, bulk density was significantly increased; infiltration rate was first decreased, and then returned to values

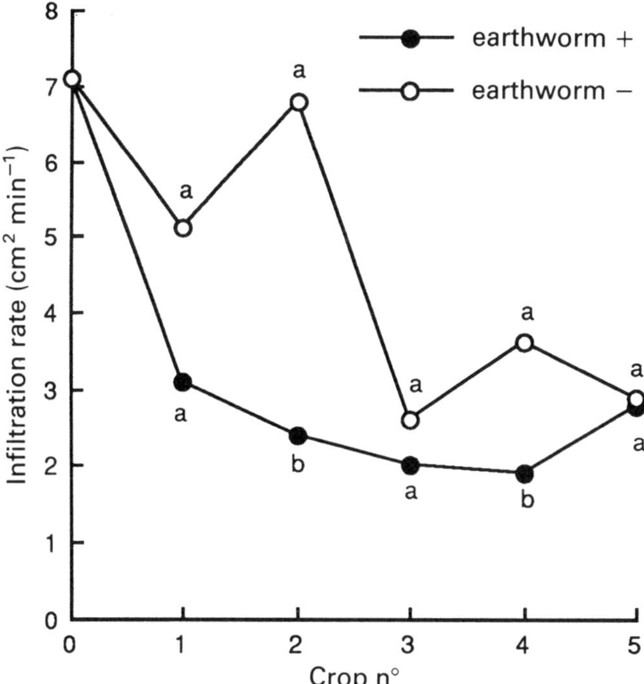

Fig. 18.9. Changes of infiltration with time. Data with different letters indicate significant differences at a given time. Bulk density was significantly higher than the initial value at time 3 in the earthworm treatment. Organic treatments had no effect.

similar to those of control treatments, after the earthworms had increased their production of surface casts and created a macropore system at the soil surface (Fig. 18.9).

Discussion

Invertebrates are major determinants of soil processes in tropical ecosystems. Whereas insect pests have been actively fought, the potential for beneficial use of insect activities has not been taken fully into consideration in the design of management practices. Current research demonstrates that practices which annihilate the activities of soil fauna are unlikely to be sustainable in the long term.

Soil fauna communities have contrasting reactions to changes induced by human land management. Their abundance and diversity are indicators of the quality of soils expressed in terms of soil organic matter, nutrient content, and physical properties such as bulk density, porosity and water regime. Annual crops generally support depleted communities, with especially low earthworm populations, whereas termites are usually less affected. Perennial cultures (e.g. sugar cane), pastures, and plantations (e.g. oil palm with a legume cover) generally have less diverse communities than the original ecosystem, but biomass is often higher due to colonization by peregrine earthworm species, and persistence of key native species. However, as the system degrades a depletion of the soil macrofauna occurs.

The abundance of soil fauna communities is significantly affected by the availability of suitable food resources and by their diversity. This is the reason why communities are much depleted in annual crops which have non-permanent and little-developed root systems, and limited inputs of organic residues. In some cases, the activity of soil fauna may be sustained for some time at the expense of energetic sources which had been accumulated in the original ecosystem. Examples of such sources are the dead woody material in recently deforested areas or coarse particle size fractions in SOM reserves. Our observations suggest that some termites or earthworms may live on resources with high lignin contents and/or high C:nutrient ratio. Therefore, there is scope for the use in carefully designed conditions of wastes like sawdust or coir (e.g. Pashanasi *et al.*, 1992). Before using such practices, the timing and placement of these residues will need to be addressed to prevent asynchrony between nutrient availability and plant requirements.

Preliminary experiments conducted for two years on five successive cropping cycles have given some insight on the effect of monospecific populations of endogeic earthworm communities on traditional agricultural systems. During that period there was no evidence that the depletion of SOM and the nutrient reserves of the soil were decreased by introduced earthworms. Nevertheless, qualitative differences in SOM, as assessed in terms of particle size fractions, were different and indications of better use of nutrients by plants were obtained. Soil fauna activities in general might result in the long term in the appearance of a new state of equilibrium which might be characterized by lower amounts of SOM, but with a much faster turnover rate, as a result of both increased mineralization of the organic matter in the supposedly more recalcitrant fine soil fraction, and the relative protection of coarse fractions. The duration of our experiments was too short to reach a new state of equilibrium of the system.

Earthworms had major effects on soil physical properties. The species chosen significantly increased macroaggregation in the upper 10 cm. Bulk density was significantly higher in the presence of earthworms (+ 15%). Infiltration was first decreased up to the stage when earthworms started to deposit surface casts and open macropores at the soil surface. Significant changes were observed in water regime, with increased flooding or drought at very moist or dry periods. Not all earthworms will have such effects (Casenave and Valentin, 1988). In wet savannahs of Côte d'Ivoire, Blanchart (1990) has described complementary effects of large earthworm species, which can transform a significant proportion of the soil microaggregates into larger macroaggregates, and smaller-size species splitting up these large aggregates into smaller ones which are mixed with root litter. Earthworm casts and other faunal structures often have high structural stabilities and the overaccumulation of these structures may alter physical properties in the soil. Complementary biotic or abiotic processes which regulate the dynamics of aggregation are thus necessary. These may include assemblages of species from different zoological groups acting simultaneously or in temporal successions (David, 1988; Blanchart, 1990).

References

Anderson, J.M. and Flanagan, P. (1989) Biological processes regulating organic matter dynamics in tropical soils. In Coleman, D.C., Oades, T. and Uehara, G. (eds.) *Dynamics of Soil Organic Matter in Tropical Ecosystems,* NifTAL project, University of Hawaii, Honolulu, pp. 97–125.

Andrén, O., Paustian, K. and Rosswall, T. (1988) Soil biotic interactions in the functioning of agroecosystems. *Agriculture, Ecosystems and Environment* 24, 57–65.

Barois, J. and Lavelle, P. (1986) Changes in respiration rate and some physiocochemical properties of a tropical soil during transit through *Pontoscolex corethrurus* (Glossoscolecidæ, Oligochæta). *Soil Biology and Biochemistry* 18(5), 539–541.

Blanchart, E. (1990) Rôle des vers de terre dans la formation et la conservation de la structure des sols de la savane de Lamto (Côte d'Ivoire). Thèse Université Rennes I., 278 pp.

Blanchart, E. and Spain, A. (1989) Rôle des vers de terre dans l'élaboration et la conservation de la structure des sols de savane. In: Lavelle, P. (ed.) *Processus Biologiques et Fertilité du Sol dans les Savanes Humides de Côte d'Ivoire. Recherches Fondamentales et Appliquées sur le Rôle des Vers de Terre,* Ministère de l'Environnement (SRETIE), Paris, pp. 26–43.

Casenave, A. and Valentin, C. (1988) *Les États de Surface de la Zone Sahélienne. Leur Influence sur l'Infiltration.* Rapport CEE-ORSTOM, 202 pp. ORSTOM, Bondy, France.

Curry, J.P. and Boyle, K.E. (1987) Growth rates, establishment and effect on herbage yield of introduced earthworms in grassland on reclaimed cutover peat. *Biology and Fertility of Soils* 3, 95–98.

Darwin, C. (1881) *The Formation of Vegetable Mould Through the Action of Worms with Observations on their Habits.* Murray, London.

Dash, M.C., Senapati, B.K. and Mishra, C.C. (1980) Nematode feeding by tropical earthworms. *Oikos* 34, 322–328.

David, J.F. (1988) Les peuplements de Diplopodes d'un massif forestier tempéré sur sols acides. Thèse Université Paris VI, 225 pp.

Feller, C. (1979) Une méthode de fractionnement granulométrique de la matière organique des sols: application aux sols tropicaux à texture grossière, très pauvres en humus. Cahiers de l'ORSTOM. *Série Pédologie* 17(4), 339–346.

Gilot, C., Lavelle, P., Kouassi, Ph. and Guillaume, G. (1992) Biological activity of soils in Hevea sands of different ages in Côte d'Ivoire. *Acta Zoologica Fennica* (in press).

Hoogerkamp, M., Rogaar, H. and Eijsackers, H.J.P. (1983) Effect of earthworms on grassland on recently reclaimed polder soils in the Netherlands. In: Satchell, J.E. (ed.) *Earthworm Ecology: from Darwin to Vermiculture.* Chapman and Hall, London, New York, pp. 85–105.

Huhta, V., Setälä, H. and Haimi, J. (1988) Leaching of N and C from birch leaf litter and raw humus with special emphasis on the influence of soil fauna. *Soil Biology and Biochemistry* 20, 875–878.

Kouassi, P.K. (1987) Etude comparative de la macrofaune endogée d'écosystèmes guinéens naturels et transformés de Côte d'Ivoire. Doctorat de 3ème cycle, Université d'Abidjan, 115 pp.

Lavelle, P. (1988) Earthworm activities and the soil system. *Biology and Fertility of Soils* 6, 237–251.

Lavelle, P., Barois, I., Cruz, C., Hernandez, A., Pineda, A. and Rangel, P. (1987) Adaptative strategies of *Pontoscolex corethrurus* (Glossoscolecidæ, Oligochæta), a peregrine geophagous earthworm of the humid tropics. *Biology and Fertility of Soils* 5, 188–194.

Lavelle, P., Alegre, J., Barois, I., Fragoso, C., Gilot, C., Gonzalez, C., Kanyonyo ka Kajondo, Martin, A., Melendez, G., Moreno, A and Pashanasi, B. (1992a) *Conservation of Soil Fertility in Low Input Agricultural Systems of the Humid Tropics by Manipulating Earthworm Communities.* CCE-STD2 programme. Final report project n° TS2 * 0292-F (EDB).

Lavelle, P., Spain, A.V., Blanchart, E., Martin, A. and Martin, S. (1992b) The impact of soil fauna on the properties of soils in the humid tropics. In: *Myths and Science of Soil of the Tropics.* Soil Science Society of America, special publication, 29, 157–185.

Lavelle, P., Dangerfield, M., Fragoso, C., Eschenbrenner, V., Lopez-Hernandez, D., Pashanasi, B. and Brussaard, L. (1993) The relationship between soil macrofauna and tropical soil fertility. In Swift, M.J. and Woomer, P. (eds) *Tropical Soil Biology and Fertility*, John Wiley, New York (in press).

Lee, K.E. (1985) *Earthworms: Their Ecology and Relationships with Soils and Land Use.* Academic Press, New York.

Lewis, D.H. (1985) Symbiosis and mutualism: crisp concepts and soggy semantics. In: Boucher, D.H. (ed.) *The Biology of Mutualism,* Croom Helm, Beckenham, p. 29.

Pashanasi, B., Melendez, G., Szott, L. and Lavelle, P. (1993) Effect of inoculation with the endogeic earthworm *Pontoscolex corethrurus* (Glossoscolecidae) on N availability, soil microbial biomass, and the growth of three tropical fruit tree seedlings in a pot experiment. *Soil Biology and Biochemistry* 24, 1655–1660.

Persson, T., Baath, E., Clarholm, M., Lundkvist, H., Söderström, B.E. and Sohlenius, B. (1980) Trophic structure, biomass dynamics and carbon metabolism of soil organisms in a scots pine forest. *Ecological Bulletins* 32, 419–459.

Petersen, H. and Luxton, M. (1982) A comparative analysis of soil fauna populations and their role in decomposition processes. *Oikos* 39, 287–388.

Piearce, T.G. and Phillips, M.J. (1980) The fate of ciliates in the earthworm gut: an in vitro study. *Microbial Ecology* 5, 313–319.

Rouelle, J. (1983) Introduction of amoebæ and *Rhizobium japonicum* into the gut of *Eisenia foetida* (Sav.) and *Lumbricus terrestris* L. In: Satchell, J.E. (ed.) *Earthworm Ecology*. Chapman and Hall, London, pp. 375–381.

Rouelle, J., Pussard, M., Randriamamonjizaka, J.L., Loquet, M. and Vinceslas, M. (1985) Interactions microbiennes (Bactéries, Protozoaires), alimentation des vers de terre et minéralisation de la matière organique. *Bulletin of Ecology* 16, 83–88.

Seastedt, T.R.T.T.C. and James, S.W. (1987) Experimental manipulations of arthropod, nematode and earthworm communities in a north American tallgrass prairie. *Pedobiologia* 30, 9–17.

Setälä, H. and Huhta, V. (1991) Soil fauna increase *Betula pendula* growth: laboratory experiments with coniferous forest floor. *Ecology* 72, 665–671.

Setälä, H., Tynismaa, M., Martikainen, E. and Huhta, V. (1991) Mineralization of C, N and P in relation to decomposer community structure in coniferous forest soil. *Pedobiologia* 35, 285–296.

Stork, N.E. and Eggleton, P. (1992) Invertebrates as determinants and indicators of soil quality. *Agriculture, Ecosystems and Environment* (in press).

Swift, M.J. (1986) Tropical soil biology and fertility (TSBF): inter-regional research planning workshop. Report of the third Workshop of the decade of the Tropics/ TSBF program. *Biology International,* Special Issue 13, 68 pp.

Swift, M.J. and Boddy, L. (1985) Animal–microbial interactions in wood decomposition. In: Anderson, J.M. Rayner, A.D.M. and Walton, D.W.H. (eds) *Animal-Microbial Interactions*. Cambridge University Press, Cambridge, pp. 89–131.

Syers, J.K. and Springett, J.A. (1984) Earthworms and soil fertility. *Plant and Soil* 76, 93–104.

Trofymow, J.A. and Coleman, D.C. (1982) The role of bacterivorous and fungivorous nematodes in cellulose and chitin decomposition in the context of a root (rhizosphere) soil conceptual model. In Freckman, D.W. (ed.) *Nematodes in Soil Ecosystems*. University of Texas Press, Austin, Texas, pp. 117–137.

Veeresh, G.K., Rajagopal, D. and Viraktamath, C.A. (1991) *Advances in Management and Conservation of Soil Fauna.* Oxford & IBH Publishing, New Dehli, Bombay, Calcutta.

Villalobos, F.J. and Lavelle, P. (1990) The soil coleoptera community of a tropical grassland from Laguna Verde, Veracruz (Mexico). *Revue d'Ecologie et de Biologie du Sol* 27, 73–93.

Yeates, G.W. (1981) Soil nematode populations depressed in the presence of earthworms. *Pedobiologia* 22, 191–196.

Chapter 19

Interrelationships between Biological Activities, Soil Properties and Soil Management

L. Brussaard

DLO-Institute for Soil Fertility Research, PO Box 30003, 9750 RA Haren and Agricultural University, Dept. of Soil Science and Geology, PO Box 37, 6700 AA Wageningen, The Netherlands. Present address: Agricultural University, Department of Plant Ecology and Soil Biology, Bornsesteeg 69, 6708 PD Wageningen, The Netherlands

Introduction

One of the most important processes in soil, both in terms of its role as part of the terrestrial ecosystem and in terms of its use by man, is the decomposition of organic materials, associated with humification and mineralization of elements. Decomposition is largely a biological process, with three major determinants of the rate of the process: soil organisms, the physical environment and the resource quality (Swift *et al.*, 1979).

One objective of the present chapter is to analyse the interrelationships between biological activities, soil properties and soil management, starting from the biological, physical and chemical controls on the decomposition process and, in particular, the associated mineralization of nitrogen. This is done by reference to three research projects, covering climatic conditions in tropical and temperate regions and agroforestry, permanent pastures, and arable crops. In the first project the resource quality is the starting point and effects on the soil physical environment and soil organisms will be dealt with (Tian, 1992). In the second study the emphasis is on the physical environment of the soil organisms, which influences both the biota and, possibly, the quality of organic matter in soil (Hassink *et al.*, 1993). In the third programme the focus is on quantification of the contribution of soil organisms to the mineralization of nitrogen and the functioning of the soil–crop ecosystem as affected by soil management (Brussaard *et al.*, 1988, 1990; Kooistra *et al.*, 1989; Lebbink *et al.*, 1994; Moore *et al.*, 1990; Van Faassen and

Lebbink, 1990; Moore and De Ruiter, 1991; De Ruiter *et al.*, 1993; Van Faassen and Lebbink, 1994).

Another objective of the present chapter is to derive recommendations from these three studies for both research and for practical use to make current agriculture sustainable in terms of non-declining soil fertility and crop yields over a meaningful period of time. Some elements to be incorporated in an 'index of soil fertility' and in a definition of 'meaningful period of time' will be argued for. To meet the second objective the chapter covers only field studies. Wherever appropriate, reference is made to studies under controlled conditions that provide evidence for the causal nature of the relationships found in the field.

Resource Quality and Biological Activities

Among the resource quality parameters that affect the rate of decomposition and nutrient mineralization are lignin content, polyphenol content and C:N ratio. The effects of these parameters on the decomposition rate were investigated by Tian *et al.* (1992) for two crop species (maize stover and rice straw) and three woody species (*Acioa barteri, Leucaena leucocephala* and *Gliricidia sepium*), in a litterbag study under humid tropical conditions (Table 19.1). Each of the parameters contributed significantly to the decomposition rate, and the effect of litterbag mesh-size (0.5, 2 and 7 mm) also indicated that the soil fauna were important (Table 19.2). The effects may be attributed to the different palatabilities of these materials to the soil biota, but a combination with effects of the residues on the soil microclimate is likely. Table 19.3 shows the mean soil temperature reduction and the mean soil moisture increase as compared to a control (no residues) over the first month after the application of plant residues (the 'prime effect') in a maize field and the period during which these effects exceeded the mean of $LSD_{0.05}$ ('duration of effect') over the entire experiment (15 weeks). In a separate experiment the

Table 19.1. Regression coefficient and partial correlation of effects of selected parameters on decomposition constant (week^{-1}) of plant residues. (From Tian *et al.*, 1992.)

	Regression coefficient	*F*	Probability > *F*	Partial correlation
C:N	−0.0035	13.30	0.004	−0.755
Lignin	−0.0023	8.90	0.014	−0.686
Polyphenols	−0.0188	4.82	0.053	−0.570
Mesh-size	0.0068	4.63	0.057	0.562
Constant	0.2736			

Table 19.2. Decomposition constant ($week^{-1}$) of prunings of woody species and crop residues as affected by mesh-size of litterbags. (From Tian *et al.*, 1992.)

	Mesh size (mm)			
Plant residues	7	2	0.5	Mean
Acioa	0.011	0.012	0.010	0.011
Gliricidia	0.255	0.194	0.127	0.192
Leucaena	0.166	0.147	0.062	0.125
Maize stover	0.118	0.134	0.085	0.113
Rice straw	0.124	0.157	0.106	0.129
Mean	0.135	0.129	0.078	

LSD (0.05): for species means, 0.038; for mesh-size means, 0.022; for different mesh-sizes for same species, 0.050; for different mesh-sizes for different species, 0.056.

significance of such effects on the soil microclimate for the contribution of the soil fauna to the decomposition of plant residues was demonstrated (Table 19.4). As a corollary, effects of the soil fauna on the decomposition of the chemically least-resistant and microclimatically least-influential materials were absent (*Leucaena*) or short-term only (*Gliricidia*), whereas effects on the most resistant residue species (*Acioa*) only became apparent after more than 4 weeks, with the intermediate species (maize and rice) showing intermediate results (Table 19.5). Similar results were obtained for effects of the soil fauna on N release from the plant residues. The resource quality- and the soil microclimate-mediated changes in the soil mineral N amount over the growing season of 15 weeks, after the application of plant

Table 19.3. 'Prime effect' and 'Duration of effect' of mulching on soil temperature and moisture of an Alfisol with maize in Nigeria during the cropping season of 1990. (From Tian *et al.*, 1994.)

Plant materials	Prime effect	Duration of effect (days)
	Temperature (°C)	
Acioa prunings	−2.6	90
Gliricidia prunings	−0.6	20
Leucaena prunings	−0.8	22
Maize stover	−1.7	60
Rice straw	−2.2	38
	Moisture (% v/v)	
Acioa prunings	5.5	63
Gliricidia prunings	2.1	21
Leucaena prunings	2.5	28
Maize stover	3.6	37
Rice straw	4.9	38

Table 19.4. Effect of incubation temperature and soil moisture on contributions* of earthworms and millipedes to percentage degradation of maize leaves over the control (no fauna) in a growth chamber experiment using top soil from an Alfisol, Nigeria. For each group of fauna, figures within the same column (upper case) or on the same line (lower case) indicated by the same letter are not significantly different ($P < 0.05$) according to Duncan's test. (From Tian *et al.*, 1994.)

Temperature (°C)	Soil water potential (MPa)	
	0.1	0.01
Earthworms		
32	−4.5 a A	−2.2 a A
28	1.0 a B	3.1 a B
24	0.8 a B	10.6 b C
Millipedes		
32	4.0 a A	8.6 a B
28	21.4 a B	30.9 a B
24	30.3 a B	44.8 b B

* 'maize leaf disappearance with fauna (%)' – 'maize leaf disappearance in control (%)'.

residues, were significantly correlated with changes in the maize nitrogen uptake (Tian *et al.*, 1994).

Physical Environment and Biological Activities

Protozoa and nematodes are predators of soil bacteria, and protozoa are also eaten by nematodes. Most protozoa are smaller than nematodes and can enter soil pores with neck-sizes smaller than the diameters of nematodes. Bacteria living in pores with neck-sizes smaller than the diameters of protozoa may, hence, escape predation by protozoa and nematodes. Fig. 19.1 illustrates this principle for rhizobial cells and protozoa. Once inside pores containing bacteria, it is reasonable to assume that the protozoa will increase in number. Subsequently, many may move to larger pores where they can be predated by nematodes. Various microcosm studies have given evidence for this phenomenon (Elliott *et al.*, 1980; Heynen *et al.*, 1988; Postma and Van Veen, 1990; Heijnen *et al.*, 1991). Other studies have shown that the activity of the microflora is increased under grazing pressure as long as the density of the grazers is not too high (e.g. Anderson and Ineson,

Table 19.5. Increase* in percentage degradation of plant residues with earthworms and millipedes over the control (no fauna) during 10 weeks of incubation in a growth chamber experiment using topsoil from an Alfisol, Nigeria. Figures carrying the same letter within a column are not significantly different PP < 0.05), according to Duncan's test. (From Tian *et al.*, 1994.)

	Animal groups		
Plant residues	Earthworms	Millipedes	Millipedes + earthworms
For first 4 weeks			
Acioa prunings	3.4 b	7.2 a	9.0 a
Gliricidia prunings	9.7 b	13.3 b	21.3 b
Leucaena prunings	−4.3 a	4.9 a	6.7 a
Maize stover	19.4 c	30.8 d	42.0 c
Rice straw	7.2 b	21.8 c	19.9 b
For 10 weeks			
Acioa prunings	4.6 a	23.6 b	31.9 b
Gliricidia prunings	1.3 a	10.0 a	10.6 a
Leucaena prunings	4.2 a	11.0 a	15.0 a
Maize stover	9.7 a	23.8 b	25.5 b
Rice straw	3.1 a	21.4 b	31.8 b

* 'plant residue disappearance with fauna (%)' – 'plant residue disappearance in control (%)'.

1984; Kuikman and Van Veen, 1989; Brussaard *et al.*, 1994). The concept is illustrated in Fig. 19.2. The hypothesis that these phenomena also occur under field conditions was tested by Hassink *et al.* (1992), who investigated permanent pastures on clayey, sandy and loamy soils. The total pore space, which ranged from 42 to 62%, was lower in sandy soils than in loamy and clayey soils. Based on the dimensions of the most numerous soil bacteria, assessed by direct microscopy, most bacteria were expected to occupy pores with neck-sizes of 0.2–1.2 μm. A good correlation was found between the bacterial biomass and the soil volume made up by those pores, as derived from the moisture retention characteristics (Fig. 19.3). Based on the dimensions of the most numerous nematodes a good correlation was found between their diameters and the volume of pores with neck-sizes of 30–90 μm (Fig. 19.4). The pores with diameters between 0.2 and 1.2 μm were more abundant in the loamy and clayey soils, whereas the pores with diameters between 30 and 90 μm were more abundant in the sandy soils. No pore size correlated well with the biomass of protozoa, which may be due to the large range in species and sizes occurring naturally and/or to the limited reliability of the most probable number method to enumerate them.

The mineralization rates of carbon and nitrogen are given in Table 19.6. To get an index of bacterial activity, the C mineralization rate was divided by the bacterial biomass; no differences were found among the soils studied.

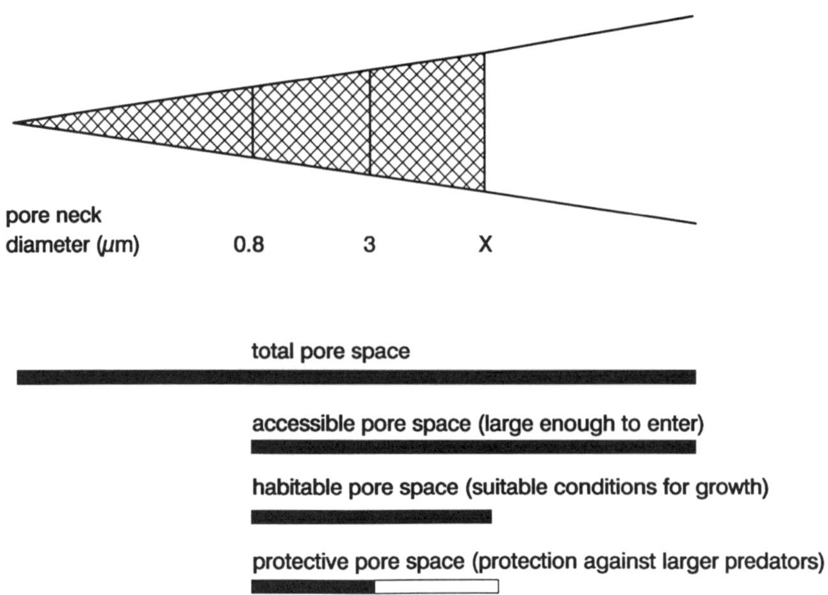

Fig. 19.1. Schematic representation of total, accessible, habitable and protective pore space for rhizobial cells. Hatched areas represent pores filled with water, and X is the pore neck diameter that is still water-filled at the water potential used (X is 30 μm at −10 kPa). Closed and open bars indicate if pore volume or pore surface area, respectively, of a certain pore diameter class are expected to be important. (From Postma and Van Veen, 1990.)

The ratio of nitrogen mineralization rate to bacterial biomass was in general higher in the sandy soils than in the loams and clays. Assuming that the ratio of biomass of grazers to biomass of bacteria is an index of grazing intensity, Hassink *et al.* (1992) investigated the relation between grazing intensity and mineralization rate of C and N per unit of bacterial biomass. This was done for amoebae, flagellates and bacterivorous nematodes. No clear relationship was found for flagellates or amoebae, the numbers of which fluctuated widely between sampling dates. However, the maximum ratio of flagellate to bacterial biomass was much higher in sandy soils than in loams and clays and coincided with the highest N mineralization rate per unit of bacterial biomass. In addition, the average grazing intensity by flagellates was highest in soils with the highest N mineralization rate per unit of bacterial biomass. Also for the bacterivorous nematodes no relationship was found between the ratio of nematode to bacterial biomass and mineralization rate of C per unit of bacterial biomass. In contrast, the mineralization rate of N per unit of bacterial biomass was much higher in soils with a high grazing intensity than in soils with a low grazing intensity by bacterivorous nematodes (Fig. 19.5). An increase in the ratio of nematode biomass to bacterial biomass by a factor 10 led to an increase in N mineral-

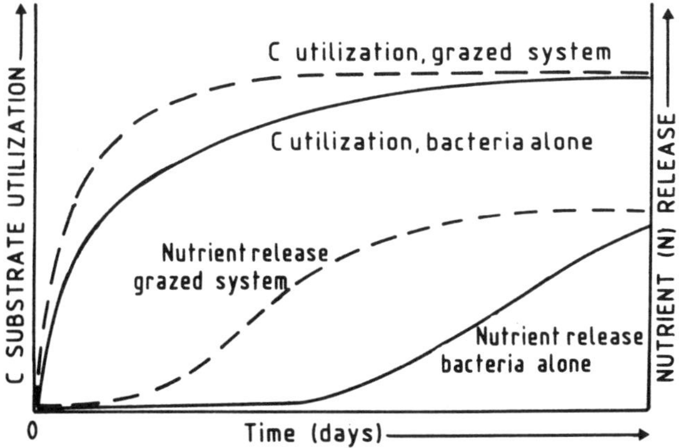

Fig. 19.2. Conceptual model of substrate utilization and nutrient mineralization with and without grazers. (From Anderson *et al.*, 1981.)

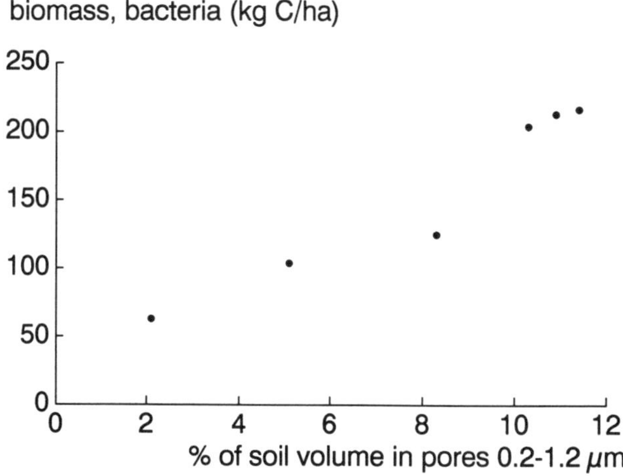

Fig. 19.3. Relation between the soil volume of pores with diameters between 0.2 and 1.2 μm and the biomass of bacteria (R^2= 0.91). (From Hassink *et al.*, 1993.)

ization rate per unit of bacterial biomass of a factor 4.

It is not clear why an effect of grazing on bacterial activity (carbon mineralization rate) was not found. It can be concluded, however, that in soils with a high grazing pressure the amount of N mineralized per bacterium is much higher than in soils with a low grazing pressure. The grazing pressure is apparently higher in sandy soils than in loamy and clayey soils, which is correlated with the accessible soil pore volume for the grazers.

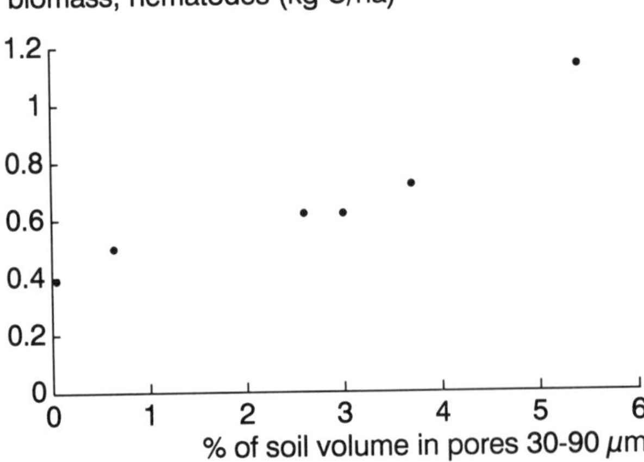

Fig. 19.4. Relation between the soil volume of pores with diameters between 30 and 90 μm and the biomass of nematodes (R^2= 0.84). (From Hassink *et al.*, 1993.)

If the lack of effect of grazing intensity on C mineralization and the substantial effect on N mineralization is a general phenomenon, it can be expected that the C:N ratio of the soil organic matter in sandy soils will be

Table 19.6. Mineralization rates of C (period 10–20 days after the start of the incubation) and N (period 0–84 days after the start of the incubation) at 20°C (kg ha^{-1} day^{-1}) and the mineralization rates of C and N divided by the bacterial biomass × 100 (kg C ha^{-1}). (From Hassink *et al.*, 1993.)

	Carbon		Rate of C mineralization per bacterial biomass (× 100)		Nitrogen		Rate of N mineralization per bacterial biomass (× 100)	
	Oct 90	May 91	Oct 90	May 91	Oct 90	May 91	Oct 90	May 91
Sand								
Heino	23.3	11.6	29.0	18.4	1.28	1.21	1.59	1.92
Achterberg	28.6		21.0		1.77		1.30	
Tynaarlo	25.8	19.6	26.5	18.9	1.00	1.85	1.03	1.78
Loam								
Swifterbant	26.2	24.6	28.7	19.7	0.45	0.67	0.49	0.54
Aduard	41.6		24.3		1.08		0.63	
Burum	52.9	51.0	37.6	23.6	1.46	2.57	1.04	1.19
Clay								
Vliert		28.8		13.5		1.77		0.83
Haskerdijk		41.6		20.4		3.05		1.50

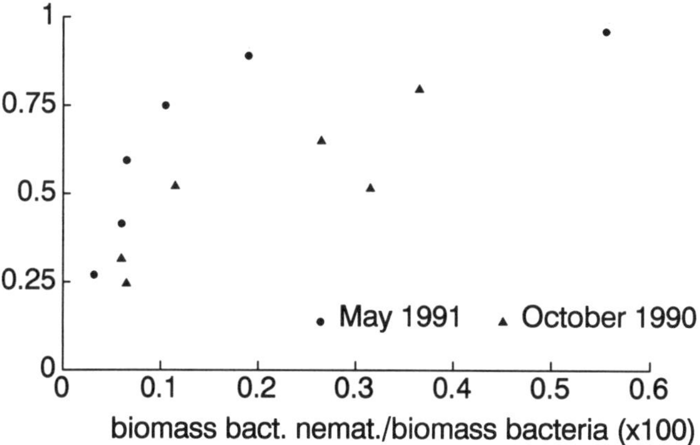

Fig. 19.5. Relation between the biomass of bacterivorous nematodes per unit of bacterial biomass and the N mineralization rate per unit of bacterial biomass. (From Hassink *et al.*, 1993.)

higher than that in loamy and clayey soils after similar management, as observed for the permanent pastures on sandy soils (C:N 17–18) as compared with loamy and clayey soils (C:N 9–12).

Soil Management and Biological Activities

Soil organisms are the central focus of research in an ecological comparison of 'integrated' and 'conventional' arable cropping systems in The Netherlands (Brussaard *et al.*, 1988; Kooistra *et al.*, 1989). In this programme, a four-year rotation of winter wheat, sugar beet, potatoes and spring barley is practised on a calcareous silt loam (Typic Fluvaquent with pH-KC1 of 7.3–7.7; organic matter 2.1–2.8%; total N 0.10–0.15%; $CaCO_3$ 9%; sand 12%, silt 68%, clay 20%; average annual rainfall 740 mm). The integrated system differs from the conventional system in the depth of soil tillage (12–15 vs. 20–25 cm), the use of pesticides (based on observations vs. calendar; no soil fumigation vs. nematicides against potato cyst-nematodes) and fertilization (manures in addition to inorganic fertilizer and crop residues vs. inorganic fertilizer and crop residues only; nitrogen fertilizer in integrated: 50–65% of conventional, depending on crop; C input on average in integrated 2400, in conventional 1600 kg ha^{-1} $year^{-1}$). The integrated and conventional management have been applied since 1985 on fields that had received 3270 or 1856 kg C ha^{-1} $year^{-1}$ during the previous 32 years. The resultant integrated organic matter contents (Fig. 19.6) show a slight increase since 1985, whereas the conventional management led to a decrease. Anticipating that the 1985 high and low levels of organic matter would approximate the equilibrium levels of the integrated and conventional management, respectively, detailed studies of various groups of soil organisms and biological processes were made on the fields with the initial high organic matter level, under integrated management, and on the fields with the low organic matter level under conventional management. These farming systems will henceforth be referred to as IF and CF, respectively.

Both systems are bacteria-dominated. The average fungal biomass (0.11 kg C ha^{-1} cm^{-1} depth) is about 100-fold less than the average bacterial biomass (9.47 kg C ha^{-1} cm^{-1} depth), based on direct counts of samples taken in 1990 (Bloem *et al.*, unpublished; Table 19.7). Generally the highest values of organism biomass and rates of processes were obtained in the integrated field (Table 19.7). Significant effects of management were found on moisture content, bacterivorous nematodes, potential O_2 consumption (in laboratory-incubated soil at 20°C) and *in situ* N mineralization (in cores incubated during 6 weeks in the field) (Table 19.7). A higher biomass of bacterivores suggests a higher turnover of the bacterial biomass (Table 19.7). The biomasses of those groups of soil organisms involved in the food web (Fig. 19.7), were determined in 1986 and used to simulate the nitrogen mineralization in the field in that year, and compared with the observed values (De Ruiter *et al.*, 1993). The two closely matched (Fig. 19.8).

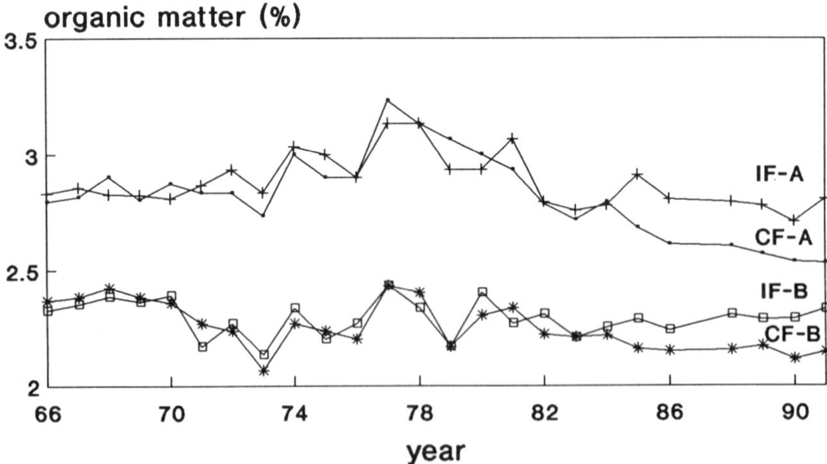

Fig. 19.6. Soil organic matter in the 0–25 cm layer of a calcareous silt loam soil, The Netherlands. Each point represents the average value for three fields. IF-A = integrated farming and CF-A = conventional farming (both since 1985) on fields that had been under an eight-year rotation including two years of ley and addition of farmyard manure from 1953–1985. IF-B = integrated farming and CF-B = conventional farming (both since 1985) on fields that had been under a six-year rotation without ley and without organic manuring from 1953–1985. (After Van Faassen and Lebbink, 1994.)

Table 19.7. Effect of management system on some abiotic and biotic soil parameters of a calcareous silt loam soil in The Netherlands in 1990. IF = integrated farming, CF = conventional farming. For differences between IF and CF see text. *P* values relate to analysis of variance. (After Bloem *et al.*, 1994.)

	Ratio IF : CF	*P*
Temperature	1.00	0.960
Moisture	1.08	0.021
Bacteria	1.08	0.396
Fungi	1.54	0.196
Bacterivorous nematodes	1.22	0.004
Potential Q_2 consumption	1.18	0.035
In situ N mineralization	1.33	0.024

Bacteria, fungi, amoebae, and bacterivorous nematodes contributed most to the mineralization of nitrogen. The contribution of the soil fauna estimated by simulation was between 25 and 30% in all cases, disregarding any effects of earthworms which did not occur in the incubated soil. When earthworms were taken into account in the simulation (Marinissen and De Ruiter, 1993), the contribution of the soil fauna to the mineralization of nitrogen rose to a maximum of 40% in IF, depending on conversion efficiencies. Earthworms do not occur in the conventional fields.

Whether or not the higher biomass and activity of soil organisms in IF than CF is associated with a tighter cycling of nitrogen was investigated by modelling the carbon and nitrogen turnover during the cropping cycle and calculating the nitrogen balance (Van Faassen and Lebbink, 1990; Van Faassen and Lebbink, submitted). The model used is a modification of that of Jenkinson and Rayner (1977) for the turnover of soil organic matter (Van Faassen and Smilde, 1985). Fig. 19.9 shows the modelled fluctuations in the organic N pools during the rotation and the resultant net accumulation of mineral N, available for plant uptake or lost to the environment; uptake and loss themselves were not modelled. Most of the mineral N is available during the crop growing seasons. After harvest of three of the four crops in the rotation N is immobilized by green manures or compost. This nitrogen will be protected from loss to the environment during winter when there is no crop uptake. After the potato crop, however, the amount of mineral N in soil increases (Fig. 19.9). This N will be prone to denitrification or leaching.

The total amounts of N circulating during the 1988–1991 cropping cycle (Fig. 19.10) showed cumulative N loss to the environment to be more than 289 kg ha^{-1} in CF and less than 181 kg ha^{-1} in IF. The N use efficiency for the IF cropping cycle was 82%, that for CF 72%. The N loss in CF is an underestimation and the N use efficiency an overestimation, depending on the extent to which the young humus pool is still decreasing and thus releasing N (Fig. 19.11). The loss in IF is an overestimation and the N use

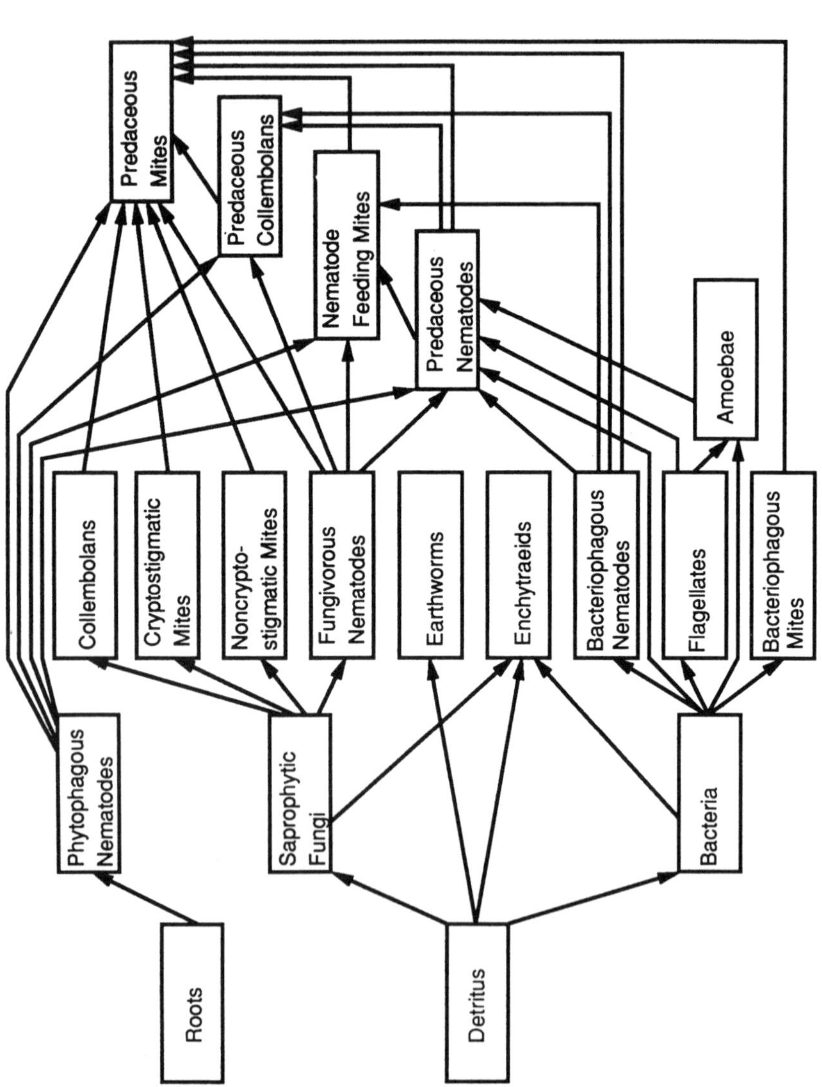

Fig. 19.7. The soil food web of a calcareous silt loam soil, The Netherlands. Arrows from each box to the detritus pool omitted. (From De Ruiter *et al.*, 1993.)

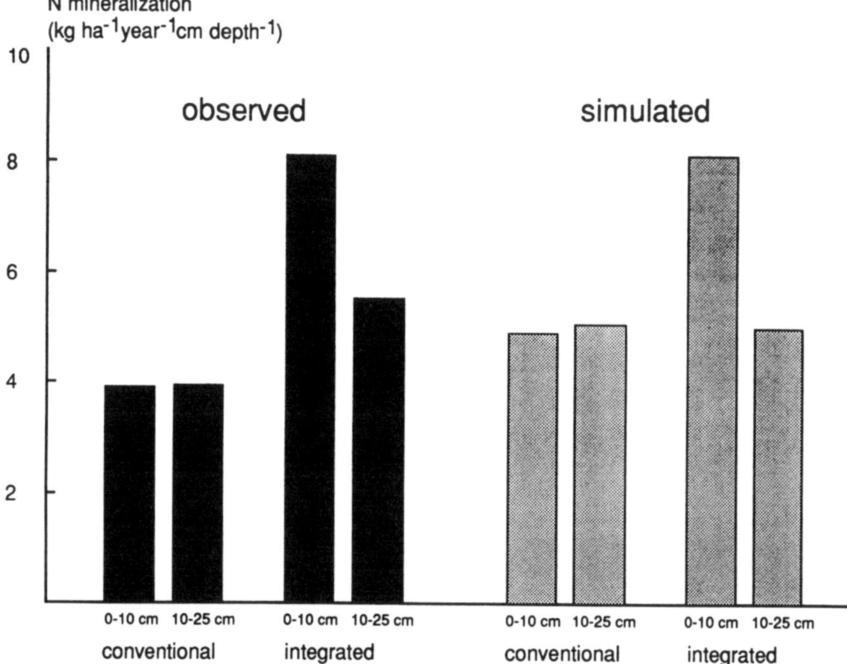

Fig. 19.8. Nitrogen mineralization under conventional and integrated farming in a calcareous silt loam soil, The Netherlands. Observed values from incubated soil, simulated values from a food web model. (From De Ruiter *et al.*, 1993.)

efficiency an underestimation, depending on the extent to which the young humus pool is still increasing and thus accumulating N (Fig. 19.11). In any case IF is conserving considerably more N than CF.

The higher deficit in the N balance and the lower N use efficiency in CF may be due to the higher input of mineral N, increasing the risk of denitrification during the growing season. Under the steady-state conditions assumed in the model (and in reality in the field given sufficient time), only small amounts of mineral N have to be applied to supplement the soil N supply in IF, but considerably larger amounts are needed in CF.

Crop yields in IF during the 1988–1991 cropping cycle were on average between 84 and 102% of those in CF (Table 19.8). A moderate yield loss with lower inputs may be acceptable and economic under surplus production of some crops in Western Europe. Given that farmers' experience in optimizing management inputs is less well developed for IF than for CF, yields under IF may be expected to improve.

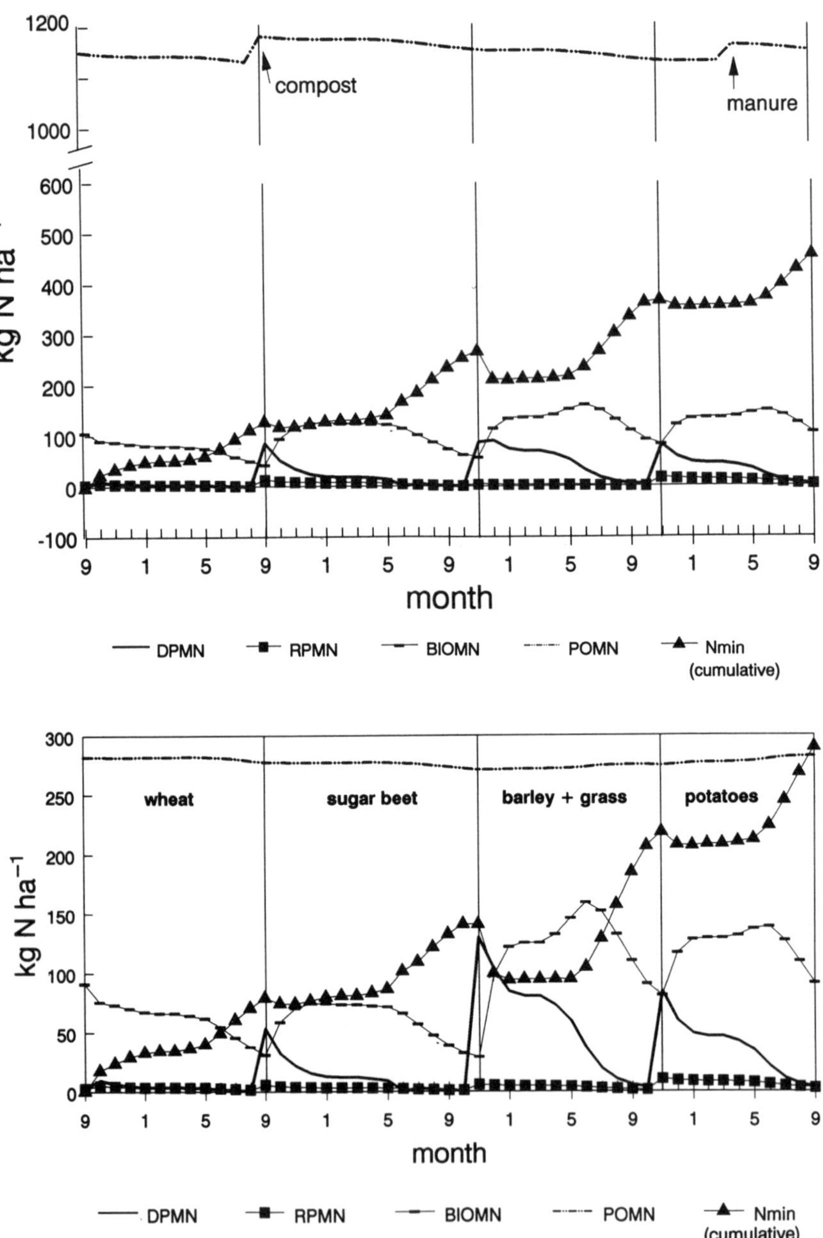

Fig. 19.9. Modelled fluctuation in crop residue and soil organic N pools and cumulative mineral N for a four-year rotation under integrated (top) and conventional (bottom) management on a calcareous silt loam soil, The Netherlands. DPMN= decomposable plant material nitrogen; RPMN= resistant plant material nitrogen; BIOMN= soil microbial biomass nitrogen; POMN= young humus nitrogen. (From Van Faassen and Lebbink, 1994.)

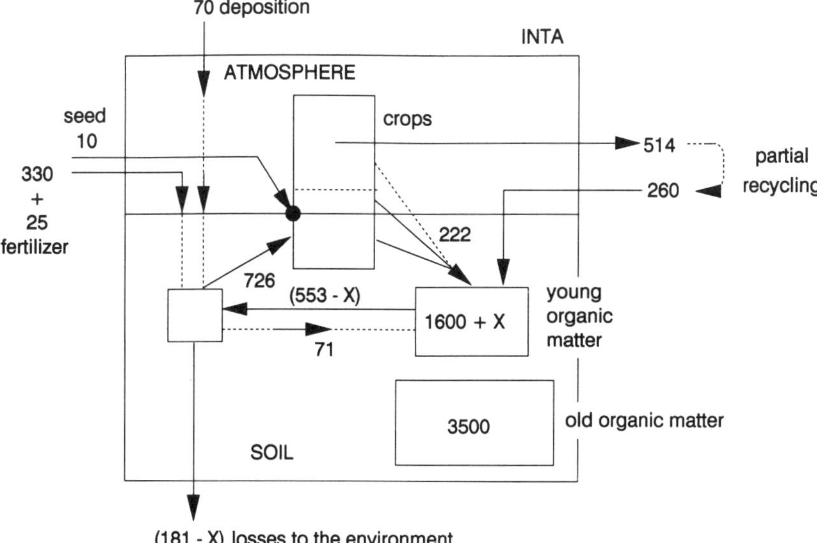

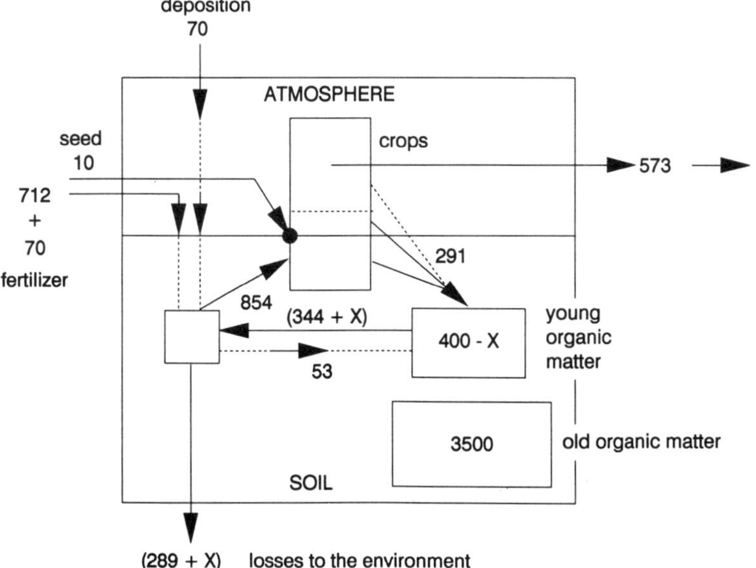

Fig. 19.10. N budget of a four-year rotation under integrated (top) and conventional (bottom) management on a calcareous silt loam soil. The Netherlands. (From Van Faassen and Lebbink, 1994.)

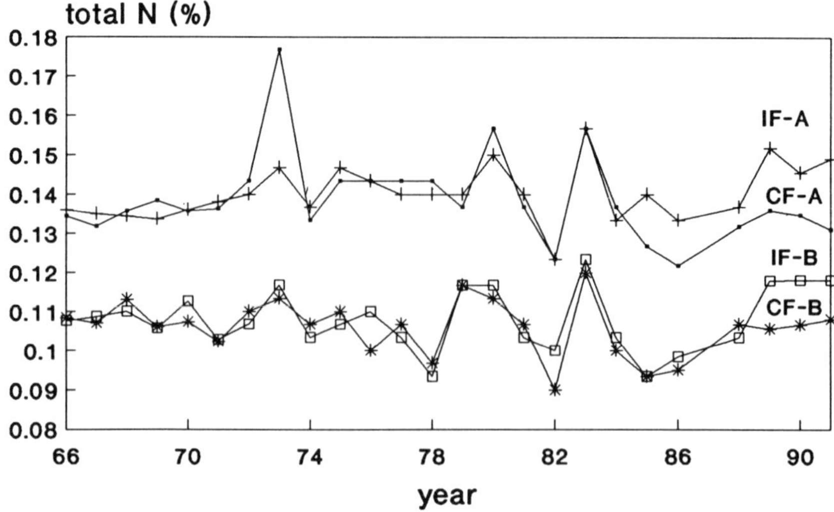

Fig. 19.11. Total N in the 0–25 cm layer of a calcareous silt loam soil, The Netherlands. For further explanation see Fig. 19.6 (After Van Faassen and Lebbink, 1994.)

Conclusions from the Case Studies

Case study number 1:

1. Plant residues influence both the soil biota and the soil microclimate and, thereby, the rate of decomposition, mineralization and crop nitrogen uptake.

Case study number 2:

2. Soil texture affects the soil biota, in particular bacteria and nematodes, to the effect that the bacterial biomass is lower, but the rate of nitrogen mineralization per bacterium higher in coarser textured soil than in finer textured soil under the same management.

Table 19.8. Average crop yields for the rotation 1988–1991 (Mg ha^{-1}) ± SD ($n = 4$) on a calcareous silt loam soil under integrated (IF) or conventional (CF) farming in The Netherlands. (After Van Faassen and Lebbink, 1994.)

	IF	CF
Winter wheat	5.6 ± 0.8	6.7 ±0.8
Sugar beet	12.9 ± 1.2	14.5 ± 1.6
Spring barley	4.8 ± 0.7	5.3 ± 0.6
Potatoes	58.4 ± 7.1	57.4 ± 7.5

Case study number 3:

3. Nitrogen mineralization in soil is related to biomass and activity of soil organisms.

4. The contribution of the soil fauna (excluding earthworms) to the mineralization of nitrogen is between 25 and 30%.

5. In the presence of earthworms this figure may rise to about 40%, depending on conversion efficiencies.

6. Total soil organic matter attains a higher equilibrium value in integrated farming than in conventional farming on the soil and in the environment studied.

7. Considerably less nitrogen is present as inorganic N in IF than in CF.

8. The nitrogen use efficiency in IF is higher than in CF (more than 82% vs. less than 72%).

9. Crop yields in IF are between 84 and 102% of those in CF.

Soil Fertility, Soil Resilience and Soil Management

A sustainable agricultural system has been defined as one 'in which output trend is non-declining and resistant, in terms of yield stability, to normal fluctuations of stress and disturbance' (Spencer and Swift, 1992). Since agriculture itself is one of the largest sources of stress and disturbance of the environment (e.g. by erosion and nutrient losses) it would seem that another important criterion for the sustainability of agriculture is 'minimal effects on the environment'. Keeping these criteria for sustainability in mind (i.e. yield stability, resilience and closed cycles), what are the attributes of soil fertility important for sustainable agriculture?

Soil fertility has been defined as 'the status of a soil with respect to its ability to supply elements essential for plant growth without a toxic concentration of any element' (Foth and Ellis, 1988). The ability to supply elements essential for plant growth may be a useful quantity parameter to define soil fertility, but the definition does not include reference to the intensity factors determining the ability of the soil to supply elements. Hence, it would seem to be of limited use in deterimining the ability of the soil to supply elements in the long run, the ability of the soil to resist disturbances, and the management action needed to maintain the quantity and intensity factors of nutrient supply.

Soil microbiological attributes have recently been reviewed by Visser and Parkinson (1993), who conclude that substrate-induced respiration (Anderson and Domsch, 1978) appears to have the greatest potential for assessing and monitoring 'soil quality'. Basal respiration measurements followed by SIR measurements on the same soil, allow the calculation of the time until the respiratory response of the microorganisms, the metabolic quotient $qCO_2 = C_{resp}/C_{micro}$ (C evolved as $CO_2 = C_{resp.}$; microbial biomass C=

C_{micro}) and the C_{micro}/C_{org} quotient (*sensu* Insam and Domsch, 1988; Insam and Haselwandter, 1989) can also be determined. These are characteristic within certain limits for soils with a certain vegetation (history) in a certain climate region, which means that deviations from expectation may provide the necessary incentive for research on causes, consequences and solutions. Based on the study by Hassink *et al.* (1993) it is hypothesized that a 'mineralization quotient' $N_{min\ rate}/C_{bact}$ may also be a useful attribute of soil fertility. This quotient is dynamic but, under normal circumstances, appears to be associated with the pore size classes that protect bacteria and are habitable for nematodes. These pore sizes and, ideally, the biomass of bacteria and nematodes, may also be useful to evaluate the presence and effects of stress and disturbance. Further research is needed to confirm these relationships.

In addition to parameters relating soil microorganisms and nematodes to soil organic matter and nitrogen mineralization, the biomass of earthworms appears to be a useful attribute of soil fertility, not only because of the well-known effects of earthworms on soil porosity and related transport of water and solutes (e.g. Edwards *et al.*, 1990) and on the dynamics of soil organic matter (e.g. Martin, 1991), but also because the studies by Tian *et al.* (1994) and Marinissen and De Ruiter (1993) indicate their influence on amount and pattern of nitrogen mineralization in soil. Moreover, earthworms are sensitive indicators of stress (e.g. pesticides and heavy metals) and disturbance (e.g. ploughing).

The studies by Tian *et al.* (1992, 1994) and Van Faassen and Lebbink (1994) show that the amount, timing and quality of organic inputs to the soil constitute the most important management tools for maintaining and improving soil fertility in terms of supply and efficiency of nitrogen use, through their effects on soil organic matter and soil biota. Organic inputs are essential to the interrelationships between biological activities and soil properties regarding the long-term supply of nutrients and closing of element cycles. Compared to high external input agriculture, such agriculture may result in somewhat lower annual yields as long as the high input systems are economically feasible. For farmers that have no access to high external inputs, however, organically based agriculture appears to be the only option if sustainability is to be achieved.

Research Priorities and Recommendations

Research priorities

1. To assess quantitatively the relationships, under 'normal' vegetation and climate conditions, between soil type and the following soil fertility attributes: $N_{min\ rate}/C_{bact}$ ('mineralization quotient'); volume of pores protective for bacteria and volume of pores habitable for nematodes; biomass of earthworms; metabolic quotient C_{resp}/C_{micro}; C_{micro}/C_{org} quotient; and time till response of microflora following substrate amendment.

2. To assess the resilience of those attributes to soil management practices (tillage, fertilization, pesticide application, harvesting) and other types of stress and disturbance for different soils under various types of land use.

3. To analyse further the interrelationships between biological activities, soil properties and soil management.

Practical recommendations

1. To prevent damage to the soil biota by physical and chemical factors.

2. To raise the nutrient use efficiency of cropping systems to improve yields and to reduce nutrient losses to the environment by judicious application of organic materials that make the soil biota immobilize and mineralize nutrients according to the presence and demand of the crop.

Acknowledgements

Thanks are due to J. Bloem, G. Tian and H.G. Van Faassen for provision of unpublished data and to J. Hassink, P.C. de Ruiter, G. Tian, H.G. Van Faassen and M. Van Noordwijk for useful comments on an earlier draft of the manuscript.

References

Anderson, J.M. and Ineson, P. (1984) Interactions between microorganisms and soil invertebrates in nutrient flux pathways of forest ecosystems. In: Anderson, J.M., Rayner, A.D.M. and Walton, D.W.H. (eds) *Invertebrate–Microbial Interactions.* Cambridge University Press, Cambridge, pp. 59–88.

Anderson, J.P.E. and Domsch, K.H. (1978) A physiological method for quantitative measurement of microbial biomass in soils. *Soil Biology and Biochemistry* 10, 215–221.

Anderson, R.V., Coleman, D.C. and Cole, C.V. (1981) Effects of saprotrophic grazing on net mineralization. *Ecological Bulletins* 33, 201–215.

Bloem, J., Lebbink, G., Zwart, K.B., Bouwman, L.A., Burgers, S.G.L.E., De Vos, J.A. and De Ruiter, P.C. (1994) Dynamics of microorganisms, microbivores and nitrogen mineralization in winter wheat fields under conventional and integrated management. *Agriculture, Ecosystems and Environment,* in press.

Brussaard, L., Van Veen, J.A., Kooistra, M.J. and Lebbink, G. (1988) The Dutch programme on soil ecology of arable farming systems. I. Objectives, approach and some preliminary results. *Ecological Bulletins* 39, 35–40.

Brussaard, L., Bouwman, L.A., Geurs, M., Hassink, J. and Zwart, K.B. (1990) Biomass, composition and temporal dynamics of soil organisms of a silt loam soil under conventional and integrated management. *Netherlands Journal of Agricultural Science* 38, 283–302.

Brussaard, L., Noordhuis, R., Geurs, M. and Bouwman, L.A. (1994) Nitrogen mineralization in soil in microcosms with or without bacterivorous nematodes and nematophagous mites. *Acta Zoologica Fennica*, in press.

De Ruiter, P.C., Moore, J.C., Zwart, K.B., Bouwman, L.A., Hassink, J., Bloem, J., De Vos, J.A., Marinissen, J.C.Y., Didden, W.A.M., Lebbink, G. and Brussaard, L. (1993) Simulation of nitrogen mineralization in the belowground food webs of two winter wheat fields. *Journal of Applied Ecology* 30, 95–106.

Edwards, W.M., Shipitalo, M.J., Owens, L.B. and Norton, L.D. (1990) Effect of *Lumbricus terrestris* L. burrows on hydrology of continuous no-till corn fields. *Geoderma* 46, 73–84.

Elliott, E.T., Anderson, R.V., Coleman, D.C. and Cole, C.V. (1980) Habitable pore space and microbial trophic interactions. *Oikos* 35, 327–335.

Foth, H.D. and Ellis, B.G. (1988) *Soil Fertility*, Wiley, New York.

Hassink, J., Bouwman, L.A., Zwart, K.B. and Brussaard, L. (1993) Relationships between habitable pore space, soil biota and mineralization rates in grassland soils. *Soil Biology and Biochemistry* 25, 47–55.

Heijnen, C.E., Hok-A-Hin, C.H. and Van Veen, J.A. (1991) Protection of *Rhizobium* by bentonite clay against predation by flagellates in liquid cultures. *FEMS Microbiology Ecology* 85, 65–72.

Heynen, C.E., Van Elsas, J.D., Kuikman, P.J. and Van Veen, J.A. (1988) Dynamics of *Rhizobium leguminosarum* biovar *trifolii* introduced into soil; the effect of bentonite clay on predation by protozoa. *Soil Biology and Biochemistry* 20, 483–488.

Insam, H. and Domsch, K.H. (1988) Relationship between soil organic carbon and microbial biomass on chronosequences of reclamation sites. *Microbial Ecology* 15, 177–188.

Insam, H. and Haselwandter, K. (1989) Metabolic quotient of the soil microflora in relation to plant succession. *Oecologia* 79, 174–178.

Jenkinson, D.S. and Rayner, J.H. (1977) The turnover of soil organic matter in some of the Rothamsted classical experiments. *Soil Science* 123, 298–305.

Kooistra, M.J., Lebbink, G. and Brussaard, L. (1989) The Dutch programme on soil ecology of arable farming systems. 2. Geogenesis, agricultural history, field site characteristics and present farming systems at the Lovinkhoeve experimental farm. *Agriculture, Ecosystems and Environment* 27, 361–387.

Kuikman, P.J. and Van Veen, J.A. (1989) The impact of protozoa on the availability of bacterial nitrogen to plants. *Biology and Fertility of Soils* 8, 13–18.

Lebbink, G. Van Faassen, H.G., Van Ouwerkerk, C. and Brussaard, L. (1994) The Dutch programme on soil ecology of arable farming systems: farm management, monitoring programme and general results. *Agriculture, Ecosystems and Environment*, in press.

Marinissen, J.C.Y. and De Ruiter, P.C. (1993) Contribution of earthworms to carbon and nitrogen cycling in agro-ecosystems. *Agriculture, Ecosystems and Environment* (in press).

Martin, A. (1991) Short- and long-term effects of the endogeic earthworm *Millsonia anomala* (Omodeo) (Megacsolecidae, Oligochaeta) of tropical savannas, on soil organic matter. *Biology and Fertility of Soils* 11, 234–238.

Moore, J.C. and De Ruiter, P.C. (1991) Temporal and spatial heterogeneity of trophic interactions within below-ground food webs. *Agriculture, Ecosystems and Environment* 34, 371–397.

Moore, J.C., Zwetsloot, H.J.C. and De Ruiter, P.C. (1990) Statistical analysis and simulation modelling of the belowground food webs of two winter wheat management practices. *Netherlands Journal of Agricultural Science* 38, 303–316.

Postma, J. and Van Veen, J.A. (1990) Habitable pore space and survival of *Rhizobium legumonosarum* biovar *trifolii* introduced into soil. *Microbial Ecology* 19, 149–161.

Spencer, D.S.C. and Swift, M.J. (1992) Sustainable agriculture: definition and measurement. In: Mulongoy, K., Gueye, M. and Spencer, D.S.C. (eds) *Biological Nitrogen Fixation and Sustainability of Tropical Agriculture.* Wiley, Chichester, pp. 15–24.

Swift, M.J., Heal, O.W. and Anderson, J.M. (1979) *Decomposition in Terrestrial Ecosystems,* Blackwell Scientific Publications, Oxford.

Tian, G. (1992) Biological effects of plant residues with contrasting chemical compositions on plant and soil under humid tropical conditions. PhD thesis, Wageningen Agricultural University.

Tian, G., Kang, B.T. and Brussaard, L. (1992) Biological effects of plant residues with contrasting chemical compositions under humid tropical conditions – decomposition and nutrient release. *Soil Biology and Biochemistry* 24, 1051–1060.

Tian, G., Kang, B.T. and Brussaard, L. (1994) Mulching effects of plant residues with chemically contrasting compositions on maize growth and nutrient accumulation. *Plant and Soil,* in press.

Van Faassen, H.G. and Lebbink, G. (1990) Nitrogen cycling in high-input versus reduced-input arable farming. *Netherlands Journal of Agricultural Science* 38, 265–282.

Van Faassen, H.G. and Lebbink, G. (1994) Organic matter and nitrogen dynamics in conventional versus integrated arable farming. *Agriculture, Ecosystems and Environment,* in press.

Van Faassen, H.G. and Smilde, K.W. (1985) Organic matter and nitrogen turnover in soils. In: Kang, B.T. and Van der Heide, J. (eds) *Nitrogen Management for Farming Systems in Humid and Subhumid Tropics.* Institute for Soil Fertility / International Institute of Tropical Agriculture, Haren / Ibadan, pp. 39–55.

Visser, S. and Parkinson, D. (1993) Soil biological criteria as indicators of soil quality: soil microorganisms. *Journal of Alternative Agriculture* 7, 33–37.

Chapter 20
Biological Indicators of Soil Health and Sustainable Productivity

C.E. Pankhurst

CSIRO, Division of Soils and the Cooperative Research Centre for Soil and Land Management, PMB 2, Glen Osmond, SA 5064, Australia

Introduction

The use of animals, plants and microorganisms as bioindicators of environmental impact is a well established concept (Paoletti *et al.*, 1991). In most instances, loss of species diversity has been the most obvious casualty accompanying the evolution of agroecosystems. Monoculture, cultivation, artificial fertilizers and agrichemicals all contribute to a reduction in species diversity and biomass (Dindal, 1989; Stinner and House, 1990; Paoletti *et al.*, 1991, 1992). As a consequence, these practices are now being re-evaluated as part of a concerted worldwide effort to promote conservation and sustainability in agriculture and to promote productivity and profitability in terms of reduced input and maximal use of natural resources.

Definitions and Concepts

There are many different concepts of sustainability, but none is generally accepted. In many respects, sustainability should be considered dynamic because, ultimately, it will reflect the changing needs of an increasing global population. Recognizing this broader concept, a working definition of sustainable agriculture (at the farm level) proposed by Campbell (1989) is one which 'maintains the productive capacity of the land and economic viability, while minimising energy and resource use, and optimising the rate of turnover and recycling of organic matter and nutrients'.

Soil health and sustainability necessarily go hand in hand. A healthy soil will be a productive soil. It will possess many characteristics, chief among which will be its capacity to support plant growth and soil organism activity appropriate for the particular soil type and climate. The presence of organic matter, essential for the provision of nutrients and maintenance of soil structure, the absence of physical and chemical constraints to plant growth such as acidity, salinity, sodicity, waterlogging, compaction and toxicant accumulations, and the absence of biological constraints to plant growth such as root diseases, are all synonymous with a healthy soil. Bioindicators of soil health therefore will be biological entities such as living plants, animals, microorganisms, the processes they carry out, or the end product of their activity, that can be used as a measure of the soil condition.

Types of Bioindicator

Bioindicators of soil health, could be divided into three broad categories. These are: bioindicators of soil productivity (indicators that are positive or negative measures of plant growth); bioindicators of soil stability and sustainability (indicators reflecting the long-term conservation of the soil resource); and bioindicators of soil pollution (indicators of contamination). However, it could be argued that it is not appropriate to distinguish between indicators of productivity and sustainability, as they are too closely interrelated. For the purposes of this discussion therefore they will be considered together.

Bioindicators of Soil Productivity and Sustainability

Living soil organisms

The soil is a habitat for a vast number of diverse organisms, many of which are yet to be identified. In fact, we should be reminded that the soil is a living entity with a diverse flora and fauna and that when devoid of its biota, the uppermost layer of earth ceases to be soil (Lal, 1991). Representatives of all groups of microorganisms and fungi, green algae and cyanobacteria, and all but a few exclusively marine phyla of animals, make up the soil biota. An indication of the biomass of soil organisms, divided on the basis of their size into microfauna, mesofauna, macrofauna, and microflora, and comparing the biota of temperate with tropical regions is provided in Table 20.1. Perhaps the most striking aspect of these statistics is the extremely large soil biomass that is present in a fertile soil. This biomass is roughly equivalent to the aboveground biomass of plants supported by fertile soils in a pasture, or woodland agroecosystem (Paoletti *et al.*, 1992).

Table 20.1. Representative composition of soil biomass in temperate and tropical zones (Lee, 1990).

Organisms classified by size	Biomass (kg ha^{-1} live weight)	
	Temperate zone	Tropical zone
Microfauna (< 2 mm) (protozoa, nematodes)	50	0.5
Mesofauna (2–10 mm) (enchytraeids, microarthropods)	20	20
Macrofauna (> 10 mm)		
Earthworms	900	300
Termites	–	15
Microorganisms		
Bacteria, fungi, and other microorganisms	20,000	?

Bacteria and fungi

Bacteria and fungi and the activities they perform are integral to the development and maintenance of a healthy soil. Bacteria are particularly important in soil because their diverse metabolic capabilities enable them to exploit many sources of energy and carbon. Unique metabolic features of bacteria include anaerobic respiration, chemolithotrophic growth, fixation of molecular nitrogen and utilization of methane (Lynch, 1983), making soil bacteria important agents for the global cycling of many inorganic compounds, especially nitrogen, sulphur, and phosphorus. Soil fungi are numerically less abundant than bacteria in soil. There are typically between 10^4 and 10^6 fungal propagules per gram of soil compared to 10^6–10^9 bacteria per gram of soil (Lynch, 1983). However, fungi may account for as much as 70% by weight of the soil biomass (Lynch, 1983). Most fungi are opportunistic, becoming active when environmental conditions are favourable. Soil fungi are active in the transformation of cellulose and are the principal agents for the transformation of lignins produced by plants. The breakdown of these polymers releases molecules that are subsequently used by other soil organisms, particularly bacteria.

Total populations of bacteria and fungi in the soil are sensitive to and respond differently to soil management practices. Where soils have been subjected to conservation management practices (minimum tillage, crop rotations, stubble retention on the soil surface etc.) and thus have a good structure and are high in organic matter, the microbial population, especially in the surface soil layers, tends to be fungal dominated (Hendrix *et al.*, 1986; Holland and Coleman, 1987; Gupta and Germida, 1988) (Table 20.2). In contrast, cultivation and incorporation of plant residues into the soil, tends to create conditions more favourable for bacterial-based food webs and faster decomposition of organic matter (Hendrix *et al.*, 1986).

Table 20.2. Microbial properties of soil aggregate size classes from a native grassland soil and an adjacent soil subjected to cultivation for 69 years. (From Gupta and Germida, 1988.)

			Fungal biomass		Microbial biomass ($\mu g\ g^{-1}$)		
Size classes (mm)	Bacteria ($10^7 \times$ cfu g^{-1})	Fungi ($10^4 \times$ cfu g^{-1})	Length ($m\ g^{-1}$)	Biovolume ($mm^3\ g^{-1}$)	C	N	S
Native							
> 1.00	1.3	11.7	874	3.74	1538	139	10.6
0.50–1.00	1.9	9.9	1276	12.89	1862	155	6.8
0.25–0.50	0.9	19.5	2163	10.85	1463	133	8.6
0.10–0.25	0.7	9.3	508	2.06	1161	116	6.1
< 0.10	2.9	1.2	511	3.59	1106	124	8.4
Cultivated (69 year)							
> 1.00	1.9	2.8	180	0.98	886	86	6.2
0.50–1.00	4.7	2.5	166	1.12	946	92	11.0
0.25–0.50	0.8	6.5	543	3.63	859	90	5.5
0.10–0.25	6.7	0.5	144	1.70	655	82	4.7
< 0.10	18.7	0.4	66	0.75	645	80	4.6

Increased fungal abundance in soils subjected to minimum tillage practices can be attributed to their ability to translocate nutrients from the soil into surface residues (fungal hyphal bridges) and to their tolerance of the low water potential often occurring in surface residues (Holland and Coleman, 1987). In addition, mycelial fungi are well adapted to penetrating and using large detritus particles, and can tolerate low pH; these features make them well suited to undisturbed soils (Doran, 1980; Cooke and Rayner, 1984). Networks of fungal hyphae, particularly those of vesicular-arbuscular (VA) mycorrhizal fungi (Foster, 1988; Tisdall, 1991), are also important in binding microaggregates in the soil into stable macroaggregates. Work by Gupta and Germida (1988) showed that a reduction in macroaggregates in a grassland soil as a result of continuous cultivation for 69 years was correlated with loss of fungal hyphae normally associated with the macroaggregates (Table 20.2).

There is an extensive list of soil bacteria and soil fungi that are associated with either beneficial or detrimental activities in the soil. An assessment of their importance to soil health will undoubtedly be related to the perceived impact that these organisms have on soil productivity and in particular the impact they have on plant growth. Soil bacteria identified as beneficial include N-fixing root nodule bacteria (*Rhizobium*/*Bradyrhizobium*), free-living N-fixing bacteria (e.g. *Azotobacter, Azospirillium*), plant growth promoting rhizobacteria (Kloepper *et al.*, 1991) and bacteria associated with the biocontrol of plant root pathogens (Campbell, 1989). On the other hand, deleterious rhizobacteria (Schippers *et al.*, 1987) and root pathogenic bacteria are undesirable and measures to exclude them from the soil are sought. In a similar fashion, soil fungi active in the decomposition of cellulosic and lignified plant residues and mycorrhizal fungi are seen as beneficial whereas root pathogenic fungi are seen as undesirable. The presence or absence of these organisms in the soil can be used as indicators of soil health and soil productivity as they are all capable of having a direct effect on plant growth. However their presence or absence may be quite independent of other soil factors that contribute to soil health and sustainability. For example, conservation soil management practices, such as minimum tillage, favour the growth of the root pathogen *Rhizoctonia solani* in many Australian soils (Neate, 1987). Apart from studies of the effect of soil conservation management practices on populations of soil bacteria and fungi generally (Hendrix *et al.*, 1986; Dick, 1992) and of root pathogenic fungi (Rovira *et al.*, 1990), there is little known of the effects of various cropping and tillage practices on populations of the other economically important soil bacteria or soil fungi.

Algae, protozoa and nematodes

Population densities of algae and protozoa in soil have been estimated to be between 10^1 and 10^6 g^{-1} of soil for algae and between 10^4 and 10^5 g^{-1} for protozoa (Atlas and Bartha, 1987). The abiotic environmental parameters regulating the growth of algae and protozoa in soils include sunlight and CO_2 for algae and O_2 for protozoa. Protozoa are important predators in soil and help to regulate the size of bacterial populations (Alexander, 1977). Algae contribute to the organic carbon, and in the case of blue–green algae, nitrogen plus carbon, and they also contribute to soil structure and erosion control (Alexander, 1977; Roger *et al.*, 1991).

Nematodes are among the most abundant soil fauna. Population densities may vary widely with season and vegetation differences; estimates for agroecosystems range from 10^4 to 10^7 m^{-2} of soil surface (Stinner and Crossley, 1982). Many are parasitic on higher plants and animals. Free-living nematodes are voracious feeders on bacteria, fungi and protozoa, and are common rhizosphere organisms. They are important as indirect regulators of decomposition and nutrient release in the soil. However, they contribute little to direct mineralization of organic matter, accounting for less than 1.0% of the total soil respiration (Yeates and Coleman, 1982).

Protozoa and nematodes have a major role to play in soil health as a consequence of their predation activities. Their interactions with soil bacteria and fungi form an essential link in the detritus food webs that are developed in soils (Ingham *et al.*, 1985). Their population size and species composition will therefore reflect the population of bacteria and fungi present (Hendrix *et al.*, 1986). Thus all undisturbed soil with a fungal dominated microflora will have a high population of fungivorous nematodes

Table 20.3. Numbers and estimated biomass of soil fauna in conventional-tillage (CT) and no-tillage (NT) agroecosystems at Horseshoe Bend. (From Hendrix *et al.*, 1986.)

	Numbers m^{-2}			mg dry wt m^{-2}	
	CT		NT	CT	NT
Nematodes					
Bacterivores	1836	*	909	237	117
Fungivores	227	*	500	14	31
Herbivores	945		1064	93	104
Total	3008		2473	344	252
Microarthropods					
Mites	41,081	*	78,256	118	303
Collembola	6244	*	14,684	17	40
Insects	2105		2548	–	–
Total	49,430		95,488	135	343
Macroarthropods					
Ground Beetles	7	*	33	6	30
Spiders	1	*	17	1	14
Others	6	*	28	–	–
Total	14		78	7	44
Annelids					
Earthworms	149	*	967	3129	20,307
Enchytraeids	1837		520	59	17
Total	1986		1487	3188	20,324
Grand total	54,438		99,526	3674	20,962

* For numbers of organisms, tillage treatments differ significantly at $P = 0.05$.

(Hendrix *et al.*, 1986) (Table 20.3), whereas a cultivated soil with a bacterial dominated microflora will have a high population of bacterivorous nematodes. It is not surprising therefore that several species of protozoa and nematodes have been shown to have potential as biocontrol agents of root disease bacteria and fungi (Chakraborty *et al.*, 1983; Nicholas, 1984). However, on the debit side, about 500 species of plant-parasitic soil nematodes are known (Nicholas, 1984) and their presence is thus detrimental to soil productivity.

Soil mesofauna

Microarthropods (principally collembola and mites) are usually the most obvious of the mesofauna. Population densities of microarthropods range from 10^1 to 10^7 m^{-2} of soil surface (Lal, 1991). Microarthropods include detritus feeders that contribute to the physical fragmentation of litter, grazers of fungi that may influence the balance between fungi and bacteria, and a wide range of predators on microfauna and mesofauna. Many are feeders in the rhizosphere, and may make an important contribution to nutrient cycling.

Soil mesofauna have been considered to be sensitive bioindicators of soil disturbance (Loring *et al.*, 1981; Hendrix *et al.*, 1986; Perdue and Crossley, 1990; Crossley *et al.*, 1992; Koehler, 1992). Populations of both collembola and mites tend to increase in size and species abundance in soils subjected to conservation management practices (Table 20.3). Analysis of species response to soil disturbance shows that some species, e.g. the oribatid mites, decline rapidly when soils are cultivated whereas others, e.g. the prostigmatid mites, may increase in numbers seasonally, following soil cultivation (Perdue and Crossley, 1990). In a study of the effect of the insecticide Aldicarb on the succession of collembola and mites in a controlled field experiment, Koehler (1992) showed major differences in the rate of return of individual species and in the diversity of species finally present after 3 years. Thus, whereas the relationships between microarthropod activities and agroecosystem processes are necessarily complex (Crossley *et al.*, 1992), it does appear that some species of mites in particular have the potential to be sensitive indicators of soil condition.

Soil macrofauna

The principal groups among the macrofauna are the annelids, which comprise the earthworms, the larger enchytraeids, and other families, including tubificids; the macroarthropods, which include a range of insects, spiders, crustaceans and other groups, and the molluscs, including snails and slugs. They include herbivores, detritivores, predators, and other trophic groups. The macrofauna, especially earthworms, termites and ants, have a special significance in that they are very widely distributed on a global scale and they are able extensively to rework and shape the soil, with far-reaching effects on soil structure, water infiltration and gas exchange with the atmosphere. Earthworms make up the major proportion of the biomass, with populations ranging from $<10\ m^{-2}$ to several hundreds and occasionally up to 2000 m^{-2}, with biomass ranging from 1–2 g m^{-2} to >300 g m^{-2} (Lee, 1985). Termites and ants are social insects, most common in tropical, subtropical and warm temperate regions. Their populations are highly aggregated; at any one time their effects on soils are patchy, concentrated around nests and feeding territories, but over long periods of time they affect very large areas more uniformly. The biomass of termites and ants is generally much less than that of earthworms (Lobry de Bruyn and Conacher, 1990).

Earthworms are sensitive indicators of soil condition (Edwards and

Lofty, 1977; Lee, 1985). The most important factor that influences earthworm populations is the amount of organic matter that is available as food. Populations are thus favoured by mulch farming practices and reduced soil disturbance (Hendrix *et al.*, 1986; Rovira *et al.*, 1987; El Titi and Upach, 1989) (Table 20.3). Earthworm species also vary in their tolerance of low soil pH and agrochemicals (Heimbach, 1985; Lee, 1985; Lal, 1988) and could thus be used as an indicator of pH and levels of certain pesticides in the soil. However, it is the indirect effects on soil properties (humification and mineralization of organic matter, soil mixing and turnover, macroporosity) that make earthworms and in the more arid environments, termites and ants, so important for the maintenance of soil productivity and thus ecosystem stability. Their presence in large numbers is usually regarded as a sign of soil health and productivity.

The impact of soil management practices on other soil macrofauna, notably arthropod pests of food and forage crops has been extensively studied (Stinner and House, 1990; Villani and Wright, 1990; Paoletti *et al.*, 1991). In a survey of 45 studies that documented influences of reduced tillage on 51 arthropod pests and their damage to crops, Stinner and House (1990) found that 28% of the species and the damage they caused increased with minimum tillage, 29% showed no significant influence of tillage, and 43% decreased with minimum tillage. These studies also illustrated the importance of monitoring the effect of the soil and crop management practice not only on the ecology of the arthropod pest but also on the ecology of its natural enemies. One of the most frequent observations is the increase that occurs in populations of soil- and litter-inhabiting predatory arthropods, especially ground beetles (Carabidae) and spiders, as tillage is decreased (Stinner and House, 1990; Booij and Noorlander, 1992). The reasons for this are unclear, but could be related to the increase in and diversity of food resources available in conservation tillage systems (Stinner and House, 1990).

Soil organism biodiversity

The possibility of using the biodiversity of soil organism populations as a tool for monitoring soil health needs to be explored (Lee, 1991; Crossley *et al.*, 1992; Paoletti *et al.*, 1992). Significant loss of soil organism biodiversity or species richness has been found to accompany conventional agricultural systems based on intensive cultivation, high fertilizer inputs and crop monoculture (Paoletti *et al.*, 1992). Conversely, the development of ecologically based conservation practices, based on reduced cultivation, reduced fertilizer inputs and crop rotations has seen a return of soil organism biodiversity in many instances (Paoletti *et al.*, 1992).

One of the major difficulties in developing biodiversity as an indicator

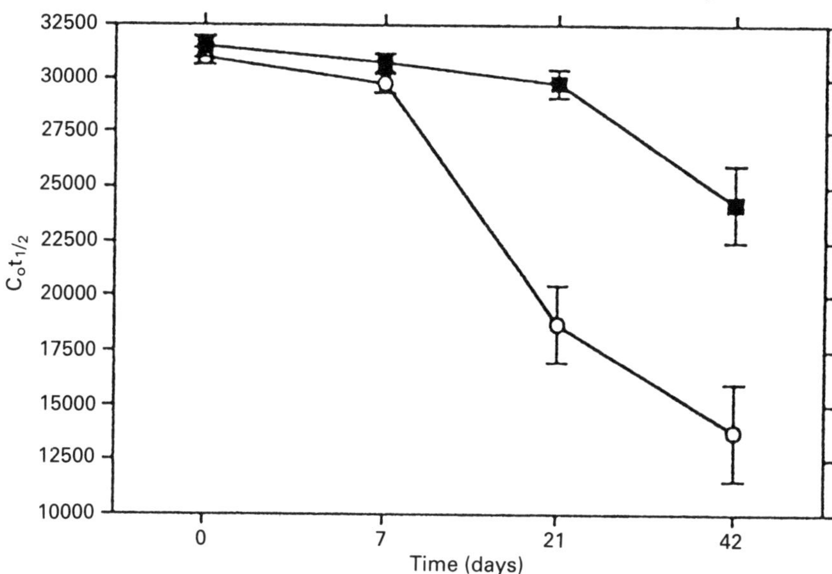

Fig. 20.1. Graph showing changes in genetic diversity (C_0t values for 50% reannealing of DNA in moles $l^{-1}s$) following 2,4,5-T treatment (open circles) and in untreated control (solid squares) soil microbial communities. (From Atlas *et al.*, 1991.)

per se is that only a limited number of soil organisms are well known taxonomically. The resources required for adequate sampling of the soil biota would also make it difficult to monitor more than a small fraction of the total biota. However, it may be possible to monitor the biodiversity of a subset of easily recognized micro-, meso- and macrofauna, and correlate changes in their biodiversity with changes in the biodiversity of other soil organisms. An example of such a subset of species would be the oribatid and prostigmatid mites and possibly various collembolan families, which appear to be very sensitive to changes in the soil environment (Crossley *et al.*, 1992). Earthworms, because of the small number of species usually encountered in agricultural lands and because they not only reflect variations in the soil environment, but also influence soil fertility and plant growth (Lee, 1991), could be another indicator group. However, the relationship between the presence or absence of species of a particular organism group and biodiversity as a whole is not understood, and it may be that other, less conspicuous species, are better indicators.

An alternative approach could be to measure soil organism biodiversity based on statistical analysis of taxonomic groupings of organisms randomly retrieved from soils, or by genetic diversity analysis of DNA extracted from soils (Atlas *et al.*, 1991). Using these approaches, Atlas *et al.* (1991) were able to show that the taxonomic and genetic diversities of microbial communities disturbed by chemical pollutants (2,4,5-T, petroleum) were lower than in undisturbed reference communities. The dominant populations within the disturbed communities had enhanced physiological tolerances and substrate utilization capabilities, indicative of their adaption to the soil conditions associated with the disturbance. The ability of 2,4,5-T to reduce the genetic diversity of a microbial community was shown by a reduction in the diversity of the DNA extracted from the soil (Fig. 20.1).

Organic matter

The organic matter content of the soil changes only slowly with time following a change in land use or management (Ladd, 1989). Changes in organic matter content are therefore difficult to detect until sufficient years have elapsed for the changes to be larger than analytical variability (Rasmussen and Collins, 1991). Despite this, it has been reliably shown that cultivation tends to increase the rate of organic matter loss in soils, principally by accelerated microbial decomposition (Blevins *et al.*, 1984) whereas reduced tillage usually increases organic C and N levels in the top 5–15 cm of the soil (Gupta and Germida, 1988; Rasmussen and Collins, 1992) (Table 20.4). The slower rate of organic matter decomposition in soils subjected to minimum tillage is related to the development of a fungal-dominated microflora in these soils (Hendrix *et al.*, 1986). In view of this, the organic matter content of the soil can be used as a coarse measure of soil health as high levels will correlate with other desirable attributes of soils eg. high soil biota, good soil structure.

Table 20.4. Some general characteristics of a native grassland soil and an adjacent soil subjected to cultivation for 69 years. (Adapted from Gupta and Germida, 1988.)

Soil characteristics	Native (grassland)	Cultivated (69 year)	Difference
pH	6.50	6.20	−0.30
Organic C (mg g^{-1})	38.60	20.40	−18.20
Total N (mg g^{-1})	3.37	2.01	−1.36
Total S (μg g^{-1})	455.00	309.00	−146.00
Total P (μg g^{-1})	871.00	620.00	−251.00
Microbial Biomass-C (μg C g^{-1})	902.00	510.00	−392.00
Microbial Biomass-N (μg NH_4-N g^{-1})	136.18	85.49	−50.69
Microbial Biomass-S (μS g^{-1})	10.65	6.50	−4.15
Microbial C:N:S ratio	85:13:1	78:13:1	Narrowed
Arylsulphatase activity (μg *p* NP g^{-1} h^{-1})	260.00	90.00	−170.00
Acid phosphatase activity (μg p NP g^{-1} h^{-1})	2100.00	1059.00	−1041.00

The quality of the soil organic matter may also have diagnostic value. For example, Arshad *et al.* (1990) demonstrated that the organic matter in undisturbed soil had a higher content of C and N than the organic matter in cultivated soil. The whole soil and the C-enriched fraction from the undisturbed soil were higher in carbohydrates (10% and 18% higher, respectively) and in amino acids (11% and 15% higher, respectively) than the whole soil and C-enriched fractions from the cultivated soil. The undisturbed soil was also higher in aliphatic C (paraffins) and contained less aromatic C than the cultivated soil. Thus adoption of conservation tillage practices may result in both quantitative and qualitative improvements in soil organic matter.

Microbial biomass

Microbial biomass has been found to be a sensitive indicator of management-induced changes in soil biological properties (Carter, 1986; Doran, 1987; Powlson *et al.*, 1987). Changes in biomass are induced by tillage practice, incorporation of crop residues, N fertilization, crop rotation sequence and changing soil moisture regimes (Carter, 1986; Doran, 1987; Powlson *et al.*, 1987; Saffigna *et al.*, 1989). For example, Gupta and Germida (1988) found that microbial biomass C, N, and S were all lower in cultivated

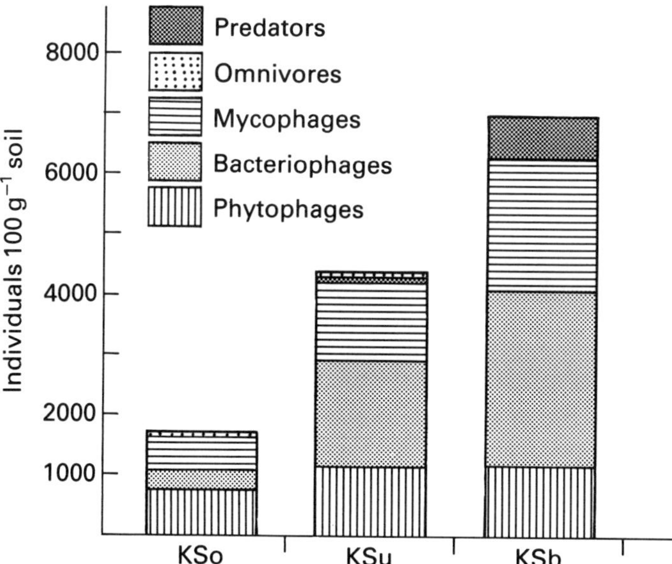

Fig. 20.2. Mean individual numbers of nematodes in control plots (*KS*o), plots treated with sewage sludge (*KS*u), and plots treated with sewage sludge plus heavy metals (*KS*b). (From Weiss and Larink, 1991.)

soil as opposed to undisturbed soil (Table 20.4). However, although it is generally assumed that microbial biomass and activity measurements are correlated with soil organic C because soil biomass depends on the quantity of degradable C sources present in the soil, this relationship is not always found (Insam *et al.*, 1989). Climatic factors, in particular the combined variables of precipitation and evaporation may have a major impact on the relationship. It seems clear, however, that shifts in biomass C measured over relatively short time periods do give an early indication of changes in soil organic matter levels long before they could be detected by other methods (Powlson *et al.*, 1987) (Table 20.5). For this reason, measurement of microbial biomass C in the top 10 cm of the soil has the potential to be used as a sensitive indicator of change in soil biological properties due to soil and crop management differences.

Potentially mineralizable C and N

Potentially mineralizable C and N in the soil may also provide sensitive parameters to assess changes in soil organic matter induced by tillage or other soil management strategies (Campbell *et al.*, 1989; Wood and Edwards, 1992). Thus higher organic C and N concentrations present in the 0–10 cm layer of soils subjected to minimum tillage was paralleled by higher respiration (C mineralization) and potential N mineralization. Below 10 cm, however, heterotrophic microbes in the minimum tillage soils appeared more N-limited than those in ploughed and harrowed soils, suggesting a greater potential for N immobilization with minimum tillage.

Nitrification and denitrification

Nitrification and denitrification are additional microbial processes for which an effect of tillage and crop management has been indicated. As is the case for N mineralization, N immobilization and N leaching, nitrification and denitrification may have a significant effect on the N fertility of the soil (Lynch, 1983). There is some evidence that nitrification proceeds more slowly in minimum-tilled soils (Doran, 1980), but this is not clearly established as a general effect of tillage (Blevins *et al.*, 1986). However, denitrification does proceed at a greater rate in minimum-tilled soils (Doran, 1980; Rice and Smith, 1982), and populations of denitrifiers may be several times greater in minimum-tilled soils than in cultivated soils. Although it has not been possible to relate population counts of individual or groups of denitrifiers to denitrifying activity in soils, there does appear to be a close relationship between microbial biomass C, organic C content and background (unamended) denitrification activity (Drury *et al.*, 1991). Higher organic matter content and higher moisture levels near the soil surface

Table 20.5. Effect of straw incorporation on total soil organic C, soil CO_2-C evolution and microbial biomass C. (From Powlson *et al.*, 1987.)

Site	Straw treatment	Total soil organic C		CO_2-C evolved by unfumigated soil		CO_2-C evolved by $CHCl_3$-fumigated soil ($\mu g\ g^{-1}$ soil)	Biomass C		Biomass C as a proportion of total soil organic C (%)
		%	t ha^{-1}	0–10 days	10–60 days	0–10 days	$\mu g\ g^{-1}$	kg ha^{-1}	
Studsgaard	Burned	2.46	59.3	35	151	60	55	132	0.2
	Incorporated	2.57	61.9	57	232	93	80	192	0.3
	% Increase	5		61*	54*	54*	45*		
Ronhave	Burned	1.16	28.7	35	116	85	110	273	1.0
	Incorporated	1.22	30.2	59	212	127	151	374	1.3
	% Increase	5		68*	83*	50*	37*		

* Indicates burned and incorporated treatments within a site significantly different at 5% level.

favour the growth and activity of denitrifiers on minimum-tilled soils (Blevins *et al.*, 1984).

Soil enzymes

Soil enzymes are the biological catalysts of innumerable reactions in soils. Although some enzymes (e.g. dehydrogenase) are only found in viable cells most soil enzymes can also exist as exoenzymes secreted by microorganisms or as enzymes originating from microbial debris and plant residues that are stabilized in complexes of clay minerals and humic colloids (Ladd, 1978; Dick, 1992). Since it is difficult to extract enzymes from soils, enzymes are studied indirectly by measuring their activity via assays done *in vitro* under controlled conditions (temperature, buffers, excess substrate, etc.). It is difficult to relate activities *in vitro* to those occurring *in situ*. None the less, soil enzyme activities provide insight into biochemical processes in soils and can therefore be useful as an indicator of biological activity (Frankenberger and Dick, 1983).

Soil urease, amidase, arylsulphatase, acid phosphatase, phosphatase, protease, dehydrogenase, and invertase have all been shown to be significantly higher in minimum-tilled soils than in cultivated soils (Klein and Koths, 1980; Doran, 1980; Dick, 1984; Gupta and Germida, 1988) (Table 20.4). Similarly, soil enzyme activities were higher in soils that had been cropped with rotations than soils that were either continuously monocultured or had more limited crop rotations (Dick, 1984). In another study, Bolton *et al.* (1985) showed that addition of a green manure crop to a wheat-based rotation over a 30-year period caused a significant increase in urease, phosphatase, and dehydrogenase activities, and soil biomass. An important aspect of the latter study was the observation that soil enzyme activities and soil biomass measurements detected changes between the management systems whereas direct microbial counts did not. Thus it would appear that in addition to microbial biomass, the measurement of certain soil enzyme activities could be used as bioindicators to detect changes in soil biological properties.

Plant growth

The ultimate measure of soil health will be the ability of the soil to support and sustain plant growth. Putting climatic factors to one side, poor plant growth or patchy growth invariably indicates the presence of some soil-based abiotic or biotic constraint. Plant growth has been used in many instances as an indicator or 'bioassay' for the presence of low levels of herbicide residues in soil (Banin and Kafkafi, 1980) and for assessment of the level of disease and beneficial soil organisms in soils (Rovira *et al.*, 1990).

Plant growth has also been used as an indicator of mineral deficiencies or toxicities in soils and for the detection of heavy metal toxicities. For example, the level of boron in barley grain was a sensitive indicator of boron-deficient soils in South Australia (Nable *et al.*, 1990). The development of an ecologically based sustainable agriculture is geared towards removal of these physical, chemical and biological constraints through adoption of a range of soil management practices, including conservation tillage, crop rotation, residue management, improved drainage, water conservation, and use of organic fertilizers (Stewart *et al.*, 1991).

Bioindicators of Soil Pollution

Currently there are two main sources of pollution of agricultural soils. Both are a consequence of our efforts to conserve soils and make better use of natural resources. First, adoption of conservation management practices in agriculture has increased the need for farmers to use herbicides to control weeds and other unwanted plant growth as an alternative to control via cultivation. Thus, in addition to the continued use of pesticides to control pests and diseases, some soil problems may be associated with the increased use of herbicides. Second, due to the growing concern over the disposal of sewage sludge in the oceans and the high cost of incineration, land application is becoming more common. This can be beneficial because sludge contains plant nutrients (especially N and P) and organic matter which can be of agronomic benefit especially in light-textured soils. However, sludges often contain appreciable amounts of heavy metals, eg. Zn, Cu, Ni, Cd and Pb, which can persist in soil for tens of thousands of years (McGrath, 1986).

Pesticide residues and heavy metals can have a significant effect on the composition and the activity of the soil biota. Consequently, there have been several reports of the potential of using soil organisms as bioindicators of toxicants in the soil environment (Domsch *et al.*, 1983; Paoletti *et al.*, 1990; Weiss and Larink, 1991; Koehler, 1992). In a study of the influence of sewage sludge and heavy metals on nematode populations in an arable soil, Weiss and Larink (1991) found that total nematode numbers were highest on the treated sites (Fig. 20.2). The increase in nematode numbers in the sludge-treated sites was due to a much higher abundance of bacterial-feeding (e.g. *Rhabditidae*) and fungal-feeding (e.g. *Apelenchoides*) nematodes and to the presence of predatory nematodes not found in the control sites (Fig. 20.2). In contrast the abundance of omnivorous nematodes declined with the addition of sewage sludge and were absent in the presence of both sludge and heavy metals. Of interest here was the finding that one omnivorous nematode, the genus *Ecumenicus,* was found in equal abundance in all sites except where heavy metals were present, suggesting that this genus could be used as a sensitive bioindicator for the presence of heavy metals in the soil.

An alternative strategy that can be used to detect the presence of heavy metals or pesticide residues in soil is to look for the development of microbial populations (bacteria and fungi) that are resistant to the contaminant (Summers, 1985). This method was used successfully by Radford *et al.* (1981) to demonstrate the presence of mercury in a mercury-contaminated soil. The number of mercury-resistant bacteria in the thus affected soil was significantly higher than the number in a non-contaminated soil. The detection of plasmid-mediated resistances to a wide range of heavy metals (e.g. arsenate, chromate, cadmium, cobalt, copper, lead, nickel) in bacteria (Silver and Misra, 1988), suggests that this kind of bioassay could be further developed.

Soils contaminated with heavy metals (produced by long-term applications of contaminated sewage-sludge) have been found to contain reduced levels of microbial biomass compared to soils receiving equivalent levels of nutrient from farmyard manure (Brookes and McGrath, 1984; Chander and Brookes, 1991). This appears to be due to the reduced capacity of the biomass in the metal-contaminated soil to utilize substrates for biomass synthesis. Also no relationship was found between biomass C and total soil organic C in the contaminated soils (Chander and Brookes, 1991). In contaminated soils therefore, the ratio of biomass C to soil organic C may provide a useful indicator of changing soil conditions due to chemical pollution long before such changes can be detected in soil organic matter as a whole.

Role of Bioindicators in Sustainable Land Use

In the foregoing discussion the impact of current agricultural practices on the soil biota and a range of biological processes carried out by the soil biota have been considered. This has shown which soil organisms or groups of organisms and which biological processes are sensitive to changes in soil condition and could therefore be useful as bioindicators of soil health.

Bioindicators of soil productivity and sustainability include:

- Healthy plant growth
- Low levels of soil-borne disease
- High populations of beneficial soil organisms (e.g. earthworms, mycorrhizal fungi, biocontrol bacteria, heterotrophic protozoa and nematodes, predatory arthropods)
- High soil organism biodiversity
- High organic matter
- Active microbial biomass

- High levels of potentially mineralizable C and N
- High levels of hydrolase and oxidoreductase enzymes

Bioindicators of soil pollution include:

- Low microbial biomass
- Presence or absence of some soil organisms (e.g. nematodes, mites, earthworms)
- Presence of microbial populations resistant to the toxicant
- Plant growth and presence of toxicant in plant tissues

The bioindicators of soil health categorized above as indicators of soil productivity and sustainability and indicators of soil pollution, could be developed in conjunction with other indicators of agricultural sustainability as a monitoring system to guide future sustainable land use. With the possible exception of soil organic matter, none of the bioindicators are sufficiently robust to have general applicability. Rather, individual bioindicators will need to be evaluated for their suitability for different agroecosystems, soil types and climates. In addition, their usefulness as part of a monitoring system to guide sustainable use of soils will be just as important as their usefulness as part of a monitoring system for the early detection of signs of soil and land degradation.

Research Needs

Further research is needed to refine in detail the most relevant and sensitive bioindicators of soil health and sustainability, and to determine if they can also be used as indicators of soil resilience. Some recommendations are:

1. Further studies of the impact of conservation soil management practices on soil organism populations, with the development of technology for the rapid measurement of populations of 'keystone' species.
2. Definition of the importance of soil organism biodiversity in sustainable farming systems, with the development of new methods (e.g. DNA methods) for rapid measurement of biodiversity.
3. Further refinement of measurements of soil biomass and assessment of how this can be used more effectively and conveniently as a bioindicator.
4. Further development of soil enzyme assays (bioassays) as bioindicators.
5. Further development of soil organisms as indicators of soil pollution.

References

Alexander, M. (1977) *Introduction to Soil Microbiology*, 2nd edn. Academic Press, New York.

Arshad, M.A., Schnitzer, M., Anders, D.A. and Ripmeester, J.A. (1990) Effects of till vs no-till on the quality of soil organic matter. *Soil Biology and Biochemistry* 22, 595–599.

Atlas, R.M. and Bartha, R. (1987) *Microbial Ecology: Fundamentals and Applications.* Cummings, Menio Park, Carlifornia.

Atlas, R.M., Horowitz, A., Krichevsky, M. and Bej, A.K. (1991) Response of microbial populations to environmental disturbance. *Microbial Ecology* 22, 249–256.

Banin, A. and Kafkafi, U. (1980) *Agrochemicals in Soils.* Pergamon Press, Oxford, 448 pp.

Blevins, R.L., Smith, M.S. and Thomas, G.W. (1984) Changes in soil properties under no-tillage. In: Phillips, R.E. and Phillips, S.H. (eds) *No-Tillage Agriculture: Principles and Practices.* Van Nostrand Reinhold, New York.

Bolton, H., Elliott, L.F., Papendick, R.I. and Bezdicek, D.F. (1985) Soil microbial biomass and selected soil enzyme activities: Effect of fertilization and cropping practices. *Soil Biology and Biochemistry* 17, 297–302.

Booij, C.J.H. and Noorlander, J. (1992) Farming systems and insect predators. *Agriculture, Ecosystems and Environment* 40, 125–135.

Brookes, P.C. and McGrath, S.P. (1984) Effects of metal toxicity on the size of the microbial biomass. *Journal of Soil Science* 35, 341–346.

Campbell, C.A. (1989) Bridging the gap between sustainable and conventional agriculture. *Australian Journal of Soil and Water Conservation* 2 (2), 43–46.

Campbell, C.A., Biederbeck, V.O., Schnitzer, M., Selles, F. and Zentner, R.P. (1989) Effect of 6 years of zero tillage and N fertilizer management on changes in soil quality of an Orthic Brown Chernozem in Southwestern Saskatchewan. *Soil and Tillage Research* 14, 39–52.

Carter, M.R. (1986) Microbial biomass as an index for tillage-induced changes in soil biological properties. *Soil and Tillage Research* 7, 29–40.

Chakraborty, S., Old, K.M. and Warcup, J.H. (1983) Amoebae from a take-all supressive soil which feed on *Gaeumannomyces graminis* var. *tritici* and other soil fungi. *Soil Biology and Biochemistry* 15, 17–24.

Chander, K. and Brookes, P.C. (1991) Microbial biomass dynamics during the decomposition of glucose and maize in metal-contaminated and non-contaminated soils. *Soil Biology and Biochemistry* 23, 917–925.

Cooke, R.C. and Rayner, A.D.M. (1984) *Ecology of Saprophytic Fungi.* Longman, New York.

Crossley, Jr, D.A., Mueller, B.R. and Perdue, J.C. (1992) Biodiversity of microarthropods in agricultural soils: relations to processes. *Agriculture, Ecosystems and Environment* 40, 37–46.

Dick, R.P. (1992) A review: long-term effects of agricultural systems on soil biochemical and microbial parameters. *Agriculture, Ecosystems and Environment* 40, 25–36.

Dick, W.A. (1984) Influence of long-term tillage and crop rotation combinations on soil enzyme activities. *Journal of the Soil Science Society of America* 48, 569–574.

Dindal, D. (1989) *Soil Biology Guide.* John Wiley, New York, 1349 pp.

Domsch, K.H., Jagnow, G. and Anderson, T.H. (1983) An ecological concept for the measurement of side-effects of agrochemicals on soil microorganisms. *Residue Review* 86, 65–105.

Doran, J.W. (1980) Soil microbial and biochemical changes associated with reduced tillage. *Journal of the Soil Science Society of America* 44, 765–771.

Doran, J.W. (1987) Microbial biomass and mineralizable nitrogen distributions in no-tillage and plowed soils. *Biology and Fertility of Soils* 5, 68–75.

Drury, C.F., McKenny, D.J. and Findlay, W.I. (1991) Relationship between denitrification, microbial biomass and indigenous soil properties. *Soil Biology and Biochemistry* 23, 751–755.

Edwards, C.A. and Lofty, J.R. (1977) *Biology of Earthworms.* Chapman and Hall, New York, NY, 333 pp.

El Titi, A. and Ipach, U. (1989) Soil fauna in sustainable agriculture: results of an integrated farming system at Lautenbach, F.R.G. *Agriculture, Ecosystems and Environment* 27, 561–572.

Foster, R.C. (1988) Microenvironments of soil microorganisms. *Biology and Fertility of Soils* 6, 189–203.

Frankenberger, W.T. and Dick, W.A. (1983) Relationships between enzyme activities and microbial growth and activity indices in soil. *Journal of the Soil Science Society of America* 47, 945–951.

Gupta, V.V.S.R. and Germida, J.J. (1988) Distribution of microbial biomass and its activity in different soil aggregate size classes as affected by cultivation. *Soil Biology and Biochemistry* 20, 777–786.

Heimbach, F. (1985) Comparison of laboratory methods, using *Eisenia foetida* and *Lumbricus terrestris,* for the assessment of hazard of chemicals to earthworms. *Journal of Plant Disease Protection* 92, 186–193.

Hendrix, P.F., Parmelee, R.W., Crossley, D.A., Coleman, D.C., Odum, E.P. and Groffman, P.M. (1986) Detritus food webs in conventional and no-tillage agro-ecosystems. *Bioscience* 36, 374–380.

Holland, E.A. and Coleman, D.C. (1987) Litter placement effects on microbial and organic matter dynamics in an agroecosystem. *Ecology* 68, 425–433.

Ingham, R.E., Trofymow, J.A., Ingham, E.R. and Coleman, D.C. (1985) Interactions of bacteria, fungi and their nematode grazers: effects on nutrient cycling and plant growth. *Ecology Monographs* 55, 119–140.

Insam, H., Parkinson, D. and Domsch, K.H. (1989) Influence of macroclimate on soil microbial biomass. *Soil Biology and Biochemistry* 21, 211–221.

Klein, T.M. and Koths, J.S. (1980) Urease, protease, and phosphatase in soil continuously cropped to corn by conventional or no-tillage methods. *Soil Biology and Biochemistry* 12, 293–294.

Kloepper, J.W., Zablotowicz, R.M., Tipping, E.M. and Lifshitz, R. (1991) Plant growth promotion mediated by bacterial rhizosphere colonizers. In: Keister, D.L. and Cregan, P.B. (eds) *The Rhizosphere and Plant Growth.* Kluwer Academic, Dordrecht, The Netherlands, pp. 315–326.

Koehler, H.H. (1992) The use of soil mesofauna for the judgement of chemical impact on ecosystems. *Agriculture, Ecosystems and Environment* 40, 193–205.

Ladd, J.N. (1978) Origin and range of enzymes in soil. In: Burns, R.G. (ed.) *Soil Enzymes.* Academic Press, London, pp. 51–96.

Ladd, J.N. (1989) The role of the soil microflora in the degradation of organic matter. In: Hattori, T., Ishida, Y., Maruyama, Y., Morita, R.Y. and Uchida, A. (eds) *Recent Advances in Microbial Ecology.* Japan Scientific Societies Press, pp. 169–174.

Lal, R. (1988) Effects of macrofauna on soil properties in tropical ecosystems. *Agriculture, Ecosystems and Environment* 24, 101–116.

Lal, R. (1991) Soil conservation and biodiversity. In: Hawksworth, D.L. (ed.) *The Biodiversity of Microorganisms and Invertebrates: Its Role in Sustainable Agriculture.* CAB International, Wallingford, pp. 89–103.

Lee, K.E. (1985) *Earthworms: their Ecology and Relationships with Soils and Landuse.* Academic Press, Sydney, 411 pp.

Lee, K.E. (1991) The diversity of soil organisms. In: Hawksworth, D.L. (ed.) *The Biodiversity of Microorganisms and Invertebrates: Its Role in Sustainable Agriculture.* CAB International, Wallingford, pp. 73–87.

Lobry de Bruyn, L.A. and Conacher, A.J. (1990) The role of termites and ants in soil modification: a review. *Soil Biology and Biochemistry* 28, 55–93.

Loring, S.J., Snider, R.J. and Robertson, L.S. (1981) The effects of three tillage practices on Collembola and Acarina populations. *Pedobiologia* 22, 172–184.

Lynch, J.M. (1983) *Soil Biotechnology: Microbiological factors in crop productivity.* Blackwell Scientific Publications, Oxford, London, 191 pp.

McGrath, S.P. (1986) Long-term studies of metal transfers following application of sewage sludge. In: Coughtrey, P.J., Martin, M.H. and Unsworth, M.H. (eds) *Pollutant Transport and Fate in Ecosystems,* British Ecological Society Special Publication No. 6. Blackwell Scientific, Oxford, pp. 301–317.

Nable, R.O., Lance, R.C.M. and Cartwright, B. (1990) Uptake of boron and silicon by barley genotypes with differing susceptibilities to boron toxicity. *Annals of Botany* 66, 83–90.

Neate, S.M. (1987) Plant debris in soil as a source of inoculum of *Rhizoctonia solani* in wheat. *Transactions of the British Mycological Society* 88, 157–162.

Nicholas, W.I. (1984) *The Biology of Free-Living Nematodes* 2nd edn. Clarendon Press, Oxford, 235 pp.

Paoletti, M.G., Favretto, M.R., Stinner, B.R., Purrington, F.F. and Bater, J.E. (1991) Invertebrates as bioindicators of soil use. *Agriculture, Ecosystems and Environment* 31, 341–362.

Paoletti, M.G., Pimentel, D., Stinner, B.R. and Stinner, D. (1992) Agroecosystem biodiversity: matching production and conservation biology. *Agriculture, Ecosystems and Environment* 40, 3–23.

Perdue, J.C. and Crossley, Jr, D.A. (1990) Vertical distribution of soil mites (Acari) in conventional and no-tillage agricultural systems. *Biology and Fertility of Soils* 9, 135–138.

Powlson, D.S., Brookes, P.C. and Christensen, B.T. (1987) *Soil Biology and Biochemistry* 19, 159–164.

Radford, A.J., Oliver, J., Kelley, W.J. and Reanney, D.C. (1981) Translocatable resistance to mercuric and phenylmercuric ions in soil bacteria. *Journal of Bacteriology* 147, 1110–1112.

Rasmussen, P.E. and Collins, H.P. (1991) Long-term impacts of tillage, fertilizer, and crop residue on soil organic matter in temperate semiarid regions. *Advances in Agronomy* 45, 93–134.

Rice, C.W. and Smith, M.S. (1982) Denitrification in plowed and no-tilled soils. *Journal of the Soil Science Society of America* 46, 1168–1173.

Roger, P.A., Heong, K.L. and Teng, P.S. (1991) Biodiversity and sustainability of wetland rice production: role and potential of microorganisms and invertebrates. In: Hawksworth, D.L. (ed.) *The Biodiversity of Microorganisms and Invertebrates: Its Role in Sustainable Agriculture.* CAB International, Wallingford pp. 117–136.

Rovira, A.D., Smettem, K.R.J. and Lee, K.E. (1987) Effect of rotation and conservation tillage on earthworms in a red-brown earth under wheat. *Australian Journal of Agricultural Research* 38, 829–834.

Rovira, A.D., Elliott, L.F. and Cook, R.J. (1990) The impact of cropping systems on rhizosphere organisms affecting plant health. In: Lynch, J.M. (ed.) *The Rhizosphere.* John Wiley, Chichester pp. 389–436.

Saffigna, P.G., Powlson, D.S., Brookes, P.C. and Thomas, G.A. (1989) Influence of sorghum residues and tillage on soil organic matter and soil microbial biomass in an Australian vertisol. *Soil Biology and Biochemistry* 21, 759–765.

Schippers, B., Bakker, A.W. and Bakker, P.A.H.M. (1987) Interactions of deleterious and beneficial rhizosphere microorganisms and the effect of cropping practices. *Annual Review of Phytopathology* 25, 339–358.

Silver, S. and Misra, T.K. (1988) Plasmid-mediated heavy metal resistances. *Annual Review of Microbiology* 42, 717–743.

Stewart, B.A., Lal, R. and El-Swaify, S.A. (1991) Sustaining the resource base of an expanding world agriculture. In: Lal, R. and Pierce, F.J. (eds) *Soil Management for Sustainability.* Soil and Water Conservation Society, Iowa, pp. 125–144.

Stinner, B.R. and Crossley, D.A. (1982) Nematodes in no-tillage agroecosystems. In: Freckman, D.W. (ed.) *Nematodes in Soil Ecosystems.* University of Texas Press, Austin, pp. 14–28.

Stinner, B.R. and House, G.J. (1990) Arthropods and other invertebrates in conservation-tillage agriculture. *Annual Review of Entomology* 35, 299–318.

Summers, A.O. (1985) Bacterial resistance to toxic elements. *Trends in Biotechnology* 3, 122–125.

Tisdall, J.M. (1991) Fungal hyphae and structural stability of soil. *Australian Journal of Soil Research* 29, 729–743.

Villani, M.G. and Wright, R.J. (1990) Environmental influences on soil macroarthropod behaviour in agricultural systems. *Annual Review of Entomology* 35, 249–269.

Weiss, B. and Larink, O. (1991) Influence of sewage sludge and heavy metals on nematodes in an arable soil. *Biology and Fertility of Soils* 12, 5–9.

Wood, C.W. and Edwards, J.H. (1992) Agroecosystem management effects on soil carbon and nitrogen. *Agriculture, Ecosystems and Environment* 39, 123–138.

Yeates, G.W. and Coleman, D.C. (1982). Role of nematodes in decomposition. In: Freckman, D.W. (ed.) *Nematodes in Soil Ecosystems.* University of Texas Press, Austin, pp 55–80.

Chapter 21
Biodiversity and Soil Resilience

L.F. Elliott[1] and J.M. Lynch[2]

[1]*USDA ARS, National Forage Seed Production Research Center, Oregon State University, 3450 SW. Campus Way, Corvallis, Oregon, 97331–7102:* [2]*Microbiology and Crop Protection Department, Horticulture Research International, Littlehampton, West Sussex, BN17 6LP, UK. Present address: School of Biological Sciences, University of Surrey, Guildford, Surrey, GU2 5XH, UK*

Introduction

Energy is fixed into soil mainly from photosynthesis and then partitioned among various components of the biota. A crucial question is whether the ecosystem benefits from the energy utilization being accomplished by a diverse or a more restricted community, especially in terms of the sustainability of agroecosystems.

In the proceedings of WEFSA I, the need for conceptual clarity in sustainable agriculture was identified (Hawksworth, 1991). The following was the first statement of findings:

> The needs of human populations for food, fuel and fibre have historically been supported, directly or indirectly, by genetic diversity among the microorganisms and invertebrates. These organisms perform functions which prime and fuel the metabolism of soils, plants and animals. The development of sustainable agroecosystems and the enhancement of agricultural productivity will depend increasingly on the maintenance of such diversity for:
>
> The improvement of soil structure and fertility, through decomposition of organic material added to the soil, and the detoxification of pesticides and other pollutants ...

Somewhat ironically biodiversity was not defined and we would choose the following definition:

Biodiversity: richness of life as indicated by the variety of biota and interrelated biochemical processes in a habitat.

From *Soil Resilience and Sustainable Land Use*
(eds D.J. Greenland and I. Szabolcs). CAB INTERNATIONAL Wallingford

Various mathematical indices have been used to measure population biodiversity, but none should be regarded as absolute. The definition is complicated by the species concept, particularly as it is now recognized that genes can be fluid among microorganisms. Eventually we may wish to define biodiversity as a series of interrelated processes rather than in terms of populations. It should also be noted that no correlation has ever been made between biodiversity of populations and stability of an ecosystem. Indeed there is ample evidence that small interactive communities are very stable.

Soil resilience: we are happy with the definition in the Recommendations presented in this volume: the ability of the soil to recover after disturbance.

Soil quality has been defined as the capacity of the soil to produce healthy and nutritious crops, resist erosion, and reduce the impact of environmental stresses on plants (Papendick and Parr, 1992). It seems likely that soil quality and soil resilience may be related. Here we will consider approaches for assessing biodiversity and relating it to processes or approaches which maintain or improve soil quality.

Soil degradation is the antithesis of soil resilience and quality, and is defined as the loss of a soil's capacity to produce crops, often a consequence of soil erosion. It is disturbing to note that at present good estimates are not available of the extent and degree of soil degradation but it is reassuring that there is an effort under way to determine this (Oldeman, Chapter 7). It is estimated that 2 billion hectares of once-productive land have been rendered unproductive since settled agriculture began. Estimates of land subject to desertification are equally alarming. However, these statistics may be inaccurate and efforts are underway to improve these estimates. Regardless, these figures show the problem is large and threatening. We also do not have firm knowledge of sustainable agricultural practices which must be linked to the soil environment, climate, and food quality. Other evidence of soil degradation includes decreased nutrient content and recycling capacity, decreased size of the soil microbiota and fauna, decreased soil organic matter content and quality, and poor physical qualities such as crusting, reduced water-holding capacity, poor aggregation, and increased compaction. However, soils do have the resilience to tolerate continuous use for extended periods without total loss of productivity as evidenced by some soils in Asia and the Near East (Lal, Chapter 4).

Soil resilience can be restored and biodiversity protected by several methods such as recycling agricultural and organic wastes, use of forests and cover crops for enhancing soil organic matter content, content controlling soil erosion, reducing tillage, increasing crop rotations, and in irrigated areas developing better water management practices. Improved land use systems must be developed to preserve and enhance soil productivity to maintain environmental quality, ecological stability, and soil resilience. The relationship of these qualities to biodiversity should then be investigated.

A Relationship of Biodiversity and Soil Resilience

The relationship of biodiversity to the resilience process can appear to be different. For example if a centimetre of topsoil is lost due to erosion, replacement occurs as a function of the soil formation process which is more dependent on weathering than on biodiversity. However, biodiversity will affect the rate of soil loss by erosion. Management of those biodiversity factors that increase a soil's resistance to erosion will decrease the rate of erosion. Biological factors or processes that increase soil stability have been identified (Lynch and Bragg, 1985).

In another example indicators of biological activity in soil have been identified. Soil from an organic farm had a higher dehydrogenase level than soil from a conventional farm (Bolton *et al.*, 1985). The question is, if the soil on the conventional farm is managed similarly to that on the organic farm, how long will it be before the soil dehydrogenase levels are equal? This would provide a biochemical index for managing to increase soil resilience. It can be assumed that the increased dehydrogenase activity would correlate with increased resilience because the soil also had a higher soil organic matter and microbial biomass content. Can a series of measurements such as dehydrogenase activity, resistance to erosion, etc., be integrated so as to predict resilience? More importantly we must be able to determine how management and associated biodiversity will improve resilience.

Soil biodiversity might also be defined as a specific grouping of macro- and microbiota and their associated functions to sustain selected soil properties and processes that sustain plant production, reduce erosion, resist environmental perturbations, or some other important function under given soil and climatic conditions. A goal for soil biodiversity should not only mean maximizing the number of species in an ecosystem. More likely it means the ability to retain the current macro- and microflora and to emphasize certain species and processes to accomplish specific objectives. Obviously the broad main objective is to manage biodiversity for maximum soil resilience. Initially it may mean emphasizing more easily understandable components of resilience such as resistance to erosion.

Responses to Management

Soils of natural ecosystems generally contain more organic matter and nutrients and have better physical characteristics than the same soils which are cropped. The native soil organic matter content of a soil is largely dictated by rainfall and temperature (Stevenson, 1982). The organic matter content of a soil correlates with its potential productivity, tilth, and fertility and likely resilience (Smith and Elliott, 1990) and is dependent on the climatic zone. When these soils are disturbed by tillage, erosion, or a decrease in crop rotation, soil quality decreases. However, with conservation tillage practices, serious degradation does not have to occur. Studies of native prairie cropped by no-till and conventional-till seeding methods in western Nebraska showed that the particulate organic matter fraction in the no-till seeded soil was larger than that fraction in soil from the conventional tillage treatments (Cambardella and Elliott, 1992). Much earlier, in western Nigeria, Lal (1974) showed no-till conserved soil N and organic matter when compared with conventional tillage.

It is generally conceded that reduced tillage protects soil organic matter, nutrients, soil microbial biomass and greatly decreases soil erosion (Doran, 1980a, b; Papendick, Chapter 14). This is especially true in the surface horizons (Lynch and Panting, 1980). It has been shown that a legume, grass, a legume–grass mix used as a cover crop, or green manure in the rotation reduced erosion, increased soil enzymes, and increased the microbial biomass (Bolton *et al.*, 1985; Elliott *et al.*, 1987; Reganold *et al.*, 1987). Increased crop rotations also increased the size of the microbial biomass (Bolton *et al.*, 1985) and decreased the incidence of soil-borne diseases (Rovira *et al.*, 1990). Anderson (Chapter 17) similarly showed management practices were related to the size of the soil microbial biomass and Lavelle *et al.* (Chapter 18) showed similar relationships for the soil fauna.

There are questions regarding the effect of management practices on biodiversity and on soil resilience. Diversity of the soil community might be reduced by land management practices such as intensive land clearing and tillage, monoculture, fertilizer and pesticide use. Alternative practices such as minimum tillage, mixed cropping, rotations, and agroforestry systems improve soil resilience and could preserve soil biodiversity when compared with the previous detrimental practices. Decreases in soil resilience could result from decreases in the number or types of species involved in a beneficial process and in this case that level of resilience may not be restorable. Hassink *et al.* (1991a, b) showed the rhizosphere community was affected most by management practices. Approaches are being developed for studying root–microbial relationships (Lam *et al.*, 1990). They prepared insertion mutants from root-colonizing pseudomonads to determine effects on colonizing abilities. These and related approaches provide the ability to determine how we might be able beneficially to affect these relationships. Definition of these interactions will also help to define the interaction of biodiversity and soil resilience. Many studies have shown the potential for improving the structure of cultivated soils through manipulation of the microbiota (Lynch and Bragg, 1985; Elliott and Lynch, 1984; Elliott and

Wildung, 1992). Biotechnology such as the use of DNA probes (Lynch, Chapter 5) will most certainly play a role in defining the role of biodiversity in soil resilience.

There is a perception that fertilizers and pesticides harm soil resilience. Earthworms are especially susceptible in this regard. Generally the damage occurs when these materials are used as a salvage operation or when they are used to allow monoculture or very limited rotation farming. Studies have not been conducted that separate the beneficial and harmful effects of fertilizers and pesticides on soil quality and the soil community. The value of crop rotations, and of green manure in the rotation, on soil biological properties and erodibility was shown by Bolton *et al.* (1985) and Reganold *et al.* (1987). The soil that had an increased crop rotation including green manure had significantly greater microbial biomass, selected soil enzymes, and less erosion. These differences were not likely to have been caused by fertilizer and pesticide use. More careful studies of the effect of degree of fertilizer and pesticide use on the soil biodiversity must be conducted.

Soil Degradation

As discussed by Oldeman (Chapter 7) the International Soil Reference and Information Centre prepared and published in October 1990 the world map of the status of human-induced soil degradation. It is now possible to make some global and continental interpretations of the extent of soil degradation, severity, and causative factors. Two categories of human-induced soil degradation processes are recognized: (i) wind and water erosion; and (ii) chemical and physical degradation. The processes are then subdivided into removal of topsoil and terrain deformation, loss of nutrients, salinization, acidification, pollution, compaction, sealing and crusting, waterlogging, and subsidence of organic soils. Four degrees of soil degradation are recognized. The causative factors were listed as deforestation, overgrazing, agricultural activities, over-exploitation of the vegetation for domestic use, and industrial and bioindustrial activities. The map provides an overview of the consequences of land exploitation and assists policy and decision-makers in establishing strategies for remediation. The role of the soil biota has not yet been addressed. This is a pressing need and we must respond with effective approaches for assessment.

Factors Affecting Soil Biodiversity

Soil organic matter levels have declined or are declining mainly because of intensive tillage practices that stimulate the microbiological decomposition of crop residues and residual soil organic matter (i.e. humus) and accelerate erosion. Erosion is a severe problem in many developing countries for other reasons. Use of monocultures, lack of crop residues for soil protection, and erosion-promoting management have also lowered biodiversity and soil resilience. With a specific tillage management system soil organic matter levels will usually reach a steady state. Soil organic matter loss is a function of degree of tillage and crop residue management (Smith *et al.*, 1946; Collins *et al.*, 1992). In many situations, previous crop residues are burned to facilitate cropping which is very detrimental to soil organic matter content. Loss of soil organic matter results in poorer soil structure, decreased microbial biomass, reduced infiltration rates, increased crusting, decreased water-holding capacity, increased resistance to root penetration, decreased nutrient availability, and accelerated soil erosion by both wind and water.

Reduced tillage systems that will maintain and improve soil quality components can be adapted to most cropping systems. With reduced tillage, soil organic matter content and microbial biomass levels will increase. Soil physical properties and soil resilience will also improve. For severely eroded soils, reduced tillage may not be completely adequate for restoring soil productivity and beneficial soil properties. In these cases, some type of soil fracturing and deep placement of phosphoric fertilizer may be needed. Additionally, the introduction of a drought-tolerant legume in the crop rotation system can be beneficial. Legumes can provide significant quantities of N to other crops in the system thereby increasing their yield potential with a minimum input of N fertilizer. Legumes will also help to maintain soil organic matter levels and improve or restore soil resilience. These cropping practices benefit the size and activity of the microbial biomass which will be reflected in biodiversity. Thus benefits are not relegated to only soil physical properties and nutrient content. Some crop rotations may promote better beneficial microbial–rhizosphere relations than others. Crop and residue management affects the soil microbial biomass and undoubtedly biodiversity. Microbial carbon represented 4.3, 2.8, and 2.2% of the total soil carbon under grass pasture, annual cropping, and wheat-fallow, respectively (Collins *et al.*, 1992). Additionally they found that annual cropping significantly reduced declines in soil organic matter and soil microbial biomass. This study illustrated methods for improving soil biodiversity.

That the components of biodiversity in the soil are changed by management has been established by other studies. Stroo *et al.* (1989) showed that the macrofauna contributes less than 5% of the CO_2 evolved from winter wheat residue during decomposition. Lavelle *et al.* (Chapter 18) showed the macrofauna can be responsible for as much as 30% of the CO_2 respired from organic matter turnover. These large differences reflect differences in management and climate.

Earlier it was mentioned that biodiversity is affected by tillage systems.

Doran (1980a, b) and Lynch and Panting (1980) surveyed tillage systems and found that microbial biomass, soil organic matter content, and biochemical activity increased near the soil surface when soil planted without tillage was compared with soil conventionally tilled and planted. Soil microbial biomass and selected enzyme levels also responded similarly to tillage in the subarctic (Cochran *et al.*, 1989). When the soils from an organic farm were compared with those from a conventional farm, it was shown that the organic farm soil generally had higher microbial biomass, greater enzyme activity, higher soil organic matter content, and less erosion. The conventional farm soil had lost almost 16 cm more soil to erosion (Bolton *et al.*, 1985; Reganold *et al.*, 1987). The differences in the soils were probably caused by the greater crop rotation and the use of green manure for the organic farm. After two years, Brussaard *et al.* (1990) compared an integrated farming system with a conventional system and found total biomass of soil organisms averaged 907 kg C ha^{-1} and 690 kg C ha^{-1}, respectively. They found bacteria constituted more than 90%, fungi approximately 5% and protozoa less than 2% of the total biomass. However, nutrient use efficiency in the integrated system was not improved compared to the conventional system (Brussaard, Chapter 19). Hassink *et al.* (1991a) found that the size and activity of the microbial biomass from the soil of the reduced-input field was greater than that for the conventional. In another report Hassink *et al.* (1991b) showed that microbial and fungal populations in the two soils changed during the study with the greatest changes occurring in the rhizosphere. At the end of the growing season the fungal counts in the root zones were significantly different, and the ability of the fungi from the conventional field to utilize specific substrates had decreased when compared to the reduced-input field. The reasons for these changes are not readily obvious but must be related to cropping sequences and schemes. They are most likely related to differences in kind and amounts of substrate inputs. Undoubtedly further studies are needed to establish the role of biodiversity in soil resilience.

Components of biodiversity which are targets to reduce are the plant pathogens and pests. The role of crop rotation in soil-borne disease control is well established. Normally a three-year rotation will alleviate most soil-borne disease problems (Rovira *et al.*, 1990). The rotations positively affect biodiversity so that soil-borne diseases are kept in check. Crop diseases will have a large effect on soil resilience. Decreased disease incidence increases crop productivity so that more photosynthate is produced and rhizodeposition is increased. Crop residue return is increased as disease incidence decreases because plant growth is increased. This will result in increases in the soil macro- and microfauna when compared with the limited rotation soil-borne diseased situation.

A case can be made that plant root health reflects the effects of biodiversity and that types of root colonists or root response may be the most sensitive indicator of biodiversity, at least as it relates to plant productivity and environmental response. This area has been largely overlooked in this regard. Root colonization by beneficial organisms is increased by crop rotations, which decrease soil-borne disease incidence. Also soil microbial biomass and enzyme activity are increased and erosion is decreased. Deleterious rhizobacteria can subclinically reduce plant root growth and may be a substantial crop yield constraint (Schippers *et al.*, 1987; Elliott and Fredrickson, 1987). These organisms have been associated with reduced-tillage cropping systems for winter wheat production (Elliott and Lynch, 1984b, 1985). They may also be associated with reduced crop rotation. Hassink *et al.* (1991b) indicates that the greatest microbial and fungal population changes associated with management occur in the rhizosphere. The concept that root health may be associated with biodiversity or specific patterns of biodiversity is interesting and worthy of investigation.

Tracking Soil Biodiversity

Many biological measurements have been used to link management with changes in the macro- and microbiota. As mentioned earlier, measurements of species will likely fail because of the complexities involved. Generally measurements of biodiversity lack precision. Many biological measurements which can be used to assess changes in soil biodiversity were summarized by Parkinson and Coleman (1991). There is a need for more precise and less complicated measurements that will track populations and processes. Studies developing these techniques and using them on similar soils managed quite differently such as organic versus conventional management may provide the optimum situation for studying biodiversity and soil resilience. Our knowledge in these areas must be improved so that more solid approaches are developed.

The problem is to determine how integrated land management and cropping practices are affecting soil resilience, biodiversity, and economic sustainability. This must be determined both for systems with high and low off-farm inputs and for farms in developing countries. Only then will we know which practices are most beneficial. Most importantly those relationships must be determined on a cropping systems basis.

Techniques that are readily agreed upon for studying soil biodiversity directly or indirectly include: soil organic matter content, measurement of the soil organic matter active fraction, water infiltration, resistance to penetration, water-holding capacity, aggregate stability, energy of wetting, microbial biomass, soil animals, soil enzyme activity, soil crusting, nutrient content, nutrient cycling, disease content, and microbial–faunal composition. Improved methods for following some of these processes will become available. These will include genetic probes and possibly the use of DNA digests to follow specific processes or indicators (Lynch, Chapter 5). These approaches might even involve techniques for following mRNA that initiate processes.

Integrating Soil Biodiversity and Resilience

Soil biodiversity and resilience are qualitative terms and to define them we must integrate the quantitative terms that we can measure concerning the soil flora and associated processes using a hierarchial approach. With this approach, levels of knowledge are accumulated and integrated at each level, which works progressively towards definitions. Swaminathan in the Foreword to this volume offers a similar approach for defining sustainable soil health management.

The interaction of soil biodiversity with soil properties and processes, as regulated by crop rotation, cropping, system tillage, crop residue management, pollutant input, soil amendments, and climate will determine soil resilience and sustainability of the cropping system. These factors must be integrated to provide predictive capabilities of how the system might be improved. To arrive at this goal, first-order quantitative inputs would include measurements such as rainfall, heat units, soil classification and composition, water-holding capacity, quantities of soil enzymes, soil organic matter content, specific biota-regulated processes, genetic probes to follow specific biota or biota–economic inputs and outputs, soil aggregate stability, infiltration properties, microbial biomass and composition, crop yield, soil loss rate, pH, soil respiration and other easily quantifiable measurements must be identified and measured. In other words, just about every measurement possible is included initially. These measurements are then integrated to provide correlations with population and/or process diversity (micro- and macrofauna), tillage resistance, compactability from traffic, disease incidence, and crop vigour, for example, to name some characteristics that might be considered second order and not so easily quantifiable. The integration is then used to determine higher level qualitative interactions until soil biodiversity and resilience can be defined by the model. This approach will rapidly identify redundant and non-critical measurements showing which should fall out and it will identify measurements that should be made. This approach provides an excellent research approach to the development and improvement of sustainable cropping systems. As new ideas and approaches emerge, they can be incorporated into the system.

A different approach will most likely be required to inform policy and decision-makers. An approach such as the **Prevention Index** formulated by Rodale Press (1992) might be used. In this case, subjective qualitative parameters can be identified with the help of the research model. Examples could be soil colour, smell, pH, crust formation, and crop yield to name some. There are many other possibilities but they must be simple subjective measurements. The parameters are placed into the model based on a general description. These terms are then integrated to develop a soil resilience index that, although not extremely sensitive, can be employed for rapid readout, for setting general course directions, and for assisting decision-making. As the research model is developed, the technology transfer model will be improved.

The initial approach for measurement and model development should be on a very restricted basis with measurements taken at three to four sites only. Selection of the sites should be based on rainfall, soil type, and agro-ecosystem. The reason is to develop the greatest precision possible with the minimum number of measurements. Many measurements will fall out rapidly, and it will be apparent that some not being taken will be required. There must be user involvement to determine if the approach is achievable, realistic, acceptable, and if it fits the real world. The efforts must involve integrated systems; for example, an independent modelling effort, which is a main tool, will likely fail. Questions such as short-term versus long-term sustainability must be asked. In other words, the cost of purchasing and applying the chemical or remedial action may be minimal; the resultant need for eventual environmental remediation may be the main cost. A goal of increased biodiversity is very likely unreal. It may be most desirable for biodiversity to remain relatively constant with changes in population and processes differentially responding to input pressures. The goal is to describe those pressures and management activities that most benefit the desired biodiversity and resultant soil resilience needed. Our comparison of the ideal situation is usually with the pristine environment. This may be incorrect for several reasons. One is that as improvements are developed, soil resilience should be a moving target and we should be able to improve it as knowledge increases. Another example is that the pristine situation may not be optimum for erosion or disease control. Eventually we may have the ability to design the system for specific needs. Integration of soil biodiversity with soil resilience is imperative for research.

Recommendations

We identify the following needs for study.

1. Identify key species or key assemblages necessary for the maintenance of the beneficial functions under the spatial constraints that determine the biotic colonization of a range of soils and plants.

2. Analyse critically the biodiversity of communities necessary for organic matter decomposition, particularly in terms of metabolic processes which lead to improved soil conditions. Similar considerations should be applied to soils being used in bioremediation processes. In all processes fauna should be considered with microorganisms as potential members of structured communities.

3. Use holistic approaches in soil management practices by considering the role of the biota in plant production systems.

References

Bolton, Jr, H., Elliott, L.F., Papendick, R.I. and Bezdicek, D.F. (1985) Soil microbial biomass and selected soil enzyme activities: effect of fertilization and cropping practices. *Soil Biology and Biochemistry* 17, 297–302.

Brussaard, L., Bouwman, L.A., Geurs, M., Hassink, J. and Zwart, K.B. (1990) Biomass, composition and temporal dynamics of soil organisms of a silt loam soil under conventional and integrated management. *Netherlands Journal of Agricultural Science* 38, 283–302.

Cambardella, C.A. and Elliott, E.T. (1992) Particulate soil organic matter changes across a grassland cultivation sequence. *Soil Science Society of America Journal* 56, 777–783.

Cochran, V.L., Elliott, L.F. and Lewis, C.E. (1989) Soil microbial biomass and enzyme activity in subarctic agricultural and forest soils. *Biology and Fertility of Soils* 7, 283–288.

Collins, H.P., Rasmussen, P.E. and Douglas, Jr, C.L. (1992) Crop rotation and residue management effects on soil carbon and microbial dynamics. *Soil Science Society of America Journal* 56, 783–788.

Doran, J.W. (1980a) Microbial changes associated with residue management and reduced tillage. *Soil Science Society of America Journal* 44, 518–524.

Doran J.W. (1980b) Soil microbial and biochemical changes associated with reduced tillage. *Soil Science Society of America Journal* 44, 765–771.

Elliott, L.F. and Fredrickson, J.K. (1987) Plant microbe interactions in the rhizosphere. In: Boersma, L.L. (ed.) *Future Developments in Soil Science Research.* Soil Science Society of America Inc., Madison, WI, pp. 145–156.

Elliott, L.F. and Lynch, J.M. (1984a) The effect of available carbon and nitrogen in straw on soil and ash aggregation and acetic acid production. *Plant and Soil* 78, 335–343.

Elliott, L.F. and Lynch, J.M. (1984b) Pseudomonads as a factor in the growth of winter wheat (*Triticum aestivum* L.). *Soil Biology and Biochemistry* 16, 69–71.

Elliott, L.F. and Lynch, J.M. (1985) Plant growth-inhibitory pseudomonads colonizing winter wheat (*Triticum aestivum* L.) roots. *Plant and Soil* 84, 57–65.

Elliott, L.F. and Wildung, R.E. (1992) What biotechnology means for soil and water conservation. *Journal of Soil and Water Conservation* 47, 17–20.

Elliott, L.F., Papendick, R.I. and Bezdicek, D.F. (1987) Cropping practices using legumes with conservation tillage and soil benefits. In: Power, J.F. (ed.) *The Role of Legumes in Conservation Tillage Systems.* Soil Conservation Society of America, Ankeny, IA, pp. 81–89.

Hassink, J., Leffink, G. and van Veen, J.A. (1991a) Microbial biomass and activity of a reclaimed-polder soil under a conventional or a reduced-input farming system.

Soil Biology and Biochemistry 23, 507–513.

Hassink, J., Oude Voshaar, J., Nijhuis, E.H. and van Veen, J.A. (1991b) Dynamics of the microbial populations of a reclaimed-polder soil under a conventional and or a reduced-input farming system. *Soil Biology and Biochemistry* 23, 515–524.

Hawksworth, D.L. (ed.) (1991) *The Biodiversity of Microorganisms and Invertebrates: Its Role in Sustainable Agriculture.* CAB International, Wallingford.

Lal, R. (1974) No-tillage effects on soil properties and maize (*Zea mays* L.) production in western Nigeria. *Plant and Soil* 40, 321–331.

Lam, S.T., Ellis, D.M. and Lion, J.M. (1990) Genetic approaches for studying rhizosphere colonization. *Plant and Soil* 129, 11–18.

Lynch, J.M. and Panting, L.M. (1980) Variations in the size of the soil biomass. *Soil Biology and Biochemistry* 12, 547–550.

Lynch, J.M. and Bragg, E. (1985) Micro-organisms and soil aggregate stability. In: *Advances in Soil Science.* Vol. 2. Springer-Verlag, New York, pp. 133–170.

Papendick, R.I. and Parr, J.F. (1992) Soil quality – The key to a sustainable agriculture. *American Journal of Alternative Agriculture* 7, 2–3.

Parkinson, D. and Coleman, D.C. (1991) Methods for assessing soil microbial populations, activity, and biomass. Microbial communities, activity and biomass. *Agriculture, Ecosystems and Environment* 34, 3–33.

Reganold, J.P., Elliott, L.F. and Unger, Y.L. (1987) Long-term effects of organic and conventional farming on soil erosion. *Nature* 330, 370–372.

Rodale Press (1992) *The Prevention Index.* 33 E. Minor Street, Emmaus, PA. p. 36.

Rovira, A.D., Elliott, L.F. and Cook, R.J. (1990) The impact of cropping systems on rhizosphere organisms affecting plant health. In: Lynch, J.M. (ed.) *The Rhizosphere.* John Wiley, Chichester, pp. 389–436.

Schippers, B., Bakker, A.W. and Bakker, P.A.H.M. (1987) Interactions of deleterious and beneficial rhizosphere microorganisms and the effect of cropping practices. *Annual Review of Phytopathology* 25, 339–358.

Smith, J.L. and Elliott, L.F. (1990) Tillage and residue management effects on soil organic matter dynamics in semiarid regions. In: *Advances in Soil Science.* Springer-Verlag, New York, pp. 69–88.

Smith, V.T., Wheeting, L.C. and Vandecaveye, S.C. (1946) Effects of organic residues and nitrogen fertilizers on a semiarid soil. *Soil Science* 61, 393–410.

Stevenson, F.J. (1982) *Humus Chemistry.* John Wiley, New York, p. 443.

Stroo, H.F., Bristow, K.L., Elliott, L.F., Papendick, R.I. and Campbell, G.S. (1989) Predicting rates of wheat residue decomposition. *Soil Science of America Journal* 53, 91–99.

Part V

Methodologies for the Study of Soil Resilience and Sustainable Land Use

Chapter 22

Long-term Field Experiments: their Importance in Understanding Sustainable Land Use

D.S. Powlson and A.E. Johnston

AFRC Institute of Arable Crops Research, Rothamsted Experimental Station, Soil Science Department, Harpenden, Herts., AL5 2JQ, UK

Introduction

> Sustainable agriculture is not a luxury ... When an agricultural resource base erodes past a certain point, the civilisation it has supported collapses ... There is no such thing as a post-agricultural society.
>
> (Holmberg *et al.*, 1991)

Any definition of sustainability related to the managed use of land must include physical, environmental and socioeconomic aspects. No agricultural system will be sustainable if it is not economically viable both for the farmer and the society of which he is a part. But economic sustainability cannot be bought at the cost of environmental damage, which is ecologically, socially or legally unacceptable, or physical damage which leads to irreversible soil degradation or uncontrollable outbreaks of pests, diseases and weeds. Future studies must attempt to relate all these aspects although not all are amenable to research in long-term experiments.

It is essential to have an indication of the sustainability or otherwise of an agricultural system well before the catastrophic consequences of non-sustainability become apparent. In relation to soil resilience and sustainable land use, future research must concentrate on finding indicators of soil change which accurately reflect likely changes in sustainability before adverse effects on yield become serious. Currently, yield is often our only measure of sustainability. Thus there is a need to determine critical values for relevant soil parameters and this can only be done in long-term experiments. Although indispensible, these are not perfect; the maintenance of crop production in small, well-managed field plots is no guarantee that the

(eds D.J. Greenland and I. Szabolcs)

system will be robust enough when practised by farmers on a large scale on a wide range of soils. However, long-term experiments do provide the best practical approximation to a test of sustainability. Provided they have contrasted treatments they offer the best opportunity to assess sustainability, explain why a specific system fails and seek modifications to improve sustainability. The long-term aspect is essential because many soil properties change slowly over many decades and cannot be detected in experiments lasting only a few years. Another important aspect is that the data can be used to construct mathematical models to describe the system in whole or in part. These models can then be used to extrapolate the results to different soil types or environments provided that the controlling factors have been identified and measured using independent, quantitative data.

Although it might seem desirable to work towards a single, worldwide index of soil resilience and sustainable land use, in practice this is likely to be difficult to achieve and it might not even be very helpful. This is because any agricultural system is as weak as its weakest link which may be any one of a wide range of variables such as soil type, soil fertility, climate, management skill or financial viability; these and other variables will not always rank in the same order. For example Rothamsted's present Classical Experiments have continued for more than 130 years because yields, the measure of sustainability used, have been maintained or improved. At least over this time scale the husbandry systems tested have been sustainable. However, some experiments started in the middle of the last century with identical cropping and manuring but on other soils they were stopped because yields failed. In these and other experiments where yields also failed, soil management and/or cropping systems were inappropriate for the soil and climate, although identical husbandry might have succeeded on other soils (Johnston and Powlson, Chapter 23). Thus a critical level set for any parameter controlling yield will invariably differ between soils, farming systems and climate. This chapter discusses some topics relevant to soil resilience and sustainable land use using examples from research on soil fertility.

Parameters Affecting Soil Fertility

Soil fertility develops slowly over many decades. Whereas inept management can destroy soil fertility rapidly, small, less obvious, but insidious changes can be equally damaging in the long term. Soil fertility arises from complex interactions between the biological, chemical and physical properties of soil each of which, in turn, is affected by the interaction of its constituent parameters. Little is known about the complexity of some of these interactions or their rates of change, especially in relation to our ability to predict changes in soil fertility. Soil fertility is not synonymous with agricultural productivity. Indeed, it is attempts to increase productivity in some countries since the 1950s that have led to concern about the impact of agriculture on the environment, for example, the movement of nitrate, phosphate and pesticide residues to potable water. In consequence Johnston (1991a) proposed that there should be a clear distinction between soil fertility and soil productivity and that each should be considered separately. Soil fertility can be defined in relation to those properties of a soil, like acidity, P and K status and soil organic matter level which change slowly over time. For each it should be possible to define a critical level below which yield declines rapidly. Provided that soil fertility factors are optimum, the level of production at any one time can be set by the quantity of nitrogen applied in fertilizers and manures and the use of agrochemicals, like weedkillers, fungicides and insecticides, to protect the yield potential.

This chapter discusses some parameters affecting soil fertility and uses them to provide illustrations of how a knowledge of the way they change, as determined by long-term experiments, provides insights into sustainable land use. The examples are mainly from long-term experiments in the temperate region, and are from five experimental stations in South East England now managed by Rothamsted. The Rothamsted experiments are the oldest. They were started by J.B. Lawes and J.H. Gilbert between 1843 and 1856 on a silty clay loam (Chromic Luvisol) with 700 mm rainfall. The other experimental stations, together with their year of commencement, soil type and approximate annual rainfall are: Woburn, 1876, sandy loam (Cambic Arenosol), 600 mm; Saxmundham, 1899, sandy clay loam (Eutric Gleysol), 600 mm; Brooms Barn, 1962, sandy loam to clay loam developed in glacial drift (Haplic and Chromic Luvisols), 500 mm; Brimstone Farm, 1978, clay (Typic Haplaquept), 680 mm. Some details of the history and development of the experimental programme at each station are given in Chapter 23, which also discusses some of the practical aspects of the conduct of long-term experiments.

Evidence for Sustainable and Non-sustainable Production at Rothamsted and Woburn

Perhaps the two best-known and most remarkable examples of the continuous and sustained production of arable crops are at Rothamsted. Winter wheat has been grown on all or part of Broadbalk field each year since 1843; spring barley has been grown continuously on Hoosfield since 1852. Also, herbage production has been maintained on the permanent grassland experiment on Park Grass since 1856.

Yields of both spring barley on Hoosfield and winter wheat on Broadbalk changed little for many decades. Then, with the introduction of

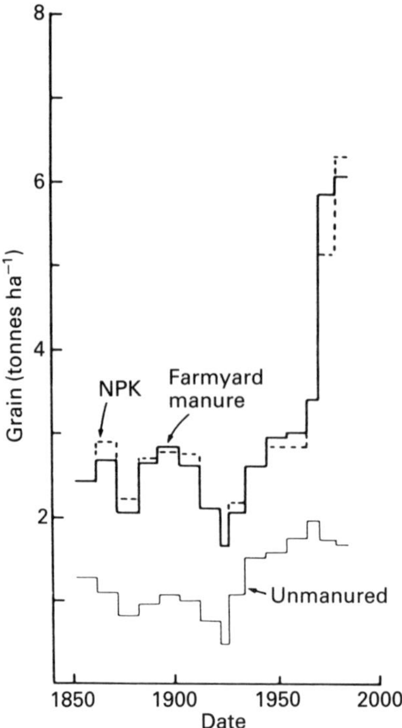

Fig. 22.1. Yields of winter wheat on three plots of the Broadbalk experiment, 1852–1986. Each year inorganic fertilizers supply 144 kg N, 33 kg P, 90 kg K, 12 kg Mg ha^{-1}; farmyard manure, 35 t ha^{-1}, supplies about 3000 kg C, 225 kg N, 40 kg P and 210 kg K.

cultivars with a high yield potential and the use of agrochemical inputs (herbicides, fungicides, pesticides) to protect that potential, yields of both cereals have increased appreciably since the late 1960s. Fig. 22.1 shows the yields of wheat grown on Broadbalk when 35 t ha^{-1} of animal manure (farmyard manure, FYM) was applied each year and when NPK fertilizers, supplying 144 kg N, 33 kg P and 90 kg K ha^{-1} were used and compares yields with those on the unmanured plot. The decline in yield in the 1920s is considered to be because of difficulties in controlling weeds, previously controlled by hand hoeing. In 1925 a bare fallow one year in five was used for weed control; since 1957 weedkillers have been used. Grain yields have not decreased on the unmanured plot. On plots given fertilizers or FYM yields have remained essentially equal throughout the whole period of the experiment and have increased appreciably in recent years because plant breeders have succeeded in improving the grain:straw ratio in modern cultivars. On soils given only fertilizers, soil organic matter declined a little at the start of the experiment but has remained largely unchanged during the last 100 years. Where FYM has been applied each year the organic matter content of the soil is now about two and a half times that on fertilizer-only plots (Johnston, 1969). The similarity of the yields on these two plots shows that in some conditions yields can be maintained by the use of fertilizers and lime in the absence of organic manures. From the data it has been assumed by some that the organic matter content of a soil is not very important; recent results suggest that this is not always the case even at Rothamsted (Johnston and Powlson, Chapter 23). It is also unfortunate that many have sought to extrapolate this result unthinkingly to other soils and other climates.

In marked contrast to the sustainability of cereal yields at Rothamsted, yields of barley, and to some extent wheat, grown continuously on the sandy loam soil at Woburn began to decline about 15 years after the start of experiments there in 1876 (Table 22.1). Lawes and Gilbert were instrumental in starting these experiments in which winter wheat and spring barley were grown each year, much as on Broadbalk and Hoosfield at Rothamsted. The manurial treatments were similar to those at Rothamsted and included a comparison of ammonium sulphate and sodium nitrate. The decline in yield was most marked on plots getting N as ammonium sulphate and tests of lime ($CaCO_3$), the first in England, were started in 1898 (Johnston, 1975). Similar tests to correct what we now term soil acidity were started inde-

Table 22.1. Yields of wheat and barley grain (t ha^{-1}) grown continuously and effect of calcium carbonate on plots given ammonium sulphate, Continuous Wheat and Barley experiments, Woburn 1877–1926. (Adapted from Johnston, 1975.)

Crop and treatment[a]	Period				
	1877–86	1887–96	1897–1906	1907–16	1917–26
Winter wheat					
Unmanured	1.08	0.83	0.61	0.66	0.46
NPK – no chalk	2.04	1.94	1.68	1.11	0.64
– chalk[b]	–	–	–	1.25	0.66
FYM	1.76	1.83	1.69	1.38	1.20
Spring barley					
Unmanured	1.56	0.98	0.60	0.60	0.49
NPK – no chalk	2.57	2.10	0.19	0.19	0.30
– chalk	–	–	1.39	1.39	0.90
FYM	2.39	2.30	1.87	1.87	1.54

[a]46 kg N ha^{-1} as ammonium sulphate, FYM 17.6 t ha^{-1} per year on average.
[b]2.5 t CaO ha^{-1} to winter wheat, half in 1905, half in 1918.
10.0 t CaO ha^{-1} to spring barley, half in 1898, half in 1912.

pH, in water, of soils sampled in 1927 were:
Wheat, no chalk 4.6; chalk 5.0
Barley, no chalk 4.8; chalk 5.8
Applying chalk would have raised soil pH more soon after application.

Table 22.2. Yields of potatoes grown continuously during 1876–1901, Exhaustion Land, Rothamsted.[a]

	Mean annual yield of total tubers (t ha^{-1})		
Treatment[b]	1876–1881	1882–1887	1888–1901
Unmanured	5.7	4.2	2.1
NPK	18.9	14.9	10.4
FYM	14.0	10.7	11.6

[a]Selected treatments, for full details see Johnston and Poulton (1977).
[b]Rate per ha: N, 96 kg; P, 33 kg; K, 135 kg; FYM 35 t fresh weight.

pendently about the same time in the United States. However, liming did not fully restore yields to the levels obtained in the early years of the experiment (Table 22.1). With hindsight it might be conjectured that there was also a build-up of cereal cyst nematode on this light-textured soil because yields also declined on plots getting sodium nitrate, which had little effect on soil pH (Johnston and Chater, 1975), but did not decline where cereals were grown in rotation on adjacent experiments. At Rothamsted, the silty clay loam soil had as much as 5% free calcium carbonate when the experiments started, and the use of ammonium sulphate, supplying up to 144 kg N ha^{-1}, caused no serious problem with soil acidification until the late 1940s and early 1950s. At that time remedial action was taken to adjust soil acidity to near pH 7 and a scheme of regular liming was introduced to prevent further problems developing (Rothamsted Experimental Station, 1955). This prevented a serious decline in yields of wheat and barley which would probably have occurred in the 1970s.

There were problems, however, at Rothamsted, when other crops were grown in monoculture. Beans (*Vicia faba*) and clover could not be grown for more than a few years at a time. Clover was probably affected by *Sclerotinia* and beans either by root fungal pathogens or possibly stem nematodes which have been particularly devastating in recent years.

Potatoes, first planted in 1876 on the Exhaustion Land experiment, were grown in consecutive years on the same plots until 1901. By 1888 yields on fertilizer-treated plots were about half those at the start (Table 22.2). Yields did not appear to decline as dramatically as on FYM-treated soils, because in the first six years the FYM supplied too little N (Johnston and Poulton, 1977). The experiment was stopped in 1901 because potato monoculture was not economically sustainable. It was thought that the physical condition of the soil not treated with FYM was poor and unsuitable for potatoes. However, yields in Table 22.2 show that during 1888–1901 the FYM-treated plot yielded little more than that given fertilizers only. Thus a factor other than soil structure was probably involved, possibly potato cyst nematodes. This pest was not recognized in the 1890s, and although soils sampled in

1903 were assayed in the 1970s, only a few remnants of recognizable cyst cases could be found. This illustrates the need for a multidisciplinary approach to long-term experiments now that many more factors affecting crop growth have been recognized.

Soil Acidity

Soils are acidified through a number of processes which include biological activity and other natural sources in soil, atmospheric deposition of acidifying pollutants, additions of fertilizers and, to a minor extent, the removal of calcium by crops. Soil acidification is a worldwide problem which can have a major effect on sustainable land use. An example has been given above of how, even in the comparatively short term, acidification of soil, due to the use of ammonium sulphate, so decreased yields of spring barley that the usual beneficial effects of nitrogen were nullified. Kang (1990) gave data for acidification of the surface 15 cm of an Oxic Paleustalf under experiment for 11 years in south western Nigeria. Compared to a fallowed soil (pH 6.63) cropping caused soil pH to decline to 5.40 and 5.83 with or without fertilizers respectively. The form in which fertilizer nitrogen was applied was not given but the amounts were 120 and 160 kg N ha^{-1} in the first 5 and second 6 years respectively. Further examples of the way that sustainable land use can be affected by soil acidity follow in which the time scales for the changes induced by soil acidity ranged from 40 to 90 years.

Table 22.3. Yields (t ha^{-1}) of turnips and winter wheat in the Agdell Rotation experiment, Rothamsted 1848–1951. (Adapted from Johnston and Penny, 1972.)

	Crop and treatment			
	Turnip roots		Wheat grain	
Years	None	NPK	None	(NPK)[a]
1848–51	1.31	2.92	1.91	1.93
1852–83	0.24	3.47	1.46	1.96
1884–99	0.13	5.09	1.60	2.47
1900–19	0.11	3.98	0.97	1.37
1920–35	0.08	1.61	0.98	0.91
1936–51	0.04	0.54	1.27	2.07
Soil pH 1953	8.2	5.6		

[a]NPK fertilizers were applied only once every four years to the turnips. Wheat followed a one year clover ley which would have left a nitrogenous residue.

Effect of soil acidity on the yields of turnips

On the heavier textured soil at Rothamsted, with an initial pH of about 8.0, soil acidification due to ammonium sulphate occurred over a much longer period than at Woburn. The Four-course Rotation experiment on Agdell field started in 1848. Turnips, spring barley, clover or beans and winter wheat were grown in rotation. Nitrogen was tested only on the turnips and it was applied as a mixture of ammonium sulphate and rape cake. Increasing soil acidity after 80 years allowed club root (caused by the fungus *Plasmodiophora brassicae*) to flourish and decrease turnip yields to such an extent (Table 22.3) that the experiment had to be stopped in 1951. Yields of winter wheat were not affected so seriously (Table 22.3). Soil pH was raised subsequently by applying lime but once established, *Plasmodiophora brassicae* persists in soil for long periods affecting yields of most brassica crops.

Effect of vegetation on rate of soil acidification

One plot on the Park Grass experiment at Rothamsted which started in 1856 has been unmanured since then. Initially the 0–23 cm depth of soil had a pH about 5.6. Less than 1 km away is the Geescroft Wilderness. This was part of an old arable field until it was fenced off in 1886 when the top 23 cm of soil had a pH about 7.1. Since then the site has been untended and a mature deciduous woodland, mainly dominated by oak, has developed. Table 22.4 shows that surface soil pH is now about 4.8 on the Park Grass plot and 4.3 under the Wilderness. Soil acidification, arising at least in part from aerial pollutants, has been quicker and more intense on Geescroft than on Park Grass: about 3 pH units in 100 years on Geescroft and about 1 pH unit in 140 years on Park Grass. The inference must be that the tree canopy has been more efficient in trapping aerial pollutants than low growing grass herbage. Under the trees the 23–46 and 46–63 cm subsoils have also acidified appreciably (Table 22.4). Such intense acidification to such a depth has implications not only for sustainable land use, but also for drainage water quality. Once soil pH has fallen to about 4.0 at the depth of which water moves into drains then it is to be expected that this drainage water will contain appreciable amounts of aluminium, iron and manganese.

Yield and species composition of grassland

The Park Grass experiment tests the effects of N, P, K singly and in various combinations on the yields of permanent grassland cut for hay in June and silage in autumn. The soil initially had a pH (water) of about 5.6 but it was not until 1903 that many of the plots were halved for a test of liming once every four years. Yields and the number of species present in the sward now

Table 22.4. Effect of both 'natural' and additional fertilizer acidifying inputs on soil pH at different depths under woodland and grassland at Rothamsted. (Adapted from Johnston *et al.* (1986) with additional unpublished data.)

	Year and experiment									
	Geescroft Wilderness Woodland					Park Grass Grassland				
Horizon, cm	1883	1904	1965	1983	1991	1876	1923	1959	1984	1991
Natural inputs										
0–23	7.1	6.1	4.5	4.2	4.3	5.4	5.7	5.2	5.0	4.8
23–46	7.1	6.9	5.5	4.6	5.1	6.3	6.2	5.3	5.7	5.4
46–69	7.1	7.1	6.2	5.7	6.0	6.5	–	–	–	5.7
Additional fertilizer inputs										
0–23	–	–	–	–	–	4.2	3.8	3.7	3.4	3.2
23–46	–	–	–	–	–	6.3	4.4	4.1	4.0	3.8

reflect both soil pH and nutrient status after 60 years of treatment (Thurston *et al.*, 1976). Supplying nitrogen together with phosphorus and potassium greatly increased yield but appreciably decreased the number of species present in the sward. Both soil pH and nutrient status have had a major impact on the dominance of different grass species (Warren and Johnston, 1964). Equally important a number of species have become both morphologically and physiologically adapted to the level of soil fertility on the plots on which they grow, such that they do not thrive when moved to other plots (Thurston *et al.*, 1976). The liming test was modified in 1965 by dividing all plots into four subplots to establish soils with pH 5, 6 and 7, plus one unlimed subplot, for each manurial treatment (Warren *et al.*, 1965). Changing the pH of soils on plots where a mat of only partially decomposed organic material overlaid the mineral soil took many years. This was because much calcium was absorbed by the organic matter at the surface. It was not until this began to decompose, once microbial activity was enhanced by raising pH, that calcium was released and moved down into the underlying mineral soil and its pH rose (Johnston, 1972).

One aspect of inappropriate technology transfer may well be attempts to use cultivars tested on neutral or calcareous soils, and herbicides whose efficiency was proven on such soils, to acid soils in those parts of the world where limestone is not freely available to raise soil pH.

Soil Organic Matter

Humus content of soil

In any farming system, soil organic matter or humus content changes towards an equilibrium value that depends on: (i) the quantity of added organic material and its rate of decomposition; (ii) the rate of breakdown of existing soil humus; (iii) soil texture and pH; and (iv) climate. Changes in both %C and %N in the top 23 cm of soil have been used to assess changes in the humus content of soil in long-term experiments at Rothamsted and Woburn.

Fig. 22.2(a) shows that on Hoosfield at Rothamsted, soil humus content has been constant for about 100 years on both the unmanured plot and that given NPK fertilizers. The quantity is a little larger in the fertilized soil because larger crops have been grown and, although straw is removed each year, there have been larger residues from stubble, leaves and roots returned to the soil. Annual additions of 35 t ha^{-1} fresh FYM have increased soil humus, rapidly at first and then more slowly as the appropriate equilibrium value for this system is approached. But it is important to note that the timescale over which this change has occurred is more than

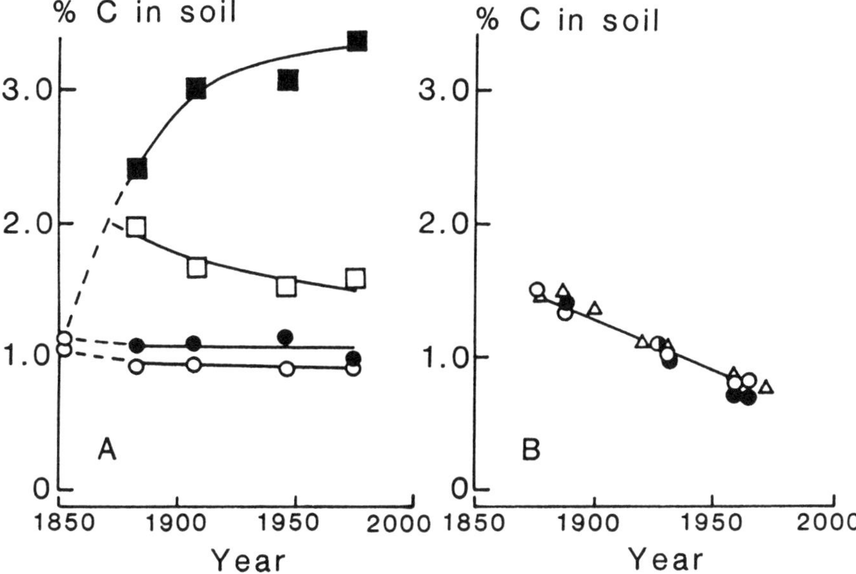

Fig. 22.2. Change in % carbon in soil with time under all-arable cropping systems on: A, a silty clay loam at Rothamsted and B, a sandy loam at Woburn. A. Barley grown each year, annual treatments continuous since 1852: ○, unmanured; ●, NPK fertilizers, 48 kg N ha^{-1}; ■, FYM 35 t ha^{-1}; □, FYM 1852–71, none since. B. Cereals each year: ○, unmanured; ●, NPK fertilizers; Δ, manured four course rotation 36.

130 years for this medium-textured soil in a temperate climate.

Figure 22.2 well illustrates the importance of soil texture on equilibrium levels of soil humus. The sandy loam at Woburn has about 10% clay whereas the silty clay loam on Hoosfield has about 20%. At the start of the long-term experiments at the two sites, soil humus levels were higher at Woburn than at Rothamsted (for explanation see Johnston, 1991b) but under continuous arable cropping there is now less humus in Woburn soil than Rothamsted (cf. Fig. 22.2A and B). Johnston (1991b) gives further examples of the effects of soil texture for other farming systems at the two sites.

The slow rate of change in humus content of a soil in a temperate climate is also well demonstrated by the changes in old arable soils, with about 0.1% N, put down to permanent grass at Rothamsted. It took 100 years to achieve the apparent equilibrium level of humus, corresponding to about 0.28% N, characteristic of permanent grassland in these conditions. The half-way stage was reached in 25 years (Johnston, 1991b).

The humus content of soil changes more rapidly in the tropics. For example Jenkinson and Ayanaba (1977) showed that the decomposition of uniformly ^{14}C-labelled ryegrass was four times faster under field conditions at IITA (Nigeria) than at Rothamsted (Fig. 22.3). Such a contrast between the rate of decomposition of soil organic matter in England and Nigeria is now

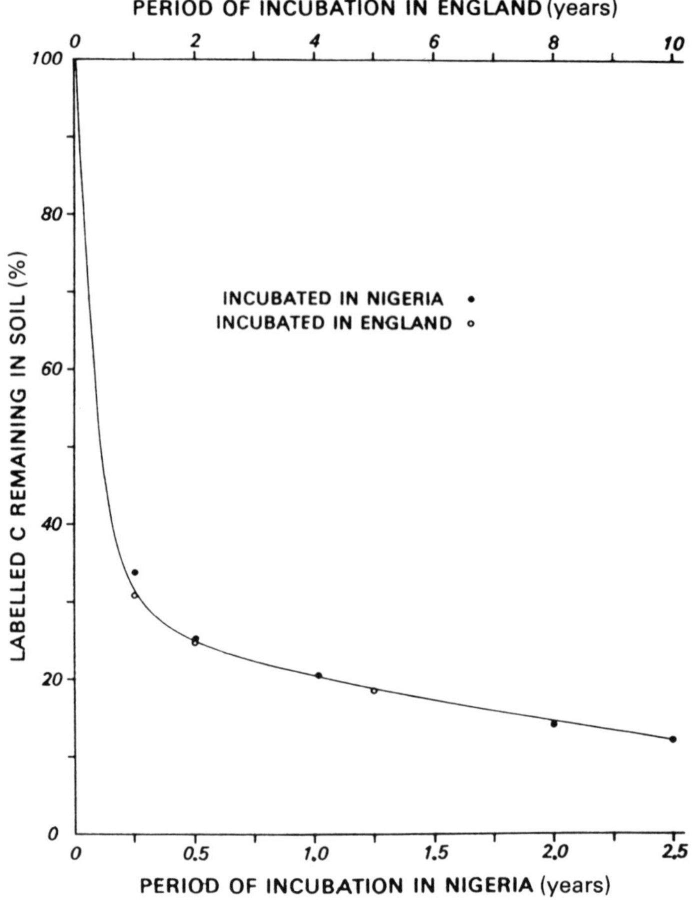

Fig. 22.3. Decomposition of uniformly ^{14}C-labelled ryegrass in England (Rothamsted) and in Nigeria (IITA).

well documented and understood. But many years ago the well-founded Rothamsted results on the lack of crop response to soil humus level and the slow changes in soil organic matter were promulgated widely without the rider that they needed to be confirmed for other soils, under different farming systems in other parts of the world, and this led to problems which would have been avoided if data from other long-term experiments had been available.

Modelling changes in soil humus content

The large amount of data on changes in soil organic matter in different Rothamsted experiments made it possible to develop models to describe

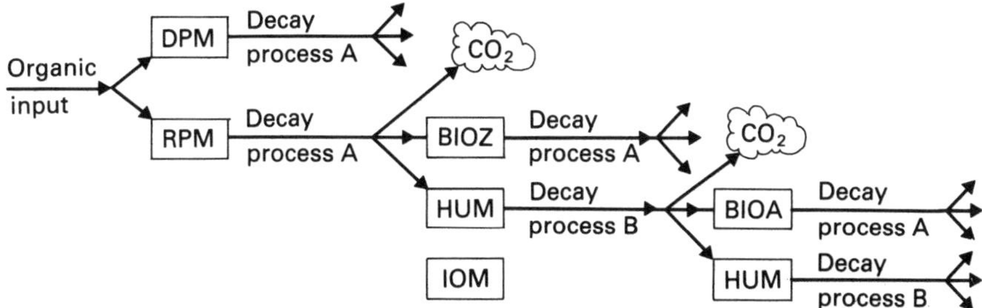

Fig. 22.4. Flow of C through the model. DPM is decomposable plant material, RPM is resistant plant material, BIOZ is zymogenous soil microbial biomass, and BIOA is autochthonous biomass: all these compartments decay by process A. HUM is humified organic matter which decays by process B. These processes differ only in the proportions of CO_2, biomass and HUM formed. IOM is biologically inert organic matter.

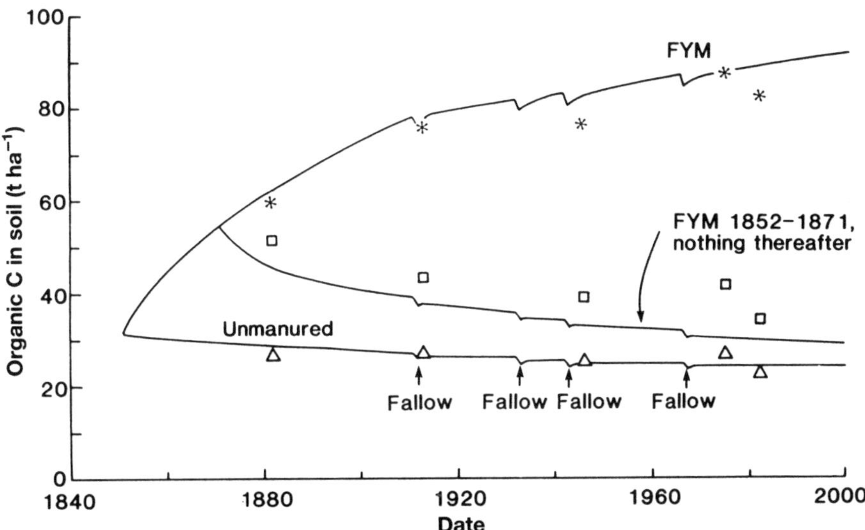

Fig. 22.5. Organic C in the top 23 cm of a soil under continuous spring barley (Hoosfield). Data, corrected for changes in bulk density, are from Jenkinson and Johnston (1977), except for results for 1982 (unpublished). Plot 1-0 is unmanured (△); the FYM plot (7-2) receives 35 t FYM annually (*); the FYM residues plot (7-1) received 35 t FYM annually between 1852 and 1871 and nothing since (□). The FYM, applied in late autumn (after 1916), was assumed to be equivalent to 60% of the plant material from which it came and to contain DPM, RPM and HUM (but no biomass) in the proportions 0.53, 0.38 and 0.09 respectively. Before 1916, FYM was applied in spring and was assumed to be equivalent to 45% of the original plant material, the corresponding proportions being 0.34, 0.49 and 0.17. The C inputs used, all in t C ha^{-1} year^{-1}, were: unmanured plot, 1.1; FYM plot, 1.5 (plant debris) + 3.0 (FYM); FYM residues plot, as FYM plot during 1852–1871, 1.5 during 1872–1876, thereafter, 1.1.

such changes (Jenkinson and Rayner, 1977). Data from other experiments could then be used to validate and extend the model to soils and climates other than those at Rothamsted (Jenkinson *et al.*, 1987). The current model considers the flow of carbon entering the system to pass through six compartments (Fig. 22.4). The fit of the model to the observed changes in soil humus in the Hoosfield Barley experiment is good (Fig. 22.5); Jenkinson *et al.* (1987) gave other examples. It should be emphasized that Fig. 22.5 is a true test of the model because no data from the Hoosfield experiment were used to set the model parameters and no adjustments were made to improve the fit. This approach to modelling from, as it were, the bottom up, i.e. a single site studied in great detail, has the benefit that the parameters in the model can be determined on the basis of well-estimated data and then, as other long-term data sets become available, the parameters can be adjusted so that the model adequately describes a wider range of soils, farming systems and climates.

Effect of soil organic matter on crop yield

Soil organic matter can affect the yield of arable crops through at least three mechanisms: release of nutrients, improved soil structure and improved water-holding capacity. Although these mechanisms can be identified they cannot be quantified readily. In part this is because in temperate climates it takes many years to establish soils with different levels of humus (Fig. 22.2).

Effects of soil structure

In one experiment at Rothamsted it took 12 years to establish soils with 0.87 and 1.40% organic carbon and with a range of sodium bicarbonate soluble P levels but with exchangeable K values above those at which arable crops would respond to K. Only then did the main experiment start; it aimed to measure the response of arable crops to readily soluble soil P in the presence of adequate N and K. The results (Table 22.5) show, however, that there were large differences in yields between soils with different levels of organic matter when the soils were grouped by the level of soluble P they contained. This particular soil was among the most difficult to cultivate at Rothamsted yet by many standards it was not intractable and yields were good. This effect of extra soil organic matter was probably through improved soil structure, allowing better root exploration of the soil. This is the most likely explanation because after the soil from each plot was sampled, air dried and ground to pass a 2 mm sieve, it was mixed with half its weight of inert quartz, to give a good rooting medium, and the mixture cropped with ryegrass in pots in the glasshouse. Irrespective of soil organic matter content, when ryegrass yields were plotted against soluble P there was a

Table 22.5. Yields (t ha^{-1}) of potatoes, sugar from sugar beet and spring barley, on soils with different amounts of readily soluble P and at two levels of soil organic matter.

Crop[a]	% organic carbon in soil	P soluble in 0.5 M $NaHCO_3$ (mg kg^{-1})				
		0–9	9–15	15–25	25–45	45–70
Barley grain	0.87	–	2.71	3.29	4.21	4.61
	1.40	3.18	4.78	5.20	5.00	5.46
Potato tubers	0.87	–	31.5	36.7	39.3	44.0
	1.40	26.6	43.0	45.3	46.4	47.8
Sugar	0.87	–	5.03	5.91	6.66	6.80
	1.40	2.45	5.92	7.06	6.73	6.85

[a]Barley grown in 1970, 1971; potatoes 1971, 1972; sugar beet 1970, 1972.

single response curve, the shape of which was closely similar to that for the arable crops grown in the field on the higher organic matter soil.

Effects on water-holding capacity

It is sometimes difficult to envisage how small differences in organic matter in soils in temperate climates can affect yield through changes in water-holding capacity. The reason may be that although the difference in available water is often small, perhaps equivalent to only 3 or 4 days' use by the crop, this could be sufficient to prevent the onset of severe water stress before the next rainfall. Such effects are likely to be of most benefit to shallow-rooting, spring-sown crops. When spring barley, potatoes, winter barley and winter wheat were grown on the same experiment at Woburn with different rates of nitrogen only the spring-sown crops benefited from extra soil organic matter (Table 22.6). The winter-sown crops, having deeper rooting systems, probably used much more water stored in the subsoil than the spring-sown crops and were, therefore, less sensitive to short periods of water shortage in the surface soil.

Other effects of soil organic matter

Soil organic matter can affect yields indirectly. For example beans (*Vicia faba*) were grown at Rothamsted on soils with different levels of organic matter but well supplied with P and K; simazine was applied to the surface for weed control. On the soil with least organic matter yield was reduced considerably (Table 22.7) presumably because, in the absence of enough organic matter, simazine was leached down from the surface and taken up

Table 22.6. Yields of potatoes, winter wheat and winter and spring barley at Woburn 1973–80. (Adapted from Johnston, 1986.)

% organic C in soil	Fertilizer N applied N0	N1	N2	N3	Average April–July rainfall and deviation from long-term average[a]
Potatoes, tubers (t ha^{-1}) 1973 and 75					
0.76	25.7	35.6	41.7	43.2	
2.03	27.1	40.6	50.7	59.0	266 (−14)
Spring barley, grain (t ha^{-1}) 1978					
0.76	2.19	5.00	6.73	7.05	
1.95	2.58	5.12	6.85	7.81	222 (+20)
Winter wheat, grain (t ha^{-1}) 1979					
0.76	3.54	7.32	8.05	7.82	
1.95	4.81	7.21	8.09	8.08	247 (+45)
Winter barley, grain (t ha^{-1}) 1980					
0.76	3.05	6.01	7.32	7.83	
1.95	3.57	5.92	7.00	7.98	258 (+56)

N0 N1 N2 N3: 0, 100, 200, 300 kg N ha^{-1} for potatoes; 0, 50, 100, 150 kg N ha^{-1} for cereals.
[a]April–August rainfall for potatoes.

by the bean roots to the detriment of top growth and grain production. Darker-coloured soils with more organic matter may also warm more readily in spring enhancing germination and early growth.

Quantifying the separate factors which contribute to the 'organic matter effect'

Recently, in a number of experiments at Rothamsted, yields of arable crops have been larger on soils with more organic matter irrespective of the amount of fertilizer nitrogen applied. By fitting response curves of the same form it is possible to estimate both maximum yield and its associated

Table 22.7. Effect of simazine on yield of field beans (*Vicia faba*) at three levels of soil organic matter.

% organic matter	Yield grain (t ha^{-1}) No added simazine	Simazine added (0.94 kg ha^{-1})
1.2	2.5	1.6
2.0	2.6	2.4
3.5	3.1	2.9

nitrogen application. It is then possible to bring the response curves into coincidence by bringing the maximum yields into coincidence. This usually requires a diagonal shift which has both horizontal and vertical components. The horizontal shift can be equated to a spring nitrogen application, the vertical shift to an effect of organic matter other than nitrogen. This has been done for the effects of extra organic matter on yields of wheat on Broadbalk (Johnston, 1987). Nitrogen released from soil organic matter increased yield as much as a spring application of 70 kg N ha^{-1}; other non-spring N effects gave a yield increase of 1.4 t ha^{-1} grain. This procedure has also been used to apportion the effects of different rotations on yields of wheat on Broadbalk (Dyke *et al.*, 1983) and to estimate the benefits of grass clover leys of varying duration on the yields of arable crops which followed (Johnston *et al.*, 1993). In the latter case the nitrogen effect was large in the first year but insignificant in the following two years when much of the benefit could perhaps be ascribed to better soil physical conditions of which increased water-holding capacity could be important. The importance of soil physical conditions are also dealt with in Chapters 4, 14 and 25.

Soil Microbial Biomass

The proportional change in soil microbial biomass content of a soil in response to a change in organic input, or in factors influencing its rate of breakdown, is much greater than the corresponding change in total soil organic matter. For example, in an experiment with sorghum in a sub-tropical climate in Queensland, Australia, crop residues were either returned to or removed from a soil over a 5-year period. The increase in biomass carbon of about 15% was much larger than that (9%) in total organic carbon (Saffigna *et al.*, 1989). Similarly, returning or removing barley straw over an 18-year period had a larger effect on biomass C (about 40%) than on total carbon (about 5%) in an experiment in Denmark (Powlson *et al.*, 1987). Thus changes in soil microbial biomass meet an important criterion for a short-term indicator of change in a soil property which itself changes only slowly but where change could lead to a serious decline in soil productivity. Future research must concentrate on finding similar indicators.

Phosphorus and Potassium

In long-term experiments at Rothamsted residues of P and K have accumulated in soils receiving both fertilizers and organic manures. The availability of these residues to plants and possible losses of both P and K from soil have been reviewed for P by Johnston and Poulton (1992) and for K by

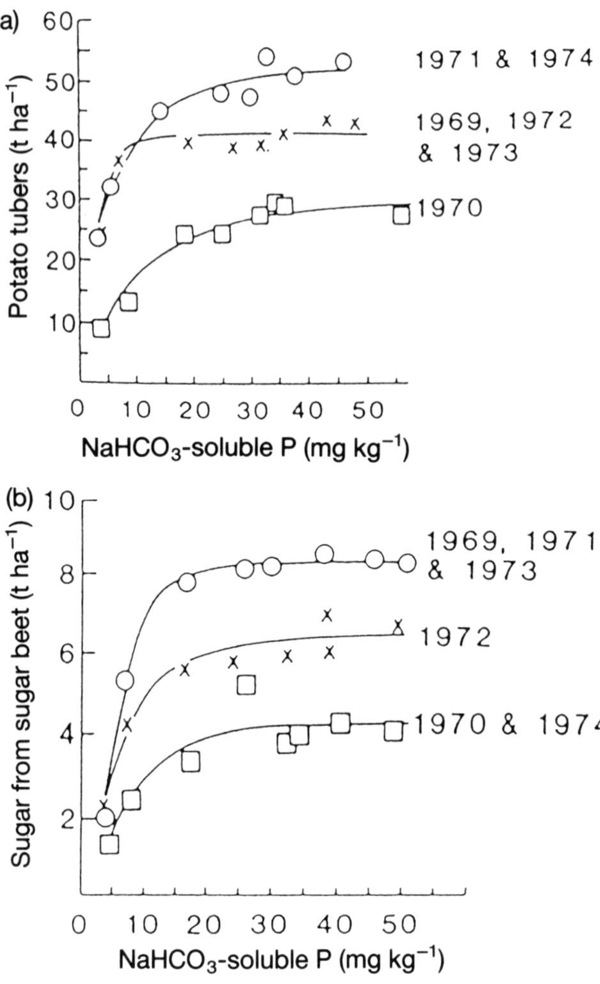

Fig. 22.6. Asymptotic regression of crop yield on $NaHCO_3$ soluble P for (a) potatoes and (b) sugar from sugar beet. Data for six years averaged for three groups of years when the asymptotic yeild was closely similar.

Johnston and Goulding (1990). These authors showed that the value of such residues can best be studied by developing plots with a wide range of both readily soluble P and K on the same soil type and under the same management. However, it takes time for both P and K to equilibrate between the various soil categories or pools into which the total P and K in soil are arbitrarily divided. But such equilibration must be allowed to happen; experience at Rothamsted suggests it takes 4–12 years. So such tests require long-term experiments. Fig. 22.6 shows the relationship between the yield of potatoes and sugar (from sugar beet) and readily soluble P. The twofold difference in yield between groups of years was because of differences in summer rainfall (Johnston *et al.*, 1985). The readily soluble P level at which yield approached the asymptote was not appreciably different for the different asymptotic yields. There was therefore a soluble P level which it would not be necessary or economic to exceed. The soil should be maintained just above the critical value for each crop as determined for each soil type. But in terms of sustainable production it is very important to realize that when soluble P levels decline below the critical value yields begin to fall catastrophically.

Some plots in the experiment from which the data in Fig. 22.6 were obtained were not given P fertilizer for 16 years. The total amount of P removed and its relation to the decline in soluble P is shown in Table 22.8. In addition, soluble P levels in soil were determined every two years (Fig. 22.7a) and the rate of decline in soluble P depended on the initial level. The individual decline curves could be brought into coincidence by suitable horizontal shifts to determine the rate of decline in soluble P over a 60-year period (Fig. 22.7b). The half-life of the soluble P was 9 years.

Compared to phosphorus it has been less easy to determine critical levels of readily soluble K in Rothamsted soils because they contain micaceous clays which release K. Fig. 22.8 shows that spring barley yields were much the same between 75 and 200 mg kg^{-1} readily soluble K but potatoes responded up to 200 mg kg^{-1}. The scatter of points in both relationships is probably because the crops got variable amounts of K from soil horizons below 23 cm (Johnston and Goulding, 1990). Also, potassium which is not readily soluble in reagents used for soil analysis, may have widely varying availability when soils are stressed to provide K (Johnston, 1988). For some soils it may be many years before mineral or matrix K becomes the only source of K. Currently there is no simple, rapid laboratory method to estimate non-readily soluble but plant available K residues in soil.

Table 22.8. Relationship between P balance and decline in $NaHCO_3$-soluble P in a sandy clay loam, Saxmundham 1969–82.

$NaHCO_3$-soluble P (mg kg^{-1}) 1969	3	7	21	28	39	44	54	67
P removed in crops (kg ha^{-1}) 1969–82	94	153	217	237	253	256	263	263
Decrease in soluble P (kg ha^{-1}) 1969–82	8	12	27	50	65	78	87	120
Change in soluble P as a % of crop uptake	8	8	12	21	26	30	33	46

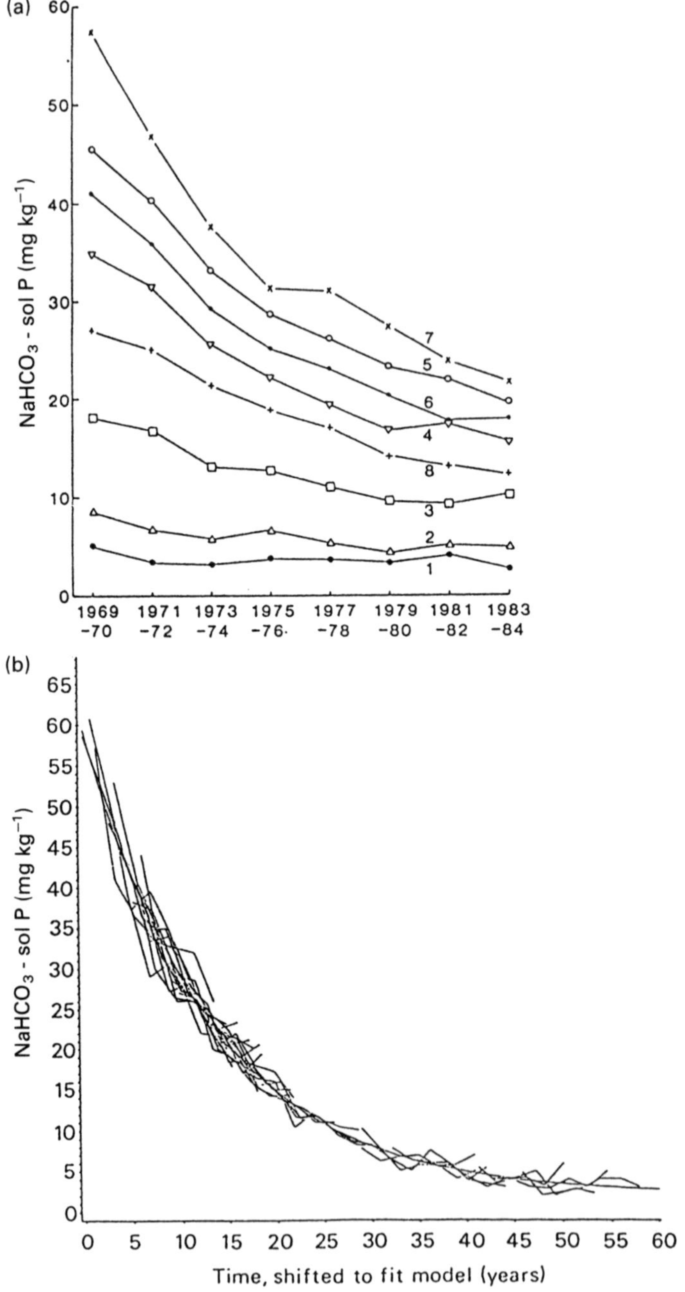

Fig. 22.7. Decrease in $NaHCO_3$-soluble P in soil given no phosphate but cropped with arable crops. (a) eight individual soils; (b) computer printout showing all data for individual plots shifted horizontally to bring curves into coincidence.

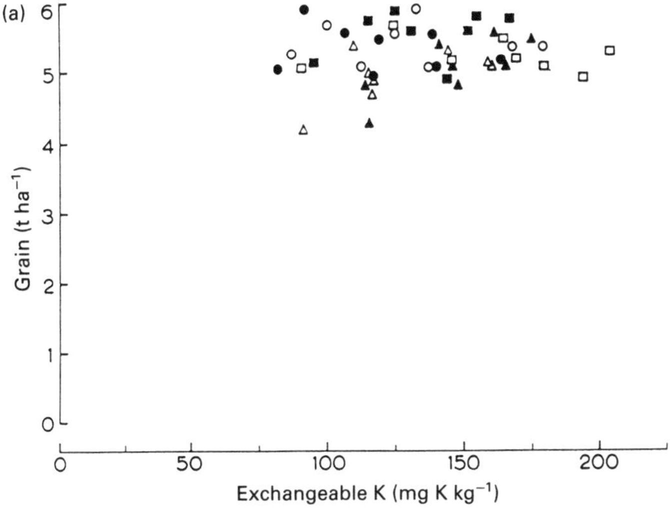

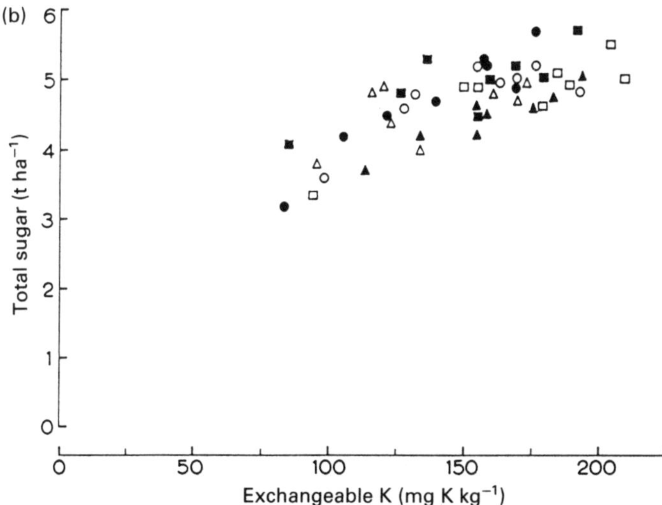

Fig. 22.8. Relationship between yields of spring barley (a) and sugar from sugar beet (b) and exchangeable K in soil. Crops grown on soils with 0.87 and 1.40% organic carbon (not shown) and manured from 1848–1951 with: NPK, circles, PK, squares; unmanured, triangles and when cropping was a 4-course rotation: turnips, barley, fallow, wheat, open symbols; turnips, barley, clover, wheat, closed symbols.

Although time consuming, the incremental release of K to cation exchange resins is perhaps the most promising method (Johnston and Goulding, 1990).

Trace Elements

Some soils may be inherently deficient in trace elements whereas others may contain amounts toxic to either plants or animals which feed on them. Deficiency or toxicity can develop in soils over a long period and such changes can only be monitored in long-term experiments. For example boron toxicity occurred after about 15 years in long-term experiments at IRRI in the Philippines where water of marginal quality was used for irrigation.

Long-term Experiments and Environmental Concerns

Pollution

Sustainable land use is being increasingly threatened by pollution causing not only soil acidification but also accumulation of heavy metals and organic pollutants, the effects of which are still only imperfectly understood. Measuring rates of accumulation, assessing bioavailability and setting critical limits for such pollutants requires long-term experiments. Only from such experiments will there be reliable data required by policy makers seeking to set limits for the amounts of pollutants in soil. The speciation of metals in waste materials added to soil may be different from that in which they are taken up by plant roots; speciation may change in soil and long periods may be required for added metals to equilibrate with other soil constituents. Also, concentrations of metals which do not affect crop growth may affect other parameters contributing to soil fertility. Brookes and McGrath (1984) showed that concentrations of heavy metals below those at which crop growth was affected had adverse effects on some components of the soil microbial population. The amount of microbial biomass in metal-contaminated soil was only half that in soil which had received FYM low in metals. The effects of the metals on nitrogen-fixing bacteria were of particular agronomic concern. Only one strain of *Rhizobium* survived in the contaminated soils and this was not able to form effective nodules on the roots of clover (McGrath *et al.*, 1988). These observations on clover were made 20 years after the last application of contaminated sewage sludge to a sandy loam soil to which sludge had been applied for 20 years in an attempt to increase the organic matter content of the soil and hence its fertility.

Point sources of pollution like those from sewage sludge can be readily identified and remedial action taken. Equally damaging to sustainable land use can be the very slow build up in soil of pollutants from atmospheric deposition. Jones *et al.* (1987) analysed soils from the Rothamsted archive to

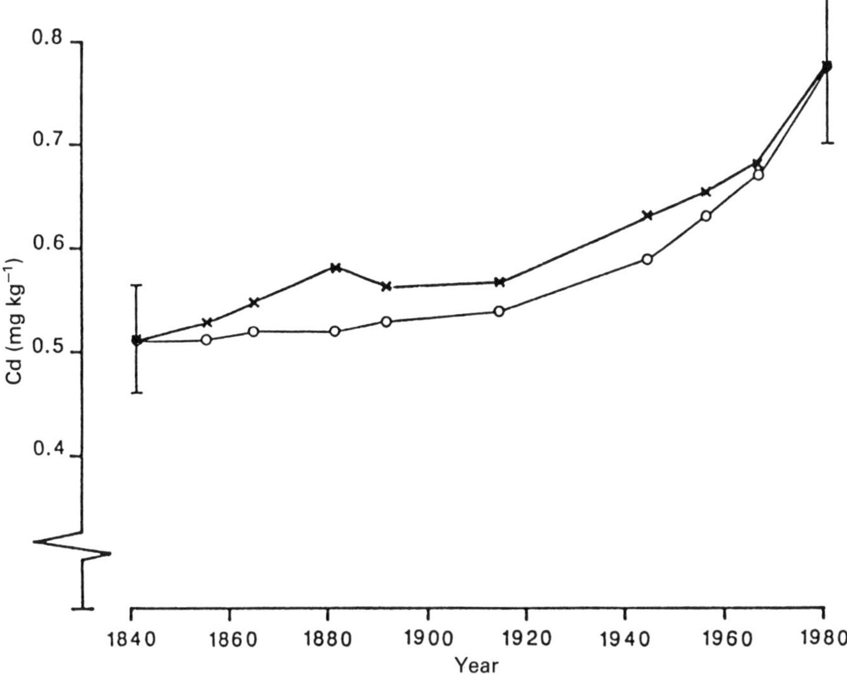

Fig. 22.9. Changes in soil Cd levels in the untreated Broadbalk plot between 1846 and 1980. Measured soil Cd dated (x—x). Predicted soil Cd concentration (o—o) based on assumed global emissions of Cd to the atmosphere. Error bars relate to analyses of eight separately digested samples for both 1846 and 1980. All other data points are the mean of duplicate analyses.

estimate changes in concentration of 12 metals over periods of up to 140 years. For eight of these metals there were no consistent changes which could be associated with either atmospheric or treatment inputs over this time scale. There was a small (~ 15%) increase of Pb from atmospheric deposition and increases in Cu, Pb and Zn associated with the large (35 t ha^{-1}) annual applications of FYM.

Johnston and Jones (1992) have recently summarized changes in the cadmium burden in soils and crops at Rothamsted and discussed their results in relation to others throughout the world. These authors undertook their detailed study because cadmium can be added to soil in superphosphate and there is concern for the health hazards it poses. Figure 22.9 shows that at Rothamsted there has been a rapid increase in soil cadmium since the 1940s on soils without any agricultural amendment, an increase which must be due to aerial deposition. It was also observed that there had been no additional accumulation of cadmium in soils low in organic matter and with pH above 6.5, to which 0.4 t ha^{-1} superphosphate had been added

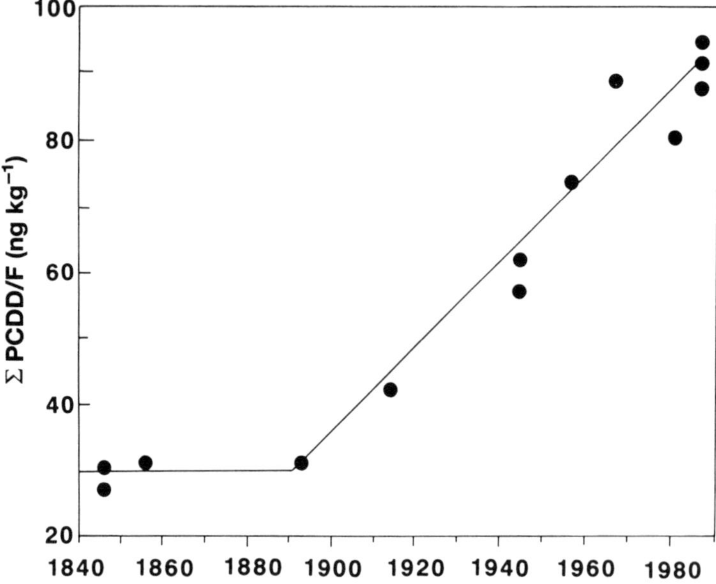

Fig. 22.10. Trends in the concentration of polychlorinated dibenzo-*p*-dioxin and -furan (ΣPCDD/F) in the 0–23 cm depth of soil on the unmanured plot of Broadbalk between 1846 and 1986.

each year for 100 years; there was some retention of cadmium added in superphosphate, however, on acid soils with more organic matter. There was no change in the cadmium concentration in cereal grain over time but that in herbage had increased. The latter could be due to increasing deposition on the surface (Johnston and Jones, 1992).

Organic pollutants like polynuclear aromatic hydrocarbons (PAHs), polychlorinated biphenyls (PCBs) and dioxins, which are of concern today, have been shown by analysis of the same archived samples to have accumulated in soil at Rothamsted, and changes in crop uptake patterns have also been determined (Jones *et al.*, 1989a, b, 1992; Kjeller *et al.*, 1991; Wild *et al.*, 1992). As an example Fig. 22.10 shows that dioxins have been increasing in soil since the beginning of this century. The value of archived samples may not be fully realized until many decades after their acquisition.

Soil erosion

Large-scale losses of fertile topsoil by wind or water erosion are a disaster ensuring loss of productive capacity. The causes of, and remedies for, rapid soil erosion are well understood and preventing such erosion is both a matter of education and of removing the pressures which bring susceptible soils into agricultural production.

Although many experiments have been conducted to establish erosion rates, there is a need to estimate the effect of less obvious soil losses on productivity. This requires carefully monitored long-term experiments, whose results need to be related to laboratory studies (cf. Chapter 4).

Climate change

Long-term data sets obtained in well-managed long-term experiments allow the construction and validation of models to describe various processes. The models can then be used to test a range of scenarios related to climate change. Jenkinson *et al.* (1991) have used Jenkinson's carbon model to estimate likely changes in the stock of soil organic carbon in a range of climatic zones if soil temperatures increase by varying amounts because of global warming. Assessing effects of climate change on crop productivity is more difficult, not least because changes in cultivars and husbandry practices, controlling pests and diseases, have allowed larger yields to be produced, but the effects of these factors are often difficult to quantify. Squarehead's Master winter wheat and closely related cultivars were grown on Broadbalk from the 1850s to 1967 but Yates (1969) considered it impossible to relate variation in yield to weather because other factors controlling yield could not be quantified or discounted.

Jenkinson *et al.* (unpublished) have recently examined yields over 100 years on five plots of the Park Grass experiment, where species composition has changed little over time. They found for both the last 30 years, when atmospheric CO_2 concentrations have been increasing, and in the preceding 70 years, that sunshine and rainfall combined account for more than 60% of the variation in annual yield. This is probably because yield potential in this system is limited by water and nitrogen and the same will probably be true in many other situations where soils support either natural vegetation or rain-fed crops.

Conclusions

The yields of winter wheat, spring barley and herbage from permanent grassland have been maintained or improved over more than 150 years in the world's oldest agricultural field experiments at Rothamsted in England. However, identical cropping on other soil types within a 40 km radius, and other husbandry systems at Rothamsted, especially those where legumes were grown frequently, became unsustainable within 30–80 years. This emphasizes the risk attached to extrapolating results from one soil type or climatic zone to another. But it is impossible to have long-term experiments on all relevant combinations of soil type, climate and cropping systems. So, inevitably, there will be a considerable amount of extrapolation but risks must be minimized. There is an urgent need to establish, in a carefully chosen range of environments, a network of long-term field experiments with husbandry systems appropriate to the soil and climate. The sustainability of each system can then be determined while changes in soil parameters affecting soil fertility and productivity are monitored. The data from individual sites can be used to construct mathematical models to describe various aspects of the system and tested with data from the other experiments to ascertain the wider applicability of the models. This greater degree of testing will minimize the risk associated with extrapolating data from one or a very few experiments.

References

Brookes, P.C. and McGrath, S.P. (1984) Effects of metal toxicity on the size of the soil microbial biomass. *Journal of Soil Science* 35, 341–346.

Dyke, G.V., George, B.J., Johnston, A.E., Poulton, P.R. and Todd, A.D. (1983) The Broadbalk Wheat experiment 1968–78. Yields and plant nutrients in crops grown continuously and in rotation. *Rothamsted Experimental Station Report for 1982* Part 2, 5–44.

Holmberg, J., Bass, S. and Timberlake, L. (1991) In: *Defending the Future.* Earthscan/IIED, London, p. 13.

Jenkinson, D.S. and Ayanaba, A. (1977) Decomposition of carbon-14 labelled plant material under tropical conditions. *Journal Soil Science Society of America* 41, 912–915.

Jenkinson, D.S. and Johnston, A.E. (1977) Soil organic matter in the Hoosfield Continuous Barley experiment. *Rothamsted Experimental Station Report for 1976* Part 2, 87–101.

Jenkinson, D.S. and Rayner, J.H. (1977) The turnover of soil organic matter in some of the Rothamsted Classical Experiments. *Soil Science* 123, 298–305.

Jenkinson, D.S., Hart, P.B.S., Rayner, J.H. and Parry, L.C. (1987) Modelling the turnover of organic matter in long-term experiments at Rothamsted. INTECOL *Bulletin* 15, 1–8.

Jenkinson, D.S., Adams, D.E. and Wild, A. (1991) Model estimates of CO_2 emissions from soil in response to global warming. *Nature* 351, 304–306.

Johnston, A.E. (1969) Plant nutrients in Broadbalk soils. *Rothamsted Experimental Station Report for 1968* Part 2, 93–112.

Johnston, A.E. (1972) Changes in soil properties caused by the new liming scheme on Park Grass. *Rothamsted Experimental Station Report for 1971* Part 2, 177–180.

Johnston, A.E. (1975) Experiments made on Stackyard Field, Woburn, 1876–1974. I. History of the field and details of the cropping and manuring in the Continuous Wheat and Barley experiments. *Rothamsted Experimental Station Report for 1974* Part 2, 29–44.

Johnston, A.E. (1986) Soil organic matter, effects on soils and crops. *Soil Use and Management* 2, 97–105.

Johnston, A.E. (1987) Effects of soil organic matter on yields of crops in long-term experiments at Rothamsted and Woburn. *INTECOL Bulletin* 15, 9–16.

Johnston, A.E. (1988) Potassium fertilization to maintain a K-balance under various farming systems. In: *Nutrient Balances and the need for Potassium,* International Potash Institute, Berne, pp. 199–226.

Johnston, A.E. (1991a) Potential changes in soil fertility from arable farming including organic systems. *Proceedings No. 306,* The Fertiliser Society of London, Peterborough, 38 pp.

Johnston, A.E. (1991b) Soil fertility and soil organic matter. In: Wilson, W.S. (ed.) *Advances in Soil Organic Matter Research: the Impact on Agriculture and the Environment.* Royal Society of Chemistry, Cambridge, pp. 299–314.

Johnston, A.E. and Chater, M. (1975) Experiments made on Stackyard Field, Woburn, 1876–1974. II. Effects of the treatments on soil pH, P and K in the Continuous Wheat and Barley experiments. *Rothamsted Experimental Station Report for 1974* Part 2, 45–60.

Johnston, A.E. and Goulding, K.W.T. (1990) The use of plant and soil analyses to predict the potassium supplying capacity of soil. In: *Development of K-fertilizer Recommendations.* International Potash Institute, Berne, pp. 177–204.

Johnston, A.E. and Jones, K.C. (1992) The cadmium issue – long-term changes in the cadmium content of soils and the crops grown on them. In: Schultz, J.J. (ed.) *Phosphate Fertilizers and the Environment,* International Fertilizer Development Centre, Muscle Shoals, USA, pp. 255–270.

Johnston, A.E. and Penny, A. (1972) The Agdell Experiment, 1848–1970. *Rothamsted Experimental Station Report for 1971* Part 2, 38–67.

Johnston, A.E. and Poulton, P.R. (1977) Yields on the Exhaustion Land and changes in the NPK contents of the soils due to cropping and manuring, 1852–1975. *Rothamsted Experimental Station Report for 1976* Part 2, 53–85.

Johnston, A.E. and Poulton, P.R. (1992) The role of phosphorus in crop production and soil fertility: 150 years of field experiments at Rothamsted, United Kingdom. In: Schultz, J.J. (ed.) *Phosphate Fertilisers and the Environment.* International Fertilizer Development Centre, Muscle Shoals, USA, pp. 45–64.

Johnston, A.E., Lane, P.W., Mattingly, G.E.G., Poulton, P.R. and Hewitt, M.V. (1985) Effects of soil and fertilizer P on yields of potatoes, sugar beet, barley and winter wheat on a sandy clay loam soil at Saxmundham, Suffolk. *Journal of Agricultural Science, Cambridge* 106, 155–167.

Johnston, A.E., Goulding, K.W.T. and Poulton, P.R. (1986) Soil acidification during more than 100 years under permanent grassland and woodland at Rothamsted. *Soil Use and Management* 2, 3–10.

Johnston, A.E., McEwen, J., Lane, P.W., Hewitt, M.V., Poulton, P.R. and Yeoman, D.P. (1993) Effects of one to six year old ryegrass–clover leys on soil nitrogen and on subsequent yields and fertilizer nitrogen requirements of the arable sequence winter wheat, potatoes, winter wheat and winter beans. *Journal of Agricultural Science, Cambridge* (in press).

Jones, K.C., Symon, C.J. and Johnston, A.E. (1987) Retrospective analysis of an archived soil collection. I. Metals. *The Science of the Total Environment* 61, 131–144.

Jones, K.C., Grimmer, G., Jacob, J. and Johnston, A.E. (1989a) Changes in the polynuclear aromatic hydrocarbon content of wheat grain and pasture grassland over the last century from one site in the UK. *The Science of the Total Environment* 78, 117–130.

Jones, K.C., Stratford, J.A., Waterhouse, K.S., Furlong, E.T., Giger, W., Hites, R.A., Schaffner, C. and Johnston, A.E. (1989b) Increases in the polynuclear aromatic hydrocarbon content of an agricultural soil over the last century. *Environmental Science & Technology* 23, 95–101.

Jones, K.C., Sanders, G., Wild, S.R., Burnett, N. and Johnston, A.E. (1992) Evidence for a decline of PCBs and PAHs in rural vegetation and air in the United Kingdom. *Nature* 356, 137–139.

Kang, B.T. and Balasubramanian, V. (1990) Long-term fertilizer trials on Alfisols in West Africa. *Transactions of 14th International Congress of Soil Science* Volume IV, Commission IV, pp. 20–24.

Kjeller, L.-O., Jones, K.C., Johnston, A.E. and Rappe, C. (1991) Increases in the polychlorinated dibenzo-*p*-dioxin and -furan content of soils and vegetation since the 1840s. *Environmental Science & Technology* 25, 1619–1627.

McGrath, S.P., Brookes, P.C. and Giller, K.E. (1988) Effects of potentially toxic metals in soil derived from past applications of sewage sludge on nitrogen fixation by *Trifolium repens* L. *Soil Biology and Biochemistry* 20, 415–424.

Powlson, D.S., Brookes, P.C. and Christensen, B.T. (1987) The measurement of soil microbial biomass provides an early indication of changes in total soil organic matter due to straw incorporation. *Soil Biology and Biochemistry* 19, 159–164.

Rothamsted Experimental Station (1955) Liming programme for the Classical Experiments. *Rothamsted Experimental Station Report for 1954* pp. 146–148.

Saffigna, P.G., Powlson, D.S., Brookes, P.C. and Thomas, G. (1989) Influence of sorghum residues and tillage on soil organic matter and soil microbial biomass in an Australian Vertisol. *Soil Biology and Biochemistry* 21, 759–765.

Thurston, J.M., Williams, E.D. and Johnston, A.E. (1976) Modern developments in an experiment on permanent grassland started in 1856; effects of fertilisers and lime on botanical composition and crop and soil analyses. *Annales Agronomique* 27 (5–6), 1043–1082.

Warren, R.G. and Johnston, A.E. (1964) The Park Grass Experiment. *Rothamsted Experimental Station Report for 1963* 240–262.

Warren, R.G., Johnston, A.E. and Cooke, G.W. (1965) Changes in the Park Grass Experiment. *Rothamsted Experimental Station Report for 1964* 224–228.

Wild, S.R., Jones, K.C. and Johnston, A.E. (1992) The polynuclear aromatic hydrocarbon (PAH) content of herbage from a long-term grassland experiment. *Atmospheric Environment* 26, 1299–1307.

Yates, F. (1969) Investigations into the effects of weather on yields. *Rothamsted Experimental Station Report for 1968* Part 2, 46–49.

Chapter 23

The Setting-up, Conduct and Applicability of Long-term, Continuing Field Experiments in Agricultural Research

A.E. Johnston and D.S. Powlson

AFRC Institute of Arable Crops Research, Rothamsted Experimental Station, Soil Science Department, Harpenden, Herts., AL5 2JQ, UK

Introduction

The world's oldest long-term agricultural field experiments were started between 1843 and 1856 by J.B. Lawes and J.H. Gilbert at Rothamsted, 40 km north of London in South East England. In 1895, Lawes and Gilbert attempted a brief summary of how they came to be started. They stated that,

> the general plan was to grow some of the most important arable crops in rotation, each separately, year after year, for many years in succession on the same land, without manure, with farmyard manure and with a great variety of chemical manures (fertilizers), the same description of manure being as a rule applied year after year on the same plot. ... field experiments, conducted as above described, must of themselves throw much light on the characteristic requirements (for nutrients) of the particular crop ...
>
> (Lawes and Gilbert, 1895)

Considering the lack of scientific knowledge about many aspects of crop production in the 1840s, this approach was very successful. Their results clearly demonstrated the source from which all plants, other than legumes, get their nitrogen, and the relative importance of phosphorus and potassium. The quotation stresses the importance of continuity of treatment because by the late 1840s, Lawes and Gilbert were regretting their frequent modifications to the treatments on the Broadbalk experiment in the first few years following 1843.

(eds D.J. Greenland and I. Szabolcs)

Eight of the field experiments started by Lawes and Gilbert still continue. In recent years there have been various, carefully considered modifications which have increased their agronomic and scientific value. These experiments, together with other long-term experiments demonstrate that they can be used to measure and evaluate the effects of both farming practices and non-agricultural anthropogenic activities on soil fertility and water quality and hence the sustainability of crop production. Today more such experiments are needed because it has become increasingly obvious that although the principles of crop growth and soil processes have wide applicability they interact and respond to other external factors, like weather, in very different ways in different parts of the world. Provided such experiments are both well designed and executed their value increases with time, as emphasized from the 50-year experience of Lawes and Gilbert (1895). However, longevity makes experiments ever more costly, but their cost effectiveness can be increased if they serve a number of different objectives, and if they are conducted on well-characterized sites so that the results can be extrapolated as widely as possible. Among the more easily identified objectives are:

- To test the sustainability of a particular husbandry system over a long time span and determine what changes in husbandry are needed to enhance productivity and maintain sustainability.
- To provide data of immediate value to farmers to improve best husbandry practices.
- To provide a resource of soil and plant material to further scientific research into soil and plant processes which control soil fertility and crop production.
- To allow a realistic assessment of non-agricultural anthropogenic activities on soil fertility and crop quality.
- To provide long-term data sets which can be used to develop mathematical models to predict the likely effects of management practices and of climate change on soil properties, on the productive capacity of soils, and on the wider environment.

The value of long-term experiments is enhanced when crop and soil samples are archived in appropriate conditions. This allows samples to be retrospectively analysed using new or improved techniques for elements or compounds about which there is a developing scientific interest or environmental concern and which may not have been estimated previously.

This chapter discusses and gives examples of some of the points, which need to be considered when setting up long-term experiments in agriculture. It is not intended as a detailed 'how-to-do' manual, but rather as a check list of some aspects which need to be considered. Among these are flexibility to respond to events which might bring the experiment to an end and the need to be able to introduce new factors to increase the usefulness of the data.

The Need for Flexibility

Within the biological population of soil there are organisms which are beneficial, such as *Rhizobium*, and harmful, such as parasitic nematodes. Populations of both are affected by the farming system and the chemical and physical properties of soil over long time scales and in ways that differ between soils. For example, soil-inhabiting nematodes, as opposed to seed-dispersed ones, are often a major problem on the sandy loam at Woburn but rarely so on the heavier textured soil at Rothamsted; too many pores in the latter soil are too small for nematodes to move freely through the soil mass. Again, soil acidity can encourage an increase in soil-borne fungal pathogens and interactions between fungal pathogens can become more important. It is probably very unusual for a single factor to cause a farming system to collapse, but identifying interactions is difficult. In addition, husbandry systems are constantly evolving and often it is innovative farmers, especially in developed economies, who introduce new practices at the farm level, e.g. direct drilling. Therefore there is a need for long-term experiments both to explain successes and failures in terms of measurable parameters, and to assess their applicability to a wide range of soils and climates. Thus the design of long-term experiments must allow flexibility to introduce changes although frequent, mindless changes to accommodate ephemeral fads and fashions must be avoided.

Two types of change can be identified. First there are changes which have to be made in response to factors which would otherwise jeopardize the continuity of an experiment. The second are changes made to test new ideas or hypotheses. Both types of change have been made in Rothamsted experiments and both have enhanced their value considerably. Some examples follow.

Changes to Ensure Continuity

Combined effects of chemical and biological changes

A complicated, but excellent example of slow changes in chemical and biological properties and their effect on sustainability comes from the Woburn Ley Arable experiment. Because the experiment had contrasted treatments it was in some cases possible to decide why different husbandry systems became non-sustainable. Started in 1938, the experiment initially compared four contrasted cropping sequences, each lasting three years. They were: (A) a grass-clover ley grazed by sheep; (B) lucerne (alfalfa) cut three or four times each year; (C) potatoes, wheat, kale; (D) potatoes, wheat, one-year ley. The subsequent effects of these sequences on two arable crops, initially potatoes then barley, were measured. The five-year rotations were, therefore, three years of ley and two arable crops or five years of arable crops; each phase of the rotation was present each year on one of the five blocks which constituted the experiment. The design of the experiment was very ambitious. Half the plots always had the same rotation, designated continuous rotation plots, on each plot there were four replicates of each rotation in 20 years. The other half of the plots tested a practice which might well have occurred in commercial farming. Each arable rotation was followed by a ley rotation and *vice versa*, e.g. sequences were C A D B or D A C B. Designated 'alternating rotation' plots there was least chance of a build-up of soil-borne pests and diseases, but on each plot it took 20 years to complete a cycle.

At the beginning the same total amounts of phosphorus and potassium were applied over the 5 years of each rotation. However, about 1952 it became obvious that the yields of potatoes following lucerne were less than those following grazed ley and, unexpectedly, less than those in the all-arable rotations. In other words, there was no benefit from the lucerne ley, which was contrary to the perceived wisdom. It was shown subsequently that the lucerne was removing more potassium than was being applied and more than the crops in any of the three other rotations; this was depleting soil reserves. A test of additional P and K first to potatoes, and then to sugar beet, showed this to be so. The P and K manuring was increased and in 1962 altered again so that each rotation was manured according to its estimated needs (Boyd *et al.*, 1968).

The experiment had been running for 18 years before the first biological problem became apparent. In 1955 there were large differences in the yields of potatoes grown in the treatment phase of the arable rotations between plots which were in the Alternating and Continuous rotation cycles (Table 23.1). Potatoes grown on these plots in 1955 were the sixth and eighth potato crops respectively since the start of the experiment 18 years previously and there had been a greater build-up of potato cyst nematodes (*Heterodera rostochiensis*) where potatoes were grown more frequently. Yields of winter rye were not affected. Sugar beet replaced potatoes as first test crop and the other arable crops in the all-arable rotations were changed, so that no crop was grown more than once in 5 years, to diminish the effects of soil-borne pathogens.

In 1957, lucerne on some of the plots, particularly in the second and third year, was seen to be damaged by stem eelworm (*Ditylenchus dipsaci*). Fumigation of the soil and the use of fumigated seed did not resolve this problem and lucerne was eventually replaced by clover.

Sugar beet was grown for 12 years. During the last three, the best yields with optimal amounts of fertilizer N were grown following lucerne, which yielded about 0.95 t ha^{-1} more sugar than the crop following the grazed ley (Table 23.2). This large difference was unexpected and opposite to that

Table 23.1. Yields (t ha^{-1}) of potatoes and winter rye in arable rotations following different previous rotations. Woburn Ley Arable 1955. (Adapted from Rothamsted Experimental Station, 1955.)

	Rotation cycle and previous rotation			
	Alternating		Continuous	
	Ley	Lucerne	Arable (Hay)	Arable (Roots)
Potato tubers	19.4	14.7	4.8	3.8
Rye grain	4.54	4.07	4.29	4.15

Rotations were:	Treatment phase	Test phase
Ley	3 year grass ley grazed by sheep	Potatoes, barley
Lucerne	3 year lucerne ley cut for conservation	Potatoes, barley
Arable (Hay)	Potatoes, rye, hay	Potatoes, barley
Arable (Roots)	Potatoes, rye, sugarbeet	Potatoes, barley

For definition of alternating and continuous rotations see text. These rotations, with some minor modifications in the arable crops, had been maintained for 18 years before the results quoted here.

Table 23.2. Yields of sugar beet with optimum nitrogen in different rotations. Woburn Ley Arable 1965–1967. (Adapted from Boyd *et al.*, 1968.)

	Rotation			
	Ley	Lucerne	Arable (Hay)	Arable (Roots)
Total sugar (t ha^{-1})	8.36	9.31	8.70	9.14

Rotations were:	Treatment phase	Test phase
Ley	3 year grass ley grazed by sheep	Sugar beet, barley
Lucerne	3 year lucerne ley cut for conservation	Sugar beet, barley
Arable (Hay)	Potatoes, rye, hay	Sugar beet, barley
Arable (Roots)	Potatoes, rye, carrots	Sugar beet, barley

These rotations, with some minor modifications in the arable crops, had been maintained for 28 years before the results quoted here.

recorded earlier for potatoes. Eventually it was shown that there was a much larger population of free-living nematodes (*Longidorus* and *Trichodorus*) in the grazed ley plots. Yields in the arable rotations were similar to those following lucerne, provided that the nitrogen applications were adjusted correctly.

After 40 years much had been learned about the problems of doing the experiment and maintaining sustainability of production. It was necessary to manure with P and K to meet the needs of the crop; in 1938, uniform P and K manuring to all rotations had followed the then fashionable rigorous statistical approach. It was essential to be aware of the role played by pests and diseases and to try either to eliminate them or estimate their effect on yield. None of these problems could have been foreseen at the outset. Only in the last few years or so has it been possible to measure the effects of extra organic matter accumulated in soil by growing leys, which was the original aim of the experiment.

Effects of acidity

Soil acidity, particularly when it results from some of the treatments tested, can mask treatment effects. Soil acidity, arising from the use of ammonium sulphate, decreased yields of spring barley on the sandy loam at Woburn after only 15 years, whereas it took 80 years for the effects of soil acidity to decrease yields of turnips at Rothamsted (Chapter 22). These results highlighted the need to maintain soil pH in the UK, but liming many tropical soils, which decreases the effects of acidity, can induce zinc deficiency. Extrapolating without testing can lead to disaster.

At the start of long-term experiments it is important to remember that soil acidity can also affect yields through factors which are not being tested. These factors include: the mobilization of Al, Fe and Mn in very acid soils; varying the populations of fungal pathogens, e.g. the severity of *Guaemannomyces graminis* appears to depend on soil pH (Glynne, 1935); and varying the efficiency of some soil-applied herbicides.

Effects of weeds

Although many farmers appreciate the need for weed control, quantifying the benefits to justify the cost can only be done in long-term experiments because the seed bank in both treated and untreated soils takes many years to stabilize. During the first 70 years of its history, weeds in the winter wheat crop on Broadbalk were controlled by hoeing, either with horse hoes or by hand. After about 1915 labour was no longer readily available for this task and weed competition gradually decreased yields (Chapter 22, Fig. 22.1). In 1926 the plots were divided into five sections which were periodically bare fallowed until, in 1931, a regular cycle of bare fallowing, 1 year fallow, 4 years wheat, was introduced in turn on each section (Johnston and Garner, 1969). This led to some interesting research on weed biology and the persistence of weed seeds (Thurston, 1969); on the effects of a bare fallow on yields of wheat (Garner and Dyke, 1969), and on wheat bulb fly (Johnson *et al.*, 1969). Regular bare fallowing continued until 1964 when, after some preliminary trials starting in 1957, weedkillers were used on all but one half section. This half section has continued to be without weedkillers, and the effects of weeds can now be estimated. Weed competition on the higher yielding plots decreased yield in 1985–90 by as much as

1.8 t ha^{-1} grain (Table 23.3); whereas on the plot with farmyard manure (FYM) plus fertilizer N the lost yield exceeded 3.0 t ha^{-1}. On the plot which received PK fertilizers but no N, two annual legumes, *Vicia sativa* and *Medicago lupulina* flourished. The nitrogen they fixed, either directly or indirectly through ploughed-in residues, increased wheat yield by more than 1.0 t ha^{-1} compared to where these legumes were killed by spraying (Table 23.3).

In areas prone to soil erosion, bare fallowing is, of course, not an option for weed control and weeds, together with crop residues after harvest, are one way of minimizing the risk of erosion. Such a need must be identified at the start of any long-term experiment and a decision taken whether to leave weeds or, if possible, to measure soil erosion in the presence or absence of crop and weed residues which act as a surface mulch.

Availability of cultivars

Whether to change cultivars of annual crops, and if so how frequently, is perhaps one of the more difficult decisions for sponsors of long-term experiments. This is especially so now that there is often a bewildering array of new cultivars to choose from. The benefits to yield of changing cultivars are well illustrated by results from Rothamsted (Chapter 22 and Fig. 23.1). On the Broadbalk wheat experiment there was a long period when Squareheads Master was grown and it was hoped that it would be possible to investigate effects of weather on wheat yields but Yates (1969) found this impossible largely because there were many sources of variation, only some of which could be quantified. Current practice for long-term experiments at Rothamsted is to review cultivars regularly. On Broadbalk for example, this is done every five years when a five-course rotation has completed a cycle.

Besides effects on yield, one important observation from changing

Table 23.3. Effect of using weedkillers on yields of winter wheat grown continuously during 1985–90, Broadbalk, Rothamsted. The comparison with and without weedkillers started in 1963 after 120 years of continuous wheat cropping with occasional bare fallows to control weeds. Since 1963 the section without weedkillers is bare fallowed occasionally when weed populations become excessive.

Treatment	Mean annual yield of grain (t ha^{-1}) at 85% dry matter								
	PK plus N (kg ha^{-1})							FYM plus N (kg ha^{-1})	
	0	48	96	144	192	240	288	0	96
No weedkillers	2.9	3.9	4.4	4.9	5.4	4.5	5.1	5.0	4.9
Weedkillers	1.4	4.0	5.3	6.6	6.2	6.3	6.9	6.6	8.2

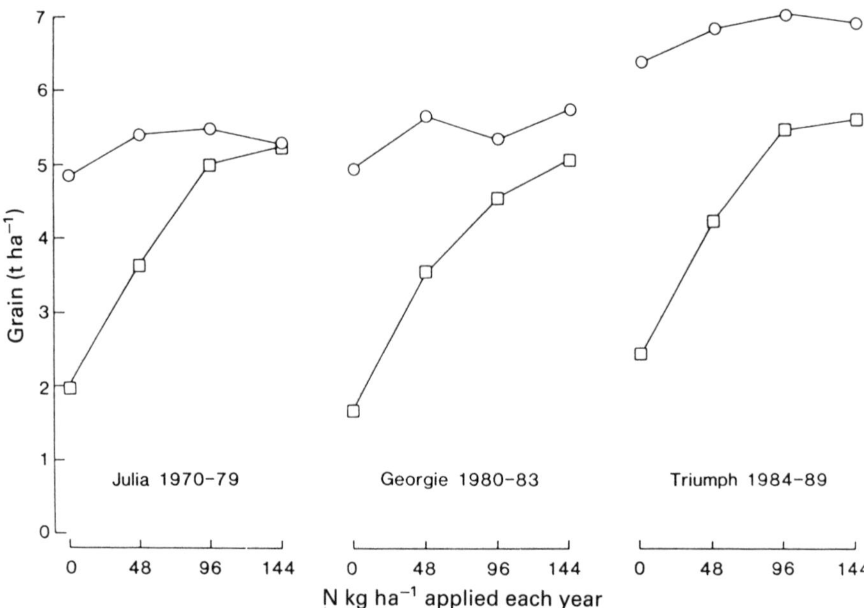

Fig. 23.1. Mean annual yields of three cultivars of spring barley grown continuously on soils which have received either PK fertilizers (□) or farmyard manure (○) each year since 1852, Hoosfield Barley experiment, Rothamsted. (Adapted from Johnston, 1991.)

cultivars on Broadbalk has been the improvement in recovery of fertilizer nitrogen (Table 23.4). From values of about 30% up until the 1970s, recovery now exceeds 70% because present cultivars yield more grain which contains most nitrogen. Within each period recovery of nitrogen varied little with the amount applied. No matter how little fertilizer nitrogen was used there was always some which could not be accounted for. Such information obtained on one site over many years is invaluable in discussions with those concerned with the environmental aspects of fertilizer N use because it gives reliable long-term data on the amounts of N not accounted for by crop removal and changes in organic N in soil.

The need to change cultivars may also focus attention on other soil factors which affect yield. Figure 23.1 shows grain yields of three cultivars of spring barley grown on the Hoosfield Continuous Barley experiment since 1970. In this experiment, started in 1852, two of the treatments compare annual applications of PK fertilizers, 33 kg P, 90 kg K ha^{-1}, and animal manure (FYM), 35 t ha^{-1}. By 1968 the FYM-treated soils contained two and a half times as much soil organic matter as fertilizer-treated soils, and in that year both plots were divided to test four amounts of N as inorganic fertilizer, 0, 48, 96, 144 kg ha^{-1}. In the first period, 1970–1979, yields of cv Julia on fertilizer-treated soils given 96 kg N ha^{-1} were the same as on FYM-

Table 23.4. Percentage recovery of fertilizer nitrogen applied to winter wheat grown continuously on Broadbalk, Rothamsted. (Adapted from Johnston and Jenkinson, 1989.)

	% recovery[a]			
Period	48 kg N ha^{-1}	96 kg N ha^{-1}	144 kg N ha^{-1}	192 kg N ha^{-1}
1852–1871	32	33	32	29
1966–1967	32	39	36	–
1970–1978	56	63	59	52
1979–1984	69	83	76	69
1985–1987	67	77	67	57

[a]In grain plus straw and determined using N in crop grown on plot where no N was applied as control.

treated soils. This equivalence of the yields with fertilizers and FYM had been an unchanging feature of the results since the experiment started. In 1980–1984 yields of cv Georgie on fertilizer-treated soils were again equal to those on FYM-treated soil, but 144 kg N ha^{-1} was needed. More importantly, however, on FYM-treated soils yields were further increased by giving extra fertilizer N. In the third period, 1985–1990, cv Triumph showed an even greater response to extra fertilizer on FYM-treated soil. Spring sown crops, with a high yield potential, have to grow quickly to achieve good yields and this requires good soil physical conditions for rapid root growth to explore the soil for nutrients and water. A similar benefit of having extra organic matter in soil is also seen for autumn sown winter wheat on Broadbalk (Table 23.5). In recent years yields have always been largest when extra fertilizer N was given to crops grown on FYM-treated soil. On these plots the readily available N from the annual application of FYM and the nitrate mineralized from the soil organic matter is not sufficient for the yield potential of current cultivars. It is thought that improved soil structure and the slightly enhanced water-holding capacity in the FYM-treated soil is the explanation for the difference in yields now observed between the two soils when sufficient fertilizer N is applied; it is also possible that part of the benefit on FYM-treated soil comes from mineralization of organic N late in the growing season although this has a greater effect on grain %N than on yield (Johnston and Jenkinson, 1989).

Plot size and soil movement

Decisions about plot size in long-term experiments must take account of the need for soil cultivation, especially ploughing and harrowing, which can move soil across plot boundaries. That soil had been moved between plots

Table 23.5. Yields of winter wheat (grain t ha^{-1}) given by fertilizers, farmyard manure and farmyard manure plus fertilizer N, Broadbalk, Rothamsted.

	Cultivar grown			
	Flanders 1979–1984		Brimstone 1985–1990	
Treatment	Continuously	In rotation	Continuously	In rotation
NPK[a]	6.93	8.09	6.69	8.61
FYM	6.40	7.20	6.17	7.89
FYM + N[b]	8.13	8.52	7.92	9.36

[a]Best yields of cv Flanders were given by 192 kg N ha^{-1} and of cv Brimstone by 288 kg N ha^{-1}.
[b]FYM plus 96 kg ha^{-1} fertilizer N.

during 100 years of cultivation was recognized on the Hoosfield Continuous Barley experiment at Rothamsted in the 1960s (Warren and Johnston, 1967) and the harvested area of the plot was truncated to allow for this. Sibbesen (1985) introduced the concept of a diffusion process to describe soil movement and modelled it in one dimension using data from an experiment at Askov in Denmark. McGrath and Lane (1989) developed a more complex two-dimensional model using data for concentrations of heavy metals in soils across transects of plots to which very different amounts of metals had been added in sewage sludge over a period of 20 years. Fig. 23.2 shows the

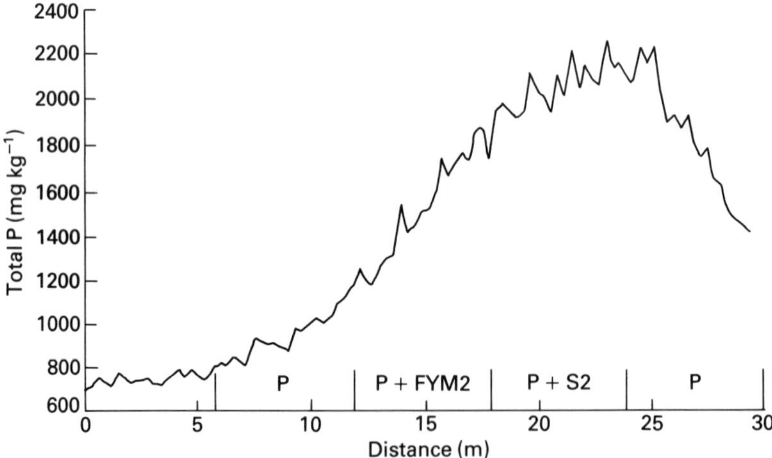

Fig. 23.2. Change in the total P content of soil along a transect crossing plots which had received varying amounts of P. All plots received superphosphate (supplying 1.1 t P ha^{-1}), in addition farmyard manure (FYM 2) supplied 7.7 t P ha^{-1} and sewage sludge (S2) supplied 13.1 t P ha^{-1}. (Adapted from Johnston, 1989.)

phosphorus concentration along a transect across four plots. Each plot was only 6.04 m wide × 8.38 m long and ploughing was always at right angles to the length of the plot. Such movement is difficult to observe unless crops respond to the soil nutrient gradient with a gradient in crop growth. However, the effect of cultivation, together with that of soil movement due to soil erosion processes, can seriously jeopardize the useful life of an experiment. If plots have to be small, soil movement between plots must be prevented, e.g. by concrete barriers or grass guard strips, and on larger plots every effort should be made to minimize soil movement.

Changes to Test New Ideas

Effect of soil-borne pathogens

In their experiments on arable crops, Lawes and Gilbert grew the crops in monoculture with one exception. On Broadbalk and Hoosfield this has been spectacularly successful. This is especially so when considered against our present knowledge of fungal pathogens and root diseases. Glynne (1969) discussed the detailed records of crop growth on Broadbalk up to the 1930s and related accounts of poor growth to possible fungal diseases. She pointed out that diseases had been thought to be no worse on Broadbalk than on fields growing cereals in rotation and that growing cereals continuously led to no accumulation of any disease. However, since the 1930s our knowledge of plant pathology has developed rapidly and two major diseases of cereals, eyespot caused by *Cercosporella herpotrichoides,* and take-all, caused by *Guaemannomyces graminis,* have been studied extensively on Broadbalk.

Observations of the incidence of take-all on wheat on Broadbalk and other fields gradually led to the idea that when wheat was grown continuously factors inimical to take-all prevented it developing in its most severe form (Glynne *et al.*, 1956); a feature that became known as take-all decline. However, although take-all decline occurred when susceptible cereals were grown continuously, there was evidence to suggest that even with maximum take-all decline, yields could be less than in the absence of take-all. To test this it was decided in 1968 to subdivide each of the five sections on Broadbalk to create ten sections. On some sections wheat was grown continuously, on others a two-year break from cereals was introduced. Such a break was known to minimize any risk of take-all affecting the next crop. Yields during the next 15 years are shown in Fig. 23.3. From 1970 to 1978 cv Cappelle Desprez was grown. On plots given fertilizers, yields of wheat grown continuously increased up to 96 kg N ha^{-1} with little further increase to more N; when grown after a two-year break, yields peaked at 96 kg N ha^{-1} and then declined. When 96 kg N ha^{-1} was given the benefit

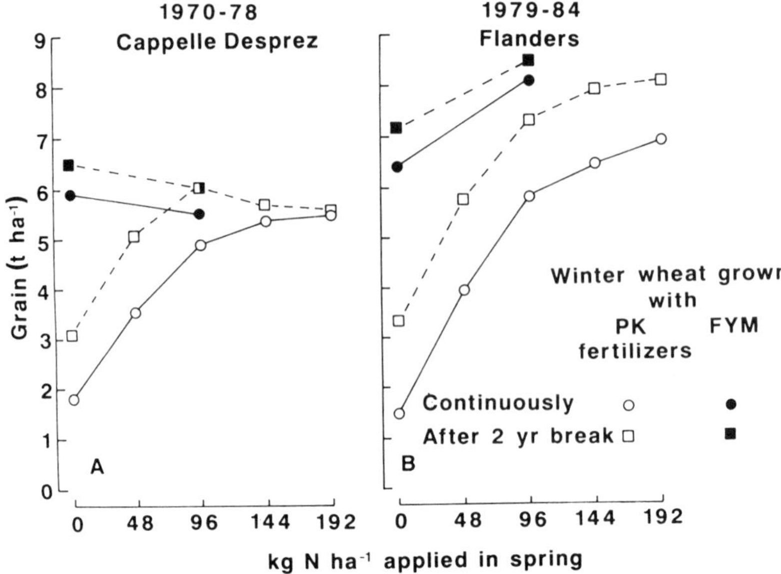

Fig. 23.3. Relationship between the mean yield of two cultivars of winter wheat, grown in two rotations and spring-applied fertilizer N during two periods on soils with two levels of organic matter because of treatment with or without FYM since 1843, Broadbalk, Rothamsted.

of the two-year break was 1.18 t ha^{-1} grain. On plots with more organic matter from repeated applications of FYM, the two-year break increased yields by only 0.63 t ha^{-1} and yields declined when extra fertilizer N was given. In 1979–1984 cv Flanders was grown and yields in all situations increased up to the maximum amount of N tested (192 kg ha^{-1}). At this level of N the benefit of the two-year break was 1.14 t ha^{-1} on fertilizer-treated soils; on FYM-treated soils the effect of 96 kg N ha^{-1} was less, only 0.14 t ha^{-1}. The different efficiency with which fertilizer N was used in the two periods was because fungicides were used to control foliar pathogens in the second period. The effect of breaking take-all decline by having a two-year break and then three consecutive wheats has been tested since 1985. With fertilizers, FYM and FYM + N, best yields of the third wheat after a two-year break were always less than those of wheat grown continuously (Table 23.6) because after a two year break from cereals the causative agent for take-all decline built up less quickly than *Guaemannomyces graminis*.

Thus, in this long-term experiment, continuous wheat production has not only been sustained but increased, the value of extra organic matter in soil has been demonstrated and the role of take-all and take-all decline shown. There is also an important message for farmers, namely that although it has been possible to grow wheat continuously on this soil with careful attention to management, there are benefits to be obtained if profit-

Table 23.6. Yields of first, second and third wheat after a 2-year break compared with those of wheat grown continuously. Broadbalk, Rothamsted 1985–1990.

	Wheat after a 2-year break			
Treatment	1st	2nd	3rd	Continuous wheat
NPK[a]	8.61	7.85	6.47	6.69
FYM	7.89	5.86	5.37	6.17
FYM + N[b]	9.36	8.64	7.59	7.93

[a]Best yield of cv Brimstone was given by 288 kg N ha^{-1}.
[b]FYM plus 96 kg ha^{-1} fertilizer N.

able crop rotations can be devised in which most wheat crops are grown after a two-year break.

Effect of P and K accumulated in soil from fertilizers and manures

Lawes and Gilbert applied large amounts of P and K in their experiments; annual applications were 33 kg P ha^{-1}, and 90 kg K ha^{-1} to cereals and 200 kg K ha^{-1} to root crops and grass. These quantities far exceeded offtakes in the harvested produce and Lawes and Gilbert made some simple tests of effects of the residues accumulated from previous applications of P and K fertilizers (Table 23.7). Liebig (1872) and Dyer (1902) showed that dilute acid extractants removed more P and K from fertilized soils than from unmanured ones once fertilization had continued for some years, but only a fraction of the estimated residue remained readily soluble. Recent studies have shown that only about 13% of the increase in total soil P from applications of fertilizers and FYM remains soluble in 0.5 M $NaHCO_3$ (Johnston and Poulton, 1992) and about 40% of K residues remain exchangeable to 1M NH_4OAc (Johnston and Goulding, 1990). However, the cumulative effects of residues from many applications may be important for soil fertility and sustainable land use.

Table 23.7. Effect of residual P and K in soil on yields (t ha^{-1}) of winter wheat grain, Broadbalk, Rothamsted. See text for details. (Adapted from Johnston, 1970.)

	Annual treatment			
Period	PK	N	NPK	N with PK residues
1852–1861	1.3	1.6	2.4	2.2
1862–1871	1.1	1.8	2.6	2.2

In 1949, P and K residues from 50 applications of PK fertilizers and 26 applications of FYM applied between 1856 and 1901 in the Exhaustion Land experiment at Rothamsted gave such large increases in yields of spring barley that further studies on the evaluation of accumulated residues were started. Yields (Table 23.8) showed that, on this soil, with pH above 6.0, P and K residues from inorganic fertilizers had remained in forms available to plants for more than 50 years. This result was subsequently confirmed in the Hoosfield Barley experiment where P and K residues from 20 applications of FYM were still available to plants after 100 years (Table 23.9). In both experiments the beneficial effects on yield were from the combined effects of P and K. The evidence that P and K residues could accumulate in soil and remain available to plants over many years was indisputable. However, it must be pointed out that between 1872 and 1968 on Hoosfield and 1902 and 1941 on Exhaustion Land no nitrogen was applied to the crops. In consequence yields and the amounts of P and K removed from the soil were small. Once N was applied, and the soils were stressed to supply P and K because yields increased, the effectiveness of the residues began to decline (Tables 23.8 and 23.9).

In 1957 it was decided to test P and K separately and the plots in a number of experiments were subdivided to make these tests. Johnston and Warren (1970) gave examples of the way in which the subdivisions were made. These were only possible because the original experiments had large plots. A selection of results (summarized in Tables 23.10 and 23.11) showed, not surprisingly, that soils with residues gave larger yields than those

Table 23.8. Effect of residual P and K in soil on yields of spring barley, Exhaustion Land, Rothamsted, 1949–1974.

	Grain (t ha^{-1}) at 85% dry matter				
	Treatment[a] 1856–1901				
Period	Unmanured	FYM	NPK	P[b]	PK
1949–1953	1.59	3.03	2.87	2.72	3.05
1954–1959	1.80	3.32	3.10	2.81	3.08
1960–1962	2.02	3.09	2.62	2.59	2.73
1963	1.90	3.27	2.86	3.21	2.99
1964–1969	1.71	4.28	3.61	3.59	3.61
1970–1975	1.83	4.75	4.22	3.76	4.53

[a]For details see Johnston and Poulton (1977).
[b]K applied 1856–75.

Table 23.9. Effect of P and K residues accumulated in soil between 1852 and 1871 on yields (t ha^{-1}) of spring barley in 1970–73 and 1988–1991, Hoosfield, Rothamsted. (Adapted from Johnston and Poulton, 1992.)

	Treatment					
	1852–1991 Unmanured		1852–1871 only FYM		1852–1991 FYM	
Nitrogen, (kg ha^{-1}) since 1968	1970–1973	1988–1991	1970–1973	1988–1991	1970–1973	1988–1991
None	1.46	0.98	1.07	2.20	5.38	5.50
48	2.01	1.43	3.20	3.24	5.86	5.94
96	2.12	1.73	5.27	3.48	5.62	6.06
144	2.26	1.66	4.94	3.63	5.38	6.00

Table 23.10. Effect of residual P in soil and freshly applied P fertilizer on yields (t ha^{-1}) of three arable crops.

		Experiment and bicarbonate soluble P (mg kg^{-1})					
		Agdell		Exhaustion Land		Woburn	
Crop	P applied (kg ha^{-1})	4	13	4	12	18	42
Spring barley	0	1.54	3.41	2.03	3.11	2.62	3.34
grain	56	2.89	3.88	3.49	3.48	2.86	3.61
Potato	0	12.1	29.9	12.8	21.1	35.1	40.7
tubers	56	25.4	38.2	32.6	32.6	38.2	43.4
Sugar from	0	3.38	5.77	3.84	5.67	5.17	5.90
beet	56	4.79	6.00	5.72	6.00	5.15	6.15

For each experiment the soils with the lower levels of bicarbonate soluble P had received no P fertilizer since the start; Agdell, 1848; Exhaustion Land, 1852; Woburn, 1876.

without. More importantly, yields on impoverished soils were not increased to those on enriched soils even when generous quantities of freshly applied P or K were given. Again this result derived from long-term experiments highlights the importance of maintaining the P and K fertility of soil both to the immediate benefit of farmers and to the longer-term sustainability of crop production.

Table 23.11. Effect of residual K in soil and freshly applied K fertilizer on yields (t ha^{-1}) of five arable crops.

		Experiment and exchangeable K (mg kg^{-1})					
		Exhaustion land		Woburn		Saxmundham	
	K applied (kg ha^{-1})	83	111	61	84	113	166
Spring barley grain	0	3.34	3.54	3.14	3.32	nt	nt
	63	3.56	3.56	3.38	3.31	nt	nt
Winter wheat grain	0	nt[a]	nt	nt	nt	8.49	8.50
	52	nt	nt	nt	nt	8.54	8.60
Potato tubers	0	17.1	27.6	32.9	41.2	28.8	43.1
	125[b]	31.1	36.7	44.2	47.2	39.6	44.0
Sugar from beet	0	4.87	6.15	3.59	4.60	nt	nt
	125[c]	5.67	5.43	5.81	5.80	nt	nt
Beans[d] grain	0	nt	nt	nt	nt	2.52	4.42
	52	nt	nt	nt	nt	3.60	4.38

[a]nt, not tested.
[b]K to potatoes 208 kg ha^{-1} at Saxmundham.
[c]K to sugarbeet 375 kg ha^{-1} at Woburn.
[d]*Vicia faba.*
For each experiment the soil with the lower level of exchangeable K had received no K fertilizer since the start; Exhaustion Land, 1852; Woburn, 1876; Saxmundham, 1899.

Testing efficient use of nitrogen fertilizers

There have been a number of major concerns in the agriculture/environment debate some of which can be answered only by long-term experiments. One such concern has been with increasing levels of nitrate in potable waters which many people perceive as directly and simplistically related to the increasing use of inorganic nitrogen fertilizers. One of the main difficulties with this conclusion is that it was some years after reports of increasing levels of nitrate in water that the amounts of fertilizer N applied to, for example, cereals first exceeded the amount of N in the harvested crop (Sylvester-Bradley *et al.*, 1987). Research into the efficient use of fertilizer N was urgently needed. This could be done best using labelled ^{15}N fertilizer, preferably at sites where there was no net immobilization or release of organic N from humus. This situation only occurs in long-term experiments where soil organic matter levels are in stable equilibrium. Powlson *et al.* (1986) described experiments made during four years on Broadbalk, in which varying rates of ^{15}N-labelled inorganic N fertilizers were applied to winter wheat. Averaged over the four years, about 20% of the spring-applied N fertilizer was found in the soil after harvest mostly as organic N (Table 23.12). Less than 2% of the total applied N was present as mineral N. This N balance used measured values for N in soil, grain and straw (Table 23.12), and showed, by difference, that about 20% of the applied N was unaccounted for. In these and similar experiments there was a strong relationship between the loss of spring-applied labelled N and rainfall in the three weeks after fertilizer application (Fig. 23.4). The losses could have been by volatilization of ammonia on application, by leaching or by denitrification. In 13 out of 16 of these experiments denitrification was the most likely cause of nitrate loss because rainfall did not exceed evapotranspiration for soils already below field capacity (Addiscott and Powlson, 1992).

These results led to the suggestion that much of the nitrate which appears in soil in autumn comes from the mineralization of organic matter, an observation important for those framing legislation on N fertilizer use. The exceptions are where fertilizer N has been applied in excessive amounts relative to the yield potential of the site for a particular crop or where the crop has failed for some reason (see Johnston and Jenkinson, 1989 for examples).

The other consideration is that the long-continued use of nitrogen fertilizers can lead to the accumulation of more humus in fertilized soils compared to those which are unmanured. For example, on Broadbalk, straw has always been removed at harvest and the only return of organic matter has been in stubble, roots and fallen leaves. Shen *et al.* (1989) showed that the soil which had received 144 kg N ha^{-1} fertilizer nitrogen each year for 137 years contained 3600 kg organic nitrogen in the top 23 cm, compared to 2900 kg in soil receiving no nitrogen. In laboratory incubation studies the amount of nitrogen mineralized was the equivalent of 14 and 22 kg N ha^{-1} in the soils without and with fertilizer nitrogen, but soil-derived nitrogen

Table 23.12. Percentage distribution at harvest of fertilizer-derived nitrogen applied to winter wheat at 144 kg N ha^{-1} labelled with ^{15}N, Broadbalk, Rothamsted. (Adapted from Johnston and Jenkinson, 1989.)

	% fertilizer nitrogen in			
Year	Grain	Straw	Soil	Unaccounted for
1980	55	13	17	15
1981	37	16	20	27
1982	45	23	24	8
1983	44	13	16	27
Mean	45	16	19	19

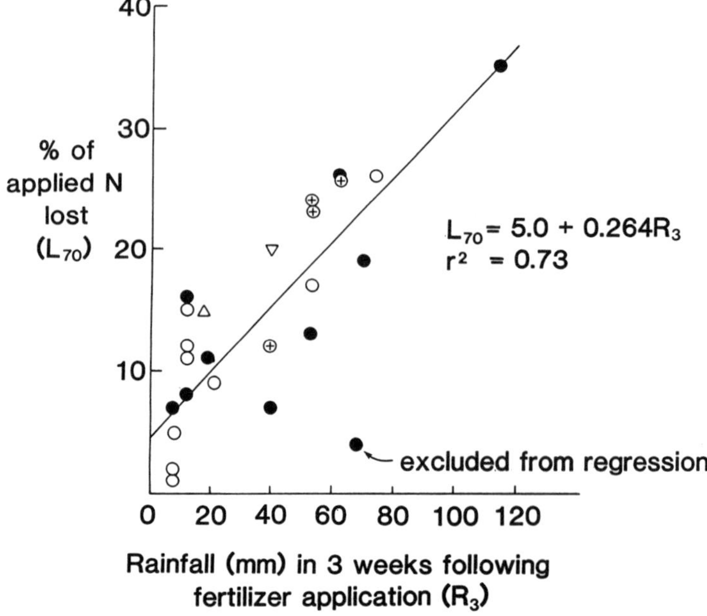

Fig. 23.4. Relationship between rainfall (R_3, mm) in the three weeks following application of ^{15}N-labelled fertilizer to winter wheat on three soil types and percentage loss (L_{70}) of fertilizer N from the crop:soil system. Loss defined as labelled N not recovered in crop or soil to a depth of 70 cm at time of harvest. $L_{70} = 5.0 + 0.264\ R_3$; $r^2 = 0.73$. Fertilizer N applied above (⊕) or below (○) that giving maximum grain yield, other results (●). Arrowed figure and data for forage maize (▽) and winter oilseed rape (△) not used in calculating the regression. (Adapted from Powlson *et al.*, 1992.)

supplied to a fully matured crop was larger (31 and 72 kg N ha^{-1} respectively), determined using ^{15}N-labelled fertilizer nitrogen. However, per unit area of land, these amounts of 'available' nitrogen are small relative to the much larger quantities of nitrate mineralized in the first autumn or spring following the ploughing of grass–clover leys (Johnston *et al.*, 1993) or the incorporation of residues from grain legumes.

The fate of nitrogen in organic manures

A large proportion of the N applied in organic manures can be unaccounted for. For 20 years, FYM, sewage sludge and two composts were each applied at 37.5 and 75 t ha^{-1} fresh manure in the Market Garden experiment at Woburn. The straight-line relationship between the accumulation of N in soil and that applied indicated that only about 37% of the applied N could be found in extra soil organic matter at the end of the period in which these large applications were made (Johnston, 1975). Similarly during the first 135 years of the Broadbalk experiment, about 125 kg N ha^{-1} of the 255 kg applied to the FYM-treated soil in the annual applications of FYM and by aerial deposition could not be accounted for (Table 23.13). Much of this loss was probably as nitrate because Powlson *et al.* (1989) found much larger concentrations of nitrate in soil throughout the winter on similar FYM-treated soils compared to those given fertilizers only. In experiments where differences in soil humus content have accumulated over long periods it is possible to show enhanced denitrification with extra organic matter (Goulding and Webster, 1989).

Long-term or Continuing Experiments

Long-term or continuing field experiments are essential to our understanding of, and opportunity to test, the sustainability of agricultural systems. It is impossible, however, to say how long an experiment should continue, but undoubtedly the opportunity for slow changes, which have adverse or beneficial effects, to manifest themselves increases with time. It is more realistic and perhaps important to suggest that experiments continue until they demonstrate that the husbandry system being tested is not sustainable. If doubts about sustainability are raised, every effort should be made to seek modifications to the system to ensure its continuity.

In 1843 Lawes and Gilbert could not have conceived that, of the field experiments they were to start over the next 14 years, eight would still be continuing in 1993, providing agronomic and scientific data of immense value. But some of the experiments they, together with others at Rothamsted, started have not continued because yields became too small. Where yields declined because of increasing soil acidity this was corrected

Table 23.13. Increase in the amounts of nitrogen in the soil and the nitrogen applied to, and removed in winter wheat from the farmyard manure-treated plot on Broadbalk, Rothamsted, at various periods. (Adapted from Johnston *et al.*, 1989.)

Period	Annual increase in soil nitrogen (kg ha^{-1})	Nitrogen in harvested crop (kg ha^{-1})	Nitrogen applied[a] (kg ha^{-1})	Nitrogen not accounted for (kg ha^{-1})
1852–1861	70	65	255	120
1892–1901	30	90	255	135
1970–1978	5	125	255	125

[a]The amount of FYM applied has remained constant at 35 t ha^{-1} each year and its nitrogen content has varied little; an average value of 225 kg ha^{-1} has been used. The total nitrogen input of 255 kg ha^{-1} includes 30 kg ha^{-1} from other sources (see Powlson *et al.*, 1989).

by liming but failure to grow legumes continuously was not explained satisfactorily.

The Introduction refers to Lawes and Gilbert's view, expressed in 1895, that long-term experiments become more valuable with time but they did not say what made them continue after the first few years when the experiments had apparently answered their original purpose. The answer may lie in the fact that results in the first year indicated the greater need to supply readily available N to crops (other than legumes) rather than mineral nutrients and this was quickly exploited and proven. However, the result was contrary to the views expressed by Liebig (1843), which led to an acrimonious debate with Liebig which lasted for more than 25 years. It may be that it was the need to show that they, rather than Liebig, were correct, that persuaded Lawes to continue to fund, from his own resources, the Rothamsted experiments (Johnston, 1991b). Having allowed their experiments to pass the first critical 'time hurdle', perhaps not by the initial concept, Lawes and Gilbert realized the ever-increasing value of the accumulating data sets. Together they published more than 130 scientific papers on topics such as the nutrient requirement of crops, nutrient composition of crops, effects of weather on yield, species composition of grassland and long-term changes in the nutrient status of soil.

In the section on the need for flexibility, the effect of treatments in the Woburn Ley Arable experiment and the sponsors responses to ensure continuity of the experiment over a 50-year period are discussed in detail. Powlson and Johnston (Chapter 22) give examples where soil acidity had adverse effects on cereal yields after 15 years on a sandy loam soil but not for a 100 years on a clay loam. These authors also showed that soil organic matter changes slowly in temperate climates; 20 years are needed for changes to be measured reliably, whereas easily detected changes in readily soluble P and K in soil might be seen after 10 years.

Experiments can be too short and the results misleading. Although not related to soil resilience, a good example is provided by the data in Fig. 23.5. Over a 30-year period (1957–1989), herbage lead concentrations declined, although there was much seasonal variation, and the decline mirrored a decline in atmospheric lead concentrations (Jones *et al.*, 1991). Viewed in isolation, however, data from the mid 1970s to the early 1980s could be used to show a slight increase in lead concentrations over about seven years which, although real, was nevertheless against the overall downward trend.

Approaches to New Long-term Experiments

A number of criteria must be satisfied before setting up long-term experiments. Two factors need to be settled before much time is spent on detailed

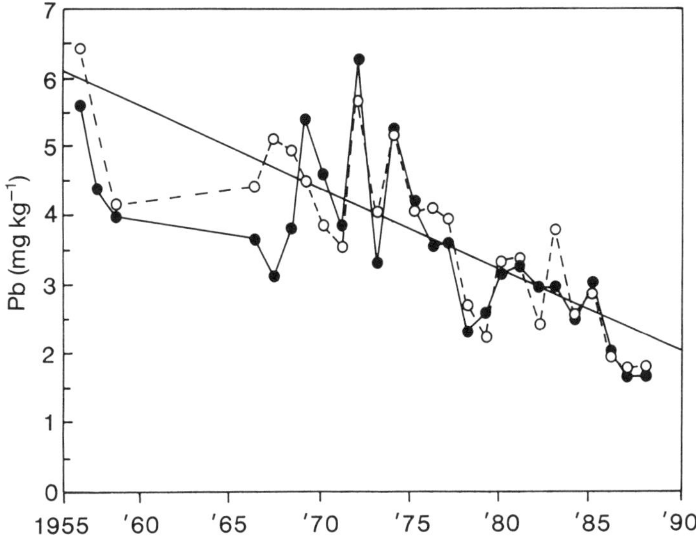

Fig. 23.5. Lead concentrations in herbage at Rothamsted 1956–1988. Soils treated with fertilizers (●) or farmyard manure (○). (Adapted from Jones *et al.*, 1991.)

planning. One is tenure of the site, the other is funding.

In no small measure the Rothamsted long-term experiments have continued through many decades because there was security of tenure of the sites; the Lawes Agricultural Trust either owns or has very long leases on them. Also employees of the Trust are responsible for the field operations. Thus scientists could be assured that if they could imbue their junior colleagues with enthusiasm for the experiments they had started, the sites would still be there for the experiments to continue. Today Rothamsted still continues eight of the very large field experiments started by Lawes and Gilbert in the 1840s and 1850s.

Perhaps in the past funding for long-term experiments was more readily available than it is today. Certainly costs were surprisingly small compared to those currently incurred. It would be realistic to consider that funding should be multi-resourced if the different objectives set out in the Introduction are to be achieved. Partial farmer-funding should be available where data are provided which help farmers reach decisions about their day-to-day management practices. But studies at the interface of agricultural practice and the maintenance of continuing or long-term productivity and environmental objectives, on which administrative decisions will be based, should be funded by an appropriate agency of Government.

Funding for field experiments must be justifiable on the basis of getting reliable and valuable data which could not be otherwise obtained. The data must be well interpreted, and then made freely available. Policy makers especially need good interpretation if they are to reach rational decisions so that future generations enjoy both adequate food and a safe and pleasant environment. The need for good quantitative data has never been more apparent. It is frightening that today so many are trying to infer too much on the possible effects of, for example, global warming, the greenhouse effect and organic farming on future food supplies on the basis of totally inadequate data. Even with well-founded data, especially those for complex biological and physicochemical processes such as those occurring in soils and plants, there is almost always scope for a variety of interpretations. Trying to infer processes from a poor database, then model it and make global predictions, leaves much to be desired. The importance of long-term, continuing experiments to global concerns is something to which not only governments but also the United Nations and other international agencies should give their attention and funds.

As long as site tenure and funding can be secured other aspects of setting up long-term experiments can then be considered. Building on current experience and some of the examples discussed above, it is essential that any new agricultural long-term field experiments should be multidisciplinary in approach. Hence the first task is to assemble a group of like-minded researchers willing to invest their time and expertise under the leadership of a chairperson with whom they can work. This group must then decide not only the questions they want to ask of the experiment but also, and much more difficult, the likelihood of their experiment providing the answers and over what time scale. In seeking compromises in design to include treatments to meet individual research needs, it is essential to ensure that the achievement of the original aims will not be jeopardized.

The questions and the treatments which might answer them are closely interwoven. Lawes and Gilbert, as shown above, asked very simple questions of the experiments they started in the 1840s and 1850s at Rothamsted. They chose simple treatments and got very clear cut answers, albeit after some years. Many years later this approach was rebutted by R.A. Fisher who said that, 'Nature will best respond to a logical and carefully thought out questionnaire'. But such a questionnaire, directed at the complexities of plant growth responding just to the interactions of the biological, chemical and physical properties of soil, let alone the effects of external factors like climate, will give answers that are strongly influenced by the way the questions are framed. The results may then reflect the intentions and preconceptions of those setting the questionnaire and statistical treatment of the results will not remove this bias. Equally the questionnaire may have been incomplete and the results will identify new research projects which it may or may not be possible to accommodate within the original experiment.

The scale of an experiment and its design will depend on the crops to be grown and the number of treatments to be tested. It is essential to consider possible cropping sequences at the start of any long-term experiment. Crop rotations should exclude wherever possible the risk of build-up of pests and diseases unless their control either chemically, biologically or by rotation can be included in the experimental design and the damaging effect of the pest or disease estimated.

There are excellent texts which deal with the statistical aspects of long-term experiments (see e.g. Dyke, 1988; Pearce *et al.*, 1988). The design should allow a robust statistical analysis of the data, but recognize that modern statistical analysis can now use years as replicates (see Dyke *et al.*, 1983 for an example) and time series analyses undertaken (Chatfield, 1985).

The effect of soil variability within an experiment can be minimized by careful siting of the experiment after detailed soil sampling and analysis. The initial state of the site and the soil should be recorded carefully. Long-term experiments will almost certainly outlive cultivars of arable crops and the agrochemicals which may be used as treatments or basal applications. The protocol adopted must be flexible enough to allow for change when change is essential. The size of plots must also be sufficient to exclude edge and soil creep effects and permit the opportunity to test additional treatments. Besides treatment effects it is also essential to monitor those factors which can affect soil fertility and plant composition or health but are not under test. Weather at the site should be monitored rigorously especially those parameters which are needed to derive other data, like evapotranspiration, which may be important in helping to understand differences in yield.

Besides yield, long-term data sets about factors controlling growth, which are of interest to the individual members of the research group, will accumulate. However, it is also essential to decide initially on a policy for collecting and archiving crop and soil samples. Here there must be a rigid protocol if results are to be compared over time. It is especially important to take a soil sample from each plot before any treatments are applied, and to sample subsoils as well as surface soils.

Three other major problems of long-term experiments must never be underestimated. One is the publication of results, especially when they take some years to accumulate. Too often publication and promotion appear inseparable to the young research worker. Senior management within research organizations must give credit for good work on long-term experiments. However, within the framework of established treatments, there is often considerable opportunity for carefully supervised projects leading to higher degrees and postdoctoral research and these possibilities should never be overlooked or dismissed lightly. Then there is the continuity of experiments beyond the active participation of the initiators. Within the framework of multidisciplinary sponsorship it is likely that there will be a range of age, experience and management skill. Thus as any one sponsor leaves the group a suitable replacement can be found and their enthusiasm and skill used to maintain or enhance the research programme. Finally, it is extremely important to recognize that all long-term experiments are likely to have periods of great interest alternating with a routineness which threatens the morale of even those closely involved. Such interludes can also lead to questions about the need to continue by those responsible for funding or wanting to make alternative use of the site. Such periods of doubt must be minimized. One of the advantages of the long-term experiments at Rothamsted is that there is more than one. Invariably there is something new and interesting in one experiment or another.

Conclusions

The Rothamsted experiments, the oldest of which have now continued for 150 years, show that long-term experiments which have continued with few, but well-chosen modifications are of immense value. To achieve such value continuing or long-term experiments in agriculture must: (i) be on sites with security of tenure; (ii) have acceptable levels of funding to achieve their aims; (iii) be multidisciplinary; (iv) have objectives that are well defined but not rigid; (v) be designed so that the data can be analysed statistically but allow for acceptable change; (vi) have large plots which can be subdivided to include additional tests; (vii) have clearly defined but flexible experimental protocols, except for crop and soil sampling which should remain unchanged; (viii) have sets of measurements that are agreed at the start; (ix) provide a continuous output of good data with well-documented interpretations.

References

Addiscott, T.M. and Powlson, D.S. (1992) Partitioning losses of nitrogen fertilizer between leaching and denitrification. *Journal of Agricultural Science, Cambridge* 118, 101–107.

Boyd, D.A. and The Ley Arable Sponsors (1968) Experiments with Ley and Arable Farming Systems. *Rothamsted Experimental Station Report for 1967*, 316–331.

Chatfield, C. (1985) *The Analysis of Time Series – An Introduction* 3rd edn. Chapman and Hall, London, 286 pp.

Dyer, B. (1902) Results of investigations on the Rothamsted soils. *Bulletin of Official Experiment Stations* No. 106, US Department of Agriculture, 180 pp.

Dyke, G.V. (1988) *Comparative Experiments with Field Crops*, 2nd edn. Charles Griffen, London, 262 pp.

Dyke, G.V., George, B.J., Johnston, A.E., Poulton, P.R. and Todd, A.D. (1983) The Broadbalk Wheat experiment 1968–78. Yields and plant nutrients in crops grown continuously and in rotation. *Rothamsted Experimental Station Report for 1982* Part 2, 5–44.

Garner, H.V. and Dyke, G.V. (1969) The Broadbalk Wheat experiment: yields. *Rothamsted Experimental Station Report for 1968* Part 2, 26–46.

Glynne, M.D. (1935) Incidence of take-all on wheat and barley on experimental plots at Woburn. *Annals of Applied Biology* 22, 225–235.

Glynne, M.D. (1969) Fungus diseases on wheat on Broadbalk, 1843–1967. *Rothamsted Experimental Station Report for 1968* Part 2, 116–136.

Glynne, M.D., Salt, G.A. and Slope, D.B. (1956) *Rothamsted Experimental Station Report for 1955,* 103.

Goulding, K.W.T. and Webster, C.P. (1989) Denitrification losses of nitrogen from arable soils as affected by old and new organic matter from leys and farmyard manure. In: Hansen, I.A. and Henriksen, K. (eds) *Nitrogen in Organic Wastes applied to Soils.* Academic Press, London, pp. 225–234.

Johnson, C.G., Lofty, J.R. and Cross, D.J. (1969) Insect pests on Broadbalk. *Rothamsted Experimental Station Report for 1968* Part 2, 141–156.

Johnston, A.E. (1970) The value of residues from long-period manuring at Rothamsted and Woburn. II. A summary of the results of experiments started by Lawes and Gilbert. *Rothamsted Experimental Station Report for 1969* Part 2, 7–21.

Johnston, A.E. (1975) The Woburn Market Garden experiment, 1942–69. II. Effects of the treatments on soil pH, soil carbon, nitrogen, phosphorus and potassium. *Rothamsted Experimental Station Report for 1974* Part 2, 102–131.

Johnston, A.E. (1989) Phosphorus cycling in intensive arable agriculture. In: Holm Tiessen (ed.) *Phosphorus Cycles in Terrestrial and Aquatic Ecosystems,* SCOPE Regional Workshop I: Europe, Saskatchewan Institute of Pedology, Saskatoon, pp. 123–136.

Johnston, A.E. (1991a) Soil fertility and soil organic matter. In: Wilson, W.S. (ed.) *Advances in Soil Organic Matter Research: the Impact on Agriculture and the Environment.* Royal Society of Chemistry, Cambridge, pp. 299–314.

Johnston, A.E. (1991b) Liebig and the Rothamsted experiments. In: Judel, G.K. and Winnewisser, M. (eds) *Symposium '150 Jahre Agrikulturchemie',* Justus Liebig-Gesellschaft zu Giessen, Giessen, pp. 37–64.

Johnston, A.E. and Garner, H.V. (1969) The Broadbalk Wheat Experiment: historical introduction. *Rothamsted Experimental Station Report for 1968* Part 2, 12–25.

Johnston, A.E. and Goulding, K.W.T. (1990) The use of plant and soil analyses to predict the potassium supplying capacity of soil. In: *Development of K-fertilizer recommendations.* International Potash Institute, Berne, pp. 177–204.

Johnston, A.E. and Jenkinson, D.S. (1989) The nitrogen cycle in UK arable agriculture. Proceedings No. 286. The Fertilizer Society of London, Peterborough, pp. 1–24.

Johnston, A.E. and Poulton, P.R. (1977) Yields on the Exhaustion Land and changes in the NPK contents of the soils due to cropping and manuring, 1852–1975. *Rothamsted Experimental Station Report for 1976* Part 2, 53–85.

Johnston, A.E. and Poulton, P.R. (1992) The role of phosphorus in crop production and soil fertility: 150 years of field experiments at Rothamsted, United Kingdom. In: Schultz, J.J. (ed.) *Phosphate Fertilizers and the Environment.* International Fertilizer Development Centre, Muscle Shoals, USA, pp. 45–64.

Johnston, A.E. and Warren, R.G. (1970) The value of residues from long-period manuring at Rothamsted and Woburn. III. The experiments made from 1957 to 1962, the soils and histories of the sites on which they were made. *Rothamsted Experimental Station Report for 1969* Part 2, 22–38.

Johnston, A.E., McGrath, S.P., Poulton, P.R. and Lane, P.W. (1989) Accumulation and loss of nitrogen from manure, sludge and compost: long-term experiments at Rothamsted and Woburn. In: Hansen, J.A.A. and Henriksen, K. (eds) *Nitrogen in Organic Wastes Applied to Soils.* Academic Press, London, pp. 126–139.

Johnston, A.E., McEwen, J., Lane, P.W., Hewitt, M.V., Poulton, P.R. and Yeoman, D.P. (1993) Effects of one to six year old ryegrass–clover leys on soil nitrogen and on subsequent yields and fertilizer nitrogen requirements of the arable sequence winter wheat, potatoes, winter wheat, winter beans. *Journal of Agricultural Science, Cambridge* (in press).

Jones, K.C., Symon, C., Taylor, P.J.L., Walsh, J. and Johnston, A.E. (1991) Evidence for a decline in rural herbage lead levels in the UK. *Atmospheric Environment* 25A, 361–369.

Lawes, J.B. and Gilbert, J.H. (1895) The Rothamsted Experiments. *Transactions of the Highland and Agricultural Society of Scotland* Fifth Series, 7.

Liebig, J. (1872) Soil statics and soil analyses. (English abstract.) *Journal of the Chemical Society* 25, 318 and 837.

Liebig, J. (1843) *Organic Chemistry in its Application to Agriculture and Physiology,* 3rd edn, Taylor and Walton, London, p. 400.

McGrath, S.P. and Lane, P.W. (1989) An explanation for the apparent losses of metals in a long-term experiment with sewage sludge. *Environmental Pollution* 60, 235–256.

Pearce, S.C., Clarke, G.M., Dyke, G.V., and Kempson, R.E. (1988) *A Manual of Crop Experimentation.* Charles Griffen, London, 358 pp.

Powlson, D.S., Pruden, G., Johnston, A.E. and Jenkinson, D.S. (1986) The nitrogen cycle in the Broadbalk Wheat Experiment – recovery and losses of ^{15}N-labelled fertilizer applied in spring and inputs of nitrogen from the atmosphere. *Journal of Agricultural Science, Cambridge* 107, 591–609.

Powlson, D.S., Poulton, P.R., Addiscott, T.M. and McCann, D.S. (1989) Leaching of nitrate from soils receiving organic or inorganic fertilizers continuously for 135 years. In: Hansen, I.A. and Henriksen, K. (eds) *Nitrogen in Organic Wastes Applied to Soils.* Academic Press, London, pp. 334–345.

Powlson, D.S., Hart, P.B.S., Poulton, P.R., Johnston, A.E. and Jenkinson, D.S. (1992) Influence of soil type, crop management and weather on the recovery of ^{15}N-labelled fertilizer applied to winter wheat in spring. *Journal of Agricultural Science, Cambridge* 118, 83–100.

Rothamsted Experimental Station (1955) *Results of the Field Experiments 1955.* Lawes Agricultural Trust, Harpenden, p. 55/Be/1.4.

Shen, S.M., Hart, P.B.S., Powlson, D.S. and Jenkinson, D.S. (1989) The nitrogen cycle in the Broadbalk Wheat Experiment: ^{15}N-labelled fertilizer residues in the soil and in the soil microbial biomass. *Soil Biology and Biochemistry* 21, 529–533.

Sibbesen, E., Andersen, C.E., Andersen, S. and Flensted-Jensen, M. (1985) Soil movement in long-term field experiments as a result of soil cultivation. I. A model for approximating soil movement in one horizontal dimension by repeated tillage. *Experimental Agriculture* 21, 101–107.

Sylvester-Bradley, R., Addiscott, T.M., Vaidyanathan, L.V., Murray, A.W.A. and

Whitmore, A.P. (1987) *Nitrogen Advice for Cereals: Present Realities and Future Possibilities.* Proceedings No. 263. The Fertiliser Society of London, Peterborough, 36 pp.

Thurston, J.M. (1969) Weed studies on Broadbalk. *Rothamsted Experimental Station Report for 1968* Part 2, 186–208.

Warren, R.G. and Johnston, A.E. (1967) Hoosfield Continuous Barley. *Rothamsted Experimental Station Report for 1966,* 320–338.

Yates, F. (1969) Investigations into the effects of weather on yield. *Rothamsted Experimental Station Report for 1968* Part 2, 46–49.

Chapter 24
Modelling Changes in Soil Properties

A. Young

Visiting Fellow, School of Environmental Sciences, University of East Anglia, Norwich NR4 7TJ, UK

Systems, Models and Simulation

Models can be a considerable aid to understanding the relations between soil resilience and sustainable land use. They allow predictions to be made of the consequences, for soil properties, of different systems of land use.

A **system** is a limited part of the real world that contains interrelated elements. A **model** is a simplified representation of a system. **Simulation**, or modelling (US spelling 'modeling'), is the art of constructing mathematical models and the study of their properties, for comparison with the properties of the systems they represent (Rabbinge and de Wit, 1989).

Systems

The boundaries of a system should be clearly defined, with a high degree of interrelation of the elements within them. Still more important is that all flows which enter and leave the system should be easily isolated and measurable. The boundaries of biophysical systems are planes in space, but those of economic systems are units of accounting, e.g. farm enterprise, farm, development project.

Possible systems of relevance to modelling changes in soil properties are:

- soil systems;
- plant–soil systems;
- cropping or crop–livestock systems (farm enterprises);
- land use systems (farm systems, forestry, etc.).

Soil and plant–soil systems are biophysical, whereas land use systems also

 Soil Resilience and Sustainable Land Use (eds D.J. Greenland and I. Szabolcs)

include socioeconomic elements. The biophysical components are necessarily always present, although some economic models take them as fixed quantities.

Soil systems are bounded above and below by the ground surface and the interface between parent material and soil. Where vertically acting processes only are considered, lateral boundaries can be arbitrary, e.g. 1 m^2, and the model is effectively one-dimensional. Where erosion and lateral seepage are included, a two-dimensional cross section of a slope must be taken. Soil systems are relevant to simulations of soil processes, such as leaching and erosion, in circumstances where the effects of plants (e.g. litter, soil cover) can be taken as external.

However, the physical intermingling of roots with soil is so intimate that for most practical purposes, **plant–soil systems** are a better basis for modelling. They are a necessary basis for modelling nutrient cycling.

A plant–soil system as defined for modelling nutrient cycling is approximately equivalent to a **cropping system** as defined for economic purposes. Much economic analysis, however, is based on the **farming system**, a combination of several crop or crop–livestock enterprises. The limits of a farming system are basically the farm boundaries.

Variables

Models are made up of state, rate, and driving variables, sources and sinks, parameters, and controls.

State variables are quantities that are internal to the system, e.g. quantities of carbon, nutrients, water, biomass, or money. **Rate variables** are flows between state variables, as a result of defined processes. The flows may be of materials, energy or money, e.g. flow of carbon between plant litter and soil humus, transfer of money between income from crop sales and investment in capital equipment. **Driving variables** are external to the system being modelled, and partly control its functioning, e.g. rainfall, prices of inputs and products.

Sources and **sinks** are stores external to the system, from which the system derives, or to which it discharges, materials. Thus, for soil nitrogen, the atmosphere is the source, for phosphorus, the weathering bedrock; sinks are the atmosphere again for nitrogen (as gaseous losses) and the groundwater (or rivers) for all nutrients.

The term 'parameters' is widely misused to mean 'variables'. Correctly, **parameters** are values which are fixed, unchanging, either for all cases or for a defined set of conditions. For example, it may have been established that it takes one hour to prune a 60 m length of hedgerow; the rate of oxidation of soil humus carbon might be set at 3% or 4% per annum for soils under fallow and cultivation respectively. The discount rate used in cost:benefit analysis is another example.

Controls are the equations which determine the values of rate variables, based on driving variables, state variables and parameters. Frequently, 'rate' = 'state' times 'constant', e.g. man-hours to prune hedgerows = length of hedgerows times 1/60. It is also common for two or more state (or driving) variables to determine a rate variable, in which case the physical real-world connections are represented in the model by transfers (flows) of information.

Static and dynamic models, and iteration

Static models provide estimates of processes, and their results, over one interval of time. Changing the driving variables produces different outputs. Dynamic models represent changes, in state and rate variables, over time.

Most dynamic models work not continuously but iteratively, based on defined time intervals, commonly one day, 10 days, one month or one year. The choice of time interval greatly affects the complexity of the model; those which make use of daily or 10-daily weather data are far more complex than those based on monthly or annual averages. The iteration is then:

1. Describe the state of the system (state variables, driving variables, parameters) at the beginning of the time interval; from this,
2. Calculate the rate variables for the time interval; from these,
3. Calculate the state of the system at the end of the time interval.
4. Repeat for the next time interval.

Outputs from such models normally include both rate of processes and the progressively changing state of the system, usually at annual intervals.

Making models useful for users

In constructing a model one is pulled between two extremes. Make it simple, and specialist scientists will deride it for over-simplifying complex natural processes. Make it complex, and no-one (other than its architect) can use it – either because they cannot obtain the data required, or cannot spare the time to understand how it works.

Another spectrum is between opaque and transparent models. Opaque models contain large 'black boxes', meaning that sets of calculations are performed to obtain outputs, but the user cannot at all easily find out how the results were obtained. This is all too common in complex models. Transparent models allow the user to see the way it works. The best example is ALES (Automated Land Evaluation System), in which it is possible to highlight any output value with the cursor, press a function key, and the equation which gave that value is displayed – and this can be done again for the values in that equation (Rossiter and van Wambeke, 1989; Rossiter, 1990).

The user-friendliness of many models is lamentably poor! In far too many:

- the screen prompts, for how to input data and operate the model, are far from easy to follow;
- inputs and outputs are shown to the user as contractions, e.g. LLPHOSPC for percentage phosphorus in leaf litter;
- the user's handbook is not well written.
- innocent mistakes (e.g. entering, or not entering, '.DAT' on a file name) can cause the model to crash.

The probable reason is that the authors are so busy designing the next versions of their models that they do not bother to make them accessible to others. Some require a training course to use, the considerable expense of which could have been avoided by a good handbook. As a consequence, most models are largely used by their authors only.

Brief, clear, sound advice to those thinking of constructing a model will be found in Warner (1991).

A Grouping of Models

In ancient Greece, Hercules was given the task of cleaning the Augean stables: each day he cleaned them, but each night the horses made them dirty again. If Hercules had been living today, he might appropriately be given the task of writing a review paper on soil models! For 1991 alone, the CABI soils database contains 1314 papers indexed for 'models'.

Many such papers are directed at modelling one set of experimental results only. Of interest here are models potentially applicable to soils in all environments, or a wide range. Even so, this list is far from comprehensive. Those cited are examples selected to indicate their relevance, direct or indirect, to modelling soil changes. The meanings of acronyms, together with references, are given in the Annex. The grouping is based on the primary objective of the model.

Estimating the rate of soil erosion

The universal soil loss equation, USLE, is the best known. Others include SLEMSA and the Rose model. These are static models with a well-defined purpose, that of estimating what erosion control measures are necessary to reduce erosion to a defined 'acceptable' level.

Estimating the effects of soil erosion

EPIC and CREAMS are US-based models for this purpose, PERFECT was based on Australia, whereas THEPROM was developed for tropical grazing land conditions. EPIC is a highly complex model, based on, and initially validated for, the US, which simulates runoff, erosion, changes in soil C, N and P, and crop growth.

Estimating soil changes

This group is the most focal to the topic of this chapter. CENTURY was initially designed and validated for the US Great Plains grasslands. It has now been extended by the provision of data sets for a wider range of environmental conditions, and taken up for data analysis by the Tropical Soil Biology and Fertility programme (Woomer and Ingram, 1990). Although primarily directed at soil changes, CENTURY also simulates plant growth. SCUAF, or Soil Changes Under Agroforestry, was designed to model the special features of agroforestry systems, with its mixture of trees and crops, and penetration of roots and transfer of litter between them. Purely agricultural or forestry systems can also be analysed, since these appear in the model as limiting cases of agroforestry, with 0% and 100% trees respectively. CENTURY and SCUAF both output changes over time in soil carbon, nitrogen and phosphorus, with effects of these changes on plant growth (and therefore production), and SCUAF also gives future trends in rate of erosion.

The soil organic matter cycle

Soil modelling can be said to have begun with the equations for soil organic matter changes of Nye and Greenland (1960). These were not computerized but were nevertheless models. They formed the basis for all subsequent attempts to model soil organic matter, represented by carbon (for examples see Table 24.1). The Rothamsted carbon model is without doubt the most soundly calibrated of all models, having over 150 years of continuous field trials as the basis for its validation.

An unsolved problem is the presence in soils of 'passive' or 'slow' carbon. This is apparently both physically and chemically protected from reactions, changes very slowly, and is the cause of the high radiocarbon ages obtained for soil organic matter. It is not even known what are its effects on soil physical properties – indeed, almost nothing is known other than that it appears in standard carbon analysis. But, 'Oh, let us never, never doubt/What nobody is sure about' (G.K. Chesterton, The Microbe).

Nutrient cycling

A large number of attempts have been made to model nitrogen cycling (for examples, see Table 24.1). Nitrogen models are necessarily linked with carbon cycling models.

There are many problems of data availability and accuracy. Direct measurements of plant nitrogen fixation are difficult (and disputed), as are those of

Table 24.1. Results of modelling on SCUAF. For abbreviations to treatments, see text. HP: soil with high phosphorus release.

Run No.	Treatments	Soil loss in 20 years (cm)	Topsoil carbon year 20 (%)	Crop yield Year 1 (kg ha^{-1})	Crop yield Year 20 (kg ha^{-1})
1	Control	2[a]	0.82[a]	1354	244[a]
2	CB	1[b]	0.83[b]	1083	147[b]
3	NF	3	0.78	3000	742
4	LF	3	0.82	2107	1366
5	HF	2	0.95	3000	2704
6	CB, LF	1	0.88	1750	1303
7	CB, HF	1	0.99	2400	2238
8	CB, HF, RC	1	0.86	2400	1915
9	HI	1	0.82	1219	281
10	HI, LF	1	0.91	2068	1517
11	HI, HF	0	1.00	2700	2701
12	CB, NF	1	0.83	2400	674
13	CB, NF, HP	1	0.91	2400	1505

[a]After 12 years/in Year 12.
[b]After 15 years/in Year 15.

gaseous losses of nitrogen, and fixation of phosphorus to clay minerals. No-one has found a way to measure release of phosphorus from rock weathering, although we know from experience that P-rich parent rocks lead to soils that permit more or less continuous annual cropping. Few experimenters are prepared to go to the considerable effort of installing lysimeters to measure and monitor leaching, and instead resort to some model-based estimate. In any case, lysimeters do not normally take account of lateral (downslope) nutrient transport.

Soil water: processes and effects

In many parts of the world, crop growth is more often limited by soil water than by nutrients. The FAO model for the effects of water on plant growth is established and usable. There are various models for soil water movement and leaching, for example SLIM. Of interest is a model for rooting depth in agroforestry systems, and the effects of this for interception of nutrients otherwise lost by leaching. A review of soil water modelling is given by Feddes *et al.* (1988).

Soil productivity

Examples are the Productivity Index (PI) and QUEFTS. These are static models for estimating the potential crop yield of soils with given properties.

They can be combined with estimates of soil changes (e.g. from erosion) to give the effects of such changes on productivity.

Simulation of plant growth

These may be considered as the aristocrats of modelling. They are necessarily complex, some using daily time intervals. The CERES/DSSAT group of models has had a vast input of effort, including calibration for crop cultivars. It is rather opaque, and the soil as such (distinct from its water and nutrient-supplying capacities) appears in a shadowy manner. Its major competitor is the Wageningen group of models, published in part in the Simulation Monographs series (PUDOC, 1972–1992). Other plant growth models include CRIES and NTRM (agricultural crops), the FAO agroecological zones system (static modelling, mainly crops but has been applied to fuelwood trees), SPUR (rangeland growth) FORECAST (plantation forests), MPTGRO (multipurpose trees), ALMANAC (a version of EPIC), and DYNAMITE (natural forest). At a different level is PLANTGRO, a static model for estimating crop and tree suitability from soil variables.

The sequence usually applied in plant growth simulation modelling is as follows.

1. Estimate potential plant growth based on radiation only, with water and nutrients not limiting.
2. Add the constraint to growth from soil water availability.
3. *Either* add the constraint to growth from nutrient availability; *or* specify the nutrient inputs required to give stated levels of growth.
4. In some models, a further constraint from pests and diseases is added.

Estimation of soil changes is not the primary objective of this group of models. However, some either contain within them values for soil properties over time which could be output, or else have the potential for development to give such results. Because of their complexity, plant growth simulation models make considerable demands for data.

Economic and socioeconomic models

There are many economic models for cost–benefit analysis, for example MULBUD, in which soils, with crop yields and other biophysical variables, are inputs which play no part in the simulation procedure. COAL is a linear programming model for land use optimization, covering mostly economic data but with some environmental factors – nitrogen and pesticide use – also included. The FAO population-carrying capacity study is a world-scale static model of soil productivity (Higgins *et al.*, 1987). There have been few attempts to integrate biophysical with socioeconomic modelling; examples are discussed below.

Modelling Soil Changes

The basic sequence

The iterative procedure outlined above can be applied to modelling soil changes in the following basic manner:

1a Describe the initial soil conditions.
1b Describe other initial conditions, such as climate, slope.
2a *Either* simulate, *or* input, crop growth in Year 1.
2b Simulate plant–soil processes (e.g. litter fall and humification) and soil processes (e.g. erosion, oxidation of humus, leaching) during Year 1.
3 Calculate soil conditions at the end of Year 1.
4 Repeat for Year 2, taking the new soil conditions as the basis.

If plant growth at the outset is input (as a driving variable), then feedback equations are needed, denoting the effect of changes in soil properties on plant growth. If plant growth is included in the simulation, this effect is automatically achieved.

Examples of modelling

To illustrate the kinds of outputs that can be obtained, the SCUAF model will be used, partly because it is familiar to the writer, but also because it is transparent and easily permits a basic operation, that of changing one input variable to obtain modified outputs. SCUAF contains a set of default values, representing typical values for specified environmental conditions. Those taken as the initial basis for the following sequence of outputs are:

Climate:	tropical subhumid (savannah)
Slope:	gentle, 5°
Soil:	medium textured, acid, well-drained; initial topsoil values: carbon 1.00%, nitrogen 0.10%, available phosphorus 8 mg kg^{-1}
Crop:	maize
Erosion control:	none
Fertilizer:	none
Crop residues:	retained on the soil (except Run 8)

In relation to crop requirements, the soil is nitrogen deficient but not, initially, phosphorus deficient. The effects of rainfall variability, which would superimpose a random variation in crop yield (and can be included in SCUAF) are omitted, in order to show the soil changes more clearly.

In successive modelling runs, one or more of the input conditions is varied. Treatments, applied singly or in combinations, are (abbreviations as in Table 24.1):

CB	Contour banks:	Bank-and-ditch structures, which reduce erosion to 10 t ha^{-1} $year^{-1}$ and occupy 20% of the land area
NF	N fertilizer:	150 kg N ha^{-1}
LF	Low fertilizer:	75 kg N, 10 kg P ha^{-1}
HF	High fertilizer:	150 kg N, 20 kg P ha^{-1}
HI	Hedgerows:	Contour-aligned hedgerow intercropping (alley cropping), hedges occupy 10% of the land area

The runs are for 20 years. The indicators of sustainability used are soil loss by erosion in 20 years; the values after 20 years of soil C, N and available P; and the crop yield in Year 1 (to indicate the initial attractiveness to the farmer) and Year 20. Results are given in Table 24.1 and Fig. 24.1.

Run 1: Control

No erosion control, no fertilizer. Year 1 crop yield is limited by nitrogen deficiency to 1354 kg ha^{-1}. With a poor plant cover, erosion is rapid, losing 0.5 cm in the first 4 years. Crop yield, initially low, declines steadily, reaching zero in Year 13. At this point, the SCUAF model is designed to 'crash', symbolizing a catastrophe situation: the farmer abandons the field, or gully erosion occurs.

This scenario is clearly non-sustainable. In practice, a farmer would abandon this land after 3–5 years, as in shifting cultivation. If this option were not available, some other management system must be found.

Run 2: Contour banks

No fertilizer is added. A bank-and-ditch system of erosion control is installed, assumed to reduce erosion to the 'tolerable' level of 10 t ha^{-1} $year^{-1}$, but with no other improvements. Initial crop yield is only 1083 kg ha^{-1}, since the control works take up 20% of the land. Erosion loss is only 1 cm in 20 years, but this does not make the system sustainable. After 15 years, crop yield is a miserable 147 kg ha^{-1} and topsoil carbon has fallen to 0.83%.

This system is very clearly uneconomic. Erosion is reduced, but the farmer gets no benefit, neither initially nor in the long term. This modelling result has been frequently observed in practice. What then happens is that the farmer – not surprisingly – fails to maintain the erosion control works.

In runs 3, 4 and 5, different levels of fertilizer are added, but without erosion control.

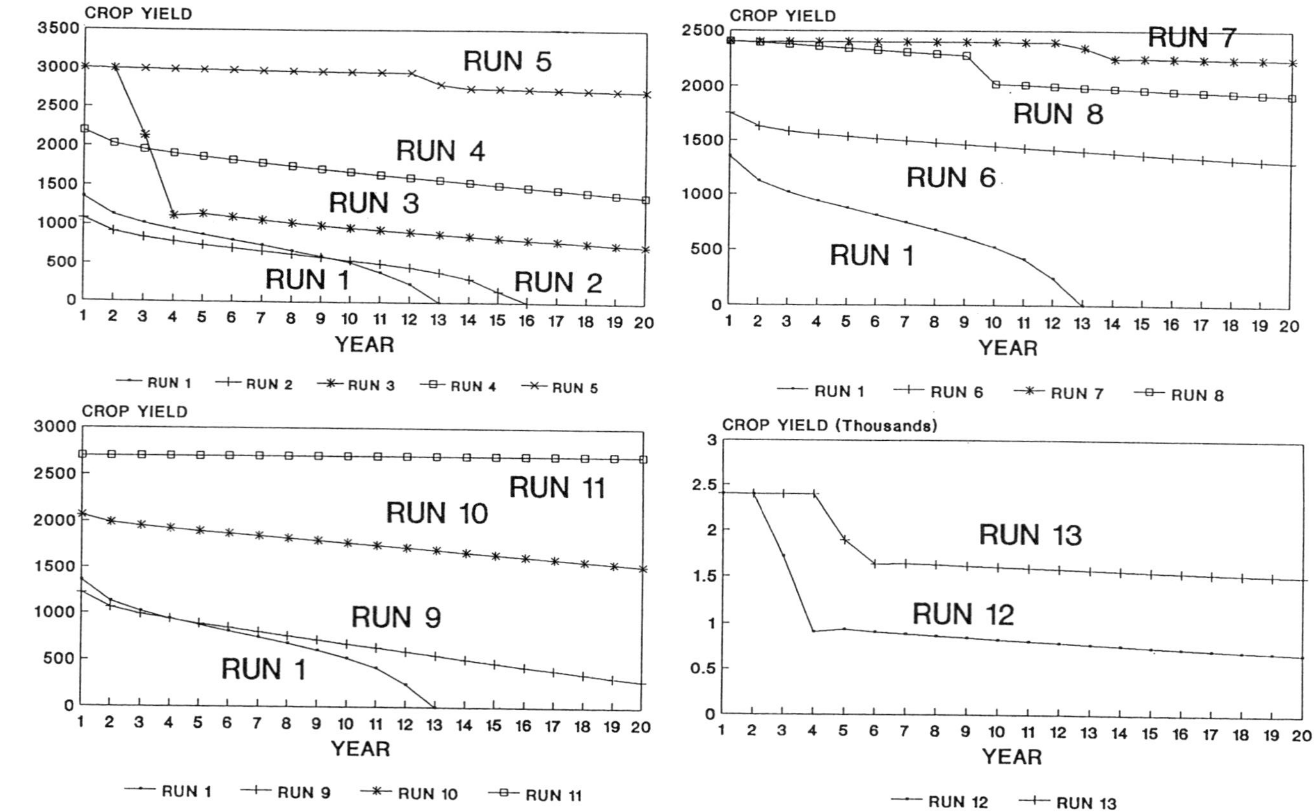

Fig. 24.1. Results of modelling on SCUAF. For treatments, see Table 26.1.

Run 3: Nitrogen fertilizer

The crop shows nitrogen deficiency, to remedy which, the obvious course is to add nitrogen fertilizer. Initial success! Crop yield in Year 1 rises to 3000 kg ha^{-1}, which is the climatically limited maximum. It remains at this level for a second year, but then in Years 3–4 collapses to 1113 kg ha^{-1}. What has happened is that the initial high rate of plant growth used up the available soil phosphorus, and a state of P deficiency set in. Yield in Year 20 is 742 kg ha^{-1}, limited by the rate of natural P inputs (rock weathering and rainfall). Erosion is substantial, and soil carbon loss is greater than with any other treatment.

Run 4: Low fertilizer

Crop yield is initially lower than in Run 3, but maintained at a higher level by the added P. This low but more balanced fertilizer treatment is 'less unsustainable' than that of high nitrogen only.

Run 5: High fertilizer

High fertilizer but without erosion control is very much more sustainable than the converse (Run 2). The good plant cover reduces erosion to 2 cm in 20 years, whilst the recycling of carbon from crop roots and residues keeps soil carbon up to 0.95%. Crop yield after 20 years has fallen only from 3000 to 2704 kg ha^{-1}. In this environment, with only a gentle slope, the conservationist might have reservations, but the farmer would consider the system to be sustainable. (However, this management system could lead to soil acidification, which is not modelled.)

Run 6: Contour banks plus low fertilizer

Now we install erosion control works plus the lower of the two fertilizer levels. This is neither attractive to the farmer nor sustainable in terms of soil properties.

Run 7: Contour banks plus high fertilizer

Not unexpectedly, this system controls soil loss and maintains crop yield at close to its initial level. Its drawback is the loss of 20% yield to the land taken up by the conservation works. The practice in Run 5, high fertilizer but without the erosion control, is more attractive to the farmer.

Run 8: As Run 7, but crop residues removed

The complete maintenance of soil organic matter in Run 7 was achieved only by return of crop residues to the soil. If these are removed, topsoil carbon declines to 0.86, and due to the poorer soil physical conditions, yield in Year 20 is also lower.

Runs 9–11 cover the estimated effects of an agroforestry system, that of contour-aligned hedgerow intercropping (alley cropping). There is experimental evidence that this management system successfully checks erosion (Young, 1989).

Run 9: Hedgerows, no fertilizer

Although erosion is checked, the N fixation from the hedgerows, transferred to the soil as prunings, is by no means sufficient to make up the N deficiency. Crop yield commences at 1219 kg ha^{-1} (1354 less 10% occupied by hedges) and falls steadily. Because of the poor plant growth, soil carbon falls. This hedgerow system, without fertilizer, is neither productive nor sustainable.

Run 10: Hedgerows, low fertilizer

Combining a low fertilizer rate with the hedgerows raises crop yield to 2068 kg ha^{-1} in Year 1, declining slowly to 1517 kg ha^{-1} in Year 20. The system is not fully sustainable but many farmers might accept it. This treatment improves upon Run 4, low fertilizer without erosion control, and also compares favourably with Run 6, the same fertilizer level plus conventional erosion control.

Run 11: Hedgerows, high fertilizer

This system is totally sustainable! Crop yield remains constant at 2700 kg ha^{-1}, and soil carbon, nitrogen and phosphorus in Year 20 are almost identical with their initial levels. It compares favourably with high fertilizer plus conventional erosion control (Run 6), because hedgerows take up less space than bank-and-ditch structures. Nevertheless, the farmer might still prefer high fertilizer without hedgerows (Run 5), which is slightly less sustainable but avoids the labour of establishing and maintaining the hedgerows.

On a moderate slope, say 10°, the gain from erosion control by the hedgerows is greater. Crop yield, initially lower by 10%, overtakes that of the without-hedgerows treatment from Year 4 onwards.

Runs 12 and 13: Comparison between soil types

In the examples above, the variables changed have been the treatments. It is equally possible to model the effect of the same treatment under different climatic conditions, or on different slopes or soils. **Run 12** is on the same soil as above, with contour banks and nitrogen fertilizer only. **Run 13** is the same treatment but on a soil derived from basic parent material, assumed to supply phosphorus by rock weathering at a higher rate (10 kg P ha^{-1} $year^{-1}$ instead of 2). Both show the same feature, a yield maintained during the early years but then dropping when phosphorus deficiency sets in. On the high-P supplying soil, however, the fall in yield commences later and the new equilibrium is reached at a higher level, about 1500 instead of 700 kg ha^{-1}.

This above sequence illustrates several features of modelling. First, once a basic set of data has been obtained, then a large number of results can be produced in a short time. Second, the outputs are much more 'neat and tidy' than in reality; the random effects that affect field experiments, such as insect attack, are excluded. In particular, microvariability in soil properties is totally excluded: every time the same treatments are modelled on the same site, the results are identical – 'replication' is perfect!

A third important feature is to show that modelling generates hypotheses. The above results were not wholly 'expected' when work began. It was indeed expected that the non-utility of erosion control without other improvements might be demonstrated, but the devastating inefficiency of this course of action was not anticipated! As an agroforestry specialist, the author 'hoped' that the hedgerow treatments would be the best and indeed, hedgerows plus fertilizer was the only combination to achieve full sustainability. But in this particular environmental situation, the high fertilizer treatment alone would be more attractive to farmers who could obtain and afford it, with erosion control achieved by high cover factor resulting from good crop growth. It should be added that this result is hardly new – it is stated, for example, by Hudson (1971)! However, erosion could still occur during the period of crop establishment.

A further feature is that to obtain the above results required the impossible situation of a complete set of data, plus perfect knowledge of processes. For example, assumptions were a knowledge of the annual biomass increase of roots, what proportion of hedgerow roots grow into the soil under crops, and what is the loss of carbon in conversion from plant residues (above-ground and root) to soil humus. Not only must the rate of leaching be known, but how it is related to root biomass, by which nutrients in soil solution are retrieved and recycled. There are also the simplifications, employed in modelling to demonstrate the behaviour of the modelled features of the system more clearly; for example, it is assumed that no elements other than nitrogen and phosphorus become limiting, and that acidity does not become a problem.

We cannot, of course, obtain complete data, nor have perfect knowledge of processes, but this fact in itself is instructive. First, it reveals the extent of our ignorance. There are far more studies of above-ground litter decay and nutrient paths, for example, than of root processes. Second, modelling shows which processes need to be known accurately, since they have large effects on the results, and which are less critical. This is analogous with sensitivity analysis in economics. To give one example, a modelling study of very long-term soil changes under beech woodland in Denmark showed that out of all the variables, that which had the greatest effect on equilibrium soil carbon levels was root biomass (Petersen *et al.*, 1985).

Linking Models to Field Experiments

Modelling can be employed in conjunction with field experiments at various stages:

- for validation of the model;
- in the planning of new experiments;
- in the conduct of experiments;
- to extend the results of experiments.

Validation of models for the environment to which they are applied is essential. Some models perform well within a restricted range of environmental conditions but give implausible results when applied to another, notably when models developed for the temperate zone are applied to the tropics. There is a danger of circular reasoning. One can 'fine tune' a model, adjusting its equations, parameters and values of variables, so that it closely simulates one set of experimental results. To claim then that it has accurately predicted these results is akin to circular reasoning! A model must show a degree of robustness, simulating results which 'it did not know about before'.

In planning new experiments, modelling can be used to investigate whether the desired results (hypotheses) may reasonably be expected from the treatments tested. For example, if the hypothesis is that hedgerow intercropping will maintain soil organic matter, modelling can indicate what rate of soil carbon input may be expected from hedgerows at various spacings. A field trial is costly, and takes a minimum of three years for annual crops, five or more for tree crops. This application is being made in current research in agroforestry.

During the conduct of field trials, modelling can indicate which measurements are required if the processes being studied are to be simulated. It was formerly common to find, for example, papers on nitrogen cycling which said, in effect, 'Then there is volatization and denitrification, which we cannot measure' – and so ignored them. Computer models will not accept missing data. Default values must be supplied, and the necessity to estimate them, at least correct to an order of magnitude, is instructive. A common result from modelling soil changes is to indicate the importance of good data on roots: growth, biomass and decay.

There are three ways to use models to extend the results of experiments:

- extrapolation in time;
- to cover field treatments not included.
- for extending the application of results in space, using geographic information systems.

Because of institutional and funding problems, there are very few long-term experiments (Chapters 22 and 23). Yet questions of sustainability often ask what will be the state of the soil in 10, 20 or even 50 years time. Can we use a well-validated five-year trial and extrapolate the results by modelling? This is tempting, but hazardous. Some quite new process, or combination of processes, may appear, although it is valuable if this is anticipated by the model; for example, it is possible to demonstrate that hedgerow intercropping with nitrogen-fixing species can maintain N supplies and crop yields for five or more years; these higher yields, however, use up the soil phosphorus reserves, and P deficiency appears in later years, a result which has been both modelled and observed.

A valuable use of models is to estimate what the results would have been from treatments not included in a field trial. Suppose an experiment is designed with three management variables: crop variety, farmyard manure, and fertilizer. This calls for, say, 27 combinations of treatments or, with control and replication, 84 plots. It is conducted for five years, and statistically valid results obtained. Someone now says, 'But you returned the crop residues to the soil; the farmers of XYZ never do this, they need them as fodder'. Rather than have the delay and expense of another trial, modelling may be able to forecast the results from this modification.

Of course, such a forecast is considerably less reliable than a field trial. Perhaps results from modelling are, say, ten thousand times less valid than those from field experiments – but it takes ten million times less cost and effort to obtain them!

A third way to use models is to extend the results of experiments to the recommendation domain to which they are applicable. Experimental results are obtained from one or more sites, which have been characterized with respect to their environmental conditions. Using a geographic information system (GIS), the area that has similar environmental conditions can be identified. Modelling can now be carried out for the whole of that area, to show, for example, the crop yield to be expected from the new treatment tested on the station. Land suitability maps for specified crops or soil management systems can be constructed in a similar way (e.g. Kassam *et al.*, 1991).

Adding Socioeconomic to Biophysical Modelling

Sustainable land use has socioeconomic as well as biophysical dimensions. Farmers can use fertilizer to maintain soil fertility only if they can afford it; they can use hedgerows only if they have the labour to prune them.

The most common way to attempt this combination is to take the output from biophysical modelling (or field trials) and apply economic analysis. A recent example is an economic analysis of bush fallowing, hedgerow intercropping and minimum tillage systems in Nigeria; this takes into account the impact of these systems on soil erosion, based on experimental results, and calculates the net present value of each management system. Not unexpectedly, the answer depends on the discount rate used (Ehui *et al.*, 1991).

True combinations of biophysical with socioeconomic modelling are infrequent but of interest. BEAM, or Bio-Economic Agroforestry Model, is a spreadsheet model which combines biophysically based estimates of crop yield over time with cost–benefit analysis (Thomas *et al.*, 1991). Spreadsheet modelling is transparent, but does not permit an iterative procedure.

In modelling of shifting cultivation systems, Trenbath (1989) added a further dimension by including population density as a variable. The biophysical element in this model is simple, taking 'soil fertility' as a single variable, lowered by cropping and increased by fallowing. With long fallows a stable equilibrium is achieved, but with reduction in the ratio of fallow to cultivation a threshold is passed which leads to irreversible decline in soil fertility. Combined models on these lines deserve further study.

The approach of environmental accounting should be applied to soil fertility. Farm budgeting always takes account of the stock of buildings and machinery, but should also cover the stock of plant nutrients and the soil physical conditions. Most attempts to apply economic analysis to soil properties have been in estimating the results of erosion, ranging from the crude early attempts based on reduction in soil depth to more recent work such as that based on available water capacity or 'soil life' (Elwell and Stocking, 1984; Biot, 1990) and the EPIC model.

Economic variables, such as prices for inputs and outputs, or assumed national discount rates, are highly variable in time. It is a valuable feature of modelling that results can be obtained for a range of such values, as in sensitivity analysis. A consequence is that as socioeconomic consequences are included in the same model as biophysical processes it should be possible clearly to distinguish them, through intermediate biophysical outputs.

Needs for Research

There should be no research into modelling! This statement requires qualification: what is meant is that it is not necessary to seek funding support for the construction of models alone, in isolation from field or laboratory experimental work.

The first reason is that there are quite enough people with salaries and computers who will construct such models in any case. More seriously, modelling should not be conducted in isolation. It should always be linked with either field trials, laboratory experiments, or both. For example, the CENTURY model was initially based on data from the US Great Plains grasslands, and is now being applied to results from the Tropical Soil Biology and Fertility programme; SCUAF is based on agroforestry experimental work by ICRAF; whereas, as already noted, the Rothamsted Carbon Model is unique in being supported by 150 years of soil data. Of course, there may be specialist modellers, working with experimentalists as a team; in physics, theoretical physicists calculate and publish consequences of hypotheses, which experimentalists seek to verify. Modelling is only valid when it is based on an interactive exchange with field data. Because changes can occur that were not anticipated, there is a particular need for long-term experiments (Chapter 23).

An important consequence of modelling research is what data are needed, and what knowledge of processes. The importance of root processes has been shown, for example, and is currently leading to an expansion of this formerly neglected branch of study.

Lastly, there is a need for better links between biophysical and socioeconomic models. In large part, these have been developed independently, and the few attempts to combine them are not yet nearly as strong. Models are quite capable of handling combined data (after all, farmers do so all the time!). The most likely way of doing this is to extend the iterative method to an annual interchange: environment – socioeconomic circumstances – action taken in farm management – biophysical consequences (crop yields, and soil conditions) – economic consequences (profit/loss), repeating the cycle on the improved or degraded soil the next year.

Conclusions and Recommendations

When properly allied to field experimental data, modelling can provide a powerful means to demonstrate the consequences, for soils, plant growth and production, of alternative systems of land use and management. It can be used to demonstrate whether soil degradation, steady state, or soil improvement will occur under specified land use practices. To achieve better impact, predictions of changes in soil properties should be accompanied by estimates of the consequences for plant growth, production, economics, and population-supporting capacity.

Recommendations given above cover, with respect to soil resilience and sustainable land use:

- the limitations of modelling, and dangers of its uncritical application;
- the proper relation between modelling and field experiments;
- the value of modelling in the generation of hypotheses, including on sustainability;
- the value of modelling in indicating needs for data collection and improved knowledge of processes;
- the potential for combining biophysical with socioeconomic modelling;
- the need to make models more accessible to users other than their authors, by improving user-friendliness;
- the use of modelling in public relations, for demonstrating the consequences of alternative courses of action.

Annex

Examples of models which estimate, or are related to, changes in soil properties and the consequences of these changes

This list is by no means comprehensive. Only more recent or accessible references are given. Foster (1988) is a secondary source which lists primary references.

Rate of erosion

'Rose' erosion and sedimentation model (Rose *et al.*, 1983; Rose and Freebairn, 1985).

SLEMSA: Soil Loss Estimator for Southern Africa (Elwell, 1981; Foster, 1988).

SOILLOSS: Soil Loss; adaptation of USLE for New South Wales, Australia, conditions (Edwards and Rosewell, 1990).

USLE: Universal Soil Loss Equation (Wischmeier and Smith, 1978).

Effects of erosion on soil productivity

ALMANAC: Agricultural Land Management Alternatives with Numerical Assessment Criteria; a version of EPIC (Jones and O'Toole, 1987).

EPIC: Erosion Productivity Impact Calculator (Cole *et al.*, 1987; Foster, 1988).

CREAMS: Chemicals, Runoff and Erosion from Agricultural Management Systems (Foster, 1988).

PERFECT: Productivity Erosion Runoff Functions to Evaluate Conservation Techniques (Littleboy *et al.*, 1989).

THEPROM: Theoretical Erosion Productivity Model (Biot, 1990).

Changes in soil properties

CENTURY (Parton *et al.*, 1987, 1989a,b; Cole *et al.*, 1989). Also simulates plant growth.

SCUAF: Soil Changes Under Agroforestry (Young and Muraya, 1990).

Soil productivity potential

PI: Productivity Index (Rijsberman and Wolman, 1985; Foster, 1988).

QUEFTS: Quantitative Evaluation of the Fertility of Tropical Soils; estimates unfertilized maize yields from soil properties (Janssen *et al.*, 1990a).

Soil organic matter cycle

Modelling of the soil organic matter cycle (e.g. Nye and Greenland, 1960; Smith, 1979; van Veen *et al.*, 1981; Bosatta and Agren, 1985; Bouwman, 1989; Chenfang Lin *et al.*, 1987; van der Linden *et al.*, 1987; Kovda *et al.*, 1990; Verberne *et al.*, 1990).

Rothamsted Carbon Model (Jenkinson *et al.*, 1987; Jenkinson, 1990).

Nutrient cycles

NCSOIL: Nitrogen and Carbon in Soil (Hades *et al.*, 1987).

Modelling of the nitrogen cycle (e.g. Rosswall, 1980; Robertson, *et al.*, 1982; Frissel and van Veen, 1982; Wilson, 1988).

Cycles of other nutrients (e.g. Frissel, 1977).

Compare also plant growth simulation models listed below.

Soil water

CROPWAT: Crop Water, a program for irrigation planning and management (Smith, 1992).

SLIM: Solute Leaching Intermediate model (Addiscott and Whitmore, 1991).

SWACROP: Soil Water–Crop Production model (Feddes *et al.*, 1978; De Jong and Kabat, 1990).

Rooting depth and nutrient uptake (van Noordwijk, 1989).

Effects of soil water on plant growth (e.g. Doorenbos and Kassam, 1979).

Simulation of Plant Growth

Agro-Ecological Zones model; agricultural crops, relative soil and climatic suitability (Higgins *et al.*, 1987).

CERES/DSSAT: Decision Support System for Agricultural Technology; agricultural crops (Jones and Kiniry, 1986; IBSNAT Project, 1990).

CRIES: Comprehensive Resource Inventory and Evaluation System; the yield model, for agricultural crops, is part of a Geographic Information System and land evaluation model (Schultink, 1987; Schultink *et al.*, 1987).

DYNAMITE (formerly NUTCYC): growth of natural forest, also crops in shifting cultivation (Janssen *et al.*, 1990b).

FORECAST (previously FORCYTE): forestry plantations (Kimmins, 1985; Kimmins *et al.*, 1990).

MPTGRO: Multipurpose Tree Growth (Adams and Cady, 1988).

NTRM: Nitrogen, Tillage and Residue Management; agricultural crops (Cole *et al.*, 1987).

PLANTGRO: Plant Growth; empirical prediction of soil and climatic suitability for crops, annual and perennial (Hackett, 1991).

'Wageningen' models: a family of models, covering many aspects of soils and plant growth, including agricultural crops (Penning de Vries and van Laar, 1982; van Keulen and Wolf, 1986; van Keulen *et al.*, 1987; PUDOC, 1972–1992).

SPUR: Simulation of Production and Utilization of Rangelands: grassland (Cole *et al.*, 1987).

Economic or partly economic models

ALES: Agricultural Land Evaluation System; economic land evaluation (Rossiter and van Wambeke, 1989; Rossiter, 1990).

BEAM: Bioeconomic Agroforestry Model (Thomas *et al.*, 1991).

FAO population carrying capacity model; soil, slope and climatic capacity to support human populations at specified input levels (Higgins *et al.*, 1987).

GOAL: General Optimal Allocation of Land use; linear programming model, economics with environmental aspects (Latensteijn and Rabbinge, 1991).

MIDAS: bioeconomic modelling of a dryland farm system (Kingwell and Pannell, 1987).

MULBUD: Multiple cropping Budget; cost:benefit analysis of agroforestry systems (Etherington, 1991).

'Trenbath' model of shifting cultivation (Trenbath, 1984, 1989).

GUMTREE: General Utilization Model of Timber Resource Economic Evaluation (Slemon, 1991); an Australian model.

References

Adams, N.A. and Cady, F.B. (eds) (1988) *Modeling Growth and Yield of Multipurpose Tree Species.* Winrock International, USA.

Addiscott, M. and Whitmore, A.P. (1991) Simulation of solute leaching in soils of differing permeabilities. *Soil Use and Management* 7, 94–102.

Biot, Y. (1990) THEPROM – an erosion productivity model. In: Boardman, J., Foster, I.D.L. and Dearing, J.A. (eds) *Soil Erosion on Agricultural Land.* Wiley, London, pp. 466–479.

Bosatta, E. and Agren, G.I. (1985) Theoretical analysis of decomposition of heterogeneous substrates. *Soil Biology and Biochemistry* 17, 601–610.

Bouwman, A.F. (1989) Modelling soil organic matter decomposition and rainfall erosion in two tropical soils after forest clearing for permanent agriculture. *Land Degradation and Rehabilitation* 1, 125–140.

Chenfang Lin, Tasng-shen Liu and Tai-Lee Hu (1987) Assembling a model for organic residue transformation in soils. In: *Soils and Fertilizers in Taiwan*, 1–16. Reprinted from *Proceedings of the National Science Council. B. Life Sciences*, 11, 175–186.

Cole, C.V., Williams, J., Shaffer, M. and Hanson, J. (1987) Nutrient and organic matter dynamics as components of agricultural production systems models. *Soil*

Science Society of America Special Publication 19, 147–166.

Cole, C.V., Stewart, J.W.B., Ojima, D.S., Parton, W.J. and Schimel, D.S. (1989) Modelling land use effects of soil organic matter dynamics in the North American Great Plains. In: Clarholm, M. and Bergström, L. (ed.) *Ecology of Arable Land.* Kluwer, Dordrecht, pp. 89–98.

De Jong, R. and Kabat, P. (1990) Modeling water balance and grass production. *Soil Science Society of America Journal* 54, 1725–1732.

Doorenbos, J. and Kassam, A.H. (1979) *Yield Response to Water.* FAO Irrigation and Drainage Paper 33. FAO, Rome.

Edwards, K. and Rosewell, C.J. (1990) Evaluation of alternative land management and cropping practices for soil conservation. *Soil Use and Management* 6, 120–124.

Ehui, S.H., Kang, B.T. and Spencer, D.S.C. (1991) Economic analysis of soil erosion effects in alley cropping, no-till, and bush fallow systems in southwestern Nigeria. *IITA Research* 3, 1–6.

Elwell, H.A. (1981) A soil loss estimation technique for southern Africa. In: Morgan, R.P.C. (ed.) *Soil Conservation: Problems and Prospects.* Wiley, Chichester, pp. 281–292.

Elwell, H.A. and Stocking, M.A. (1984) Estimating soil-life span for agricultural modelling. *Tropical Agriculture* 61/2, 148–150.

Etherington, D.M. (1991) The use of MULBUD to determine the economics of agroforestry. *Australian Systems and Information Technology Newsletter* 3(3), 13–15.

Feddes, R.A., Kowalik, P.J. and Zaradny, H. (1978) *Simulation of Field Water Use and Crop Yield.* Simulation Monograph 17. PUDOC, Wageningen.

Feddes, R.A., Kabat, P., Bakel, P.J.T. van, Bronswijk, J.J.B. and Halbertsma, J. (1988) Modelling soil water dynamics in the unsaturated zone – state of the art. *Journal of Hydrology* 100, 69–111.

Foster, G.R. (1988) Modeling soil erosion and sediment yield. In: Lal, R. (ed.) *Soil Erosion Research Methods,* Soil and Water Conservation Society, Ankeny, USA, pp. 97–117.

Frissel, M.J. (ed.) (1977) Cycling of mineral nutrients in agricultural ecosystems. *Agro-Ecosystems* 4, 1–354.

Frissel, M.J. and van Veen, J.A. (1982) A review of models for investigating the behaviours of nitrogen in soil. *Philosophical Transactions of the Royal Society* B296, 341–349.

Hackett, C. (1991) PLANTGRO. *A Software System for Coarse Prediction of Plant Growth.* CSIRO, Brisbane.

Hadas, A., Molina, J.A.E., Feigenbaum, S. and Clapp, C.E. (1987) Simulation of nitrogen-15 immobilization by the model NCSOIL. *Soil Science Society of America Journal* 51, 102–106.

Higgins, G.M., Kassam, A.H., van Velthuizen, H.T. and Purnell, M.F. (1987) Methods used by FAO to estimate environmental resources, potential outputs of crops, and population-supporting capacities in the developing nations. In: Bunting, A.H. (ed.) *Agricultural Environments. Characterization, Classification and Mapping.* CAB International, Wallingford, UK, 171–183.

Hudson, N.W. (1971) *Soil Conservation.* Batsford, London.

IBSNAT Project (1990) *Decision Support System for Agrotechnology Transfer (DSSAT v.2.1): User's Guide.* University of Hawaii, Honolulu, USA.

Janssen, B.H., Guiking, F.C.T., van der Eijk, D., Smaling, E.M.A., Wolf, J. and van

Reuler, H. (1990a) A system for quantitative evaluation of the fertility of tropical soils (QUEFTS). *Geoderma* 46, 299–318.

Janssen, B.H., Joij, I.G.A.M., Wesselink, L.G. and van Grinsven, J.J.M. (1990b) Simulation of the dynamics of nutrients and moisture in tropical forest ecosystems. *Fertilizer Research* 26, 145–156.

Jenkinson, D.S. (1990) The turnover of organic carbon and nitrogen in soil. *Philosophical Transactions of the Royal Society* B329, 361–368.

Jenkinson, D.S., Hart, P.B.S., Rayner, J.H. and Parry, L.C. (1987) Modelling the turnover of organic matter in long-term experiments at Rothamsted. *INTERCOL Bulletin* 15, 1–8.

Jones, C.A. and Kiniry, J.A. (eds) (1986) *CERES – Maize.* Texas A and M University Press, USA.

Jones, C.A. and O'Toole, J.C. (1987) Application of crop production models in agroecological characterization: simulation models for specific crops. In: Bunting, A.H. (ed.) *Agricultural Environments. Characterization, Classification and Mapping.* CAB International, Wallingford, UK, pp. 199–209.

Kassam, A.H., van Velthuizen, H.T., Fischer, G.W. and Shah, M.M. (1991) *Agroecological Land Resources Assessment for Agricultural Development Planning. A Case Study of Kenya. Resources Database and Land Productivity. Main report.* FAO, Rome.

Kimmins, J.P. (1985) Future shock in forest yield forecasting: the need for a new approach. *Forestry Chronicle,* December 503–512.

Kimmins, J.P., Comeau, P.G. and Kurz, W. (1990) Modelling the interactions between moisture and nutrients in the control of forest growth. *Forest Ecology and Management* 30, 361–379.

Kingwell, R.S. and Pannell, D.J. (eds) (1987) *MIDAS, a Bioeconomic Model of a Dryland Farm System.* Simulation Monographs 26. PUDOC, Wageningen.

Kovda, V.A., Bugrovskii, V.V., Kerzhentsev, A.S. and Zelenskaya, N.N. (1990) Model of the transformation of organic substance in soil for quantitative study of the function of soil in ecosystems. *Doklady, Biological Sciences* 312, 299–302.

Littleboy, M., Silburn, D.M., Freebairn, D.M., Woodruff, D.R. and Hammer, G.L. (1989) *PERFECT. A computer simulation model of Productivity Erosion Runoff functions to Evaluate Conservation Techniques.* Queensland Department of Primary Industries, Brisbane.

Noij, I.G.A.M., Janssen, B.H. and van Reuler, H. (1988) Modeling nutrient cycling in tropical forest areas – a case study of Tai, Côte d'Ivoire. *Proceedings of the Second International TROPENBOS Programme,* Ede, The Netherlands, 122–128.

Nye, P.H. and Greenland, D.J. (1960) *The Soil Under Shifting Cultivation.* Technical Communication 51. Commonwealth Bureau of Soils, Harpenden.

Parton, W.J., Schimel, D.S., Cole, C.V. and Ojima, D.S. (1987) Analysis of factors controlling soil organic matter levels in Great Plains grasslands. *Soil Science Society of America Journal* 51, 1173–1179.

Parton, W.J., Sanford, R.L., Sanchez, P.A. and Stewart, J.W.B. (1989a) Modelling soil organic matter dynamics in tropical soils. In: Coleman, D.C., Oades, J.M. and Uehara, G. (eds) *Dynamics of Soil Organic Matter in Tropical Ecosystems.* University of Hawaii, Manoa, USA, pp. 153–171.

Parton, W.J., Cole, C.V., Stewart, J.W.B., Ojima, D.S. and Schimel, D.S. (1989b) Simulating regional patterns of soil C, N. and P dynamics in the U.S. central grasslands region. In: Clarholm, M. and Bergström, L. (eds) *Ecology of Arable*

Land. Kluwer, Dordrecht, pp. 99—108.
Penning de Vries, F.W.T., and van Laar, H.H. (1982) *Simulation of Plant Growth and Crop Production*. Simulation Monographs 21, PUDOC, Wageningen.
Petersen, H., O'Neill, R.V. and Gardner, R.H. (1985) Use of an ecosystem model for testing ecosystem responses to inaccuracies of root and microflora productivity estimates. In: Fitter, A.H. (ed.) *Ecological Interactions in Soil*. Blackwell Scientific, Oxford, 233–242.
PUDOC (1972–1992) *Simulation Monographs* 1–32 (continuing). PUDOC, Wageningen.
Rabbinge, R. and de Wit, C.T. (1989) Systems, models and simulation. In: Rabbinge, R., Ward, S.A. and van Laar, H.H. (eds) *Simulation and Systems Management in Crop Protection*. Simulation Monographs 32. PUDOC, Wageningen.
Rijsberman, F.R. and Wolman, M.G. (1985) Effect of erosion on soil productivity: an international comparison. *Journal of Soil and Water Conservation* 40, 349–354.
Robertson, G.P., Herrera, R. and Rosswall, T. (eds) (1982) *Nitrogen Cycling in Ecosystems of Latin America and the Caribbean*. Nijhoff, The Hague.
Rossiter, D.G. (1990) ALES: a framework for land evaluation using a microcomputer. *Soil Use and Management* 6, 7–20.
Rossiter, D.G. and van Wambeke, A.R. (1989) *ALES. Automated Land Evaluation System. ALES Version 2 User's Manual*. Cornell University, Ithaca, USA.
Rosswall, T. (ed.) (1980) *Nitrogen Cycling in West African Ecosystems*. Royal Swedish Academy of Sciences, Stockholm.
Rose, C.W. and Freebairn, D.M. (1985) A new mathematical model of soil erosion and deposition processes with applications to field data. In: El-Swaify, S.A., Moldenhauer, W.C. and Lo, A. (eds) *Soil Erosion and Conservation*. Soil Conservation Society of North America, Ankeny, pp. 549–557.
Rose, C.W., Williams, J.R., Sander, G.C. and Barry, D.A. (1983) A mathematical model of soil erosion and deposition processes. *Soil Science Society of America Journal* 47, 991–1000.
Schultink, G. (1987) The CRIES resource information system: computer-aided land resource evaluation for development planning and policy analysis. *Soil Survey and Land Evaluation* 7, 47–62.
Schultink, G., Amaral, N. and Mokma, D. (1987) *User's Guide to the CRIES Agro-economic Information System: Yield Model*. Michigan State University, East Lansing, USA.
Slemon, G. (1991) General Utilization Model of Timber Resource Economic Evaluation. *Australian Systems and Information Technology Newsletter* 3(3), 66.
Smith, M. (1992) *CROPWAT: A Computer Program for Irrigation Planning and Management*. FAO Irrigation and Drainage Paper 46. FAO, Rome.
Smith, O.L. (1979) An analytic model of the decomposition of soil organic matter. *Soil Biology and Biochemistry* 11, 585–606.
Thomas, T.H., Willis, R.W., Wojtkowski, P. and Bezkorowajnyj, P. (1991) Spreadsheets for the economic modelling of agroforestry systems. *Australian Systems and Information Technology Newsletter* 3(3), 25–28.
Trenbath, B.R. (1984) Decline of soil fertility and the collapse of shifting cultivation systems under intensification. In: Chadwick, A.C. and Sutton, S.L. (eds) *Tropical Rain-Forest: the Leeds Symposium*. Leeds Philosophical and Literary Society, Leeds, UK, pp. 279–292.
Trenbath, B.R. (1989) The use of mathematical models in the development of shifting

cultivation systems. In: Proctor, J. (ed.) *Mineral Nutrients in Tropical Forest and Savanna Ecosystems.* Blackwell Scientific, Oxford, pp. 353–369.

van der Linden, A.M.A., van Veen, J.A. and Frissel, M.J. (1987) Modelling soil organic matter levels after long-term applications of crop residues, and farmyard and green manures. *Plant and Soil* 101, 21–28.

van Keulen, H. and Wolf, J. (eds) (1986) *Modelling of Agricultural Production: Weather, Soils and Crops.* Simulation Monograph 25. PUDOC, Wageningen.

van Keulen, H., Berkhout, J.A., van Diepen, C.A., van Heemst, H.D.J., Janssen, B.H., Appoldt, C.R. and Wolf, J. (1987) Quantitative land evaluation for agro-ecological characterization. In: Bunting, A.H. (ed.) *Agricultural Environments. Characterization, Classification and Mapping.* CAB International, Wallingford, UK, pp. 185–197.

van Latensteijn, H. and Rabbinge, R. (1991) *Future Changes in the Rural Areas of the EC.* Scientific Council for Government Policy, The Hague.

van Noordwijk, M. (1989) Rooting depth in cropping systems in the humid tropics in relation to nutrient use efficiency. In: *Nutrient Management for Food Crop Production in Tropical Farming Systems,* pp. 129–144.

van Veen, J.A., Paul, E.A. and Voroney, R.P. (1981) Organic carbon dynamics in grassland soils. *Canadian Journal of Soil Science* 61, 185–202; 211–224.

Verberne, E.L.J., Hassink, J., De Willingen, P., Groot, J.J.R. and van Veen, J.A. (1990) Modelling organic matter dynamics in different soils. *Netherlands Journal of Agricultural Science* 38, 221–238.

Warner, A.J. (1991) A hitch-hiker's guide to modelling. *Australian Systems and Information Technology Newsletter* 3(3), 5–6.

Wilson, J.R. (1988) *Advances in Nitrogen Cycling in Agricultural Ecosystems.* CAB International, Wallingford, UK.

Wischmeier, W.H. and Smith, D.D. (1978) *Predicting Rainfall Erosion Losses – a Guide to Conservation Planning.* US Agriculture Handbook 537. US Department of Agriculture, Washington DC.

Woomer, P.L. and Ingram, J.S.I. (eds) (1990) *The Biology and Fertility of Tropical Soils. Report of the Tropical Soil Biology and Fertility Programme (TSBF) 1990.* TSBF, Nairobi.

Young, A. (1989) *Agroforestry for Soil Conservation.* CAB International, Wallingford, UK.

Young, A. and Muraya, P. (1990) *SCUAF. Soil Changes Under Agroforestry. Computer Program with User's Handbook. Version* 2. With program on diskette. ICRAF, Nairobi.

Chapter 25
Structural Aspects of Soil Resilience

B.D. Kay, V. Rasiah and E. Perfect

Department of Land Resource Science, University of Guelph, Guelph, Ontario, Canada N1G 2W1

Introduction

Soil structure has a major influence on the ability of soil to support root development, to receive, store and transmit water, to cycle carbon and nutrients, and to resist erosion and the dispersal of agricultural chemicals. Land use practices which are sustainable must therefore maintain the structure of soil, over the long term, in a state which is optimum for a range of processes related to crop production and the maintenance of environmental quality. This chapter outlines a conceptual framework for characterizing the impact of land use practices on soil structure, outlines methodology to describe some of the variables in the conceptual framework, presents data to characterize the influence of land use practices on these variables, and concludes with a discussion on needs for future research.

A Conceptual Framework for Studying Soil Structure

Soil structure has been studied for decades by investigators concerned about the influence of land use practices on soil structure and the variation of this influence with differences in inherent soil characteristics (e.g. texture, pH, mineralogy, organic matter content) and climate (e.g. precipitation, evapotranspiration, wetting/drying cycles and freezing/thawing events). Other investigators have tried to relate soil structural characteristics to processes which influence the productivity of soils, the ability of soils to receive, store and transmit water and the movement of sediments and agricultural chemicals. In the absence of any unifying conceptual framework this wealth of information is difficult to synthesize for use in making management decisions. This difficulty is further compounded when the term soil structure means different things to different people. A clear understanding of the different aspects of soil structure and the development of a conceptual framework which links these aspects of soil structure to processes influencing plant growth is, therefore, crucial to any assessment of the effect of land use practices on sustainable productivity.

Definitions

Four different aspects of soil structure are represented in the conceptual framework which will be outlined: form, stability, resilience and vulnerability. The terms degradation and regeneration will be used to describe detrimental and beneficial changes in soil structure within this framework.

Structural form

The term structural form will be used to describe the heterogeneous arrangement of solid and void space that exists in soil at a given time. Examples of characteristics of structural form include total porosity, pore size distribution, and continuity of the pore system. Other examples are the arrangement of primary soil particles into hierarchical structural states identified on the basis of failure zones of different strengths.

Structural stability

The term structural stability describes the ability of the soil to retain its arrangement of solid and void space when exposed to different stresses (Kay, 1990). Stability characteristics are generally specific for a characteristic of structural form and the type of stress being applied. Stresses may arise from tillage or from compaction due to vehicular traffic or livestock. The nature and magnitude of the stresses are characteristic of the land use practice. The ability of the soil to retain its arrangement of solid and void spaces when exposed to stress may be characterized by the maximum deformation in structural form which is possible and the rate at which this deformation occurs.

Structural resilience

The term structural resilience has been used (Kay, 1990) to describe the ability of a soil to recover its structural form through natural processes when the applied stresses are reduced or removed. Resilience arises from a range of recovery or regenerative processes including freezing/thawing, wetting/

drying, and biological activity (e.g. root development, soil fauna). Terms such as tilth mellowing (Utomo and Dexter, 1981a) and self-mulching (Grant and Blackmore, 1991) have been used to describe specific aspects of this characteristic. Resilience is characterized by the maximum recovery in structural form which is possible and the rate at which recovery occurs. The maximum recovery in structural form that a soil in a given state can experience is defined as the resilience potential.

Structural vulnerability

The term structural vulnerability is introduced in this chapter to describe the inability of the structural form of soil to cope with stress. Structural vulnerability reflects the combined characteristics of stability and resilience. Soils which are least stable under a given stress and which do not recover when the stress is reduced or removed (i.e. have low resilience under prevailing weather conditions) are the most vulnerable; conversely, soils which have a great resistance to stress and are very resilient are the least vulnerable.

Degradation/regeneration

Qualitative terms are often used to imply value to the direction of change in different characteristics of soil structure. Implicit in the use of such terms must be recognition of the value of soil structure for some purpose, e.g. crop growth. Alternative purposes might include disposing of organic wastes, ground water recharge, etc. The influence of soil structure on crop growth is strongly dependent on climate and therefore use of the term degradation with respect to crop growth implies a change in aspects of soil structure which, under prevailing climatic conditions, will lead to decreased crop growth. Under different climatic conditions the same changes may be considered a form of regeneration.

Crop growth may not change monotonically with a change in a given direction of soil structure. Consequently, use of qualitative terms such as degradation and regeneration must be based on an understanding of the optimum structural conditions for a given climatic zone or range of weather conditions. For instance, a decrease in the magnitude of a parameter describing soil structure may not be defined as degradation in all circumstances since the decrease may lead to increased or decreased productivity depending on the magnitude of the parameter relative to the optimum value for those soil and climatic conditions.

Form/stability/resilience

Structural form, at any given time, is the structural characteristic that determines the availability of water and oxygen in soil and the resistance of the soil to penetration by roots. Stability and resilience relate to the dynamics of soil structural form, i.e. its variation over time. The concept of structural form changing in response to stress and natural processes implies a temporal dependence. The introduction of new land use practices seldom results in a step change in structural form and therefore stability, resilience and vulnerability must be defined in terms of rates of change in soil structural form.

Rates of change of structural form: a conceptual model

The time scale

Changes in soil structure arising from changes in land use practice can be considered over time scales ranging from hours to centuries. The rate of change in soil structure occurring within a time scale compatible with a farmer's planning horizon is of paramount importance. For most farmers this time scale is one of years rather than days or decades, and it is within this time scale that choices are made with regard to production practices which can influence the rate of change in soil structure. Information on rates of change in soil structure will be most useful to farmers in making management decisions if the data apply to this time scale and can be projected with some degree of confidence to a scale of decades.

Balancing degradation and regeneration: sustainable land use

The technologies available to farmers today are incorporated into some land use practices which lead to a degradation of soil structures and others which lead to a regeneration of soil structure. Farmers may, for different reasons, not have the option to pursue systems which never lead to the degradation of soil structure. Sustainable land use therefore requires balancing the extent of structural degradation over time caused by one system with the extent of regeneration by other systems such that soil structure is maintained or enhanced. The extent of degradation or regeneration under a given practice is determined by the rate of change in soil structure and the duration over which the practice occurs. Information on rates of change in soil structure under different soil, climate and land use practices can then be used to assess the impact of different choices in land use practices.

A conceptual model

The different aspects of soil structure can be represented in a conceptual model which allows assessment of the rate of change of soil structure under different land use practices and different soil and climatic conditions. The model builds on earlier work (Kay, 1990).

The variation with time, t, of a characteristic of structural form, S, for a given land use practice is given by

$$S = f(t) \tag{25.1}$$

The first hypothesis upon which the model is based is that the rate of change of a characteristic of structural form can be related to three functions that are additive: a function related to stresses arising from the land use practice, a biological-related function and a weather-related function. Each of the functions includes a variable characterizing the magnitude of the agent causing the change in structural form and a parameter which links the causal agent to the rate of change in structural form (i.e. a response characteristic).

The stress-related function includes two variables, the stress or force per unit area (F/A), applied to the soil and a stability parameter (R_{sf}) which is appropriate for a given characteristic of structural form and a given stress. The form and magnitude of the stress is determined by the land use practice and is normally related to tillage and traffic. The stability is determined by soil properties and the influence of biological factors and weather on those properties.

The biological-related function describes the rate of change in a characteristic of structural form caused by soil flora and fauna such as roots and earthworms. The biological-related function includes a parameter which describes the population/activity of organisms (b) and a resilience parameter (L_b) which relates the rate of change in structural form to the population/activity of the organism.

The weather-related function is made up of a parameter which describes the number/intensity of wetting/drying cycles or freezing/thawing cycles (w) and a resilience parameter (L_w) which relates the rate of change in structural form to w.

The parameters L_b and L_w are primarily related to soil properties whereas b is related to factors controlling biological activity (soil properties, weather and land use) and w is related to weather conditions or water management practices (irrigation/drainage).

The rate of change of a characteristic of soil structural form is hypothesized to be related to these functions as follows:

$$\frac{\partial S}{\partial t} = f(R_{SF}, \frac{F}{A}) + f(L_b, b) + (L_w, w) \tag{25.2}$$

The second hypothesis upon which the model is based is that stability is a function of the level of stabilizing materials (c) and weather (w) and that changes in stability arising from the introduction of a land use practice are due to a progressive change in the level of stabilizing materials, that is:

$$R_{\mathrm{S,F}} = f(c, w)_{\mathrm{S,F}} \tag{25.3}$$

and

$$c = f(t) \tag{25.4}$$

or

$$R_{\mathrm{S,F}} = f(t, w)_{\mathrm{S,F}} \tag{25.5}$$

The stabilizing materials may be inorganic and/or organic in form but only those materials which change in content or effectiveness with land use practices are included. Changes in the concentration of these materials may arise from changes in b, or land use practices as reflected in changes in soil pH, salinity, aeration, the extent of leaching, application of soil amendments, the amount, quality or distribution of photosynthate in the soil, and soil microbiological activity. The formulation of eqn (25.5) is based on the assumption that w does not progressively change with time and that any changes in b over time which influence R are incorporated into c.

The relative importance of the different terms in eqn (25.2) is determined largely by the magnitude of the stress applied. For instance, stress associated with tillage can readily destroy any change in structural form introduced by biological factors or weather. Under such conditions $\partial S/\partial t$ is largely determined by the stability and the stress applied (e.g. Blackwell *et al.*, 1991). If, however, stress is minimized, $\partial S/\partial t$ will be much more strongly influenced by the resilience parameters as well as b and w. Soils which are most vulnerable are those with a low stability, and a small value of $f(L_b, b) + f(L_w, w)$.

Methodology to Characterize Variables in the Conceptual Framework

The rate of change of soil structural characteristics will vary with climate, soils and land use practices. Data are needed that will enable comparison of rates of change in structural characteristics between broad soil/climate zones, between soils within a soil/climate zone, and between land use practices on a given soil. An infinite number of combinations of soil/climate/land use practices exist and therefore collection of data for each combination is impractical. Data must therefore be collected on a minimum number of combinations which will permit extrapolation to many of those combinations for which data are not available. Such extrapolation is facilitated if data are collected to characterize the soil/climate zone, the soils and climate characteristics of the sites over the period that studies are being conducted, the land use practices and specific characteristics related to soil structure.

Agroecological resource areas

Soil and climate characteristics can be grouped in different ways to create different types of soil/climate zones. If the purpose of characterizing combinations is to distinguish types of agriculture and biological productivity, the concept of agroecological resource areas appears to be useful. The concept is being utilized in many parts of the world (e.g. FAO, 1978). In Canada, broad areas of generally similar agricultural potential and management have been separated with an emphasis on natural (ecological) characteristics (Hiley *et al.*, 1989). The boundaries are based on ecoclimatic zones, soil development (classification at Order or Great Group level), physiographic subdivisions (landform), and soil texture (clay content).

Site characterization

An extrapolation of the impact of land use practices to different soils and different weather conditions within an agroecological resource area requires information on soil, biological and climate characteristics over the period in which the land use practice is being assessed. The ability to make reliable extrapolations increases with the number of soil and weather combinations but probably approaches some limit exponentially.

Soils

Soil characteristics which vary within an agroecological zone and which therefore should be documented at the outset of the study include:

- sand, silt and clay content
- organic carbon content
- pH
- carbonate content
- amorphous/crystalline iron, aluminium and silica
- electrical conductivity and exchangeable sodium percentage

These parameters should be described for each horizon together with the horizon depth and thickness.

Biology

Organisms leading to a change in structural form include roots, earthworms, termites and ants. Organisms which have the greatest influence on structural stability, however, include roots and microorganisms. The relative importance of the different agents will vary with the agroecological zone and with the land use practice. Biological parameters which should be measured regularly throughout the study period include:

- root length density (by size class if possible)
- earthworm, ant and termite populations
- microbial biomass

Weather

Structural form and stability are influenced by antecedent wetting/drying cycles and freezing/thawing cycles (Panabokke and Quirk, 1957; Benoit, 1973; Chamberlain and Gow, 1979; Utomo and Dexter, 1981b; Kay *et al.*, 1985). Structural stability can also be influenced by soil water content at time of sampling (Perfect *et al.*, 1990b). Parameters which should be measured regularly during the study period include:

- number of wetting/drying events
- number of freezing/thawing events
- water content at the time of sampling

There are few criteria to define what constitutes a wetting/drying or freezing/thawing event. The rate of change in soil water content has a strong influence on the change in structural stability (e.g. Grant and Dexter, 1989). Although measurements of soil properties are preferable to describe such events, it is often more practical to use meteorological data at the site. The criteria that are applied to the meteorological data must depend, however, on the depth in the soil profile that is of greatest concern in assessing the impact of land use practices. For instance, Kay and Dexter (1992), studying the influence of wetting and drying cycles on the tensile strength (a stability parameter) of aggregates from the top 10 cm of soil, arbitrarily defined wetting/drying events as those in which the daily precipitation was equal to or greater than one-tenth of the total amount of water which could exist in the top 10 cm of the profile when fully saturated.

Land use

Soil structure changes with the amount of stress that is applied to soil. Qualitative information on the extent of stress a soil may experience is provided by information on: (i) tillage (type, number of passes, depth of tillage); and (ii) traffic (number of passes and axle load of equipment, weight of animals and stocking rate).

Qualitative information on the moderation of the impact of rainfall or wind on the soil surface is provided by estimates of the proportion of the ground surface covered during periods of maximum exposure of soil surface to wind or water erosion.

Land use practices also determine the amount of organic materials returned to soil (and therefore the level of stabilizing materials), and may influence biological activity and the level of inorganic stabilizing materials. Documentation should be provided on:

- total annual above ground dry matter production and proportion left in field
- estimated below ground dry matter produced
- amendments (fertilizer, lime, etc.)
- water management (irrigation frequency, drainage)

Soil properties at the onset of a study are often strongly influenced by cropping history. Therefore land use practices in the preceding 25 years should be documented as fully as possible. Such information is particularly critical to assessing the suitability of a control treatment at the site (discussed later).

Soil structure

In considering the methodology to characterize soil structure five questions arise: (i) over what depth in the profile should characterization be done? (ii) which structural characteristics are most important? (iii) what is the role of a control in assessing the rate of change in soil structure due to a change in land use practices? (iv) what is the form of the functions to describe the temporal variation in soil structural characteristics? and (v) can basic soil properties such as described in routine soil surveys, and characteristics of weather such as documented in climatological records, be used for partial or complete prediction of the temporal variation in these variables under other soils or weather conditions?

Depth of profile required for characterization

If the purpose of characterizing soil structure is to assess the impact of land use practices, then the depth of sampling may be restricted to that depth in the profile that is most strongly influenced by these practices. The effects of most land use practices are greatest near the surface and are attenuated with depth. The functions describing the attenuation of the effects with depth will vary with factors such as depth and type of tillage, variation in root length density with depth, variation in soil compressibility and stress with depth and the effect of land use practices on the variation in activity of soil fauna with depth. Measurements should be made in the A horizon which, in a recently glaciated humid temperate environment, means in the upper 15 cm of the profile. This depth is most susceptible to degradation but is also the easiest to regenerate. Soil structure which becomes degraded at greater depths is often much more difficult to regenerate and could therefore have a more important impact on sustainable productivity. Consequently although the bulk of sampling can be restricted to zones near the soil surface, care should be taken always to include some measurements at greater depth (e.g. the base of the tillage zone and the B and C horizons).

The most important variables

A multitude of variables and methods exist to characterize soil structure. Methods to characterize soil structure have evolved through the efforts of successive generations of researchers, with one of the earliest compendia of methods being prepared 150 years ago (Schubler, 1840). Some of the more recent methods to describe soil structure are reviewed by Dexter (1988).

The relative importance of different structural characteristics varies with soil, climate, and cropping system. The relative importance attached to a given structural characteristic is determined by the influence of that characteristic on the soil property or process that imposes the greatest limitation on the yield of a specific crop, or the transport of sediments or agricultural chemicals, under specific climatic conditions. In the case of rain-fed agriculture in humid temperate regions, growing crops may experience limitations due to a lack of oxygen (excess water), a lack of water and/or high soil resistance to penetration at different stages within the growing season. The relative importance of the limitations on crop yield may vary considerably between seasons. The impact of each of these limitations relates to weather (precipitation and potential evapotranspiration) and soil structural form. The characteristics of pore space (e.g. total porosity, pore size distribution, and pore continuity) are the most important characteristics of structural form which control aeration, available water and penetration resistance.

Total porosity or bulk density has been related to crop yield using a quadratic function in humid temperate environments (e.g. Carter, 1990; Hakansson, 1990). The variation in yield with bulk density is, however, also dependent on climate and may vary from year to year (Raghaven *et al.*, 1990). Such variation may limit the utility of bulk density alone in relating structural form to crop yield.

The concept of a range in water contents or water potentials which incorporate limitations related to aeration, resistance to penetration and available water has evolved over the past two decades (e.g. Eavis, 1972; Letey, 1985; Boone *et al.*, 1986). The limits to the range occur at water contents or potentials at which plant growth is dramatically reduced or ceases. The range in water content between these limits was referred to by Letey (1985) as the non-limiting water range, NLWR. Although this may be an unfortunate choice of terms to describe the variable (since growth may decrease or experience some limitations well before the limit is reached), the concept of the range is important since it reduces a data set (water release curve and therefore pore size distribution and total porosity, and the penetration resistance–water content function) to a single variable. Historical soil moisture (or climatological) data can then be used to predict the probability of the frequency of soil water content falling outside of this range during critical growth periods. Research is currently underway in Canada to assess the sensitivity of soils to changes in NLWR due to different land use practices and to relate NLWR and soil water content to crop response.

The role of tillage as part of the land use practice and its impact on soil structure must be considered before a final decision is made on the most important structural parameters to measure. If tillage is used to create a seedbed each year, the opportunity exists for the farmer to apply a type and amount of stress which may create, at the beginning of each growing season, an aggregate size distribution that will provide pore characteristics which are optimal for germination and early plant growth. Before tillage, the pore characteristics of the Ap horizon may reflect the influence of the antecedent effects of biological activity and weather, but these effects may be totally removed with the tillage. Whereas land use practices may alter R, it is conceivable that this change could be offset by altering F/A such that the pore characteristics remain constant over time if only measured at the time the seedbed is created. Exceptions to this generalization include cases where land use practices have resulted in progressive changes in the composition of the soil through erosion.

The pore characteristics of a seedbed will progressively change during the growing season due to stresses arising from traffic and due to wetting and drying (Carter, 1990). Such changes may have detrimental effects on plant growth, particularly in the early stages of development. A structural form which persists through the growing season is desirable and the impact of the farming practice on the persistence of seedbed characteristics would be expected to be most accurately described by a stability characteristic (eqn 25.2).

Different stability characteristics have been used to describe different processes leading to changes in structural form when different stresses are applied, e.g. compressibility index (Angers *et al.*, 1987), and friability (Utomo and Dexter, 1981b). There have been few attempts to determine the extent of correlation between the different indices of resistance. However, the change in the structural form must relate to the deformability of aggregates which, in turn, must relate to the strength of failure zones within aggregates. It has been hypothesized (Kay and Dexter, 1992) that the strength of failure zones is related to the dispersibility of clay and one would therefore expect dispersible clay to correlate with a number of resistance parameters. This expectation is confirmed in relations between dispersible clay and modulus of rupture and friability for a soil from the B horizon of a red-brown earth in Australia (Shanmuganathan and Oades, 1982). This study needs to be extended to a range of other soil conditions and to additional resistance parameters.

If land use practices are altered to reduce stress and allow a period of regeneration, then $\partial S/\partial t$ after tillage is renewed will be strongly influenced by the magnitude of R achieved during the regeneration period. Measurements should therefore include measurements of $R = f(t, w)$.

The role of a control

The variation in some structural characteristics due to processes influenced by weather is often as large as or larger than the variation due to changes in land use practices (e.g. Perfect *et al.*, 1990a). A part of this variation may be independent of land use practice and can create inconsistent trends in the temporal variation in structural characteristics within or between seasons. Such variation increases the difficulty in obtaining statistically significant effects of a change in land use when measurements are made over a relatively small number of years and therefore necessitates longer term experiments. This effect can be minimized if the study has been established on a site on which a control treatment can be maintained. The control treatment must be part of an ecosystem that is stable within the time scale of interest, i.e. soil structural characteristics are not changing progressively with time. In the absence of a land use practice × weather interaction, subtraction of the value of the parameter for the control from the value of the land use practice provides a measure of the net change due to the land use.

Form of functions

Different forms of functions may be used to describe the temporal variations in S and R (eqns 25.1 and 25.5) when a new land use practice is introduced on a given soil. A function that may be equally applicable to characteristics of S and R is proposed which takes the form (for the case where S is increasing):

$$S(t) = S^0 + \Delta S_{max}\,[(l - e^{-k(t-t_1)})] \qquad (25.6)$$

where $S(t)$ is the value of the variable S at time, t (year), subsequent to the introduction of the new land use practice; S^0 is the value of S at the time of introducing the land use practice, i.e. at $t = 0$; ΔS_{max} is the maximum projected change in S caused by this land use practice, k is the rate constant, and t_1 is the time elapsed after introduction of the land use practice before there are any measurable changes in S.

When $t < t_1$

$$S(t) = S^0 \tag{25.7}$$

If the stress is minimized, ΔS_{max} is the resilience potential (see earlier definition of resilience potential) and k is the rate constant for the recovery processes.

Eqn (25.6) has been applied to data on the temporal variation of infiltration rate (I) subsequent to the introduction of alfalfa (*Medicago sativa* L.) (Meek *et al.*, 1990). The infiltration rate will be primarily responsive to pore characteristics (structural form). Meek *et al.* (1990) reported data from two treatments, one which experienced no traffic and one which experienced traffic over 100% of the soil surface after each harvest. The data fitted the following functions:

no traffic treatment:

$$I = 19.73 + 40.97\ (l - e^{0.669t}) \qquad R^2 = 0.90 \tag{25.8}$$

100% traffic treatment:

$$I = 13.98 + 51.01\ (l - e^{-0.391t}) \qquad R^2 = 0.91 \tag{25.9}$$

Values of t_1 were zero for both treatments. There was, unfortunately, no control treatment.

The increase in infiltration rates reported by Meek *et al.* (1990) must have arisen from biological factors (roots and perhaps soil fauna) and perhaps weather-related factors (although the soil was reported to have limited shrink/swell properties). This increase was moderated by compaction in the traffic treatment. Two features of eqns (25.8) and (25.9) are noteworthy and illustrate the value of this form of the function. First, the resilience potential of the soil in its original state (with respect to infiltration rate) is given by ΔI_{max} in eqn (25.8), i.e. 40.97 mm h^{-1}. The second feature of note is that although the rate of increase in I is smaller in the traffic treatment the values of I at $t \rightarrow \infty$, $I^0 + \Delta I_{max}$, are similar, 60.70 and 64.99 cm h^{-1}, for the no traffic and traffic treatments respectively.

Values of the half-life, i.e. the time required for the alfalfa to increase I to a point midway between the initial and maximum value, were also calculated. The half-life ($t_{½}$) is given by

$$t_{½} = t_1 - \frac{\ln 0.5}{k} \tag{25.10}$$

Values of half-lives are often more meaningful than values of t_1 and k for advisory purposes. Values of $t_{½}$ were 1.04 and 1.77 years for the no traffic and traffic treatment respectively. Rasiah and Kay (1993a,b) have used a function of the form of eqn (25.6) to describe changes in the wet aggregate stability and in the dispersibility of clay subsequent to the introduction of forages on land that had been used for continuous corn production using conventional tillage for an extended period. Data collected at monthly intervals over three growing seasons on seven different soils were used to assess the function. The characteristics of the soils are given in Table 25.1. The function used by Rasiah and Kay (1993b) was:

$$SC_f\,(t, t\theta) = SC^0 + (\Delta v_f - \Delta q_f \theta_f)\,[l - e^{-k(t-t_1)}] \qquad (25.11)$$

where SC is the amount of stabilized clay (total clay − dispersible clay), v and q are constants describing the functional dependence of SC on gravimetric water content (θ), and the subscript f refers to the forage treatment. The corn treatment was used as the control on all sites. In order to minimize the impact of weather on SC (eqn 25.5) values for the control (identified with subscript c) were subtracted from the forage treatment to give the net change due to forages, SC_{net}, i.e.

$$SC_{net} = SC_f - SC_c = (\Delta v_f - \Delta q_f \theta_f)\,[l - e^{-k(t-t_1)}] + q_c\,(\theta_c - \theta_f) \quad (25.12)$$

The magnitude of the coefficients in this function for the seven soils are given in Table 25.2 together with values of the coefficient of determination.

Analysis of the data showed that there was no lag in the stabilization of clay subsequent to the introduction of forages. This is in contrast to the observation of values of t_1 ranging from 0.72 to 1.05 year for wet aggregate stability measurements which were made on the same soil (Rasiah and Kay, 1993a). The R^2 for the best fit of eqn (25.12) ranged from 0.48 to 0.84% and

Table 25.1. Description and selected properties of the soils used by Rasiah and Kay (1993a, b).

Soil no.	Series name (US Classification)	Clay (%)	Silt (%)	Organic matter (%)	pH	$CaCO_3$ (%)
1	Brookston Clay (Typic Haplaquept)	33.8	27.3	3.3	5.77	0.00
2	Brookston Clay (Typic Haplaquept)	21.1	42.3	3.1	7.33	0.79
3	Tuscola Silt Loam (Aquic Hapludalf)	18.5	56.9	3.9	6.60	0.16
4	Conestogo Silt Loam (Aquic Eutrochrept)	18.3	52.0	3.8	7.20	1.50
5	Conestogo Silt Loam (Aquic Eutrochrept)	17.9	53.4	3.4	7.17	2.04
6	Wattford Loamy Sand (Arenic Hapludalf)	6.4	10.4	2.9	6.40	0.00
7	Fox Loamy Sand (Arenic Hapludalf)	6.4	15.7	2.2	5.84	0.10

Table 25.2. Coefficients for the parameters k, Δv, Δq, and q_c in eqn (25.12) for the different soils[a].

Soil no.	Parameters						
	k	Δv	Δq	q_c	R^2	ΔSC_{max}	$t_{1/2}$
1	0.195	2.651	−0.121	0.206	0.84	4.51	3.55
2	0.186	1.639	−0.088	0.135	0.54	3.05	3.73
3	0.152	0.621	−0.061	0.127	0.51	1.77	4.56
4	0.161	0.813	−0.053	0.099	0.49	1.80	4.31
5	0.161	1.117	−0.065	0.113	0.76	2.35	4.31
6	0.096	0.667	−0.038	0.078	0.48	1.16	7.22
7	0.076	0.358	−0.018	0.041	0.58	0.55	9.12

[a]The units for Δv and ΔSC_{max} are g clay stabilized/100 g soil, Δq and q_c are g stable clay/g water; k are year^{-1}, and $t_{1/2}$, year, respectively. ΔSC_{max} is equal to $v - \Delta q\theta$.

the model was significant at $P < 0.05$ for each soil. Values of the maximum potential increase in clay stabilization, ΔSC_{max} (calculated from Δv_f, Δq_f and the average water content for each soil) ranged from 0.55 to 4.51%. Values of ΔSC_{max} provide a measure of the regeneration potential in SC. Values of $t_{1/2}$ ranged from 3.55 year for the finest textured soil (no. 1) to 9.12 years for one of the coarsest textured soils (no. 7). Values for Δq were all negative indicating the sensitivity of the clay to disperse with increasing θ as sampling decreased over time under forages. Sensitivity analyses showed the variation in SC_{net} with time was most sensitive to q_c followed in decreasing order of sensitivity by Δv, k and Δq. The relationship between SC_{net} and all four parameters was positive. Eqn (25.12) appears to provide a particularly useful form of a function to describe $R = f(t, w)$.

Extrapolation to other soils

The capability to extrapolate the impact of a land use practice to other soils is determined by the extent to which the characteristics of soils which are available through routine surveys can be used to predict the parameters which determine the rate of change of soil structure.

The term pedotransfer function has been used by Bouma and van Lanen (1987) to describe functions in which those soil properties routinely measured in soil survey are employed to predict other soil properties which are not measured. Such functions have been used to describe the variation in pore characteristics with soil properties (e.g. McBride and Mackintosh, 1984) and some stability characteristics (e.g. Kemper and Koch, 1966; Rasiah and Kay, 1992). There are, however, few descriptions of the rates of change of soil structural characteristics arising from a specific land use employing pedotransfer functions.

The value of using pedotransfer functions to predict the rates of change in structural characteristics is illustrated in the analysis of Rasiah and Kay (1993b). The coefficients in eqn (25.12) were related to soil properties and the following functions were obtained:

$$q_2 = 0.346 - 0.358\frac{OM}{\text{clay}} - 0.0222\ \text{pH} \qquad R^2 = 0.83 \qquad (25.13)$$

$$\Delta q = 0.0092 + 0.00014\ \text{clay} \times OM \times \text{pH} \qquad R^2 = 0.71 \qquad (25.14)$$

$$\Delta v = 0.0584 + 0.00272\ \text{clay} \times OM \times \text{pH} \qquad R^2 = 0.62 \qquad (25.15)$$

$$k = 0.0663 + 0.00021\ \text{clay} \times OM \times \text{pH} \qquad R^2 = 0.93 \qquad (25.16)$$

where OM is organic matter content (%).

These functions can then be substituted into eqn (25.12) to generate estimates of the temporal variation in dispersible clay which is stabilized when forages are introduced on soils of different clay content, organic matter content and pH (Fig. 25.1 and 25.2).

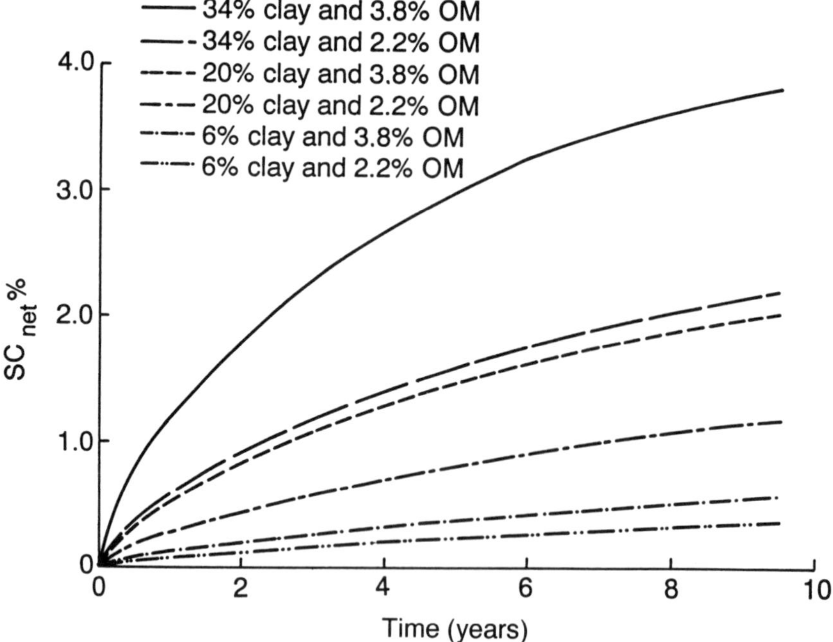

Fig. 25.1. Temporal variation in SC_{net} when forages are introduced after prolonged production of corn on soils with different clay and organic matter contents (pH = 5.77 and $\theta_c = \theta_f$).

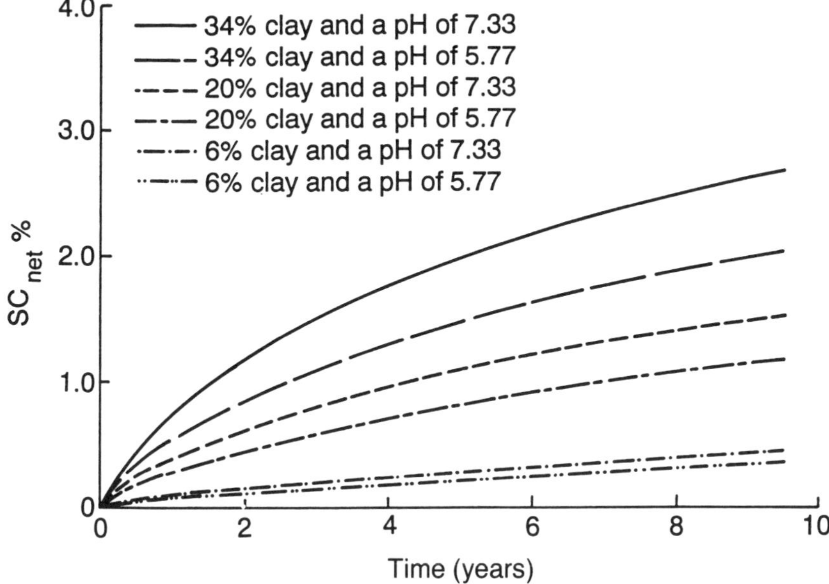

Fig. 25.2. Temporal variation in SC_{net} when forages are introduced after prolonged production of corn on soils with different clay content and pH (organic matter content = 2.22%, $\theta_c = \theta_f$).

The pedotransfer functions not only allow extrapolation to other soils but also, because of their statistical nature, identify relations which need to be evaluated experimentally to assess cause/effect. For instance, the relations imply the rate of stabilization of clay is greater on soils which initially have higher OM and pH and that this effect is greatest in soils with higher clay contents (Fig. 25.1 and 25.2). Does this mean that the effectiveness of organic cementing materials contributed by the forages is enhanced by the presence of larger amounts of organic materials and polyvalent ions such as Ca^{2+} and Mg^{2+}? If so, what are the mechanisms and what are the implications for management when forages are introduced as an ameliorative cropping practice?

Research Needs

Research is needed on topics which include the following.

- The identification of the processes within an agroecological zone which create the most frequent limitations on plant growth, contribute to the greatest dispersal of sediments and agricultural chemicals, or impose the greatest control on the recharge of ground water supplies, and thereby identify the soil structural characteristics, i.e. the limiting characteristics, which are most relevant to these processes (the use of simulation models will be crucial to meeting this need).

- The identification of characteristics of structural form and stability which may influence or be strongly correlated to as many of these limiting characteristics as possible (in this context a detailed analysis of the non-limiting water range as a limiting characteristic of soil structural form and stabilized (or dispersed) clay as a limiting characteristic of structural stability is needed).
- In the absence of quantitative data on agents causing changes in either structural form (stresses from tillage or traffic biological effects and weather effects) or structural stability (stabilizing materials, weather), the evaluation of the use of eqn (25.6) to describe the rate of change in structural form or stability arising from the introduction of a new land use practice under a range of soil and climatic conditions and determine the resilience potential for characteristics of structural form and the maximum regeneration in structural stability under these conditions.
- The development of pedotransfer functions to permit extrapolation of data to a wider range of soil conditions.
- The development of generalized approaches to quantify the agents responsible for changes in structural form, i.e. quantify the stresses due to different types of tillage and traffic so that the structural impact in terms of cumulative stress input can be determined for a given land use practice, quantify the form and activity of biological factors, and quantify the number and intensity of weather-related events responsible for altering structural form.

References

Angers, D.A., Kay, B.D. and Groenevelt, P.H. (1987) Compaction characteristics of a soil cropped to corn and bromegrass. *Soil Science Society of America Journal,* 51, 779–783.

Benoit, G.R. (1973) Effect of freeze–thaw cycles on aggregate stability and hydraulic conductivity of three aggregate sizes. *Soil Science Society of America Proceedings,* 37, 3–5.

Blackwell, P.S., Jayawardane, N.S., Green, T.W., Wood, J.T., Blackwell, J. and Hardy, H.J. (1991) Subsoil macropore space of a transitional red-brown earth after either deep tillage, gypsum or both. I. Chemical effects and long-term changes. *Australian Journal of Soil Research* 29, 141–154.

Boone, F.R., Van der Werf, H.M.G., Kroesbergen, B., ten Haq, R.A. and Boers, A. (1986) The effect of compaction of the arable layer in sandy soils on the growth of maize for silage. I. Critical matric potential in relation to soil aeration and mechanical impedance. *Netherlands Journal of Agricultural Science* 34, 155–171.

Bouma, J. and van Lanen, H.A.J. (1987) Transfer functions and threshold values from soil characteristics to land qualities. In: *Quantified Land Evaluation, Proceedings of a Workshop ISSS/SSSA,* Washington DC, ITC Publ., Enschede. The Netherlands.

Carter, M.R. (1990) Relative measures of soil bulk density to characterize compactions in tillage studies on fine sandy loams. *Canadian Journal of Soil Science* 70, 425–433.

Chamberlain, E.J. and Gow, A.J. (1979) Effect of freezing and thawing on the permeability and structure of soils. *Engineering Geology* 13, 73–92.

Dexter, A.R. (1988) Advances in the characterization of soil structure. *Soil Tillage Research* 11, 199–238.

Eavis, G.W. (1972) Soil physical conditions affecting seedling root growth. I. Mechanical impedance, aeration and moisture availability as influenced by bulk density and moisture levels in a sandy loam soil. *Plant and Soil* 36, 613–622.

Food and Agriculture Organization (1978) *Report on the Agroecological Zones Project.* Vol. 1. Methodology and results for Africa. FAO, Rome, 15 pp.

Grant, C.D. and Dexter, A.R. (1989) Generation of cracks in molded soils by rapid wetting. *Australian Journal of Soil Research* 27, 169–182.

Grant, C.D. and Blackmore, A.V. (1991) Self-mulching behaviour in clay soils: its definition and measurement. *Australian Journal of Soil Research* 29, 155–173.

Hakansson, I. (1990) A method for characterizing the state of compactness of the plough layer. *Soil and Tillage Research* 16, 105–120.

Hiley, J.C., Toogood, K.E., Pettapiece, W.W. and Wehrhahn, (1989) Description of agricultural production in Alberta using an integrated land resource/land use data base. Alberta Institute of Pedology Publ. No. M–89–1 Edmonton, Alberta, Canada, 112 pp.

Kay, B.D. (1990) Rates of change of soil structure under different cropping systems. *Advances in Soil Science* 12, 1–52.

Kay, B.D. and Dexter, A.R. (1992) The influence of dispersible clay and wetting/drying cycles on the tensile strength of a red-brown earth. *Australian Journal of Soil Research* 30, 297–310.

Kay, B.D., Grant, C.D. and Groenevelt, P.H. (1985) Significance of ground freezing on soil bulk density under zero tillage. *Soil Science Society of America Journal* 49, 373–378.

Kemper, W.D. and Koch, E.J. (1966) Aggregate stability of soils from the western United States and Canada. *Technical Bulletin No. 1355,* USDA, pp. 1–52.

Letey, J. (1985) Relationship between soil physical properties and crop production. *Advances in Soil Science* 1, 277–294.

McBride, R.A. and Mackintosh, E.E. (1984) Soil survey interpretations from water retention data: I. Development and validation of water retention model. *Soil Science Society of America Journal* 48, 1338–1343.

Meek, B.D., De Tar, W.R., Rolph, D., Rechel, E.R. and Carter, L.M. (1990) Infiltration rate as affected by an alfalfa and no-till cotton cropping system. *Soil Science Society of America Journal* 54, 505–508.

Panabokke, C.R. and Quirk, J.P. (1957) Effect of initial water content on the stability of soil aggregates in water. *Soil Science* 83, 185–189.

Perfect, E., Kay, B.D., van Loon, W.K.P., Sheard, R.W. and Pojasok, T. (1990a) Rates of change in soil structural stability under forages and corn. *Soil Science Society of America Journal* 54, 179–186.

Perfect, E., Kay, B.D., van Loon, W.K.P. Sheard, R.W. and Pojasok, T. (1990b)

Factors influencing soil structural stability within a growing season. *Soil Science Society of America Journal* 54, 173–179.

Rasiah, V. and Kay, B.D. (1993a) Characterizing rates of change in aggregate stability using the stability water content relation. *Soil Science Society of America Journal* (submitted).

Rasiah, V. and Kay, B.D. (1993b) Quantifying rates of change in clay stabilization subsequent to introduction of forages. Manuscript in preparation.

Rasiah, V., Kay, B.D. and Martin, T. (1992) Variation of structural stability with water content: Influence of selected soil properties. *Soil Science Society of America Journal* 56, 1604–1609.

Raghaven, G.S.V., Alvo, P. and McKeyes, E. (1990) Soil compaction in agriculture: A view toward managing the problem. *Advances in Soil Science* 11, 1–36.

Schubler, Professor (1840) On the physical properties of soil, and on the means of investigating them. *Journal of the Royal Agricultural Society* 1, 177–218.

Shanmuganathan, R.T. and Oades, J.M. (1982) Effect of dispersible clay on the physical properties of the B horizon on a red-brown earth. *Australian Journal of Soil Research* 20, 315–324.

Utomo, W.H. and Dexter, A.R. (1981a) Tilth mellowing. *Journal of Soil Science* 32, 187–201.

Utomo, W.H. and Dexter, A.R. (1981b) Soil friability. *Journal of Soil Science* 32, 203–213.

Chapter 26
Soil Databases for Sustainable Land Use: Hungarian Case Study

G. Várallyay

Research Institute for Soil Science and Agricultural Chemistry of the Hungarian Academy of Sciences, PO Box 35, H-1525 Budapest, Hungary

Introduction

Soils represent a considerable part of the natural resources of Hungary. Consequently, rational and sustainable land use and proper management to ensure normal soil functions have particular significance in the Hungarian national economy, and soil conservation is an important element of environment protection.

Soil Functions

The main functions of soil in the biosphere are as follows (Várallyay, 1991b):

1. Soils are the most significant, *conditionally renewable, natural resources.* During crop production they do not change irreversibly, and their quality does not decrease unavoidably and fundamentally. But their 'renewal' does not go on automatically: soil conservation, the maintenance and increase of soil fertility requires permanent activities, such as: sustainable land use, agrotechnics, remediation and reclamation.
2. Soil is *reactor, transformer and integrator* of the combined influences of other natural resources (solar radiation, atmosphere, surface and subsurface waters, biological resources) and represents the 'life-media' for microbiological activities, as well as the ecological environment for natural vegetation or cultivated crops.
3. Soil is the most important *medium for biomass production.* Water, air and available plant nutrients may occur simultaneously in soil, in this four-dimensional, three-phase, polydisperse system and soil may provide the principal requirements of plants.

 Soil Resilience and Sustainable Land Use (eds D.J. Greenland and I. Szabolcs)

4. Soil is a major *natural storage* of heat, water and plant nutrients.
5. Soil is an efficient *'natural filter'* and may prevent various pollutants reaching the deeper horizons and the subsurface waters.
6. Soils represent a *high capacity buffer medium* of the biosphere, which, to a certain limit, may moderate the various stresses caused by environmental factors (climatic droughts or too humid conditions, frost, etc.) and/or human activities (high input, fully mechanized and chemically controlled crop production; liquid manure from large-scale livestock farms; wastes and waste waters originating from industry, transport, and rural developments; etc.).

Soil Processes and Their Control

The ability of soil to fulfil these functions (*'soil value'*) is determined by the integrated impacts of various *soil properties*, which are the results of *soil processes*: mass and energy regimes, abiotic and biotic transport and transformation and their interactions. As any soil-related human activity influences the soil through these processes, their control is the main task of soil science and soil management. The Hungarian 'strategy' for the *control of soil processes* is summarized in Fig. 26.1.

Scientifically based planning and implementation of sustainable land use and rational soil management to ensure normal soil functions require adequate *soil information*: exact, reliable, 'detectable' (preferably measurable) and accurate, quantitative territorial data on well-defined soil and land properties, including characterization of their spatial (vertical, horizontal) and temporal variabilities and pedotransfer functions.

Data Sources

In Hungary, a large amount of information about ecological conditions is available as a result of long-term observations, survey and mapping activities (Hungarian National Atlas, 1989).

1. *Meteorological data*: systematic and regular measurements have been made since 1850. At present basic meteorological parameters are registered at 160 observation points; 18 stations are equipped for detailed atmospheric-chemical measurements and four EMEP stations for continuous atmospheric monitoring.
2. *Hydrological data*: regular records on the quantity and quality of *surface waters* (rivers, creeks, canals, lakes, ponds, reservoirs) are available from the first decade of the century.

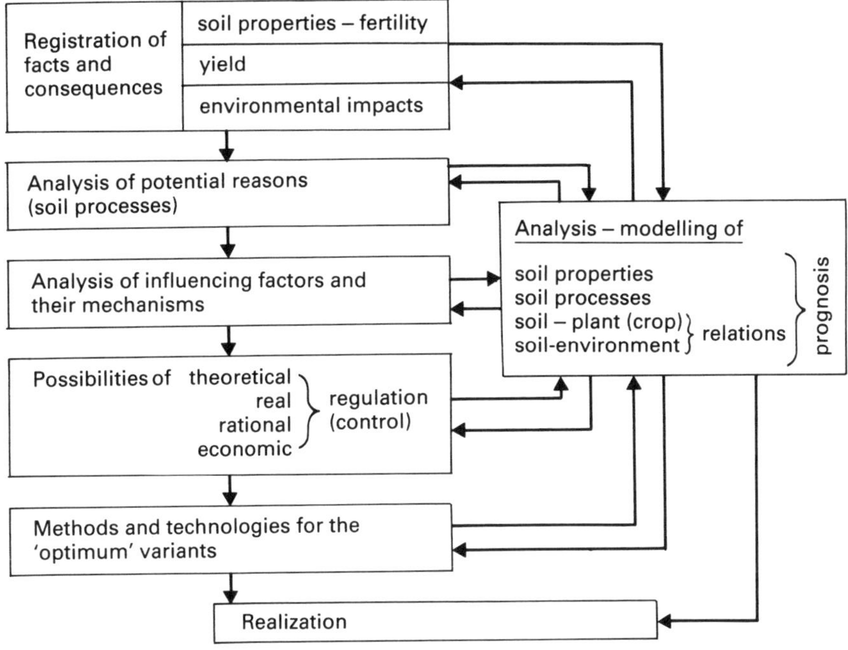

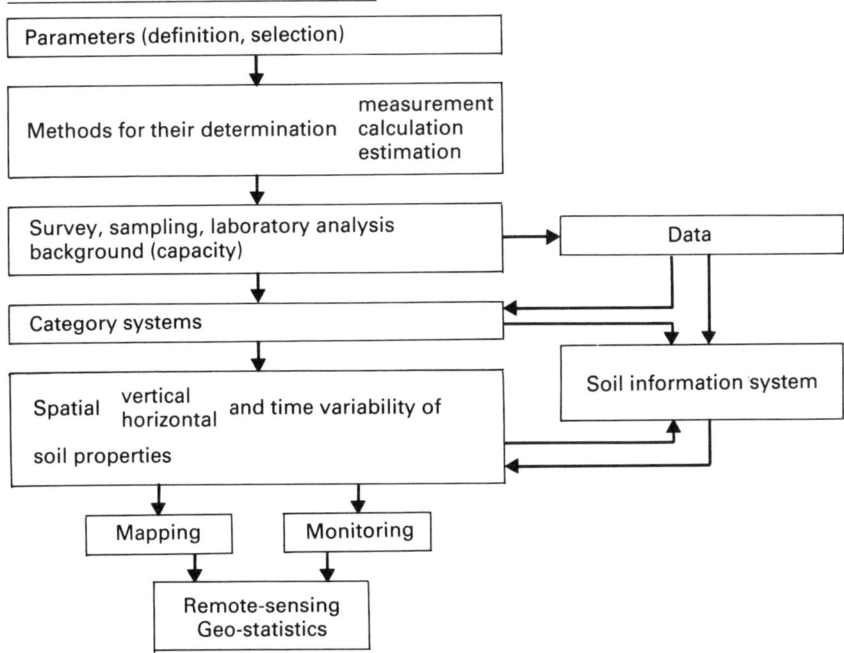

Fig. 26.1. Control of soil processes. (From Várallyay, 1989c.)

Observations of *groundwater conditions* (depth of water table; chemical composition of the groundwater) in 600–1000 groundwater testing wells are available since 1935, including 50 piezometer installations, measuring water pressure and water chemistry parameters in the deeper aquifers. On this basis the 1:200,000 scale map of average depth to the groundwater table, and the 1:100,000 scale map of groundwater chemistry (total concentration, ionic composition) have been prepared and updated; and the actual depth of water table was indicated on a 1:1 M scale map monthly.

3. *Geological data*: as a result of the 160-years of geological survey work, a 1:200,000 map for the whole country and at larger scales for various geologically important regions is available.

4. *Geomorphological data*: a 1:200,000 geomorphological map is available, and relief characteristics (slope gradient, length, complexity, exposure of the slopes) are indicated on contour maps and digital relief models.

Soil Information Sources

A large amount of soil information is available in Hungary as a result of long-term observations, and various soil survey, analytical and mapping activities on national (1:500,000), regional (1:100,000), farm (1:10,000–1:25,000) and field scales (1:5000–1:10,000) conducted during the last 60 years. Thematic soil maps are available for the whole country at a scale of 1:25,000 and for 70% of the agricultural area at a scale of 1:10,000.

There are at least three reasons why this rich soil database has been developed:

1. The small size of the country (93,000 km^2);
2. The great importance of agriculture and soils in the national economy;
3. The historically 'soil-loving' character of the Hungarian people, and particularly the Hungarian farmers.

Soil maps

Table 26.1 summarizes the most important thematic soil maps in Hungary, indicating their content, scale, author and date of preparation (Stefanovits and Szücs, 1961; Proceedings of the Hungarian–Swedish Seminar on Soil Mapping, 1989; Várallyay, 1989b).

The maps can be divided into three main groups.

Large-scale maps (nos. 1–4 in Table 26.1)

1. In the 'Kreybig – practical soil maps' (Kreybig, 1937) soil reaction, carbonate and salinity/alkalinity status is indicated by colours; physical–hydrophysical characteristics and depth of the soil by rasters; organic matter, total P_2O_5 and K_2O content, depth of the humus horizon and the groundwater table by a code number; and the soil type (according to 'Sigmond's soil classification') with Roman numbers.

2. On the 1:10,000 scale genetic soil maps (Sarkadi *et al.*, 1964; Szabolcs, 1966) the most important soil properties: soil type, subtype and local variant according to the Hungarian soil classification system; pH and carbonate status; texture; hydrophysical properties; salinity/alkalinity status; organic matter content; N, P and K status are indicated on separate thematic maps (cartograms). Recommendations are summarized in additional thematic maps for rational land use and cropping systems, soil cultivation, use of fertilizers, soil moisture control, including water conservation practices, irrigation and drainage, soil conservation practices for water- and wind-erosion control, etc.

3. The large-scale maps on the *possibilities and limitations of irrigation* (Szabolcs *et al.*, 1969), from the viewpoint of soil conditions, indicate:

- soil types, subtypes and local variants and parent material;
- physical–hydrophysical soil characteristics;
- salinity/alkalinity status of the soil (salt content, ion composition, ESP, pH);
- groundwater conditions (depth and fluctuation of water table; salt concentration, ion composition and SAR of the groundwater);

and on this basis:

- the 'critical depth' of water table and 'critical groundwater regime';
- recommendations for irrigation practices and groundwater management on separate thematic maps.

Medium scale maps (nos. 5–7 in Table 26.1)

In 1978 a national programme was initiated by the Hungarian Academy of Sciences for the 'Assessment of the agroecological potential of Hungary' (Láng *et al.*, 1983). In this programme a 1:100,000 scale map was prepared by Várallyay *et al.*, (1979, 1980a, 1985) on the soil factors determining agro-ecological potential, utilizing all available soil information. Seven soil factors were shown with an 8-digit code number:

1st and 2nd digit: Soil types (31 categories);
3rd digit: Parent material (9 categories);
4th digit: Soil reaction and carbonate status (5 categories);
5th digit: Soil texture (7 categories);
6th digit: Hydrophysical properties (9 categories);

Table 26.1 Thematic soil maps available in Hungary.

No.	Map	Scale	Date of preparation	Prepared for	Content	Author(s)
1.	Practical soil maps	1:25,000	1935–1955	the whole country per topographical map sheets	m, tm, fd, ld, e	Kreybig and coll.
2.	Large-scale genetic soil maps	1:10,000	1960–1975	70% of the agricultural land of Hungary, per farming units	m, tm, fd, ld, e	Coll.
3.	Soil conditions and the possibilities of irrigation	1:25,000	1960–1970	present and potential irrigated regions	6 thematic maps fd, ld	Coll.
4.	Large-scale maps for amelioration projects	1:10,000	1960–	amelioration projects (occasionally)	m, e	Coll.
5.	Soil factors determining the agroecological potential	1:100,000	1978–1980		m (with an 8-digit code), c	Várallyay, G. Szűcs, L. Murányi, A. Rajkai, K. Zilahy, P.
				the whole country per topographical sheets		
6.	Agrotopographical map	1:100,000	1987–1988		m (with a 10-digit code), c	Várallyay, G. Molnár, S. Szücs, L.
7.	Hydrophysical properties of soils	1:100,000	1978–1980		m, c	Várallyay, G. Szücs, L. Rajkai, K. Zilahy, P.

8.	Limiting factors of soil fertility	1:500,000	1976	the whole country	m	Szabolcs, I. Várallyay, G.
9.	Main types of moisture regime	1:500,000	1983		m, c	Várallyay, G. Zilahy, P. Murãnyi, A.
10.	Main types of substance regimes	1:500,000	1983		m, c	Várallyay, G. Szücs, L. Molnár, E.
11.	Soil erosion	1:500,000	1960–1964		m, tm, e	Stefanovits, P. Duck, T.
				the whole country		
12.	Salt-affected soils	1:500,000	1970–1974		m, e	Szabolcs, I. Várallyay, G. Mélyvölgyi, J.
13.	Susceptibility of soils to acidification	1:100,000– 1:500,000	1985–1988			Várallyay, G. Rédly, M. Murányi, A.
14.	Susceptibility of soils to physical degradation	1:500,000	1985–1988		m	Várallyay, G. Leszták, M.
15.	Soil evaluation	–	1980–1985	soil profiles	fd, ld	Coll.
		1:10,000– 1:25.000	1985–	non-mapped part of agricultural and forest land of Hungary	m, tm, fd, ld	Coll.

Key: m = soil map; tm = thematic map; fd = field description; ld = laboratory data; e = explanatory booklet; c = computer storage.

7th digit: Organic matter resource (6 categories);
8th digit: Depth of the soil (5 categories).

The territorial distribution of these categories is summarized in Table 26.2. The map was completed later by additional code numbers for two more soil characteristics:

9th digit: Clay mineral association (Stefanovits, 1989);
10th digit: Soil productivity index.

The soil contours expressing these nine characteristics were printed to a 1:100,000 basic topographical map with additional information regarding relief, land use, infrastructure, etc. Meteorological information is given on the territorial and temporal variabilities of the main climate elements on each map sheet by 'micro-maps', and monthly distribution diagrams. These *agro-topographical maps* were prepared for the whole country and are available in printed form (Várallyay and Molnár, 1989).

The soil contours were digitized and organized into a GIS-based soil information system (see later).

Small-scale maps (nos. 8–12. in Table 26.1)

From the large collection the schematic simplified version of two 1:500,000 scale maps are presented in Figs 26.2 and 26.3, on the status of soil erosion (Stefanovits, 1964) and salinity/alkalinity in Hungary (Szabolcs, 1974).

Soil susceptibility/vulnerability maps

Recently special attention has been paid to the characterization of soils from

Table 26.2. Territorial distribution of the various soil factors determining the agroecological potential in Hungary (in hectares).

Soil factors determining the agro ecological potential	Total	%
Parent material		
1 Glacial and alluvial deposits	3 433 430	37.7
2 Loess, loess-like deposits	4 374 920	48.0
3 Tertiary and older deposits	681 440	7.5
4 "Nyirok"	151 660	1.7
5 Limestone, dolomite	238 950	2.6
6 Sandstone	11 430	0.1
7 Shale, phyllite	28 530	0.3
8 Granite, porphyryt	9 740	0.1
9 Andesite, riolite, basalt	179 350	2.0

Soil reaction and carbonate status		
1 Strongly acidic soils	1 228 930	13.5
2 Slightly acidic soils	3 848 550	42.2
3 Calcareous soils (effervescence with dilute acid from the surface)	3 493 090	38.4
4 Salt affected soils, calcareous from the surface	385 260	4.2
5 Salt affected soils, non-calcareous from the surface	153 620	1.7
Soil texture		
1 Sand	1 437 230	15.8
2 Sandy loam	875 460	9.6
3 Loam	3 932 320	43.2
4 Clay loam	1 692 630	18.6
5 Clay	632 840	6.9
6 Organic soils (peat, partly decomposed peat)	117 560	1.3
7 Coarse fragments (gravel, non- or partly weathered rocks, etc.)	421 410	4.6
Soil-water management properties		
1 Soils with very high infiltration rate (IR), permeability (P) and hydraulic conductivity (HC); low field capacity (FC); and very poor water retention (WR)	957 420	10.5
2 Soils with high IR, P and HC; medium FC, poor WR	1 009 910	11.1
3 Soils with good IR, P, and HC; good FC; and good WR	2 264 230	24.9
4 Soils with moderate IR, P and HC; high FC, and good WR	1 735 640	19.1
5 Soils with moderate IR, poor P and HC; high FC; and high WR	571 080	6.2
6 Soils with unfavourable water management: low IR, very low P and HC and high WR	1 349 750	14.8
7 Soils with extremely unfavourable water management; very low IR, extremely low P and HC; and very high WR	329 210	3.6
8 Soils with good IR, P and HC and very high FC	117 560	1.3
9 Soils with extreme moisture regime due to shallow depth	774 650	8.5
Organic matter content (t ha^{-1})		
1 < 50	481 750	5.3
2 50–100	1 915 130	21.0
3 100–200	2 596 270	28.5
4 200–300	1 923 590	21.1
5 300–400	1 887 270	20.7
6 > 400	305 440	3.4
Depth of the soil (limited by solid or slightly fragmented rocks, gravel, cemented layers, pans, peat, loose sand, groundwater, etc.)		
1 < 20 cm	25 780	0.3
2 20–40 cm	445 260	4.9
3 40–70 cm	480 310	5.3
4 70–100 cm	370 630	4.0
5 > 100 cm	7 787 470	85.5
TOTAL	9 109 450	100.0
Lakes	95 900	
Towns	98 150	
Total area of the country	9 303 500	

Fig. 26.2. Map of soil erosion in Hungary. 1. Non-eroded areas; 2. slightly eroded areas; 3. moderately eroded areas; 4. severely eroded areas; 5. sedimentation territories; 6. forests.

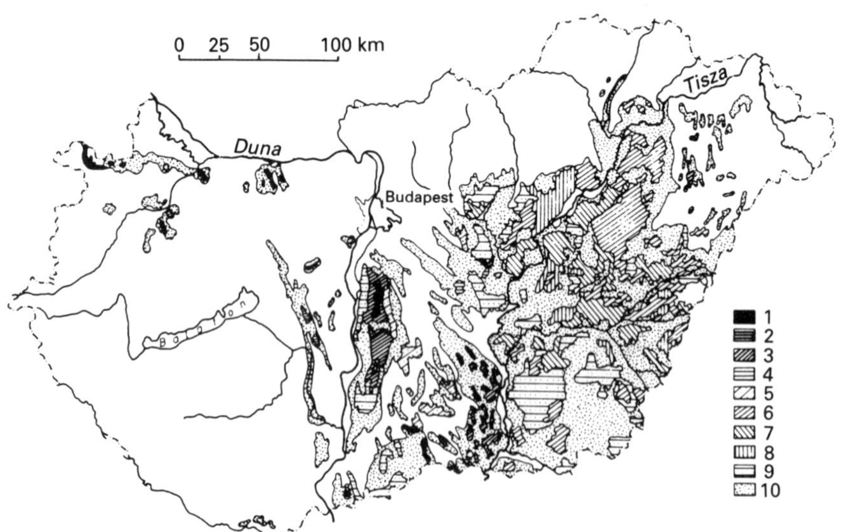

Fig. 26.3. Map of actual and potential salt-affected soils in Hungary. 1. Chloride–sulphate solonchaks; 2. soda solonchaks; 3. solonchak solonetzes; 4. calcareous meadow solonetzes; 5. calcareous solonetzic meadow soils; 6. meadow solonetzes; 7. meadow solonetzes turning into steppe formation; 8. solonetzic meadow soils; 9. soils saline and/or solonetzic in the deeper layers; 10. potential salt-affected soils (potential hazard of salinity/alkalinity).

the viewpoint of their vulnerability to various natural and human-induced stresses. This research involved three steps:

1. analysis of the mechanisms and factors influencing the impact of various stresses;
2. elaboration of category systems for the classification of the susceptibility of soils to various stresses, preferably with quantitative limits;
3. mapping of these categories using the database of the above-mentioned soil maps and GIS facilities.

For practical applications, the following steps are taken:

4. preparation of maps of the existing stress factors (e.g. atmospheric deposition; soil pollution; P load of surface waters; nitrate load of groundwaters, etc.);
5. preparation of excedance maps (comparing (**2**) and (**4**) indicating the ecologically sensitive, threatened or degraded 'hot spots').
6. action plans to prevent, stop or moderate the environmental damage due to these factors.

The following thematic soil susceptibility maps have been prepared:

1. Susceptibility of soils to water and wind erosion (1:1 M) (Stefanovits, 1964, Stefanovits and Várallyay, 1992);
2. Susceptibility of soils to acidification (1:500,000, 1:100,000) (Várallyay *et al.*, 1989);
3. Susceptibility of soils to salinization/alkalization (1:500,000) (Szabolcs *et al.*, 1969);
4. Susceptibility of soils to physical degradation, such as structural damage, compaction and surface sealing (1:500,000) (Várallyay and Leszták, 1990);
5. Vulnerability of soils to various pollutants (Molnár *et al.*, 1992).

Examples of the use of this approach include the following.

Susceptibility of soils to acidification in Hungary

Soil acidification has been increasing significantly in recent years. In Hungary the main causes are the high rate of application of mineral fertilizers (especially N fertilizers, mainly NH_4NO_3 and urea) and wet and dry acid deposition. We defined acidification as a decrease in the acid neutralizing capacity which may result in (but is not equivalent to) pH decrease. The effect of various acid loads on different soil properties have been studied in detail. To characterize the acid neutralizing capacity of soils 'titration curves' were determined for representative Hungarian soils. The influence of various soil characteristics on the titration curves was analysed by stepwise regression. Based on the titration curves, Hungarian soils were classified into six groups according to their susceptibility to acidification, and 1:500,000 and 1:100,000 scale maps of these categories were prepared. The simplified version of the 1:500,000 scale map is shown in Fig. 26.4 (Várallyay 1989).

Susceptibility of soils to physical degradation in Hungary

Large-scale, fully mechanized crop production may result in the physical degradation of soils. Compaction, structural damage, surface crusting and 'sealing' occur frequently in areas used for intensive agriculture, and lead to serious consequences, e.g. extreme soil moisture regime.

Várallyay and Leszták (1990) elaborated a comprehensive classification system for soils according to their susceptibility to physical degradation, in particular to structural damage and compaction. Based on all available data, a map was prepared at a scale of 1:500,000 and 1:100,000, showing eight categories of physical degradation. The simplified version of the 1:500,000 scale map is shown in Fig. 26.5.

Fig. 26.4. Map of the susceptibility of Hungarian soils to acidification. 1. Strongly acidic soils (13.5%); 2. highly susceptible soils due to their low buffering capacity (slightly acidic soils with light texture and low organic matter content) (14%); 3. susceptible soils due to their medium buffering capacity (slightly acidic soils with medium texture and organic matter content) (5.0%); 4. moderately susceptible soils due to their high buffering capacity (slightly acidic soils with heavy texture and/or high organic matter content) (23.2%); 5. slightly susceptible soils (salt affected) (5.9%); 6. non-susceptible soils (calcareous from the surface) (38.4%).

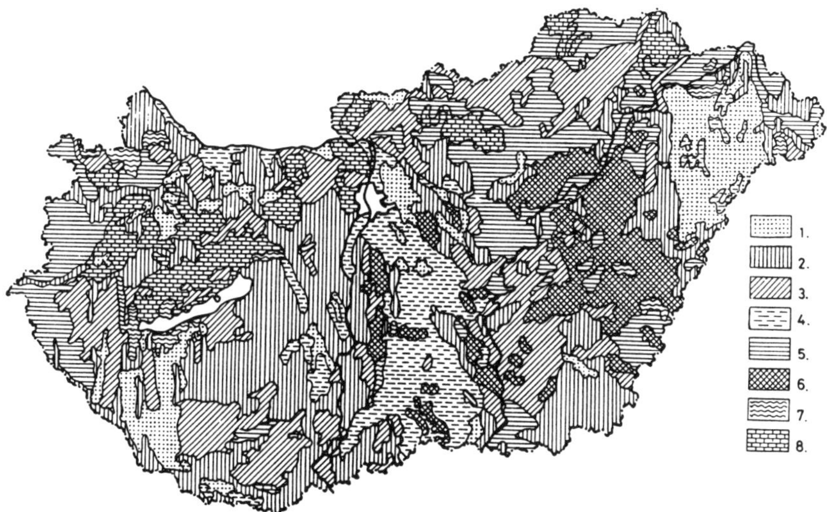

Fig. 26.5. Map of the susceptibility of soils to physical degradation. 1. Non-susceptible soils (sandy soils without structure and with a low content of cementing compounds, such as carbonates or sesquioxides) (10.5%); 2. slightly susceptible soils (medium-textured soils with well-developed structure and high aggregate stability) (23.0%); 3. moderately susceptible soils (medium-textured soils with moderately developed structure and low aggregate stability) (17.8%); 4. soils susceptible to compaction and surface crusting but not to structural damage (structureless sandy soils with large amounts of cementing compounds, mainly carbonates) (11.0%); 5. soils susceptible to structural damage and compaction (heavy-textured soils of swelling-shrinkage character and low structural stability) (12.9%); 6. soils susceptible to both structural damage and compaction due to salinity–alkalinity) (9.6%); 7. organic soils (peats) (5.7%); 8. shallow soils (solid rock or cemented layer near the surface) (9.5%).

Soil information system

Recently all existing soil data have been included in a computerized geographic soil information system (HunSIS = TIR), which is schematically illustrated in Fig. 26.6 (Csillag *et al.*, 1988; Kummert *et al.*, 1989). TIR consists of two main parts:

1. The soil data bank, including three different types of information:
   (a) basic topographic information (geodetic data standards and geographic reference systems);
   (b) point information (measured, calculated, estimated or coded data on the various characteristics of soil profiles (or borings) or their different layers, diagnostic horizons (at present 35 soil and land characteristics), and
   (c) territorial information (1:25,000 thematic maps on various physicogeographical factors (geomorphology, relief, groundwater conditions) and soil properties.

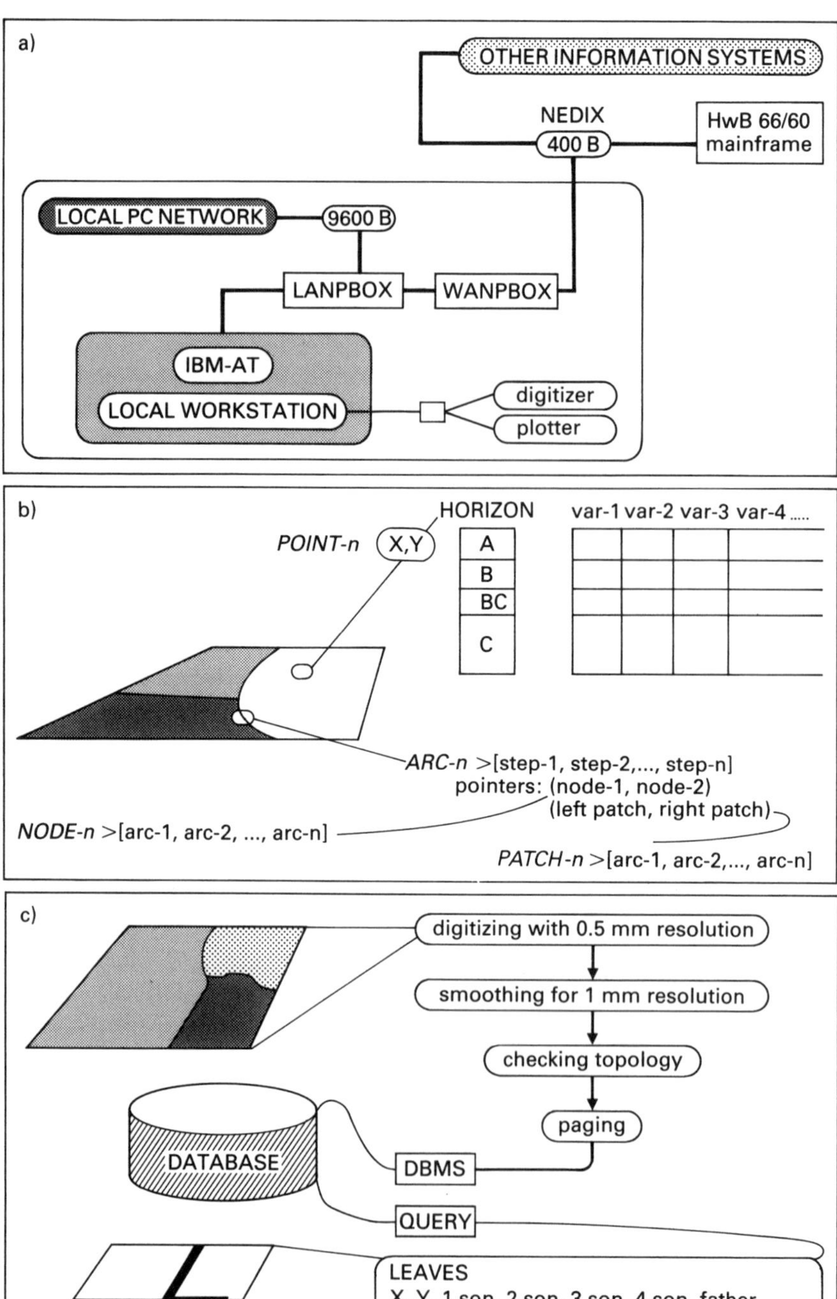

Fig. 26.6. Hungarian Soil Information System HunSIS = TIR). (a) System scheme. (b) Input data encoding. (c) Map data flow from input via database to output.

2. The information system, including models of moisture and plant nutrient regimes of soils; susceptibility of soils to various soil degradation processes, such as water and wind erosion, acidification, salinization/alkalization, structural damage and compaction; soil–water–plant relationships; status of soil pollutants and potentially toxic elements, etc.

The digitizer–computer–plotter distributed system, including adequate software, is able to search for either location or attributes and display results in digital, tabular, graphical or cartographical form (data, categorized data, results of model calculations, thematic maps, etc.).

Current system development is focused on the enhancement of local (i.e. workstation) modelling and editing functions, to make this quadtree-based thematic GIS compatible with other gridded data sources.

The simultaneous application of (**1**) and (**2**) type inputs opens new output facilities: integrated data; classification and grouping of soil according to various criteria; interpreted results; practical recommendations for sustainable land use and proper soil management.

Soil monitoring systems

In order to register soil changes three systematic monitoring systems were established.

Soil fertility control system (AIIR)

In this system the most variable soil characteristics (pH, $CaCO_3$ and organic matter content; saturation percentage; total salt content; total and 'mobile' N content; 'available' P, K and Ca content; 'soluble' Mg, S, Cu, Zn, Mn content) were measured in the topsoil (0–30 cm or ploughed horizon; later in the 30–60 cm layer, as well) of about 100,000 agricultural fields covering nearly 5 mha, in 3-year cycles. The programme started in 1978 (I: 1978–1981; II: 1982–1985; III: 1986–1989) and stopped before completing the third cycle (Baranyai *et al.*, 1987).

The data were computer-stored per agricultural field, without inner contours of the maximum 12 ha sampling sites, where mixed samples were collected in two replicates for laboratory analysis.

Microelement survey

In this system – in addition to the above-mentioned basic soil parameters – the 'total' (interpreted as a potential 'pool') and 'soluble' (interpreted as mobile and plant available) contents of Al, B, Cd, Co, Cr, Cu, Fe, Hg, Mn, Mo, Na, Ni, Pb, S, Se, and Zn were determined in the 0–30, 30–60, and

60–90 cm layers of 6000 agricultural fields representing about 5 mha in 3-year cycles (Molnár *et al.* 1992). The programme stopped during the second cycle in 1990.

Soil information and monitoring system (TIM)

TIM is an independent subsystem of the integrated Environmental Information and Monitoring System (KIM) (Várallyay, 1992). Based on soil physiographical–ecological units about 1200 'representative' observation points were selected (and exactly defined by geographical coordinates): 800 points on agricultural land, 200 points in forests and 200 points in environmentally threatened 'hot spot' regions (representing 12 different types of environmental hazards or particularly sensitive areas, such as: degraded soils; ameliorated soils; drinking water supply areas; watersheds of important lakes and reservoirs; protected areas with particularly sensitive ecosystems; 'hot spots' of industrial, agricultural, urban and transport pollution; military fields; areas affected by (surface) mining; waste (water) disposal affected spots).

During the first sampling (15 September–15 October 1992) all important soil characteristics were determined. From the 1200 basic points about 1000 will be selected for monitoring. Some soil parameters will be measured every year, others every 3 or 6 years, depending on their changeability (stability). In addition to soil maps remote sensing helps in the territorial interpretation of the registered point data. The hardware–software configuration of the database guarantees the compatibility of TIM with the other subsystems of the integrated Environmental Information and Monitoring System (KIM), which is now under elaboration.

Long-term field trials

There are 20–25 long-term field trials in Hungary. Among these the most important are:

1. National long-term field trials. The 12 sites represent various physiographical conditions (geology, relief–geomorphology, climate, hydrology), soils and land use practices: cropping pattern and crop rotation; tillage practices; fertilizer treatments, etc. The observation period in most cases is more than 30 years.
2. Long-term field experiment on non-calcareous sandy soils in Nyirség region (*Westsik trials*).
3. Long-term field experiments of various research institutes.

Application of Data

Hungarian soil science, soil survey and soil testing practices are always conducted to serve agricultural development, the planning and organization of crop production and environment control (Proceedings of the Seminar on Technologies for Sustainable Agriculture, 1985; Proceedings of the Hungarian–Swedish Seminar on Soil Mapping, 1989; Várallyay, 1989b).

Some examples from the Hungarian experiences in this respect are summarized below.

Control of factors limiting soil fertility and soil degradation processes

The main *factors limiting soil fertility* and soil degradation processes are presented in Table 26.3 (Szabolcs and Várallyay, 1978; Várallyay, 1989c).

Table 26.3. Limiting factors of soil fertility and soil degradation processes in Hungary.

No.	Limiting factor of soil fertility	Area (1000 ha)	%[a]	No.	Soil degradation processes
1.	Extremely coarse texture	746	8.0	1.	Soil erosion
2.	Soil acidity	1200	12.8		by water
	combined with erosion	(348)	(3.7)		by wind
	depth	(67)	(0.7)	2.	Soil acidification
3.	Salinity/alkalinity	757	8.1	3.	Salinization/alkalization
4.	Salinity/alkalinity in the			4.	Physical soil degradations
	deeper layers	245	2.6		structure destruction
5.	Extremely heavy texture	630	6.8		compaction
6.	Peat formation/water logging	161	1.7		surface sealing
7.	Soil erosion	1455	15.6	5.	Extreme moisture regime
	combined with acidity	(348)	(3.7)		overmoistening,
8.	Shallow depth	217	2.3		waterlogging
	combined with acidity	(67)	(0.7)		drought sensitivity
				6.	Biological degradation
					decrease of O,M.
					deteriorations of soil biota
				7.	Unfavourable changes in the nutrient regime
					leaching
					biotic and abiotic immobilization
				8.	Decrease of the buffering capacity, soil pollution, 'toxicity'

[a]Percentage of the total area of Hungary (93,000 km^2). In the case of soil acidity combined with erosion or with shallow depth only one factor was taken into account.

The necessity and rationality of the *reclamation of soils* with limited fertility depends on economic and ecological considerations. The radical improvement of salt-affected soils, sandy soils or peatlands requires expensive complex measures, and is not normally economic. At the same time saline lakes and soils, wetlands and sand regions are, in many cases, protected ecosystems, habitats of protected plants and animals, and consequently represent special environmental value. These areas must be kept in their 'original' condition, and their reclamation is not advisable (although this was pressed sometimes in the last decades). On the contrary, the improvement of soils with moderate limitations (e.g. liming of acid soils, loosening of compacted soils, etc.) can be an efficient and economic tool for agricultural development.

Soil degradation is usually a complex process in which several component features of soil deterioration can be recognized to be contributing to the loss of land or its 'productive capacity', to the limitation of normal soil functions, and/or to the decrease of soil fertility due to unfavourable changes in soil processes and, consequently, in soil properties.

Soil degradation is *not* an unavoidable consequence of intensive (but rational) agricultural production and social development because most soils are resilient, to a certain extent at least, to most of the degradation processes, and their unfavourable consequences can be prevented, eliminated or at least moderated.

These activities have always been the main task of Hungarian soil science, soil survey and advisory services. Their main results and achievements have been summarized by Várallyay (1989c).

For successful *prevention,* satisfactory *prognosis* is necessary. It can be rationally based on comprehensive *vulnerability (susceptibility) sensitivity analyses* (Várallyay, 1989c).

Control of soil moisture regime

In Hungary the soil moisture regime strongly influences (sometimes determines) the ecological potential and agricultural productivity of a given area, the biomass production of various natural and agroecosystems, and the hazard of 'nutrient pollution' of surface and subsurface waters (Várallyay, 1985; State of the Hungarian Environment, 1990).

The average annual precipitation is about 620 mm in Hungary but there is extremely high territorial and temporal variability – even on a micro-scale. Under such conditions a considerable part of precipitation is lost by surface runoff, downward filtration and evaporation. The non-uniform rainfall distribution is one reason for the extreme moisture regime: the simultaneous hazard of waterlogging or over-moistening and drought-sensitivity in extensive areas, sometimes at the same places within a short period.

The other two reasons for the extreme moisture regime are:

- the relief (in addition to the undulating surfaces the heterogeneous microrelief of the 'flat' Hungarian Plain); and
- the unfavourable hydrophysical properties of some soils.

The hydrophysical properties of soils are closely related to the limiting factors of soil fertility and soil degradation processes. This is illustrated by Figs 26.7 and 26.8.

The highly variable moisture regimes would necessitate a special *'double-faced' soil moisture control* in Hungary, i.e. ensuring (or making possible) the drainage of excess water and giving the necessary additional water when and where it is necessary, sometimes simultaneously. Both actions are costly and faced with serious limitations:

1. *drainage:* poor permeability of soil; limited capacity of drainage canals and drainwater reservoirs; salinity problems;
2. *irrigation:* relief; limited and still decreasing water available for crop production as a result of limited surface and subsurface water resources, and the increasing water demand of other sectors.

Consequently, all efforts have to be taken to *improve* agricultural *water use efficiency* by proper soil management: (i) assist infiltration; (ii) to increase water storage capacity and (iii) to improve water availability to plants.

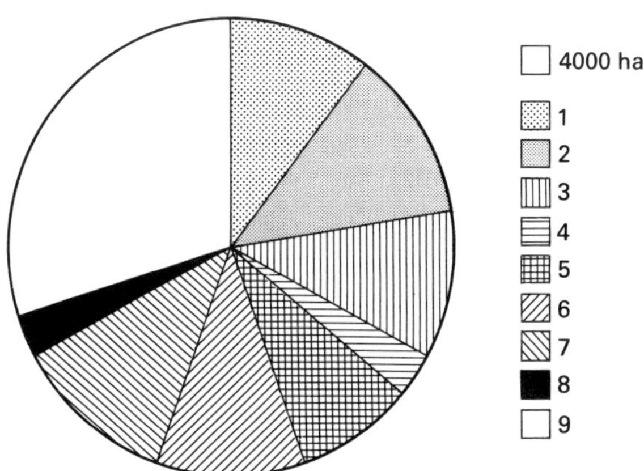

Fig. 26.7. Distribution of soils in Hungary with good, moderate and unfavourable hydrophysical properties. 1–5, Soils with unfavourable hydrophysical properties: 1, due to very coarse texture; 2, due to very heavy texture; 3, due to strong salinity–alkalinity; 4, due to peat formation; 5, due to shallow depth; 6–8, Soils with moderately unfavourable hydrophysical properties: 6, due to coarse texture; 7, due to heavy texture; 8, due to moderate salinity–alkalinity in the deeper layers. 9 Soils with good hydrophysical properties.

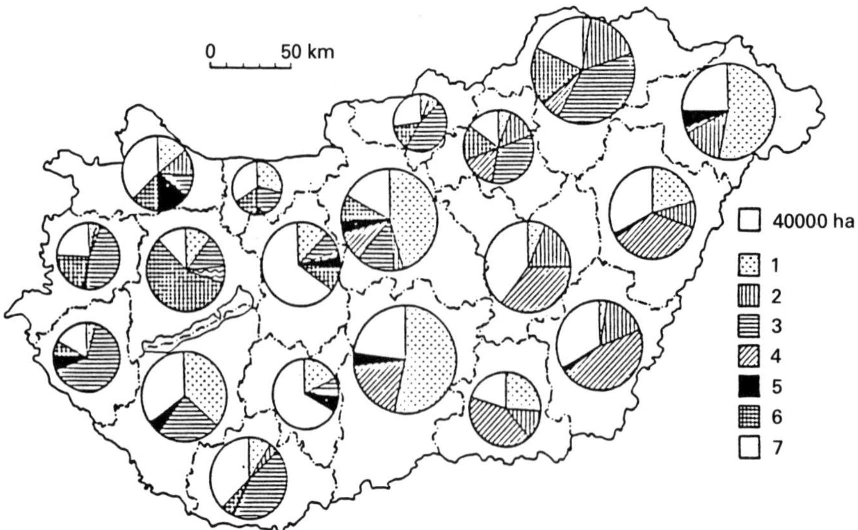

Fig. 26.8. Distribution of soils with good and unfavourable hydrophysical properties in the counties (administrative districts) of Hungary. 1–6, Soils with unfavourable hydrophysical properties: 1, due to coarse texture; 2, due to heavy texture; 3, due to clay accumulation in the B horizon; 4, due to salinity/alkalinity; 5, due to peat formation; 6, due to shallow depth. 7, Soils with good hydrophysical properties.

In Hungary almost all actions to ensure normal soil functions are closely connected with soil moisture control. On the other hand, most of the measures for soil moisture control are, at the same time, the elements of environment control, as shown in Table 26.4.

A comprehensive soil survey – analysis – categorization – mapping – monitoring system was developed for the exact characterization of hydrophysical properties, modelling and forecasting of water and solute regimes of soils (Várallyay *et al.*, 1980b; Várallyay, 1985, 1989a). The system may serve as a scientific basis for soil moisture control and is used for practical soil water management both for crop production and environmental protection.

Control of nutrient regime

Before the Second World War the plant *nutrient status* of Hungarian soils was rather poor, due to removal of more nutrients in cultivated crops than were returned in the form of organic and green manures or fertilizers (Baranyai *et al.*, 1987; Sarkadi and Várallyay, 1989; Várallyay, 1991a; Várallyay *et al.*, 1991).

Table 26.4. Elements and methods of soil moisture control with their environmental impacts.

Elements		Methods	Environmental impacts[a]	
			Favourable	Unfavourable
Surface runoff	Reduction	Increase of the duration of infiltration (moderation of slopes: terracing, contour ploughing; establishment of permanent and dense vegetation cover; tillage): improvement of infiltration; soil conservation farming system	1, 1a, 5a, 8	–
Evaporation	Reduction	Help infiltration (tillage, deep loosening)	2, 4	–
Feeding of groundwater by filtration losses	Reduction	Increase of the water storage capacity of soil; moderation of cracking (soil reclamation); surface and subsurface water regulation	5b, 7	–
Rise of water table	Reduction	Minimization of filtration losses (↲); groundwater regulation (horizontal drainage)	2, 3, 5b, 5c	–
Infiltration	Increase	Minimization of surface runoff (tillage practices, deep loosening) (↲)	1, 4, 5a, 7	–
Storage in the soil in available form	Increase	Increase of the water retention of soil; adequate cropping pattern (crop selection)	4, 5b, 6, 7	–
Irrigation		Irrigation; groundwater table regulation	4, 5c, 7	9, 10
Surface / Subsurface drainage		Surface / Subsurface drainage	1, 2, 3, 5c, 6, 7	11

[a]Environmental impacts

Favourable				Unfavourable	
Prevention, elimination, limitation or moderation of the following phenomena:					
1	Water erosion	5a	Surface runoff (→ surface waters → eutrofication) of plant nutrients	9	Overmoistening; waterlogging; peat and swamp formation; secondary salinization/alkalization;
1a	Sedimentation	5b	Leaching (→ subsurface waters) of plant nutrients	10	Leaching of plant nutrients
2	Secondary salinization, alkalization	5c	Immobilization of plant nutrients	11	Drought sensitivity
3	Peat formation, waterlogging, overmoistening	6	Formation of phyto-toxic compounds		
4	Drought sensitivity, cracking	7	'Biological degradation'		
		8	Flood hazard		

From 1955 there was a rapid increase in *fertilizer consumption*. This tendency was one of the reasons for the substantial yield increase during the same period. Another consequence was that, due to the positive nutrient balance, the nutrient status of Hungarian soils was significantly improved.

In spite of these developments there were serious problems and inadequacies in the fertilizer application technology (improper N–P–K ratio; lack of Ca, Mg and micronutrient supply; limited variety of fertilizers; problems with their storage, time of application, method of distribution; etc.). The main problem, however, was an unfavourable 'polarization' tendency in fertilizer application (Várallyay *et al.*, 1991).

1. better soil → rich farm → higher rate of fertilizer application (in spite of the lower requirements ← better nutrient status of soils) → overdosage;
2. poor soils → poor farms → lower rate of fertilizer application (in spite of the higher requirements ← lower nutrient supply of soils) → underdosage.

The overgeneralization and the imperative 'maximum-concept' led to false conclusions, decreased the effectiveness and efficiency of mineral fertilization, and resulted in undesirable environmental side-effects, like soil acidification (due to the use of inappropriate fertilizers, and the lack of simultaneous lime application) and its consequences (mobilization of toxic elements, fixation of nutritive elements), pollution of surface waters by phosphorus compounds (due to surface runoff and erosion) contamination of groundwaters by nitrates (leaching of nitrates mainly through open cracks and biological channels to the shallow water table); accumulation of harmful toxic elements in the various stages of the 'food chain': in soils, plants, animals and human organs, according to their solubility, mobility and availability (Kádár, 1991; Molnár *et al.*, 1992; Várallyay, 1991b).

These side-effects, however, are not inevitable consequences of fertilizer application. They can be prevented by the rational use of fertilizers, by taking into consideration the nutrient requirement of cultivated crops, nutrient status and other properties of soils, as well as the climatic and hydrological conditions of the site. These factors have now been taken into consideration in the development of our new plant nutrition advisory system (Sarkadi and Várallyay, 1989; Várallyay *et al.*, 1991).

Control of soil pollution

The transport and transformation of various elements in the soil are controlled by abiotic and biotic soil processes. Most elements can be found in the soil. Their quantity, quality, solubility, mobility, availability for micro-organisms, plants, animals and human-beings show an extremely wide spectrum. Most of these elements are essential for living organisms, but, over a certain 'threshold concentration', the same elements can be harmful, or even 'toxic' for the same organisms (Kádár, 1991; Molnár *et al.*, 1992; Várallyay, 1991b).

The occurrence of these elements can be due to natural sources, such as:

- air (N, etc.);
- water (B, Na, N, etc.);
- soil and geological deposits as a result of weathering (P, K, Ca, Mg, Na, Fe, Al, Mn, As, Co, Cu, Ni, Se, Zn, etc.);

and can be due to various human activities (irrigation, use of organic manure, liquid manure, mineral fertilizers, or amendments for soil reclamation; use of sewage waters, sewage sludges, solid wastes, pollution due to transport, industry, etc.).

The evaluation of the content, state, regime and ecological impact of these elements necessitates the determination, evaluation and interpretation of the following criteria:

1. *'Total content'* (interpreted as a potential 'pool').
2. *'Soluble content'*: this quantity is highly solute-specific. The 'soluble content' depends on the characteristics of the given compound (e.g. solubility, electronegativity, polarizability, rate of oxidation, ability of complex formation) and on the soil properties (e.g. soil reaction and carbonate status, texture, clay content, clay mineral association, organic matter content and quality, absorption capacity, base saturation, exchangeable cation composition, moisture content, redox potential, microbial activity, etc.). The *'in situ'* soluble content depends to a great extent on plant root activities.

Non-soluble compounds are immobile, and not available for plants, and consequently, cannot be toxic (through the food chain) for plants, animals and human beings.

The 'Chemical Time Bomb' effect means that the concentration of various elements may exceed the permissible (tolerable) concentration in soil and may become 'toxic' without any addition of the element to soil, as a consequence of changing soil properties. Thus increasing acidity helps the mobilization of numerous 'potentially toxic' (elements most of the heavy metals, etc.), so that the previously immobilized element becomes soluble and mobile, and its toxicity expressed. The 'toxic element mobilization hazard' of soil acidification can be predicted and prevented in due time by using proper neutralization measures (e.g. liming).
3. *'Mobile content'*: only the mobile fraction of the various elements (compounds) can be transported and translocated to other places (leaching, accumulation, migration), and may reach water resources and plant root surfaces.
4. *'Plant available content'*: in addition to the total, soluble and mobile content of a certain elements in the soil, its availability for a certain plant

depends on its selective ion-uptake, which is a plant-dependent characteristic (species, variety).

Consequently, the term 'toxic element' is not sufficiently specific as far as soils are concerned.

Future Tasks under New Conditions

The rational privatization of land and the market-oriented production provide possibilities for the establishment of a flexible *sustainable agriculture* (Láng and Csete, 1992). It will be based on various production systems with special roles for 'intensive' (not equal to 'big', 'high', 'large', 'much'), low input, and organic (biological–ecological) farming systems. The following are necessary for its implementation.

1. Territorial coordination of the agroecological conditions and the ecological requirements of cultivated crops, taking into consideration both the production and the environmental aspects in short-, mid- and long-term time scales: rational land use.
2. Rationalization of the structure of agricultural fields by the optimization of field size according to the given physiographic conditions, to form more homogeneous fields, so that uniform agrotechnical measures can be correctly used on each field.
3. Elaboration, adoption and implementation of scientifically based crop production technologies, including five fundamental elements:
 (a) adequate cropping pattern and crop rotation;
 (b) reduction of 'production wastes';
 (c) improvement of water-use efficiency;
 (d) rational plant nutrition, including fertilizer application;
 (e) integrated pest management with minimum use of chemicals.

The rate, direction and technologies of crop production are economy driven. In contrast, the maintenance of soil functionality as well as the quality of surface and subsurface water resources, and the protection of the natural environment (the biosphere) are not economy-dependent, but imperative tasks.

For the above-mentioned purposes:

1. the soil database (maps, data, GIS-based information system, monitoring, field trials, etc.) has to be updated according to the new requirements;
2. soil resilience has to be precisely defined and quantitative parameters have to be selected or elaborated for its evaluation;
3. the criteria of sustainable agricultural development and crop production have to be defined and quantified;

4. the necessary economic regulations have to be elaborated (such as: tax, price, credit, subsidy regulations) guaranteeing the fulfilment of these criteria;
5. the defined and quantified criteria and the economic regulations have to be formulated in various legal documents (laws and related official documents);
6. the potential possibilities and efficient methods of sustainable crop production have to be elaborated, adopted, published and demonstrated, which needs the establishment of appropriate mechanisms for research, training and education, demonstration, extension and advisory service.
7. the necessary mechanisms for continuous control have to be built up.

The effective realization of these tasks must be jointly guaranteed by the State, by the land-owner and by the land-user for the benefit of efficient, rationally privatized, market-oriented, sustainable agricultural production, harmonized with successful environment protection, ensuring a pleasant environment and a promising future.

References

Baranyai, F., Fekete, A. and Kovács, I. (1987) [Results of the Hungarian plant nutrient survey] (H) Mezõgazdasági Kiadó. Budapest.

Csillag, F., Szabó, J., Várallyay, Gy., Zilahy, P. and Vargha, M. (1988) Hungarian soil information system (TIR): a thematic geographical information system for soil analysis and mapping. *Bulletin of the Hungarian National Commission for CODATA*, no. 5. 1–13.

Kádár, I. (1991) [Heavy metal content in soils and crops in Hungary] (H, e) KTM-MTA TAKI Kiadása. Budapest, pp. 1–104.

Kreybig, L. (1937) [Soil survey, analysis and mapping methodology of the Royal Hungarian Geological Institute] (H) *M. Kir. Földtani Intézet Évkönyve* 31, 148–244.

Kummert, Á., Csillag, F., Szabó, J., Várallyay, Gy. and Zilahy, P. (1989) A geographical information system for soil analysis and mapping: HunSIS (Concepts and functionality). *Agrokémia és Talajtan* 38, 822–835.

Láng, I. and Csete, L. (1992) [Adaptive Agriculture] (H) Agricola. Budapest, pp. 1–210.

Láng, I., Csete, L. and Harnos, Zs. (1983) [The Agro-ecological Potential of Hungarian Agriculture in 2000]. (H) Mezõgazd. Kiadó. Budapest, pp. 1–265.

Magyarország Nemzeti Atlasza (1989) [The National Atlas of Hungary] Akadémiai Kiadó. Budapest.

Molnár, E., Németh, T. and Pálmai, O. (1992) Problems of heavy metal pollution in Hungary – 'State of the Art' – *Proc. of the SETAC Conference*, Prague, 13–17 October.

Proceedings of the Seminar on Technologies for Sustainable Agriculture. Budapest, 10 Sept. 1984 (1985) *Agrokémia és Talajtan* 34, Suppl. 127–210.

Proceedings of the Hungarian–Swedish Seminar on Soil Mapping. Budapest, 4–11

June, 1988 (1989) *Agrokémia és Talajtan* 38, 694–837.

Sarkadi, J., Szücs, L. and Várallyay, Gy. (1964) [Large-scale genetic farm soil maps.] (H,e,g,r) OMMI Genetikus Talajtérképek. Ser. 1. No. 8. Budapest.

Sarkadi, J. and Várallyay, Gy. (1989) Advisory system for mineral fertilization based on large-scale land-site maps. *Agrokémia és Talajtan* 38, 775–789.

State of the Hungarian Environment (1990) Hungarian Academy of Sciences – Ministry of Environment and Water Management – Hungarian Central Statistical Office, Budapest.

Stefanovits, P. (1964) [The map of soil erosion in Hungary] (H,e,g,r) OMMI Genetikus Talajtérképek. Ser. 1. No. 7. Budapest.

Stefanovits, P. (1989) Map of clay mineral associations in Hungarian soils. *Agrokémia és Talajtan* 38, 790–799.

Stefanovits, P. and Szücs, L. (1961) [Genetic soil map of Hungary] (H,e,g,r) OMMI Genetikus Talajtérképek. Ser. 1. No. 1. Budapest.

Stefanovits, P. and Várallyay, Gy. (1992) State and management of soil erosion in Hungary. In: *Proc. Soil Erosion Prevention and Remediation Workshop,* Budapest, Vol. I. 79–95.

Szabolcs, I. (ed.) (1966) [Large-scale genetic soil mapping] (H,e,g,r) OMMI Genetikus Talajtérképek. Ser. 1. No. 9. Budapest.

Szabolcs, I. (1974) *Salt Affected Soils in Europe.* Martinus Nijhoff, The Hague and the Research Institute for Soil Science and Agricultural Chemistry of the Hungarian Academy of Sciences, Budapest.

Szabolcs, I. (ed.) (1989) Ecological impact of acidification. *Proc. Symp. 'Environmental Threats to Forest and Other Natural Ecosystems',* Oulu, Finland, 1–4 Nov. 1988. Budapest. pp. 1–166.

Szabolcs, I., Darab, K. and Várallyay, Gy. (1969) Methods for the prognosis of salinization and alkalization due to irrigation in the Hungarian Plain. *Agrokémia és Talajtan* 18, Suppl. 351–376.

Szabolcs, I. and Várallyay, Gy. (1978) [Limiting factors of soil fertility in Hungary] (H,e,g,r). *Agrokémia és Talajtan* 27, 181–202.

Várallyay, Gy. (1985) [Main types of water regimes and substance regimes in Hungarian soils] (H,e,g,r). *Agrokémia és Talajtan* 34, 267–298.

Várallyay, Gy. (1988) Land evaluation in Hungary – scientific problems, practical applications. In: *Land Qualities in Space and Time.* Proc. Symp. ISSS, Wageningen, 22–26 August, PUDOC. Wageningen, pp. 241–252.

Váralyay, Gy. (1989a) Soil water problems in Hungary. *Agrokémia és Talajtan* 38, 577–595.

Várallyay, Gy. (1989b) Soil mapping in Hungary. *Agrokémia és Talajtan.* 38. 696–714.

Várallyay, Gy. (1989c) Soil degradation processes and their control in Hungary. *Land Degradation and Rehabilitation* 1, 171–188.

Várallyay, Gy. (1991a) History, present status and future use of soil tests and plant analysis in Eastern Europe. *Abstracts of the International Symp.* on 'Soil Testing and Plant Analysis in the Global Community' Orlando, Florida, USA, 22–27 August, p. 22.

Várallyay, Gy. (1991b) Environmental problems of soils and land use in Hungary. *Proceedings Swedish–Hungarian Seminar on Environmental Problems in Agriculture,* 11–15 June, 1990, pp. 129–168.

Várallyay, Gy. (1992) [Soil information and monitoring system] (H) *Proc. 1st Agro-*

environm. Protection Conference. Budapest, 23–25 November, 51–62. KVIK. Budapest.

Várallyay, Gy. and Leszták, M. (1990) Susceptibility of soils to physical degradation in Hungary. *Soil Technology* 3, 289–298.

Várallyay, Gy. and Molnár, S. (1989) The agro-topographical map of Hungary (1:100,000 scale). Hungarian Cartographical Studies. 14th World Conference of ICA-ACI. Budapest. pp. 221–225.

Várallyay, Gy., Rédly, M. and Murányi, A. (1989) Map of the susceptibility of soils to acidification in Hungary. In: Szabolcs, I. (ed.) *Ecological Impact of Acidification* Proc. Symp. Environmental Threats to Forest and Other Natural Ecosystems, Oulu, Finland, 1–4 Nov. 1988. Budapest. pp. 79–94.

Várallyay, Gy., Szücs, L., Murányi, A., Rajkai, K. and Zilahy, P. (1979) [Map of soil factors determining the agro-ecological potential of Hungary, 1:100 000 I.] (H,e,g,r). *Agrokémia és Talajtan* 28, 363–384.

Várallyay, Gy., Szücs, L., Murányi, A., Rajkai, K. and Zilahy, P. (1980a) [Map of soil factors determining the agro-ecological potential of Hungary, 1:100 000, II] (H,e,g,r). *Agrokémia és Talajtan* 29, 35–76.

Várallyay, Gy., Szücs, L., Rajkai, K., Zilahy, P. and Murányi, A. (1980b) [Soil water management categories of Hungarian soils and the map of soil water properties, 1:100 000] (H,e,g,r). *Agrokémia és Talajtan* 29, 77–112.

Várallyay, Gy., Szücs, L., Zilahy, P., Rajkal, K. and Morányi, A. (1985) Soil factors determining the agro-ecological potential of Hungary. *Agrokémia és Talajtan* 34, Suppl. 90–94.

Várallyay, Gy., Buzás, I., Kádár, I. and Németh, T. (1991) New plant nutrition advisory system in Hungary. Proceedings Swedish–Hungarian Seminar on Environmental Problems in Agriculture, 11–15 June, 1990. 88–110.

Chapter 27

The Role of Information Services in Sustainable Land Use

J.L. Nowland, A.E. Shaw and W.J. Reid

CAB International, Wallingford, Oxon OX10 8DE, UK

Introduction

'Sustainable development is still more of an intention than a reality' according to the International Institute for Sustainable Development (IISD, 1992). In this situation information services have a prominent role to play, especially in sustainable land use. No country has the resources to gain from their own research programmes all the knowledge they require. Developing countries are quite dependent on research done elsewhere, the more valuable if the information has been generated in an environmentally similar geographical area. For our purpose, any definition of 'sustainable land use' by a reputable authority should suffice.

The Value of Information

The value of information flow for supporting sustainable land use derives from imparting knowledge, stimulating ideas, provoking discussion, developing new perspectives, providing a basis for future action, and giving power to manage and control (Scott, 1991). Scientists need information to generate new hypotheses, new uses for their tools and research that is not repetitious. Unfortunately, scientists' use of available information may be far from the ideal.

One example is the agricultural library service of a major developing country that finds scientists ignoring what it struggles to provide. Another unhealthy sign is the unreasonably narrow national and language concentration of bibliographic citations in scientific papers. Examination of a sample of recent literature on sustainability related soil research indicates considerable indifference to, or unawareness of, 'foreign' research. Concentration of 80–100% of citations by authors to research done in their own country is commonplace. Papers written in more 'isolated' countries have a much broader reference outlook, breaking language barriers to find the most appropriate literature for their research. The most unbiased reference lists were found in papers sponsored by international agencies and projects where the author (either a national of that country or not) required natural resource data for a specific area (e.g. soil survey projects). Such papers may be the exceptions rather than the rule, typified more often by the example of a North American scientist doing infiltration research in Africa and citing mostly North American references.

Unfortunately research is sometimes driven less by information on priority needs, than by factors such as the capability of existing laboratory equipment, new techniques searching for an application, intellectual inertia, and doomsday hysteria funding. Research managers and policy makers are therefore particularly important targets for reliable 'processed' information in non-specialized terms, to help them reassess priorities and allocate scarce resources more effectively. Responsibility for drastic reorientation of scientists to new bandwagon priorities requires better information for the manager than is generally available. Whether to push staff into global change research or N-fixation in wheat were two examples of potential pitfalls in the absence of good information.

Types and Sources of Information

If a classification of 'types of information' is required, the four types recognized by Scott (1991) for addressing biodiversity might serve as a useful starting point. They are factual, interpretive, decision-making, and predictive categories of information. It is interesting that distribution maps and taxonomy were characterized as 'interpretive' for Scott's subject area, crop protection. Should soil surveys be considered likewise, or 'factual'? 'Decision-making' information, restricted to expert systems by Scott, may not be one of the strengths of soil science, as suggested elsewhere in this paper. 'Predictive' information is probably dominated by modelling, and now closely linked to field monitoring of trends. Publications prescribing approaches to the subject, or methodologies or checklists of factors to be considered, may constitute a fifth category that could be labelled prescriptive information (e.g. Eswaran, 1992).

The sources of information related to soil and land use sustainability are widespread. In 1991 for example, the CAB ABSTRACTS database drew upon 341 journals, 13 conference proceedings, and 66 books and reports for a total of 1363 records. Journals yielding upward of 15 records are listed in Table 27.1.

Table 27.1. Journal sources of soil-related abstracts in 1991.

Journal	Total no. of papers abstracted	No. of papers related to sustainability	%
Soil & Tillage Research	77	48	(62)
Soil Science Society of America Journal	280	43	(15)
Soil Biology and Biochemistry	134	41	(31)
Mitteilungen der Deutschen Bodenkundlichen Gesellschaft	105	31	(30)
Australian Journal of Soil Research	55	28	(51)
Agriculture, Ecosystems & Environment	73	23	(32)
Water, Air, and Soil Pollution	73	23	(32)
Forest Ecology and Management	29	21	(72)
Biology and Fertility of Soils	92	20	(22)
Soil Science	105	19	(18)
Zeitschrift für Pflanzenernährung und Bodenkunde	35	18	(51)
Soviet Soil Science	90	16	(18)
Journal of Soil and Water Conservation	56	15	(27)

Note. Sources published in 1991 and received and abstracted up to present by CAB International. 'Sustainability' papers are here defined as those on the following subjects: soil structure, aggregates, erosion control, conservation tillage, soil degradation, compaction, weathering, organic matter, fertility, salinity, nutrients and cycling, biological activity, pollution, pesticides and residues, soil fauna, microorganisms, and sustainability.

Narrowing down the sources to the themes of this volume illustrates the significance of the search strategy for an abstracts database. Use of the index term 'sustainability', in vogue for only five years, yields only 79 different journal sources in 1991, 22 books, four conference proceedings and 14 miscellaneous records. The highest yielding journals were *Soil & Tillage Research* (14), *Canadian Journal of Agricultural Economics* (13), *Fertilizer News* (7), *Journal of Soil and Water Conservation* (5), and *Journal of Sustainable Agriculture* (4). The papers have a strong economic and policy element.

Use of the broader selection of 'sustainability' terms in Table 27.1, gives a better idea of the journal sources available. Even these are only a fraction of the number of journals in which soil-related material was found.

One factual and interpretive source, indigenous knowledge, is singled out for special mention.

Indigenous knowledge – a neglected source of information

Knowledge consists of information, concepts, techniques, and skills which are stored, increased and improved, developed for use, disseminated, and applied in knowledge systems (Bunting, 1992). One such knowledge system that has so far remained relatively untapped is indigenous knowledge (NRC, 1991). This has been defined as 'the sum of experience and knowledge of a given ethnic group that forms the basis for decision-making in the face of familiar and unfamiliar problems' (Warren and Cashman, 1989).

Traditional farming or agroecological systems may not always be obvious to the 'western eye' since they are not conceptually limited to what the developed world calls 'crops'. It often involves management of a variety of semidomesticates, weeds, forests, wildlife and other elements, and in a majority of cases is actually managed by women. In many cultures ecological knowledge has been disguised in forms of cosmology and ritual. The knowledge has been developed over time, through a process of trial and error. For example, farmers in Latin America have conserved biodiversity in their agricultural systems by drawing upon their ethnoscientific and indigenous knowledge and have thereby maintained their food security (Altieri, unpublished). Though systems may vary greatly between different geographical and cultural regions, they usually exhibit important elements of sustainability in that they are well adapted to their particular environments, they rely on local resources, are small in scale and decentralized, and they conserve the natural resource base. In the past, attempts to transfer information from temperate zones into the tropical environment have often failed to recognize these fundamental differences.

Much of the information about methods of managing agricultural and natural resources available from local people is of significance to further the goals of sustainability. It may guide researchers in the selection of pertinent research questions and provide a base upon which more analytical, precise scientific investigations incorporating more sociological factors can be built. Special potential lies in the blending of traditional and modern knowledge to develop practical new technologies.

It is important that local indigenous knowledge be documented and scientifically validated since new technologies that duplicate such knowledge are superfluous, and ignoring local approaches to land problems is wasteful. The growing awareness of the potential for utilizing indigenous knowledge is reflected by the establishment of the Center for Indigenous Knowledge for Agriculture and Rural Development (CIKARD) at the University of Iowa in 1987. The Centre has three main functions: (i) to collect published and unpublished materials; (ii) to develop and pursue approaches that foster integration of indigenous knowledge into agriculture and extension; and (iii) to conduct training courses on techniques for documenting and using indigenous knowledge. Training and education are essential to the dissemination of knowledge, and must bridge the gap between what is already known and what has been learned from teachers, texts, research papers and monographs. However, training is not needed by nationals of developing countries alone: those wishing to work in such countries need training also, in relevant technical, social and economic topics.

Indigenous knowledge may play a key role in the design of sustainable systems, and by understanding and working with it, the likelihood that rural populations will accept, develop and maintain innovations and interventions is increased. It may also offer a solution to the problem of communicating new ideas to the farmer by establishing a two-way communication between farmers and researchers.

The concept is equally applicable, of course, in fully developed countries. The Landcare programme in Australia appears to involve 20–30% of the country's farmers, in, among other things, direct feedback from local experience into the R&D system (Allwright, 1991). The feedback is very important to ensure that research is driven by 'client demand' rather than researchers' predilections. However, the client may not know what he wants until presented with the options.

User's Needs

Researchers, research managers, land managers, policy decision-makers, and educators all have special needs. The information content, style and carrier need to be adapted accordingly. Improvement in meeting the needs over the last two decades has been spectacular, as pointed out by Scott and Gilmore (1992), but much more so in developed countries than elsewhere. The barriers to extending the same benefits to developing regions are now less than might be imagined, thanks to low-cost microcomputer networks and software.

Scientists' efforts in disseminating their contributions are now smoothed by statistical software, and software that encourages the 'expandable page' method of writing. The keyboard is showered with half-formed thoughts, experimental evidence, and hypotheses for further manipulation. This has potential for enhanced quality of presentation, and for moving manuscripts to publishers and reviewers more quickly by electronic means. The whole information turnaround time should be shortened.

The information for scientists, managers and policy formulators in each of the sciences related to sustainable land use should meet certain criteria. The core science must obviously be covered comprehensively, in addition to everything indicating potential field applications, a weakness in soil science. Timely information on recent advances is important. Less obvious is the need with modern media and databases to safeguard the value to the scientist of old-fashioned browsing. Wide browsing has very frequently provided creative stimulation from otherwise unused sources. A recent simple example of this might be the suggestion by Nielsen (1992) that scientists could profitably approach problems of ecosystem health in a similar manner to the veterinary approach to diagnosis and cure.

For both managers and scientists, there is a lot more scope for development of timely compilations of annual and medium term internal research project plans. The Inventory of Canadian Agrifood Research (ICAR), the Soil and Water Activity Inventory (SWAIN), also in Canada, the Current Research Information System (CRIS) for publicly supported agricultural and forestry research in the USA, and CARIS (FAO) are good examples of such compilations.

CRIS provides summaries of 30,000 current projects adding approximately 3000 new ones and 20,000 progress and publication reports annually. ICAR has 4000 references to projects from industry, universities and federal and provincial establishments. The success of such information systems depends on incentives for cooperation of research scientists and managers who provide the information. The current challenge is to develop a system that is comprehensive, international in scope, attractive to contributors and affordable.

Educators have special needs, starting with the confused lexicon of soil science. A maze of ill-defined and overlapping terminology is revealed in dictionaries of soil science, and for better communication of information a major effort on a soil science thesaurus is necessary. Teachers might appreciate computer and video-assisted taxonomic keys and correlations between classification systems. Validation of taxonomic criteria on screen may offer interesting possibilities, and there appears to be a lot of scope for further development of expert systems, such as those available for insect pest identification. The substantial cost of constructing them, will have to be carefully assessed in relation to the low level of interest in soil taxonomy.

The information needs of decision-makers in public administration and industry have been apparent in their lack of consensus on the definition, scope and significance of the sustainability concept. To rectify this the G7 International Forum on Environmental Information (Supply and Services Canada, 1991) called for information:

- to be presented within an ecosystem framework at national, regional and global levels;
- to assign priorities from the standpoints of human health, depletion of resources, absorption capacity, direct and indirect costs, expert and public opinion, and systematic accounting;
- to be based on an understanding of causes of degradation, impacts and cost-effectiveness of solutions, and appraisal of information uncertainties.

Four major areas were then identified for action:

1. knowledge base on links between economy and environment;
2. monitoring;
3. decision relevance of environmental reporting;
4. institutional partnerships.

Means of Information Transfer

Overview

Information from laboratory and field experiments is normally communicated through papers in journals, conferences, networks and through informal international collegiality. Information database organizations scan the literature from most of these primary sources, producing from them abstract journals, bibliographies, electronic journals, microfilm, CD-ROMs, and geographical information systems (GIS). Indigenous knowledge discussed earlier, has not yet found sufficient means of transfer, although a good deal has been written in specialized anthropological studies.

From these sources, information, concepts and ideas are passed on and can then be developed in further research, or applied in the field. This will eventually lead to the formation of new data pools. This cycle of events illustrates how information transfer facilitates the development of scientific knowledge, and the importance of investing resources to support the system must be emphasized.

Over the last 20 years we have seen the introduction of machine-readable formats. The marked shift towards electronic media has steadily increased as they can save end-users' time, and provide the opportunity for them to create and manage personal databases gleaned from several different sources. Greater flexibility is achieved, especially with the recent advances in technology such as CD-ROM, digital scanning, optical character recognition, networking, expert systems and multimedia systems, all providing better access to information.

Computer networking

Computer networking enables many users within the same organization to access one central source of software (e.g. information database), thus avoiding multiple purchases of the application software system. Although at first it may be costly to install, as each workstation requires networking software to access the central system, computer networking increases the versatility of computer use, can save resources and increase overall efficiency. The advantages and disadvantages of installating a network system are unique for every organization.

CD-ROM

It is widely accepted that CD-ROM (compact disc-read only memory) has had an enormous impact on the ease with which information can be disseminated and retrieved. It has particular potential in countries with poor telecommunications, where little online searching and networking is done.

CD-ROM offers ease of information retrieval, portability and durability. Huge amounts of data are available on each disc (650 megabytes). For example the CABI CD-ROMs can each hold over 400,000 individual records per disc, which represents 3 years' input of all subjects into the CAB ABSTRACTS database. Products aimed at a specific audience can also be created. For example the SOILCD contains a comprehensive compilation of soil-related abstracts published in *Soils and Fertilizers* and other abstract journals between 1973 and 1991.

Geographical Information Systems (GIS)

GIS is here taken to mean primarily the archiving, processing, retrieval and display of spatial data to allow overlaying areas on maps that are characterized by lists of attributes, as in a large map legend.

The usefulness of GIS to analyse and implement sustainability in land use might easily be regarded as a panacea, perhaps even the ultimate in GIS applications. The reasons are the way GIS are supposed to handle the peculiar problems of natural resources information: diverse disciplines, formats and scales; aggregated in different ways for different organizations; difficult to render down to a common comparable base; gross volume of some data, e.g. from remote sensing, but inevitably with crucial gaps. Crop risk modelling, socioeconomic factors and a whole lot of other features can be accommodated.

On the output side, executive committees might dispense with agenda papers of text and tables, in favour of live cartographic displays of options, priorities and impacts. Experience to date suggests quality output is possible at reasonable cost. For example, in 1989, $9000 bought the Great Lakes Water Quality Board in North America the ability to display the magnitude of non-point sources of each of 180 agricultural chemicals, county-by-county across six bordering US States and one Canadian province. The cost of adding existing soil maps for the entire catchment was estimated at $15,000. Share of overheads on the original half million dollar Agriculture Canada investment in ARC/INFO was negligible (Nowland, internal report). But as in many other cases the GIS impressed by its display rather than analytical capabilities. Maps enable analyses to be better based on facts, but do not perform the analysis, although facts properly displayed in spatial relationship often make the correct analysis obvious.

There are now many examples of the successful use of GIS in addressing sustainable land use, both at the local level, e.g. in Thailand (Huizing, 1991), Nepal (Schreier *et al.*, 1991), Malaysia (Moon, Vancouver, 1992, personal communication), and continent-wide, e.g. the CIAT Agro-ecology Studies. (P. Jones, Cali, 1991, personal communication). A full treatment of the classification and capabilities of GIS in relation to user needs (e.g. Raper and Maquire, 1992), along with outlines of some available systems are provided in a special issue of *Computers and Geosciences* (1992). Burrough (1992) suggests therein that there is still a long way to go before systems can cope with the complexity of natural resources data sets and the expert opinion involved in using land information.

The various methods of information transfer outlined have shown that information database services can provide a multitude of information to the end-user in many formats, at least in developed countries. In developing countries where there are growing populations and more pressing demands on the land, and where information on sustainable land use would be of most value, the means of information transfer are by no means as effective.

Common Problems

Surveys carried out in several developing countries by CAB International to ascertain the problems involved in information dissemination and how best they could be ameliorated indicate that although many of the basic needs are similar, differences in geographical and socioeconomic situations require different solutions. For example research institutes in Zambia, Guyana and India have access to the AGRIS, AGRICOLA, and CAB ABSTRACTS databases, whereas other countries are unable to purchase more than a few of what most would regard as the essential abstract journals for basic literature searches.

Problems of information dissemination common to many agricultural sciences were recently reported from China (Zhang, 1991):

- inadequate coverage of the subject matter by available information resources;
- unnecessary duplication and serious waste in spending;
- over-concentration of information resources in cities, universities, and research institutions, but a serious shortage at 'the county level';
- overdevelopment of information resources at the research level at the cost of the operations level;

- underdevelopment of the information industry in relation to other sectors, and low utilization of information resources.

The main concern of most developing countries is the lack of funding to meet information costs, especially with respect to acquiring foreign currency. It is often hard to obtain funding in information technology, because the financial returns may not be obvious. This is a problem that must be faced by both the developing countries and database producers. The difficulties of payment are not confined to currency availability, but include the complexity and delays in internal approval for individual transactions in advance, and the difficulties of aid agencies in contributing to acquisition of journals except on a one-off basis at best for 3–5-year runs.

The CABI country surveys confirmed that there is an over concentration of information materials in major cities, with very little filtering out to towns, villages and rural research stations. This was found to be due to inadequate library infrastructures and too few, insufficiently trained staff failing to 'sell their wares'. There is sometimes a lack of motivation and education of researchers in the use of literature searches which may reduce repetition and save valuable resources from repeating work done before. Some duplication is acceptable, however, to assess local relevance.

Another problem observed was that information that reached a distant research station was not always presented in a format or style that can be interpreted by field workers, whose training may not be as comprehensive as one might wish.

An interesting point noted in some countries, e.g. Guyana, which have relatively well-equipped libraries and information services, was that they could not obtain primary journals, which is important as abstracts are only intended as guides, not substitutes, for original documents. These countries might reasonably fail to see the advantages of investing further in information processing technology.

There were many complaints about paying for databases that contain mainly mainstream literature, which do not always fulfil a particular country's needs. Users want 'grey literature' specific to their requirements to be scanned. As consumers they want value for money.

Some special problems and issues face countries of central eastern Europe, as they emerge from communist dictatorship and develop market economies. Most have been largely isolated from international information networks, required to rely on Russian-language information materials and services which cannot meet their rapidly expanding needs. Economic depression, unfortunately, severely limits the capacity of most of these countries to access global information sources, without external assistance. Hungary has managed to establish the most productive links, for example through its membership of CAB International. Other countries of central and eastern Europe are a long way behind. The Ukrainian Republic, the most important agricultural country of the old Soviet Union, for the time being, has immediate access only to Russian information materials.

There has only been restricted use of electronic media in most developing countries. The reasons are varied and include poor telecommunications, lack of computer servicing facilities and financial constraints. The situation is gradually improving, especially with the introduction of CD-ROMs, which do not rely on telecommunications.

Particular Issues in Soil Science

Among the information problems specific to soil science in relation to the sustainability of land use, four are proposed to warrant special attention. The first is the cost and time required to monitor soil quality trends. Field monitoring, seemingly tedious for scientists, too prolonged for short-term funding, and vulnerable to cost cutting by programme managers, has clearly become a critical component of information for effective conservation programming. Long-term experiments that survived accusations of obsolete practices have proved invaluable in the analysis of sustainability. Long-term monitoring and development of economical techniques must be supported.

A second issue is information on the geographical limits of extrapolation of research findings and the extent to which management practices can be extrapolated to other locations. One can conceive of zones of degrees of safe extrapolation, based upon more integration of, say, technical abstracts with existing maps of soil and terrain, global ecological zones, and physiography-based national economic planning regions. Application of GIS could contribute much to the resolution of this issue.

Third, there is the need for soil science to play a more useful role, more integrated with other disciplines, in the movement for sustainable development. 'Unsoiled' administrators commonly pay lip service to some vague land ethic, but too often show they have not been given a convincing conservation message when it comes to the vote for land development. Failure to communicate may be one reason for the limited impact of soil scientists on natural resources management decisions during the past three decades. However, soil scientists might well compare their contribution quite favourably with the sadly nebulous offerings of some other movements such as bioethics and environmental ethics (Bourdeau *et al.*, 1989).

Fourth, and so important to avoid waste of resources, information must be extracted, shaped and delivered in ways that can prevent repetition of previous research. The daily diet of repetitive research can be one of the most depressing aspects of editing an abstract journal.

Fine Tuning the Solutions

Country surveys are important to establish the problems first hand and then produce the best 'package', i.e. the information the users require, possibly including 'grey literature', and how libraries and information could best be organized.

Training is probably the most important factor in any project. Training is needed to set up and run libraries, and to promulgate the principles of abstracting and indexing, so staff can collate information cost-effectively from research done in their own country. Training on all new electronic media is needed to facilitate specific searches. There is a training element for information specialists to motivate scientists to use their services. The survey of one major developing country noted a priority commitment by government agricultural library services to reach out to their own scientists, who seem reluctant to make full use of what is available. Some scientists appear to prefer informal international collegiality, sometimes called networks (in the loosest sense), but with the risk that their literature reviews are less than they might be, even allowing for difficult circumstances.

A reliable document delivery service of reprints from primary journals is essential, otherwise scientists and researchers could lose confidence in the system and may not use its services again. Pre-issue of coupons by international donors is an effective facilitator in developing regions. A case can be made for abstract services to produce fuller abstracts giving more detail on experimental results, to meet the needs of those without direct access to good libraries.

These solutions provide a cheap and effective way of increasing the awareness of the benefits of an information system and database. Once the products are widely known and used and seen to be useful, it is reasonable to install modern equipment which will make the system more effective.

Lack of funding was considered to be the root problem of information dissemination. Any project or solution will cost money, yet it is believed that a well thought out long-term package based on a realistic appraisal of a country's needs will find it easier to attract international donor funding.

If there is one overriding principle to be followed in the improvement of information flow aimed at more sustainable use of land resources for agriculture, it is to integrate soil and water research more effectively with broader aspects of national resource management, to reduce isolation (NRC, 1991). A great deal has been learned, for example, about soil processes and properties of organic compounds, but applications of such knowledge to local problems and to limitations imposed by socioeconomic complexity, leave much scope for improvement. Better communication of information will be essential, new means of doing it will be needed, and it must serve the needs of disciplines other than soil science.

Acknowledgements

The contributions of Andrea Powell and Colin Ogbourne of CAB International and suggestions of Ian Sneddon and Dennis Greenland are gratefully acknowledged.

References

Allwright, J. (1991) The environment and sustainable growth: the key role of farmers. Presentation to the International Federation of Agricultural Producers Conference, Reykjavik, October 1991.

Bourdeau, Ph., Fasella, P.M. and Teller, A. (1989) Environmental ethics. *Proceedings of the 6th Economic Summit Conference on Bioethics,* Brussels, May, CEC, 325 pp.

Bunting, A.H. (1992) New challenge in agricultural training. *Interpaks Interchange* 9, 6–8.

Burrough, P.A. (1992) Are GIS data structures too simple-minded? *Computers and Geosciences* 18 (4), 395–400.

Eswaran, H. (1992) Role of soil information in meeting the challenges of sustainable land management. *Journal of the Indian Society of Soil Science* 40, 6–24.

Huizing, H. and Bromsveld, M.C. (1991) The use of geographical information systems and remote sensing for evaluating the sustainability of land-use systems. In: *Evaluation for Sustainable Land Management in the Developing World, Vol. 2.* Technical Papers. International Board for Soil Research and Management, Bangkok, Thailand. IBSRAM Proceedings, No. 12(2).

International Institute for Sustainable Development (1992) Sourcebook on sustainable development. IISD, Winnipeg, 133 pp.

Nielsen, N.O. (1992) Ecosystem health and veterinary medicine. *Canadian Veterinary Journal* 33, 23–26.

NRC (1991) *Toward Sustainability: Soil and Water Research Priorities for Developing Countries.* Committee on International Soil and Water R&D, National Research Council. National Academy Press, Washington DC, 66 pp.

Raper, J.F. and Maquire, D.J. (1992) Design models and functionality in GIS. *Computers and Geosciences* 18(4), 387–394.

Schreier, H., Brown, S., Kennedy, G. and Shah, P.B. (1991) Food, feed and fuelwood resources of Nepal: a GIS evaluation. *Environmental Management* 15(6), 815–822.

Scott, P.R. (1991) The universal issue: information transfer. In: Hawksworth, D.L. (ed.) *The Biodiversity of Microorganisms and Invertebrates; Its Role in Sustainable Agriculture* CAB International, Wallingford, pp. 245–266.

Scott, P.R. and Gilmore, J.H. (1992) Crop Protection and Information Technology in the Year 2000 In: Abdul Aziz S.A. Kadir and Barlow, H.S. (eds), *Pest Management and the Environment in 2000.* CAB International, Wallingford, UK, pp. 371–381.

Sombroek, W. (1991) Amazon landforms and soils in relation to biological diversity. In: *Annual Report 1990,* ISRIC, Wageningen.

Supply and Services Canada (1991) Environmental information for the twenty-first century. *Proc. International Forum*, May, 1991, Montreal.

Warren, D.M. and Cashman, K. (1989) Indigenous knowledge for sustainable agriculture and rural development. Publ. SA10 in Gatekeeper Series, Sustainable Agriculture Programme, International Inst. for Environment and Development, 15 pp.

Wiley, D.L. and Powell, A. (1992) The role of information technology in providing animal health information – a publisher's point of view. Presented at the International Conference of Animal Health Information Specialists, University of Reading, July 16–19, 1992.

Zhang, Q. (1991) Improving the accessibility and availability of information in the agricultural library and information system of China. *Proceedings of the VIIth World Congress of the International Association of Agricultural Librarians and Documentalist: Information and the End User*, 28–31 May, 1990, Budapest, Hungary. Conference issue: *Quarterly Bulletin of the International Association of Agricultural Information Specialists* 36 (1–2), 55–58.

Part VI

Promoting Soil Resilience for Sustainable Land Use

Chapter 28
Using Collaborative Research Networks to Promote Sustainable Land Use

M. Latham and J.K. Syers

IBSRAM, PO Box 9–109, Bangkhen, Bangkok 10900, Thailand

Introduction

The concept of sustainable land use has been at the forefront of recent discussions between agricultural scientists, farmers, and environmentalists. The discussion has often raised major debates between production specialists and ecologists or conservationists. It has also raised more fundamental questions about the ownership of land, its value, the right to degrade, and the responsibilities for rehabilitating it.

Sustainable land use has a much wider scope than sustainable agriculture, which is often restricted to the on-site effects of its activities. Sustainable land use needs to consider both on-site and off-site effects. The Working Group on the Establishment of a Framework for Evaluating Sustainable Land Management has defined it as a set of technologies, policies, and activities aimed at integrating socioeconomic principles with environmental concerns so as to simultaneously:

- maintain or enhance production and/or services;
- reduce the level of production risk;
- achieve environmental stability;
- be economically viable and socially acceptable.

As a result, to promote sustainable land use and management implies on the one hand, at the local level, obtaining a consensus from all the land users (not only from the farmers) on how the land should be used; on the other hand, at a more general level, developing a multilocational effort to inform all members of the global village on how to use their land more sensibly.

Collaborative research networks, which involve multiple actors in scientific and sociological groups, multiple locations, and a common vision within land-use areas are important tools for addressing the sustainable land use issue. How can they assist in this new and challenging role, what are the foreseeable limitations, and how do they need to be associated with other means to achieve sustainable land use? These are some of the issues which will be addressed in this chapter.

Issues in Terms of Sustainable Land Use

The change from traditional farming to commercial farming and interest in many new forms of land use, be they urban, industrial, or recreational, has changed the concept of sustainable land use. Traditional farmers were custodians of the land. The land was theirs for production, fishing, hunting and recreation. Where land tenure was secure the precept was that 'a man should always aim to hand over his farm to his son in at least as good a condition as he inherited it from his father' (Russell, 1973). In communal lands, the role of the custodian was taken by the chief of the community, and in other uncertain land-tenure systems the state or the regional government took the lead.

This change to commercial farming has massively increased the production of food and other agricultural products but the result has often been obtained at the expense of the land capital and with strong pressures on the environment (Ragland, 1992). Water-table pollution, land degradation through erosion, salinization, or nutrient depletion are some of the consequences of this new commercial farming situation.

Another consequence is that land and natural resources are being traded as commodities without much consideration for their intrinsic values and their potential renewability. On the one hand, farmers are squeezed by low price policies and blamed for damaging the environment and on the other hand, powerful commercial interests overpay for hardwood logs. New uses of land for industrial and urban purposes and for recreation are being found, often at the expense of the environment.

Therefore sustainable land use implies an agreement between users, which can best be obtained from sound scientific results and mitigation between users' interests. Multidisciplinary research, led by users' needs, is therefore required for societies which have an overall commitment to development (Spendjian, 1992).

Another aspect which needs to be considered in the aim for sustainable land use is the diversity of environmental factors. Climate, soil, vegetation, animal life, and human life present considerable variations from one place to another. Because there is no universal recipe for sustainable land use, production/service systems need to be adapted to their local environment. The FAO framework for land evaluation (FAO, 1976) is one of the tools to better address the compatibility between land characteristics and potential crops.

In most development schemes, production has been the immediate goal without major consideration for risk due to climate or pests. Monocropping and biouniformity can lead to disasters in years different from the climatic average or when pests spread (Spain, 1992). Biodiversity is not only important for crops but it has recently been highlighted at the UNCED conference for trees, animals, insects, and other organisms. Monocropping, which has too often replaced multiple cropping with a view to increasing productivity, has also shown its limitations. Alternative agriculture groups and others now tend to question this practice (National Research Council, 1989). Seguy *et al.* (1992) have shown in Brazil that multiple cropping – cajanus, rice, beans – associated with proper tillage can considerably increase the yield of rice and reduce the pressure of weeds in comparison with a rice monocrop with surface tillage. Similarly, Sanchez and Benites (1987) have shown that a rice–cowpea sequence without tillage and fertilizer could be a viable option for seven crops in the Peruvian Amazon, in comparison with the one or two crops of rice normally grown by the farmer.

Sustainable land use is the result of acceptable and proper techniques but to be efficient it cannot be restricted to a few users. It needs to involve most of the land users, be they farmers or others. Long-term studies by the Tennessee Valley Authority have shown that a few polluters, effluent from poultry or pig farms, or land mismanagers, can jeopardize the efforts of other members of the community. Nutalaya (1992) has shown how indiscriminate salt mining by a few entrepreneurs in the northeast of Thailand has damaged the life of an extensive farming community, forcing thousands of farmers to migrate to Bangkok. Thus sustainable land management, as noted by Watkins (1992), is a community involvement matter before being a government matter.

Using Collaborative Research Networks to Achieve Sustainable Land Use

To face the multifaceted issue of sustainable land use, a sequence of research plus community involvement plus legislation is necessary. Collaborative research networks are probably the most appropriate, cost-effective way to tackle the first two aspects. According to Greenland *et al.* (1987) and Plucknet *et al.* (1990) in collaborative research networks, countries collaborate in joint planning, implementation, monitoring, testing, and validation of research and they share the results for the benefits of all participants and also for non-participating countries. The comparative advantage of a collaborative research network over classical research is the involvement of several teams in different countries which can assist with the issue of community involvement. Joint planning, implementation, and exchange of results are also one cost-effective way to tackle sustainable land-use questions. The important point in this case is to involve in the planning stage, all possible users so the right questions are tackled. In a review of reasons for non-adoption of new techniques by farmers, Fujisaka (1992) ranked among them: absence of a problem, inappropriate innovation, or incorrect identification of adoption domain. Therefore, the identification of the right question needs to be achieved at a very early stage, through a socioeconomic survey of the land users. The question needs then to be agreed by the potential network participants, as well as the type of research to be conducted and the methods to be used. Later on, the implementation needs to be conducted by a multidisciplinary team with a regular dialogue between scientists and users, to avoid discrepancies between the research directions and the objective sought. Typically, collaborative research networks are led by a steering committee which should include representatives of the users and of the participating scientists, and which should meet at least once each year to monitor, evaluate, and eventually redirect the research.

In addition to being participatory, collaborative research networks present the advantage of being multilocationally based. They can therefore partially overcome the site specificity constraint of sustainable land use research by a proper characterization of the sites. By fitting the results obtained to existing agroenvironmental maps they allow some degree of extension of the results. Their generalization will be even more accepted if various options are offered to land users, so that they can choose the most acceptable to their needs.

Networks can also be hooked in with Geographical Information Systems or to Decision Support Systems. However, the need for comprehensive experimental ground proofing is compulsory to validate the results and allow their proper extension. But networks can hardly develop by themselves. They need a catalyst – International Agricultural Research Centres, environmental or other groups – which will coordinate, support, and enhance the activities and make them work. The main question in this case concerns leadership. Too much leadership from the coordinator may discourage the participants in their collaborative efforts. On the other hand coordination which is too loose will not allow the necessary harmonization of the research and will therefore not allow the spill-over effect expected. Networks can also be a proper way for transmitting technologies to users, especially if the contact with users is maintained throughout the research process. Users will then be interested in the research and will accept the results more readily. Collaborative research networks can lead to extension and be the basis for policy making, yet they remain research activities and they need to be complemented by extension activities of a much larger magnitude and by policy decisions which can lead to the establishment of laws or government incentives.

An Example: The IBSRAM Network on the Management of Sloping Lands in Asia

This research network provides some ideas of the evolution of a network in a sensitive environmental-agricultural domain (Aneckasamphant *et al.*, 1992). Initially targeted at postclearing management to solve the major problems raised by transmigration in Indonesia and the extensive clearings in Malaysia, the network aims shifted in 1987 due to the fact that land clearing was not really a common problem for the potential participants – Indonesia, Malaysia, Nepal, Philippines, Thailand, and Vietnam. The first two countries were the only ones with extensive land-clearing programmes. Yet most of the ancient and recent land clearing in Asia has happened on sloping lands or steeplands raising production problems due to poor productivity and major soil conservation concerns on-site (such as land and forest degradation due to shifting cultivation) and off-site (siltation of the rivers and potential pollution). After a meeting of the potential collaborators and of potential donors, namely, the Asian Development Bank and the Swiss Development Cooperation, it was decided to reorient the project toward the management of sloping lands, and to develop a project proposal centered on testing conservation farming methods throughout Asia through commonly accepted methodologies. By mid-1988 the proposal was accepted and five projects were started in Indonesia, Malaysia, Nepal, Philippines, and Thailand, followed in 1989 by Vietnam and in 1990 by China. In each of these countries teams including agronomists and soil scientists were organized. Socioeconomic surveys, aimed at a baseline assessment of the sites and at the determination of the usual farming practice were undertaken. A total of eleven experimental sites were established, testing techniques such as: alley cropping using traditional shrub legumes, but also pigeon pea, bananas and pineapples which can give a supplementary revenue; grass strips, preferably using fodder grass without seeds; mulch of crop residues; agroforestry (fruit trees, coffee); and contour ditches. The experiments were located in farmers' fields and the first three years were devoted to adapting and testing the methods. Farmers were invited to visit the experimental sites and to make suggestions but there was a need to obtain trends in the results before advising them. A coordinator was appointed and he visited the different experimental sites three or four times each year. A programme review committee including the collaborator, representatives of the donors, and the coordinator, plus one or two resource persons, was formed and met once each year.

After three years of experimentation the results show that under certain conditions sustainable land use is achievable on sloping lands:

- Biological soil conservation measures have proved to be an efficient way to reduce soil erosion wherever they have been tested.
- The effect of soil conservation measures differs from site to site. On a rather resilient soil (Alfisol), a mulch cover of crop residues – a technique by which the farmer loses no land – gave as good a result as hedgerows or grass strips. On a more erosive soil (Ultisol) grass strips, hedgerows or contour ditches are necessary. Therefore the need to adapt techniques to site characteristics was confirmed.
- Yields of the main crops have been maintained at a comparable level to the farmers' practice without fertilizer input, despite the loss of the land used for the soil conservation measure.
- Minimum fertilizer input not only increased the yields significantly but also reduced erosion by promoting a better soil cover.
- Economically, whereas the first year was to the advantage of the farmers' practice, due to the investments made, the second and third years were beneficial to the conservation farming practice, especially when fertilizer was used and when a secondary crop was introduced (bananas, pineapples).

From the extension point of view the field visits or individual initiatives have resulted in some transfer of the techniques to neighbouring communities in China, Malaysia, the Philippines, and northern Thailand. The second phase of the network will now disseminate the results through pilot villages where demonstration plots at the farmers' plot scale using the most promising technique will be used.

The first three years of operation of this network were successful because of the strong commitment from the collaborators, the donor agencies, and the coordinating agency. The amount of practical results obtained in comparison with the small amount of money spent by the donors (less than 2 million dollars) is remarkable. However, one regrets that:

- the users – farmers and environmentalists – were not more involved in the process;
- the achievements may seem disproportionately small in comparison with the magnitude of the problem; and
- the socioeconomic aspects were not pushed further and extension is still starting when the problem is pressing.

However, it should be pointed out that research on sustainability is a long-term effort and that extending unvalidated techniques may have counter effects. The first three years of research have identified trends which need to be confirmed. Some of these trends were in line with what was expected, others were not. In addition to continuing with the research, the network can now assist national systems and NGOs with extension activities.

The socioeconomic aspect and the iterative dialogue with users has not been as extensive as expected. This situation is partially due to the initial technical research emphasis. It is also due to the difficulty faced by scientists in extending their research to users. Farmers' days have been organized but no user has yet been included in the programme committee of the network.

It must be emphasized that such a network can only act as a catalyst. It must be followed by much more powerful extension efforts, either made through government agencies or non-governmental organizations, or through extension networks built around communities. An interesting point at this stage is that the government policy in soil conservation matters, which in Thailand was oriented at the beginning of the 1980s toward mechanical technologies, is now shifting to biological technologies in considerable part due to the sloping lands network.

In conclusion it must be said that collaborative research networks cannot by themselves achieve sustainable land use. Only the users can do it. Collaborative research networks can be a catalyst for concentrating research and then extension efforts towards the implementation of sustainable land use. For this purpose they need to achieve their research aims by sound scientific procedures and at the same time they need to make their results available and understandable to users and policy makers. If successful on these two points, they can be extremely powerful tools in the process of getting sustainable land use applied.

References

Aneckasamphant, C., Boonchee, S. and Sajjapongse, A. (1992) Methodological issues for soil conservation measures on sloping lands: A case study in Thailand. In: Dumanski, J., Pushparajah, E., Latham, M. and Myers, R. (eds) *Evaluation for Sustainable Land Management in the Developing World,* Volume 2 IBSRAM, Bangkok, Thailand, pp. 205–217.

FAO (1976) A framework for land evaluation. FAO. *Soils Bulletin* no. 32. FAO, Rome, 72 pp.

Fujisaka, S. (1992) Thirteen reasons why farmers do not adopt innovation intended to improve the sustainability of upland agriculture. In: Dumanski, J., Pushparajah, E., Latham, M. and Myers, R. (eds) *Evaluation for Sustainable Land Management in the Developing World,* Volume 2, IBSRAM, Bangkok, Thailand, pp. 509–522.

Greenland, D.J., Craswell, E.T. and Dagg, M. (1987) International networks and their potential contribution to crop and soil management research. *Outlook on Agriculture* 16, 42–50.

National Research Council (1989) *Alternative Agriculture.* National Academy Press, Washington, 448 pp.

Nutalaya, P. (1992) Pollution and other off-site effects on sustainable land development. In: Dumanski, J., Pushparajah, E., Latham, M. and Myers, R. (eds) *Evaluation for Sustainable Land Management in the Developing World,* Volume 2. IBSRAM, Bangkok, Thailand, pp. 135–155.

Plucknett, D.L., Smith, N.J.H. and Ozgediz, S. (1990) *Networking in International Agricultural Research.* Cornell University Press, Ithaca, 224 pp.

Ragland, J. (1992) Stewardship and wealth. In: Dumanski, J., Pushparajah, E., Latham, M. and Myers, R. (eds) *Evaluation for Sustainable Land Management in the Developing World,* Volume 2, IBSRAM, Bangkok, Thailand, pp. 481–490.

Russell, E.W. (1973) *Soil Conditions and Plant Growth,* 10th edn. Longman, Harlow, UK.

Sanchez, P.A. and Benites, J.R. (1987) Low-input cropping on acid soils of the humid tropics: a transition technology between shifting and continuous cultivation. *International Board for Soil Research and Management Proceedings* no. 7, IBSRAM, Bangkok, Thailand, pp. 85–106.

Seguy, L., Bouzinac, S. and Pieri, C. (1992) An approach to the development of sustainable farming systems. In: Dumanski, J., Pushparajah, E., Latham, M. and Myers, R. *Evaluation for Sustainable Land Management in the Developing World,* Volume 2. IBSRAM, Bangkok, Thailand, 357–386.

Spain, J.M. (1992) Genetic resources and biodiversity as tools in sustainable land management. In: Dumanski, J., Pushparajah, E., Latham, M. and Myers, R. (eds) *Evaluation for Sustainable Land Management in the Developing World,* Volume 2. IBSRAM, Bangkok, Thailand, pp. 157–171.

Spendjian, G. (1992) Economic, social and policy aspects of sustainable land use. In: Dumanski, J., Pushparajah, E., Latham, M. and Myers, R. (eds) *Evaluation for Sustainable Land Management in the Developing World,* Volume 2. IBSRAM, Bangkok, Thailand, pp. 415–436.

Watkins, W.A. (1992) Environmental aspects and their impact on sustainable land management. In: Dumanski, J., Pushparajah, E., Latham, M. and Myers, R. (eds) *Evaluation for Sustainable Land Management in the Developing World,* Volume 2. IBSRAM, Bangkok, Thailand, pp. 121–133.

Chapter 29

The Work of FAO's Land and Water Division in Sustainable Land Use, with Notes on Soil Resilience and Land Use Mapping Criteria

W.G. Sombroek

Director, Land and Water Development Division, FAO, Via delle Terme de Caracalla, 00100 Rome, Italy

Introduction

FAO is a technical cooperation agency rather than a research organization. It serves its 166 member countries in promoting the sustainable development of agriculture, forestry and fisheries, ensuring sufficient food, timber, fibre, fuel and fodder, for an ever-growing world population, especially in developing countries. It endeavours to apply the results of strategic or applied scientific research by national and international entities for those purposes, and suggests policies and strategies to make these research efforts more relevant for development and conservation.

An exception to the general situation is the Joint FAO/IAEA Division of Nuclear Techniques in Food and Agriculture (AGE), located in Vienna. Its tasks include fundamental and applied research for the benefit of developing countries.

The Land and Water Development Division (AGL, popularly known as Natural Resources Division) deals with the inventory, assessment, development, management and conservation of land and water resources and plant nutrients. It cooperates with other Divisions and Departments on agro-climatic resources (climatic databases), on the biotic resources (vegetation, forestry, animals, fish) and on socioeconomic conditions of the land (human population, actual land uses).

From its creation in 1946 (originally as Land Use Branch of the Agricultural Division), AGL has been active in the inventory and assessment of land and water resources for a multitude of developing country areas, often with UN Special Fund or Development Programme financing. These early inventories contain valuable baseline information for 'backtracking'

(eds D.J. Greenland and I. Szabolcs) CAB INTERNATIONAL Wallingford long-term ecological research, as needed nowadays in climate and land use change studies of the International Geosphere–Biosphere Programme (IGBP).

In the 1960s, thanks to the pioneering work of Drs Bramao and Dudal, in cooperation with many of the world's soil scientists, the then existing soil resources information was collated into a Soil Map of the World at 1:5 M scale (FAO, 1971–1978). It is still the only document of its kind, although badly in need of updating for several regions. Unfortunately, this major effort was not accompanied by similar projects on the world's water resources or current land uses. However, an agroclimatic database for all developing countries, together with the soil database, served for another major undertaking: the Agro-Ecological Zones (AEZ) project (FAO, 1978–1991; see also Higgins *et al.*, 1987), and its offspring, the Assessment of Potential Population Supporting Capacities of Lands in the Developing World (FAO/UNFPA/IIASA, 1982). The concepts of agroecological zoning have subsequently been elaborated and applied at country level (Kenya, Mozambique, Bangladesh, Nigeria; currently China and the Amazon region). The criteria are continually being refined, including extension of their use in temperate and cold and high altitude regions. The system is also being adapted for use in research planning and application, in view of the recent decision of the international agricultural research centres supported by the CGIAR to take AEZ as a main research orientation.

The matching of development with conservation, as an integrating or 'holistic' approach, came to the fore in the early 1970s, through the development, in cooperation with a Wageningen group, of the Framework for Land Evaluation (FAO, 1976). Subsequently, it was applied for specific land uses through the development of guidelines for rain-fed agriculture (FAO, 1983), forestry (FAO, 1984), irrigated agriculture (FAO, 1985) and extensive grazing (FAO, 1990a). As a truly interdisciplinary effort, AGL together with other FAO Divisions has recently developed Guidelines for Land Use Planning (FAO, 1992a), incorporating physical, biological, social and economic aspects (Sombroek, 1992).

Throughout its history, AGL has issued bulletins and papers for practical use at country, project and farm levels, many of them aimed explicitly at long-term productivity maintenance or improvement. They include Soils Bulletins (66), World Soil Resources Reports (68), Irrigation and Drainage Papers (48), Fertilizer and Plant Nutrition Bulletins (11), Miscellaneous Papers (19) and Miscellaneous Reports, Guides, Manuals or Filmstrips (42).[1]

FAO also developed a 'World Soils Charter' and cooperated with UNEP and others on soil degradation assessment, actual and potential, with special attention to the situation in Africa. It works currently with UNEP and ISSS on the development of National Soil Policies in individual developing countries. There are also numerous papers on land resources appraisal and land management by AGL staff, presented at international meetings and scientific conferences.

Although the notion of sustainability pervaded its field programmes and many of its past publications, the Division has only recently been confronted with explicit sustainability issues. Special Action Programmes for developing countries have been formulated, such as the International Scheme for the Conservation and Rehabilitation of African Lands (FAO, 1990b) now being broadened to Asia and Latin America as LASAD; the International Action Programme for Water and Sustainable Agricultural Development (WASAD, FAO, 1990c) and the reorientation of AGL's long-term Fertilizer Programme towards a balanced and integrated approach to plant nutrient supply and maintenance, tentatively called NUSAD.

The new problem of pollution of soil, water and nutrient resources is being addressed, and the same holds for the effects of any human-induced climatic change on the quality and quantity of land and water resources for agricultural production in its widest sense.

Valuable insights on sustainability aspects were obtained through the joint FAO/Netherlands preparations for the Conference on Agriculture and the Environment in Den Bosch, April 1990 (FAO/Netherlands, 1991). The results served as background for the sustainable agricultural and rural development (SARD) element of the World Conference on Environment and Development (UNCED, Rio de Janeiro 1992) and have been well reflected in its Plan of Action called 'Agenda 21'. A good overview of FAO's gradual move towards sustainability issues is given in a White Paper of the Organization, prepared in reaction to criticism by ecologist groups (FAO, 1992b).

Soil or Land Qualities and Their Resilience

The above may sound impressive, but FAO, just like other entities, is groping for handles to establish sustainability of land use, the resilience of natural resources in general, and of soils in particular, for subsequent application in FAO's field programme. AGL is already working with other entities on a Framework for the Evaluation of Sustainable Land Management (FESLM; unpublished working paper for FAO prepared by A.J. Smyth). The recently proposed networking between a number of entities oriented to research on soil/water/nutrient management in developing countries such as IBSRAM, Tropsoils, SMSS, IFDC and TSBF*, is a most welcome development. I advocated this some time ago, as an ISRIC/ISSS

*IBSRAM = International Band for Soil Research and Management; Tropsoils = the US Universities Consortium for Research on Tropical Soils, supported by USAID; SMSS = the USDA/SCS Soil Management Support Services; IFDC = International Fertiliser Development Centre; TSBF = Tropical Soil Biology and Fertility Programme of IUBS and UNESCO.

official (for instance in Mooney and Sombroek, 1991). FAO's AGL is prepared to support such networking in any function that the recipient developing countries, the research participants and the donors may wish.

What should be the parameters by which to measure soil resilience and sustainable land use, and what should be the threshold value for each parameter? I have no ready answers, only suggestions for discussion.

There are many definitions of sustainability and sustainable land use (see for instance the Scientific Committee on the Application of Science to Agriculture, Forestry and Aquaculture, CASAFA, 1991). In the context of the needs of the strongly growing population of developing countries, one should not equate sustainability with stability of present-day land use, but rather with a sustained growth of agricultural productivity of the land without negative environmental effects – often as yet hidden – that would cause an ultimate collapse of the land use, on-site or off-site.

Resilience could be defined as the capacity of a system (soil, landscape, farming system) to return to its original equilibrium after a major disturbance (see also Chapter 11). Again in the context of developing countries' needs, one may rather define resilience as the capacity of an (agro-) ecosystem to return to a sustained growth of productivity, rather than to its original equilibrium after a major disturbance.

If soils are considered in isolation, it is obvious that they are not static; they develop. One may want to halt or modify natural soil forming processes if the end product would be a lower set of qualities, or rather a lower biological entropy level. Farmers do this al the time, but a single-minded ecologist or pedogenesis specialist may have the view that a lack of sustainability implies a disruption of the main genetic process on the basis of the prevailing soil-forming factors, excluding the influence of humans. If this disruption is temporary, then the pedologist would still speak of soil resilience. Only if it would lead to an irreversible turning point towards another process that has an ecologically inferior end product, would this be considered as non-sustainability (examples of such 'ecotonal' or threshold soil situations are given in Scharpenseel *et al.* 1990).

The above illustrates that soil resilience can be assessed in different modalities of time, function, entropy and geography (Table 19.1).

In FAO's developing countries' perspective, the main orientation will usually be: 10–15 years as regards time; agricultural productivity as regards function; physicochemical richness as regards entropy; and the landscape element or agroecological zone as regards geography. Sustained soil or land productivity should be determined in an ecological context, and the concept of soil (or land) qualities may be given central attention (see also Larsen and Pierce, 1991). It is noted that this ecological perspective is given the same weight as productivity and reclamation cost in the definition of degrees of soil degradation in the GLASOD publication (Oldeman *et al.*, 1990) namely 'original biotic functions largely intact/partially destroyed/largely destroyed/

Table 29.1. Modalities of assessment of soil resilience.

1. Assessment in respect of *time*

years:	≤1	3–4	10–15	50–100	> 1000
activity:	annual monocropping	rotation; ley farming	shifting cultivation; perennials	natural restoration after serious soil degradation	changes in geomorphopedologic processes
	soil temperature, moisture	soil organic matter	soil physicochemical changes		

2. Assessment in respect of *function*

- agricultural productivity or biomasss production
- biodiversity (natural or man-induced; in-soil or on-soil)
- greenhouse gas fluxes
- buffering and filtering of pollutants
- physical medium for man's socioeconomic activities

3. Assessment in respect of *entropy*

- chemical richness (pH, NPK-levels)
- biological richess (soil organic matter; microbial biomass)
- physical richness (depth, structure, stability: often old, weathered soils!)
- mineral reserve richness (young soils)

4. Assessment in respect of *geography*

- point or plot level
- landscape level
- (agro-) ecological zone level
- country or regional level

fully destroyed'. The notion, however, surfaced already in the early 1970s in the first descriptions of major land qualities as equating major ecological conditions (see Beek and Bennema's descriptions of such land qualities in FAO, 1974). For practical soil survey interpretation in Brazil in those days only a few, major qualities were used, for instance, 'moisture conditions' as the combination of the net readily available soil moisture capacity and rainfall conditions; 'oxygen availability to roots' (drainage condition), 'nutrient availability', 'foothold' and 'erosion hazard'. For a fully holistic evaluation of the land for all imaginable uses (or non-uses) the whole range of imaginable land qualities should be considered. FAO's land evaluation for rain-fed agriculture (1984) gives such a range, and Buringh and Sombroek (1971) gave a tentative regrouping according to the vertical compartments of the land (Table 29.2).

The land qualities concept was meant to facilitate the assessment of individual soil and land attributes in respect of suitability for ecologically sustainable potential land utilization types, through the development of two algorithms, or models leading to indices of sustainability (Fig. 29.1).

The A-type algorithms are largely use-independent, whereas the B-type ones are directly related to the envisaged use and therefore imply a differential weighting – some qualities becoming more prominent, others losing weight or becoming completely irrelevant. At the time of development of the Framework for Land Evaluation, the A-type algorithms were unfortunately never worked out in detail, except for the 'water-behaviour' aspects of the soil (combination of storage capacity, infiltration, percolation, horizontal and vertical hydraulic conductivity; Van Beers of ILRI, unpublished). Some ad hoc efforts on the quantification of 'chemical soil fertility' and other qualities were carried out, e.g. by Van de Weg and Mbuvi

Table 29.2. Land qualities.

Land evaluation for rainfed agriculture, FAO (1984)	After Buringh and Sombroek (1971)
LQ1 Radiation regime: Total radiation Day length	A. *Atmospheric qualities* • Atmospheric moisture supply: rainfall, evaporation, dew formation.
LQ2 Temperature regime	• Atmospheric energy for photosynthesis: temperature, daylength, sunshine conditions.
LQ3 Moisture availability: Total moisture Critical periods Drought hazard	• Atmospheric conditions for crop ripening, harvesting and land preparation: dry-spell occurrence. • Liability to atmospheric calamities: hazard of tornadoes, hail storms, etc.
LQ4 Oxygen availability to roots (drainage)	
LQ5 Nutrient availability	B. *Land cover qualities* • Value of the standing vegetation as 'crop' (e.g. timber).
LQ6 Nutrient retention capacity	

Table 29.2. (*Cont'd*)

Land evaluation for rainfed agriculture, FAO (1984)	After Buringh and Sombroek (1971)
LQ7 Rooting conditions LQ8 Conditions affecting germination and establishment LQ9 Air humidity as affecting growth LQ10 Conditions for ripening LQ11 Flood hazard LQ12 Climatic hazards LQ13 Excess of salts: Salinity Sodicity LQ14 Soil toxicities LQ15 Pests and diseases LQ16 Soil workability LQ17 Potential for mechanization LQ18 Land preparation and clearance requirements LQ19 Conditions for storage and processing LQ20 Conditions affecting timing of production LQ21 Access within the production unit LQ22 Size of potential management units LQ23 Location: Existing accessibility Potential accessibility LQ24 Erosion hazard LQ25 Soil degradation hazard	• Value of the standing vegetation as germplasm (biodiversity value). • Value of the standing vegetation as protection against soil degradation. • Value of the standing vegetation as protection for crops and cattle against adverse atmospheric influences. • Hindrance of vegetation at introduction of crops and pastures: the land 'development' costs. C. *Land surface qualities* • Surface receptivity as seedbed: the tilth condition. • Surface treadibility: the bearing capacity for cattle, machinery, etc. • Surface limitations for the use of implements (stoniness, stickiness, etc.): the arability. • Spatial regularity of soil and terrain pattern: the degree of freedom in determining the size and shape of fields with a capacity for uniform management. • Surface liability to deformation: the occurrence or hazard of wind and water erosion. • Accessibility of the land: the degree of remoteness from means of transport. • Surface water storage capacity of the terrain: the presence or potential of 'waterholes', on-farm reservoirs, bunds, etc. • Accumulation position of the land: degree of fertility renewal and/or crop damaging by overflow or overblow. • Biological soil toxicity: the presence or hazard of soil-borne pests and diseases. D. *Soil profile qualities* • Physical soil fertility: the net moisture storage capacity in the rootable zone. • Physical soil toxicity: the presence or hazard of waterlogging in the rootable zone (i.e. the absence of oxygen). • Chemical soil fertility: the availability of plant nutrients. • Chemical soil toxicity: salinity or salinization hazard; excess of exchangeable aluminium. • Biological soil fertility: the N-fixation capacity of the soil biomass; the microbial capacity for the transformation of fresh soil organic matter into readily available plant nutrients.

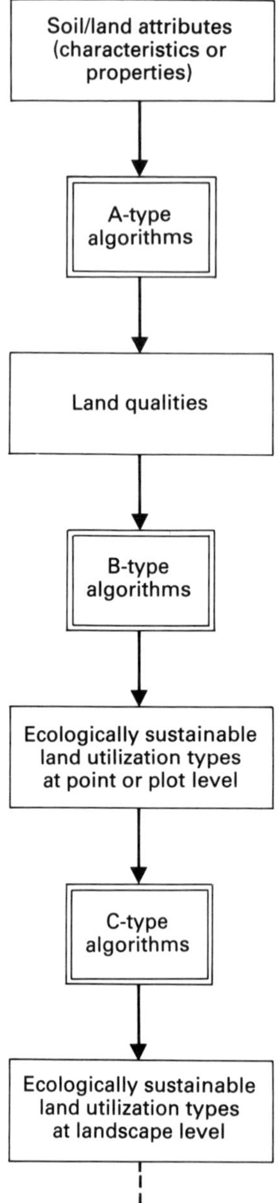

Fig. 29.1. From attributes to landscape-level sustainability.

(1975) for the Kenya Soil Survey. Only recently more systematic efforts were made through the QUEFTS system (Janssen *et al.*, 1989).

Nowadays, with the advance of computer modelling techniques, one could theoretically combine the A and B algorithms, foregoing the land qualities concept. This would allow threshold values to be given to individual soil and land attributes for sustainability and resilience purposes, such as pH, soil organic matter quantity and quality, soil bulk density and the cation exchange capacity (CEC) of the soil.[2] However, the parameters mentioned have different 'normal' levels, posing different 'normal' constraints per main pedogenetically defined soil. Full parametrizaton may therefore be elusive, and it is felt that the 'land qualities' concept as developed in the 1970s remains useful as a framework for our thinking on sustainability and resilience.

There are two reasons for using 'land' rather than 'soil':

1. One should integrate all compartments vertically; from groundwater-related qualities, through soil profile qualities, soil surface and slope position qualities, vegetative cover qualities, to overhead climatic qualities. The qualities of the vegetative cover came to the fore at AGL's recent involvement in agroecological and socioeconomic zoning of Amazon countries;
2. One should integrate all aspects horizontally at the landscape level. This is the land unit approach of physical geographers, which takes into account the typical, microgeographically repetitive elements of terrain top or plateau, scarp or upper slope, main slope, lower slope or springline, bottomland or flood plain; with their mutual influence whether natural or under current land use. This influence can be in the sense of internal hydrology (for instance rainfall moving into the soil of the plateaux and surfacing at the springline, including the lateral movement of chemical substances such as salts and hydrous oxides), or the surface transport of soil material through erosion from upper slopes and accumulation in the bottomland or flood plain.

Either of these processes can be detrimental or positive at the receiving end, depending on the rate of transport and the prevailing climatic conditions. The lateral influence also relates to chemical soil fertility: nutrients may be replaced downslope by natural processes, or on purpose in traditional farming systems. The latter implies a **nutrient mining** on plateaus or upper slopes for **nutrient enrichment** downslope to small areas near villages. This is a process of microgeographic **nutrient harvesting** that has sustained traditional land use in many places of northern Europe in the past and is still the prevalent situation in many African and Asian environments. The landscape as a whole is, or was, in a sustainable production situation. When a sudden increase of population occurs, or a sudden increase in the level of aspirations of a static population – implying the need for cash crops – the use of the landscape becomes unsustainable, unless nutrients are replenished by adding fertilizers from outside the landscape. The exclusive use of mineral fertilizers gives an initial strong improvement in the productive capacity of the degraded upper landscape parts, and less on the lower parts, but in the long run it may lead to **nutrient fatigue**, which is an unsustainable situation from the economic point of view. Without apparent reason yields do not increase further in spite of fertilizer application. Detailed research might however reveal a lack of micronutrients, a diminished activity of nitrogen-fixing bacteria or of mycorrhizal phosphorus uptake. The solution, in respect of long-term sustainability at increased population pressure, is an integrated plant nutrition system approach (IPNS). This can also be called good **nutrient husbandry**, i.e. a balanced and adequate external input of mineral fertilizers in combination with the manipulation of organic nutrients (cf. the TSBF research), the balance of which is different for upslope positions and downslope ones.

Only in rare cases, when the topography is homogeneous over large distances, e.g. the Great Plains of USA or Russia, the Cerrado lands around Brasilia, may one disregard the landscape or land unit concept. In most actual field situations, the land facet analysis of a land unit is of paramount importance for sustainability assessments. This is also the main reason behind the 'soil-and-terrain' (SOTER) approach of ISSS/ISRIC (Chapter 7) now also adopted by FAO for inventories at 1:1 M and more detailed levels.

It is realized that the landscape approach adds to the complexity of assessing soil or land sustainability. A third set of algorithms, C type, is needed to quantify the integration of sustainable soil point uses into sustainable land unit uses, as shown in the last two boxes of Fig. 29.1. (For the socioeconomic aspects a fourth set of algorithms, not elaborated here, is required).

Current Land Use Information

Any land use planning or agroecological zoning for sustainable development requires an identification of the type of current land use at two phases in the process: (i) during the defining of the locally relevant utilization types and their requirements; and (ii) for the comparison of recommended land utilization types with existing land use: what changes are involved, what are the inputs required and where precisely are they to be implemented, at country, district and at farm level. Actual land use information is, moreover, essential for the quantification of current biotic emissions of greenhouse gases, for the assessment of the impact of any climatic change on land use, and for measures to mitigate or to adapt to such change.

So far, there is no commonly agreed comprehensive system of characterization of current land use for the world at large. Every country, and at best every region (e.g. the CORINE programme of the EEC), does it on its own, which makes comparison and collation difficult. FAO is belatedly starting a programme to develop a system for land use classification and its georeferenced inventory at world scale, with in-built updating and monitoring at regular intervals, as a necessary complement to its broad country statistics, through the creation of an interdepartmental task force on the subject, with AGL as main stimulator.

Here again, a host of decisions are needed on what should be the key criteria at the different levels of any categorical system: the type of produce (crop), the degree of external inputs, the labour requirements, the land ownership situation, etc. (Table 29.3). Also ecological and sustainability considerations should enter at some level of the system, and one may think of defining them in degrees of departure from natural ecological conditions. For the purpose of inviting discussion, the following scheme is submitted (Table 29.4).

It should be stressed that the degree of departure can be negative or positive as regards sustainability aspects. Some can be interpreted as degradational, others as a definite improvement of the land resilience on a permanent basis. Stone terracing, protective hedging, liming of extremely acid soils and many other land management practices add to the intrinsic value of the land.

It may be reiterated that issues of sustainability and resilience should be considered not only in the framework of global climatic change, but also in relation to increasing or decreasing rural population pressure and the changing aspirations of the farmer communities. Food for thought on food for people.

Table 29.3. Criteria for a classification system of land use.

To be applied at categoric levels, still to be defined, in a hierarchical classification system on actual land use; units subsequently to be used, singly or in associations or complexes with other units of land use or natural vegetation, in legends for georeferenced databases (maps) at different spatial resolution (scales). Each degree to have five (or three?) quantified classes.

- □ type of produce or produce combinations/sequences, possibly to be expressed in:
  - degree of intensity of solar energy conversion (C3–C4 groupings?);
  - type of storage of conserved energy (for human use): tubers, resins, timber, fuel, fibre, grains, fruits, leaves;
- □ degree of permanence of the produce on the land (seasonal crops, annuals, perennials);
- □ degree of use of the land in time (months per year, length of fallow period);
- □ degree of farm labour (family labour, hired hands, seasonal peaks);
- □ degree of financial input, averaged per ha and per year (power and machinery, costs of fertilizers, seeds, pesticides; storage facilities);
- □ degree of parcelling and degree of cover: for linkage with remote sensing;
- □ degree of change from natural ecological conditions (see Table 29.4)

Problem:

HOW TO STRUCTURE THIS IN A MULTICATEGORICAL SYSTEM?

Table 29.4. Some concepts of ecological criteria for a typology of actual land use, with examples.

- Degree of change from the natural vegetation
  (hunting and gathering → rangeland → shifting cultivation → planted forest of one exotic species → hedged fields → fully arable fields and artificial pastures)
- Degree of change from the natural hydrological conditions
  (on-farm water catchment storage → field drainage → moveable or temporary irrigation structures → permanent third level irrigation structures → sawahs)
- Degree of change from the natural physical land surface or topographic conditions
  (superficial levelling → bunding → stone removal → camber bedding → on or off 'earthing' → stone terracing)
- Degree of change from the natural soil conditions
  (organic matter manipulation → liming, fertilization, desalinization → subsoiling or deep ploughing → decrease of soil biological diversity but control of soil-borne diseases and pests → fully controlled 'soil-less' culture)
- Degree of change from natural climatic conditions
  (increase or decrease of shading, of near-surface humidity, wind damage, or protection from frosts → temporary greenhouse conditions (plastics) → permanent greenhouse conditions (glass))

Notes

1. Most of these publications are still for sale, in several languages, through the Distribution and Sales Section of FAO and local FAO Sales Agents; some Miscellaneous Papers and Reports can be obtained free of charge upon individual requests to AGL.
2. The apparent stability over the ages of organic matter enriched soils – old arable fields in northwestern Europe, patches of terra-preta-do-indio in the Amazon, old oasis soils in arid regions – could provide a set of indicators for sustainability assessment. Stable humus–phosphorus complexing has apparently taken place in such situations, which enhanced the CEC and improved the soil structure on a permanent basis.

References

Buringh, P. and Sombroek, W.G. (1971) Ferallitic and Plinthitic Soils; syllabus of graduate lectures (cyclostyled). Dept. of Tropical Soil Science, Wageningen University, The Netherlands.

CASAFA (1991) Sustainable Agriculture and Food Security. Report prepared for UNCED. ICSU, Paris.

FAO (1971–1978) *FAO/Unesco Soil Map of the World.* 1: 5000000. Volumes II–X. UNESCO, Paris.

FAO (1974) Approaches to land classification. *Soils Bulletin* 22. FAO, Rome.

FAO (1976) A framework or land evaluation. *Soils Bulletin* 32. FAO, Rome.

FAO (1978–1991) Report on the Agro-ecological Zones Project. Volumes I–IV. *World Soil Resources Report* 48/1-4. FAO, Rome.

FAO (1983) Guidelines: land evaluation for rainfed agriculture. *Soils Bulletin* 52. FAO, Rome.

FAO (1984) Land evaluation for forestry. *Forestry Paper*. FAO, Rome.

FAO (1985) Guidelines: land evaluation for irrigated agriculture. *Soils Bulletin* 55. FAO, Rome.

FAO (1990a) Guidelines: land evaluation for extensive grazing. *Soils Bulletin* 58. FAO, Rome.

FAO (1990b) The conservation and rehabilitation of African lands, an international scheme. ARC/90/4. FAO, Rome.

FAO (1990c) An international action programme on water and sustainable agricultural development: a strategy for the implementation of the Mar del Plata action programme for the 1990s. FAO, Rome.

FAO (1992a) Guidelines for land use planning. Prepared by the Interdepartmental Working Group of Land Use Planning. *FAO Development Series* 1, FAO, Rome. (In press)

FAO (1992b) Sustainable development and the environment; FAO policies and actions: Stockholm 1972 – Rio 1992. FAO, Rome, 88 pp.

FAO/Netherlands (1991) The Den Bosch declaration and agenda for action on sustainable agriculture and rural development; report of the conference (20 volumes). FAO, Rome.

FAO/UNFPA/IIASA (1982) Potential population supporting capacities of lands in the developing world. Technical Report. FPA/INT/513, with maps at scale 1:10000000. FAO, Rome.

Higgins, G.M., Kassam, A.H., van Velthuizen, H.T. and Purnell, M.F. (1987) Methods used by FAO to estimate environmental resources, potential outputs of crops, and population supporting capacities in the developing nations. In: A.H. Bunting (ed.) *Agricultural Environments: Characterization, Classification and Mapping*. CAB International, Wallingford, UK.

Janssen, B.H., Guiking F.C.T. *et al.* (1989) A system for quantitative evaluation of soil fertility and the response to fertilizers. In: J. Bouma and A.K. Bregt (eds) *Land Qualities in Space and Time*. Pudoc, Wageningen, The Netherlands, pp. 185–188.

Larson, W.E. and Pierce, F.J. (1991) Conservation and enhancement of soil quality. In: *Proceedings of an International Workshop on Evaluation for Sustainable Land Management*, Chiang-Rai, Thailand, September 1991, IBSRAM, Banghlen, Thailand.

Mooney, H.A. and Sombroek , W.G. (1991) Terrestrial systems. In: Dooge, J.C.I. *et al. An Agenda of Science for Environment and Development into the 21st Century (ASCEND 21)*, Cambridge University Press, UK.

Oldeman, L.R., Hakkeling, R.T.A. and Sombroek, W.G. (1990) World map of the status of human-induced soil degradation: maps and explanatory note. UNEP, Nairobi, and ISRIC, Wageningen.

Scharpenseel, H., Schomaker, M. and Ayoub, A. (1990) Soils on a warmer earth. *Developments in Soil Science* 20. Elsevier, Amsterdam.

Sombroek, W.G. (1992) Land use planning and productive capacity assessment. In: *Proc. World Bank 12th Agricultural Symposium*, Washington DC, USA. 8–10

January 1992, World Bank, Washington, DC, USA.
UNCED (1992) AGENDA 21. Rio de Janeiro, June 1992.
Van de Weg, R.F. and Mbuvi, J.P. (1975) Soils of the Kindaruma area. Reconnaissance Soils Survey Report R1, Kenya Soil Survey, Nairobi.

Chapter 30

A Concept of Sustainability and Resilience Based on Soil Functions: the Role of ISSS in Promoting Sustainable Land Use

W.E.H. Blum[1] and A. Aguilar Santelises[2]

[1]*Secretary and* [2]*President, International Society of Soil Science*

The Concepts of Soil and Soil Functions

The concepts of soil and soil functions have undergone a series of changes throughout history. Two hundred years ago, the people who concerned themselves with soils had a landscape perspective and saw soils as a vital part of the environment. Since that time, there have been several dominant phases but now we are coming back to this landscape perspective (Warkentin, 1992). At the present time, scientists recognize the fundamental role of soils and soil functions on topics such as environmental quality, sustainable agriculture and global change.

With regard to these targets, and specifically to soil resilience and sustainable land use, the classical concept of soil should be enlarged to include porous sediments and other permeable rock materials, together with underground water, which these contain. Soils thus defined may reach considerable depths.

Based on such a concept (Blum, 1988, 1990a), six main soil functions can be identified in a given area which are not always complementary. Three functions are mainly ecological and three are linked to human activity.

The three ecological functions are:

1. *Biomass production,* as a base for human and animal life, ensuring the supply of food, renewable energy and raw materials. Beyond any doubt, world food production is dependent, among other factors, upon availability of agricultural land. Recent reports indicate that about 1 ha of land is lost to production from all forms of degradation about every six seconds and that

(eds D.J. Greenland and I. Szabolcs)

many countries have reached or will be reaching the limits of their arable land resource by the end of this century (IBSRAM, 1991); this situation together with the fact that the world population of about 5.2 billion will increase to about 6.2 billion at the end of this millennium, points out the need to develop new global strategies and concepts such as sustainable land use (Chapter 1). Soils as components of the resource base play a fundamental role in sustainable agriculture (Hamblin, 1992);

2. *Filtering, buffering and transforming actions* to protect the environment and particularly groundwater and the food chain from pollution. This function is explained by the porous nature of soils controlling solute transport to underground or surface waters and absorbing chemical components so that the soil acts as a filter and buffer. Similarly, soil flora and fauna are responsible for the transformation of toxic and other organic substances. When the capacities of mechanical filtration, physicochemical buffering and microbiological/biochemical transformation are exceeded, inorganic and organic compounds may be transferred to the soil solution, and from there enter the groundwater, or be taken up by plant roots. In the first case groundwater contamination occurs, in the second contamination of the food chain occurs (Blum, 1988).

3. *Providing a biological habitat and gene reserve* for many plant and animal organisms which should be protected from extinction. The conservation of this genetic heritage is one of the most important factors in man's survival.

The three functions directly linked to non-agricultural human activities are:

1. *As a physical medium,* soil serves as a spatial base for technical and industrial structures and for socioeconomic activities, e.g. housing construction, industrial development, transport and traffic systems as well as for space-consuming activities, such as sports, recreational activities, refuse-dumping and others;

2. *As a source of raw materials,* supplying water, clay, sand, gravel, minerals and others;

3. *As a cultural heritage,* forming part of our cultural development, containing palaeontological and archaeological treasures which are a unique source of information that must be protected as a testimony to the history of earth and mankind.

Each of these six soil functions plays a fundamental part in soil and land use. Awareness of the varying importance given to each of these functions and of the competition that exists between them, is extremely important in using soil resilience and devising sustainable land use strategies through rational management policies and practices.

Definition of Land Use and Sustainability

Land use can be defined as the temporarily and spatially simultaneous use of the six cited soil functions, which are of course not always complementary in a given area.

Sustainable land use can be defined as a spatial (local and/or regional) and ecological harmonization of *all* these soil or land uses. The main problem of sustainable land use by such a definition is the competition between the six soil functions (Fig. 30.1).

Fig. 30.1 shows that intensive competition exists between the three ecological soil uses such as agricultural and forest production, filtering, buffering and transformation of inorganic and organic compounds and preservation of genes on the one hand and uses linked to human activities

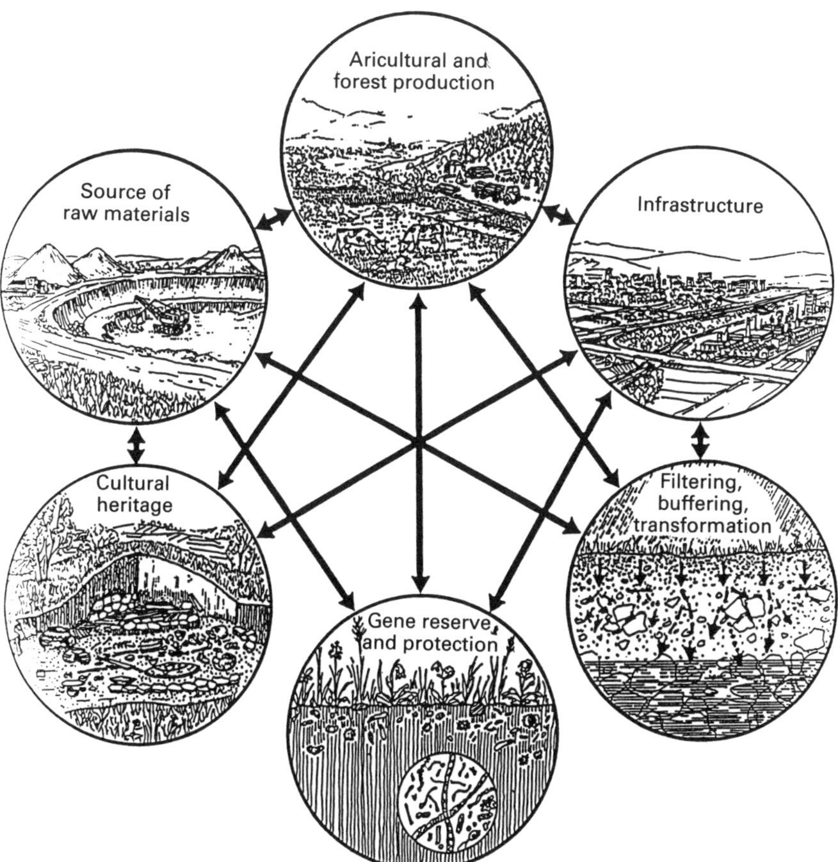

Fig. 30.1. Competition between the six main soil functions (Blum, 1992b).

such as infrastructure for housing, industrial production, transport, recreation, dumping of refuse, extraction of raw materials and the protection of cultural heritage on the other.

Within all these competitions two main problems for sustainable land use and soil resilience can be identified:

1. The use of land for infrastructural development, the extraction of raw materials and in part the protection of cultural heritage, exclude all other ecological soil uses such as agricultural and forest production, filtering, buffering and transformation action and the protection of the gene reserve.
2. From the infrastructural development and the use of fossile energy and raw materials, heavy loads are put on the three ecological land uses from linear and point sources of regional infrastructures (Blum, 1992b).

Figure 30.2 shows the three different pathways of contamination, pollution and degradation:

1. Diffuse negative influences of atmospheric origin, causing acidification, contamination or pollution by toxic substances (e.g. heavy metals, organic compounds), and contamination by radionuclides (Blum, 1990b, 1992a);
2. Negative influences by water transport, e.g. through groundwater infiltration, infiltration through river beds and irrigation;

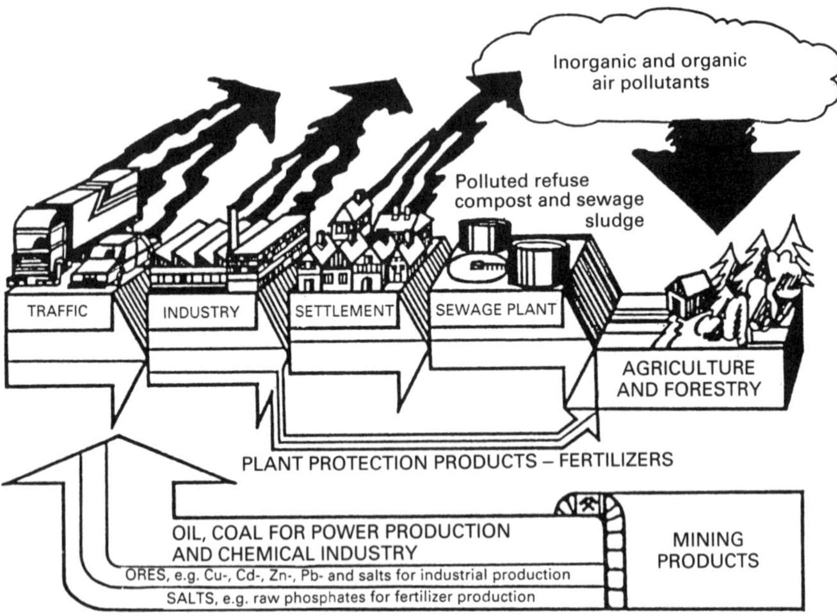

Fig. 30.2. Soil pollution by inorganic and organic compounds through the use of fossile energy and raw materials (Blum, 1988).

3. Specific negative influences of terrestrial origin which can be divided into physical influences (e.g. sealing, compacting, erosion) and chemical influences such as spreading of sewage sludge, industrial and household wastes and in part the use of fertilizers and plant treatment products.

Therefore, the spatial (local and/or regional) and ecological harmonization of all soil or land uses implies a very complex problem and can only be reached by:

- Minimizing irreversible soil losses (e.g. through sealing, excavation and others), thus allowing free land use options for future generations;
- Stopping or minimizing soil and land contamination and degradation caused by the above cited misuses of fossil energy and raw materials by infrastructural development and extraction of raw materials.

This approach can be seen as a holistic one. It is also valid for sustainable use of land for agricultural or forestry purposes in remote areas of the world where infrastructural development and the use of soil for the extraction of raw materials or the protection and conservation of the cultural heritage are of less importance. In such conditions this approach can be reduced to the influence of agricultural and forest production on the filtering, buffering and transformation capacity of soils and the protection of the gene reserve (Fig. 30.3).

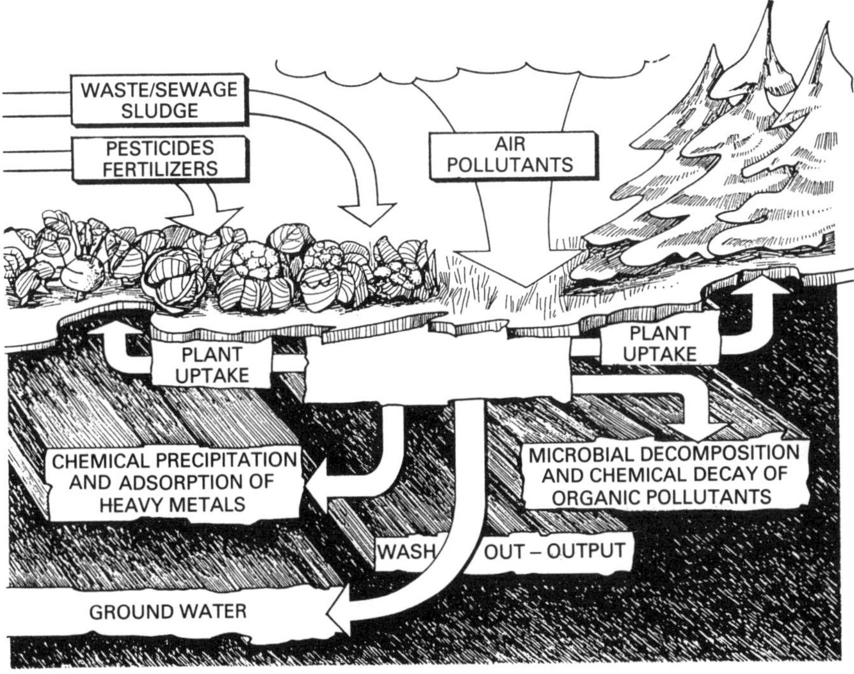

Fig. 30.3. Soil contamination by sewage sludge, fertilizers and plant protection products (Blum, 1990b).

Such a holistic approach, based on an enlarged concept of soil and the spatial and ecological harmonization of the six cited land uses is needed for:

- the definition of problems of soil resilience and sustainable land use and of methods to solve them;
- the cooperation and coordination with other sciences which are directly or indirectly concerned with soil or soil uses including biological, hydrological, geological, climatological, medical, human and many other sciences.

If soil scientists do not define their own system of land use by a broad and holistic approach, others will do it for them. Soil scientists will then not act in determining proper land use, but react to the proposals of others.

The Concept of Soil Resilience

The basic definition of resilience is the 'ability of a system to return to dynamic equilibrium after disturbance'. The problem of such a definition is that it suggests soils are more or less undisturbed systems which after disturbance could return to a more or less natural dynamic equilibrium. In reality this is not the case, with the exception of soils in very remote areas of the globe, because all our soils are more or less distributed either by global, e.g. atmospheric, disturbances or by local or regional ones caused by human activities.

Therefore, the definition of soil resilience should be 'ability of a disturbed system to return after new disturbance to a new dynamic equilibrium'.

It seems quite clear, that the term 'resilience' is much more acceptable by a broad public in the sense of soil protection and sustainable land use than terms such as disturbance, degradation and others.

Even so, it seems necessary to clarify the psychological base on which we are using or introducing this new term in the scientific discussion about sustainable land use. Moreover, it seems necessary to use the term resilience, based on the above mentioned definition only for those soil characteristics, which are influenced by biological activities. The reason for this is the fact that in soils mainly two sources of energy exist:

1. Energy derived from mineral resources inherited from the rock parent materials which reflect the energy input from the past and which is constantly diminished, e.g. through weathering of primary minerals to secondary minerals with less energy content, thus causing a constant increase of entropy in the soil system. This is a long-lasting and non-reversible process.

2. In contrast to this, the upper parts of soils receive energy from solar radiation, mainly in the form of organic compounds through photosynthesis. This energy is the base of all biological processes in soils. Through this source of energy, resilience can be actively and positively influenced by careful manipulation of the energy streams and turnover processes, e.g. by rational soil management in the sense of sustainable land use. All these processes are reversible, due to a constant flux of new energy into the system.

The Role of ISSS

The general object of the International Society of Soil Science is to foster all branches of soil science and its applications. In order to accomplish this object the Society shall:

- form commissions, subcommissions and working groups dealing with special problems of soil science;
- organize scientific congresses and conferences;
- arrange for the publication of material relevant to the activities of the Society;
- establish cooperation with other interested organizations which have related fields of work and similar objectives.

ISSS can play an important role in the promotion of soil resilience and sustainable land use through three kinds of activities:

1. more intensive cooperation with the CGIAR-International Agricultural Research Institutes, IBSRAM and others to develop the new concepts of soil resilience and sustainable land use;
2. stimulation and promotion of scientific discussion by reuniting all relevant commissions, subcommissions and working groups of the Society for the discussion of these terms which can only be done on an interdisciplinary basis, e.g. at the occasion of the 15th World Congress of Soil Science in Acapulco/Mexico 1994 with the theme 'Soil Utilization in Harmony with Nature – Learning from the Past to face the Future';
3. last but not least ISSS should develop a new paradigm of soil uses on the base of the protection of soils against misuse and the development of new techniques for the promotion of soil resilience and sustainable land use. A new paradigm can only be defined on the base of a holistic concept of soil and its functions in man's environment. On such a base it would also be possible to cooperate with many other sciences such as Geosciences, Biological Sciences, Human Sciences, Medical Sciences and others with great impact on the further development of soil science and its general acceptance within the scientific community.

References

Blum, W.E.H. (1988) Problems of soil conservation. *Nature and Environment Series* no. 39. Council of Europe, Strasbourg, 62 pp.

Blum, W.E.H. (1990a) The challenge of soil protection in Europe. *Environmental Conservation* 17, 72–74.

Blum, W.E.H. (1990b) Soil pollution by heavy metals – causes, processes, impacts and need for future actions. *Information Document,* 6th European Ministerial Conference of the Environment, Brussels, 11–12 October. Council of Europe, Strasbourg, 42 pp.

Blum, W.E.H. (1992a) Treating toxic ground. *CERES,* 24(3), 42–44.

Blum, W.E.H. (1992b) In: Schroeder (ed.) *Bodenkunde in Stichworten,* 5th edn, Hirt-Borntraeger, Berlin, Stuttgart.

Hamblin, A. (1992) Environmental indicators for sustainable agriculture. Report on a national workshop, 28–29 November, 1991. Bureau of Rural Resources, Land and Water Resource Research and Development Corporation, Grains Research and Development Corporation, Canberra.

IBSRAM (1991) International Workshop on Evaluation for Sustainable Land Management in the Developing World. IBSRAM, FAO, ISSS, Bangkok, Thailand. IBSRAM, Thailand.

Warkentin, B.P. (1992) Soil science for environmental quality. How do we know, what we know? *Journal of Environmental Quality* 21, 163–166.

Appendix
Recommendations of the Working Groups

Working Group 1: Land Management, Soil Resilience and Sustainable Agriculture

Group Leader: Professor R.S. Swift, Department of Soil Science, University of Reading, UK.
Rapporteur: Dr A.P. Hamblin, Department of Primary Industries and Energy, Canberra, Australia.

Common themes encountered in this working group were:

1. The need for *quantification* of anthropogenic impacts on soil performance, so as to distinguish exogenous from indigenous attributes and behaviour. This was sometimes linked to the need for *indicators* or *indices of status and trend* in soil quality, in assessing the sustainability of farming systems.
2. *Balance-sheets,* which provide a clear measure of biophysical sustainability; such balance sheets can be used for nutrients (inputs relative to exports from the system), water (using the accepted water balance equation), and for organic matter cycling (of particular importance in low-input systems with little or no fertilizer additions). In addition the net effect of degrading and restorative activities was identified as an area requiring further research and validation.
3. *Integration of disciplines and activities*; examples included viewing land management as a part of the total farming and socioeconomic system, analyses which integrate across spatial and temporal scales, and developing common methodologies and communications for scientists, farmers and administrators.
4. *Better information* on the soils and land uses that are *capable of restoration* after degradation, both technically and economically. There is now more information on the extent of land degradation and reasons for this than on

amelioration and restoration of previously degraded soils. The combination of technical, economic and social instruments needed for effective restoration is not often understood.

5. These themes provided the focus to *research needs,* particularly in establishing thresholds for irretrievable damage and methodologies for extrapolation of information across scales, and in using past information more effectively to assist prediction of future outcomes from courses of action in land management.

Recommendation 1

Building on past experience, interdisciplinary teams of scientists should seek better information on the relationships between land management practices and changes in key soil properties in order to identify and quantify:

- soils which can be improved and their potential capability;
- soils which are susceptible to degradation;
- the causative factors of soil degradation;
- usable indices and threshold values for these factors;
- optimum management practices to protect a given soil against degradation.

Recommendation 2

In turn, optimum soil management practices for sustainable land management should:

- simultaneously satisfy production and environmental requirements;
- be tailored to the characteristics of particular agroecosystems;
- be developed in association with socioeconomic considerations and local community interests.

Working Group 2: Biodiversity and Soil Resilience

Group Leader: Professor J.M. Lynch, Horticulture Research International, Littlehampton, UK.
Rapporteur: Dr L.F. Elliot, USDA/ARS, Corvallis, Oregon, USA.

In the workshop, the following conclusions and recommendations were agreed:

1. Biodiversity *per se* is difficult to measure in absolute terms and it is more relevant to determine any perturbations of identified components of the biota or their metabolic processes. Some biological functions are indispensable to sustain soil properties and processes that support plant production under given soil and climatic conditions.

> **Recommendation 1:** It is necessary to identify key species or key assemblages necessary for the maintenance of the beneficial functions under the spatial constraints that determine the biotic colonization of a range of soils and plants.

2. Biotic communities have positive and negative effects on plants and soils and therefore optimal, rather than maximal, biodiversity is a useful goal. It is recognized that the most successful example of modification of biotic communities is the inoculation of legumes with single species of rhizobia. Biocontrol of pests and diseases is also usually brought about by single species. By contrast, it has been demonstrated that the decomposition of xenobiotics can be most effectively accomplished by multimembered communities.

> **Recommendation 2:** It is necessary critically to analyse the biodiversity of communities necessary for organic matter decomposition, particularly in terms of metabolic processes which lead to improved soil conditions. Similar considerations should be applied to soils being used in bioremediation processes. In all processes fauna should be considered with microorganisms as potential members of structured communities.

3. Elements of scale are relevant in considering processes at the microaggregate level as well as the total field soil. Some practices may override biotic processes such as the application of N fertilizer which represses N_2-fixation.

> **Recommendation 3:** Soil management practices should be holistic in considering the role of the biota in plant production systems.

Working Group 3: Methodologies for the Study of Soil Resilience and Sustainable Land Use

Group Leader: Dr G.S. Sekhon, Department of Soils, Punjab Agricultural University, Ludhiana, India.
Rapporteur: Professor J.K. Syers, IBSRAM, Bangkok, Thailand.

Recommendation 1

To provide practical information for sound land use planning and management, it is necessary to develop quantitative measures or indices including threshold values, of those physical, chemical and biological characteristics which determine soil resilience and sustainable land use. This requires adequate attention to current practices, soil type and agroecological considerations. Models of soil resilience are of value if linked to field experiments and these should be developed.

Recommendation 2

To test the sustainability of land use management systems and to evaluate the contribution of soil resilience to sustainability, new long-term experiments are required, particularly in developing countries where pressures on soils and land are the greatest. Existing experiments which are producing useful information should be maintained. New experiments should cover major agroecological zones, be conducted by appropriate national institutions on a network basis, be coordinated by an appropriate international organisation and use the expertise of existing organisations.

Recommendation 3

To develop thinking on quantitative measures or indices of soil resilience and sustainable land use, and to facilitate integration between soil science and other disciplines, a workshop is required which should also consider the advantages of and limitations to interdisciplinary dialogue.

Working Group 4: Promoting Soil Resilience for Sustainable Land Use

Group Leader: Dr W.G. Sombroek, Land and Water Development Division, FAO, Rome, Italy.
Rapporteur: Dr H. Eswaran, World Soil Resources, USDA/SCS, Washington DC, USA.

The participants endorse the concerns and recommendations expressed in AGENDA 21 of the United Nations Conference on Environment and Development, and are acutely aware of the fact that the finite global land resource base cannot easily provide for the further rapidly increasing population and its increasing aspirations for well being. It is also recognized that the growth in production in some parts of the world has exceeded increases in population despite losses due to land degradation. However, in other areas particularly in the tropical zones, the situation is less encouraging. Consequently, the current challenge is to maintain or enhance growth in the well-endowed areas and to manage the natural resilience of soils to establish sustainable production in unimproved degraded systems.

In this context, the Symposium makes recommendations that the following activities be initiated.

Recommendation 1

Complementing the Global Database of human-induced land degradation developed by UNEP, ISRIC and FAO in collaboration with country specialists and institutions by a similar assessment showing areas with:

1. sustainable land management systems;
2. areas where lands have improved and degraded lands have been rehabilitated;
3. an analysis of the resilience of the land resource base in different ecosystems.

A simultaneous assessment of current land use globally will aid in policy and strategy development. There is also a need to quantify soil resilience and develop national and global databases of soil resilience attributes and potentials for recovery.

Recommendation 2

To implement some of the recommendations of AGENDA 21 in the area of land resource planning and management, it is recommended that an assessment of current land use should be made at national and global levels. This assessment will:

1. identify existing practices that are non-sustainable;
2. list existing practices that are not compatible with the resilience characteristics of the soil;
3. identify those areas where efforts to increase production strain the system with a potential to collapse;
4. identify those areas where efforts to increase production may have negative off-site or ecosystem impacts;
5. identify those areas that are not only agroecologically but also socio-economically similar or distinct.

Recommendation 3

To support the initiatives on ecosystem management of AGENDA 21, it is recommended that:

1. research and development activities on land management be holistic and consider the land unit (watershed, catchment, landscape unit) as a whole and not just any one component of land such as soil;
2. land users' knowledge be utilized in all stages of design and implementation of projects;
3. sustainability be monitored during the whole duration of the research and development activity.

Recommendation 4

Recognizing that sustainability is basically a 'people issue', and that consequently, social and economic factors are not only the major controls of sustainable land management but also ensure good soil husbandry, the International Society of Soil Science and interested institutions are requested to organize a Workshop involving social scientists, economists, farmers, extensionists and ecologists. Such a Workshop will integrate information on soil resilience, soil degradation and soil management from the perspective of the socioeconomic and ecological environment and will help develop a research and development agenda for sustainable land management.

Index